FUNDAMENTALS OF
carpentry
PRACTICAL CONSTRUCTION

by **WALTER E. DURBAHN**

Chairman Emeritus, Vocational Department
Highland Park High School, Highland Park, Illinois

Revised by **ELMER W. SUNDBERG**

VNR **VAN NOSTRAND REINHOLD COMPANY**
NEW YORK   CINCINNATI   TORONTO   LONDON   MELBOURNE

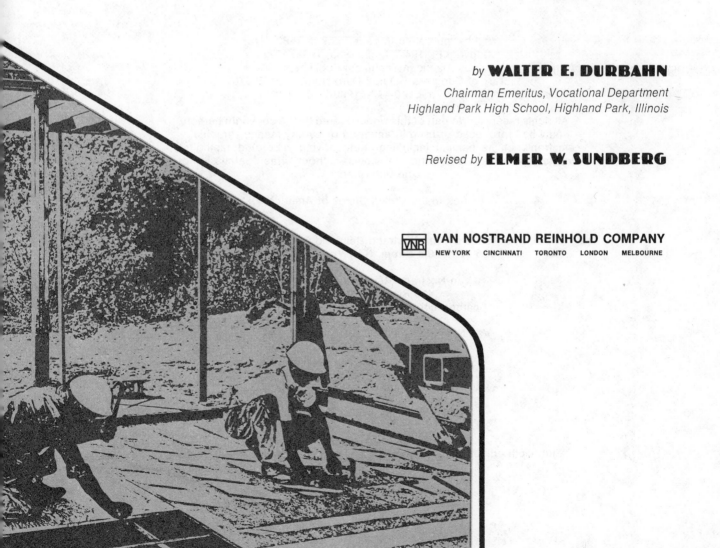

Published by Van Nostrand Reinhold Company Inc.
135 West 50th Street, New York, NY 10020

Van Nostrand Reinhold Publishing
1410 Birchmount Road
Scarborough, Ontario M1P 2E7, Canada

Van Nostrand Reinhold Australia Pty. Ltd.
17 Queen Street
Mitcham, Victoria 3132, Australia

Van Nostrand Reinhold Company Limited
Molly Millars Lane
Wokingham, Berkshire, England

1 3 5 7 9 11 13 15 16 14 12 10 8 6 4 2
Fifth cloth edition published 1977 by American Technical Publishers

PREFACE

This new edition of FUNDAMENTALS OF CARPENTRY: PRACTICAL CONSTRUCTION has been completely updated to reflect the latest techniques and methods of construction. Each step in the construction of a house is covered—from laying out the building lines to applying the final interior finish. Wherever possible, several alternative methods of solving carpentry problems are provided. Careful attention has been given to trends in building practice and variations due to regional differences. Information is often presented in step by step fashion to show how the job is actually done.

Information on new materials and how they are applied has been introduced throughout. Recent changes made in the basic structure of buildings have likewise been included. Architectural concepts which are innovative in design often present construction problems that the carpenter must be able to solve. This new edition provides the information needed to help solve many of these problems.

FUNDAMENTALS OF CARPENTRY: PRACTICAL CONSTRUCTION has greatly expanded its coverage of construction practices. New sections appear on (1) practices in site work and foundation formwork, (2) practices in wall and floor framing, and (3) practices in roof framing. New chapters have been added on industrialized building and metric measurement. A *Glossary* has been included for easy reference by the student. Hundreds of new illustrations have been added throughout this new edition to show not only the latest practices used in the field but also accepted regional variations.

Every effort has been made to present the material in a clear straightforward manner. The revising author has provided essential material that should form part of the background of every competent carpenter. Great care has been taken to assure that the illustrations accurately reflect what the young person will meet on the job. It is hoped that with the inclusion of new material, reflecting the latest accepted practices in the field, this new edition of FUNDAMENTALS OF CARPENTRY: PRACTICAL CONSTRUCTION will continue to serve as a basic foundation for young men and women entering the field of construction.

The revising author gratefully acknowledges the many people in all parts of the country who have been helpful in preparing the manuscript. Each suggestion has been given serious consideration in the task of providing an up-to-date book. He also gratefully acknowledges the cooperation and help of the many trade organizations and manufacturers who have been generous in supplying information and illustrations for this new edition.

The Publishers

CONTENTS

Wood provides an attractive finish, as shown in this view of a contemporary residence. The structure reflects the sound basic principles and quality of work provided by the carpenters. (Western Wood Products Assoc.)

CHAPTER
1
PREPARING FOR THE JOB

WHAT IS CARPENTRY?

One traditional definition of *carpentry* is "the art or science of cutting, fitting, and assembling wood or related material in the construction of buildings and other structures." Another common definition is "the occupation in which carpentry tools are used." These tools include the handsaw, the hammer, the brace and bit, chisels, and other tools generally found in a carpenter's tool box. This list of tools can be updated and expanded to include such power hand tools as the power hand saw, the router, and the electric drill. The basic definitions given above however are no longer valid.

The job of the carpenter has gone beyond working with a few select tools in a limited area of the construction industry. There is an urgent need for new housing in most areas of the country today. Of course the carpenter plays a major role in the construction of homes and apartment dwellings. But the job of the carpenter is by no means limited to this phase of the construction industry. Carpenters also figure prominently in the construction of office, apartment, and other highrise buildings, industrial complexes, sports arenas capable of seating thousands of people, giant hydroelectric dams, nuclear power plants,

and other vast public works projects. The list is long. The fact is the carpenter is involved in just about every type of construction today.

Some people are surprised that carpenters work on a wide range of construction jobs. They associate carpenters with wood and a few other building materials and a few tools. They assume carpenters only build homes and other relatively small structures. Of course this is not true. Carpenters work a wide range of construction jobs. Wood, although not always apparent on the surface, plays a very important role in many building projects. Wood is a basic material. For example, it is used to make concrete formwork for foundations as well as the formwork for floors and columns of reinforced concrete structures.

Carpenters work a variety of jobs with metals, plastics, and other synthetic materials. They install the metal framework which serves as the base for interior partitions. There are jobs which call for a wide variety of metal fastening devices, exterior prefinished metal products, metal doors and windows, and so on. Carpenters install gypsum board interior wall covering (commonly known as *drywall*) as well as exterior wall finishes made from wood, plastic, and mineral fiber. The carpentry trade includes such skilled workers as bridge, dock, and wharf builders, certain types of roofers, resilient floor coverers, cabinet makers, ship carpenters, and millwrights who prepare for and set up machinery.

Tools are no longer used to make a distinction between carpentry and other crafts. The wide range of materials used in construction require specialized tools in keeping with recommended procedures for cutting and installing. Carpenters still use various tools traditionally associated with their craft. However, they more often work with handheld electrical, pneumatic, or powder driven tools on the job. (Many of these tools are also used by other trades.) Power table or radial arm saws are set up on or near the jobsite to provide accurate cutting facilities. Carpenters have found it advantageous to know how to weld metal parts. This expansion of the trade requires obtaining new skills and familiarity with new specialized types of equipment.

Carpentry has become so diverse a field it can no longer be defined with a simple statement. It would probably take several library shelves just to hold all the books needed to adequately describe the skilled jobs and functions now ordinarily done by the carpenter. The basic question obviously is how much should any one carpenter know. How broad a range of experience should any one carpenter attempt to bring to the job?

We can look at this question more directly. Should you as a carpenter specialize in one specific area, or should you choose to acquire training in a wide range of skills and knowledge? You must make a choice. Specialization gives the worker the opportunity to concentrate all efforts and energies into one skill and become very proficient at it. However, specialization has a built-in danger. If the need for a particular skill drops off or is phased out altogether, the worker trained for it exclusively soon feels the pinch. There are less and less work days. Usually it means going for retraining in mid career.

The construction industry, as we have said, is very innovative. It is continually undergoing changes, looking for better, more efficient, more economical methods of building. Some of the changes can be called revolutionary. New jobs and new areas of skill are often created when such changes are introduced. Older established

skills are then gradually de-emphasized and eventually phased out. Workers can find themselves without employable skills. In other words, you might be topnotch at your craft, but that means nothing if there is no market for your skills.

There is an increasing demand in the construction industry for flexibility, for being experienced at doing more than one job or job function. Of course skills vary too. Some require considerable experience and know-how to master them. Others can be learned in a shorter period of time and with relatively little experience. Having a broad background doesn't have to mean you have a sketchy knowledge of various skills.

Certain types of outdoor construction work is seasonal, especially in those areas of the country which undergo dramatic weather changes. Flexibility on the job can mean you will work a greater number of days in the year. Since the construction industry is inevitably bound up with changes, it would seem that a broad background, with a wide range of skills and experience, is the most practical course to take.

Fundamentals Of Carpentry: Practical Construction covers the basic information which will help you acquire many insights useful in doing the job. This volume likewise gives you a broader outlook, the foundation on which you can build for the future and keep up with changes and growth. Sufficient detail is included to increase and fortify both your practical knowledge and skills.

Accident Prevention

As a beginning carpenter, you should be aware of the hazards of a construction job before you even start working. You should develop an attitude of safety consciousness. There is some danger in just about every occupation. There is of course greater danger in an industry in which sharp tools, power tools, and various heavy materials are frequently moved around, many workers are in the same area, each doing a different type of work, and some persons are required to work at a considerable height.

You can do carpentry work without worrying about having an accident if you keep alert, know what you are doing, and keep your working area shipshape. (Note: See *Fundamentals of Carpentry: Tools, Materials, and Practices,* Chapter 2, for a detailed treatment of accident prevention.)

MATERIALS, TOOLS AND TECHNIQUES

The construction industry is continuously at work to find better ways of building. Materials currently used are often tested and improved. Research is conducted to develop new and better building materials. Methods which provide a more efficient use of both labor and materials are likewise developed.

Plywood is a good example of how a building material can be improved to advantage. See Fig. 1-1. The development of better glues has resulted in the manufacture of plywoods with various new properties. These plywoods have been successfully introduced into many phases of the building industry. For example, plywood is used in such applications as sheathing for making concrete forms, sheathing used for exterior walls and roofs, plywood for rough flooring, and interior and exterior wall finishes.

Fig. 1-1. Plywood rough floor is applied with a pneumatic nailer. (Inland Ryerson)

Wood fiberboard, which includes particleboard and hardboard, is another example of how materials are developed to meet particular building needs. Wood fiberboard is made of wood particles bonded together and then pressed into sheets. Wood is chipped into fine particles then mixed with a liquid which serves as a bonding agent. (In some types, the liquid is extracted from the wood itself.) The mass is pressed then cut into sheets. Particleboard is made using large chips of wood and is relatively soft. It is used in such applications as cabinet construction. Hardboard is made from finer particles. It is pressed to make very dense sheets with a smooth surface. It has many applications, such as backing for prefinished interior and exterior wall covering and floor underlayment.

Metal interior framework has also been developed for use in commercial buildings. See Fig. 1-2. This framework, used for interior partitions, consists of steel studs which have steel tracks at the floor and ceiling. It has also been adapted for use in residential dwellings.

Steel clad doors have proved to be satisfactory for exterior use in all climates. Metal windows and frames are likewise popular in certain areas. They tend to be nearly the exclusive choice for new buildings in warm climates. They are used less frequently in the colder climates because the heat (or cold) transmission properties of metal cause condensation of moisture on the metal parts.

Aluminum siding, and gutter and soffit systems are increasing in popularity. Since they are prepainted with a baked on finish, they withstand weathering for many years, reducing the amount of labor and overall cost which normally goes into such exterior upkeep.

Vinyl plastics are used in various carpentry applications. In solid form, they are used as interior and exterior moldings. Such moldings are available in various colors and simulated wood grain finishes. Gutters and siding are made of solid vinyl and cut with ordinary carpentry hand tools. In sheet form, vinyl plastic is used as the facing for interior wall panels because it can be adapted to simulate wood grain or show other decorative surfaces and is very durable. Vinyl is

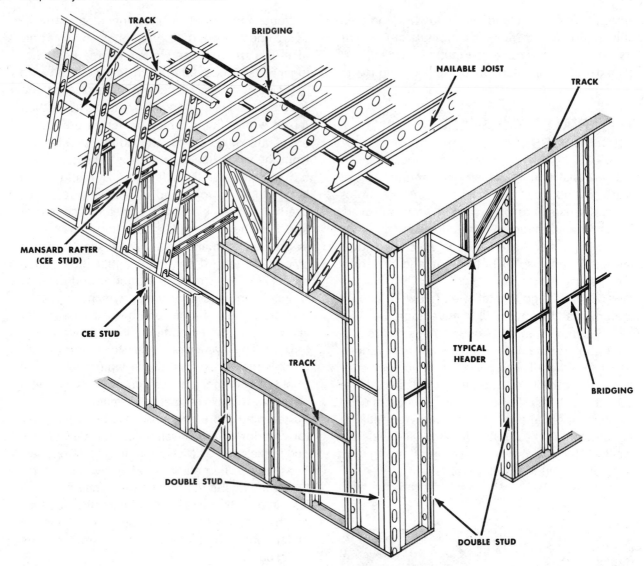

Fig. 1-2. Steel structural members are used to frame a building. The vertical members are snapped into tracks. (Wheeling-Pittsburgh Steel Corp.)

also used to shield the wood parts of windows. It provides a durable finished surface which never requires painting.

Many other materials and products are now common on the construction site. Fiberglass is widely used as insulation. Mineral Fiber (former-ly known as *cement asbestos*) is used in sheet and shingle form as covering for exterior walls. Asphalt impregnated material is applied as roof-ing. There are of course many other materials and products. These examples show that the

building industry has developed new products from discoveries made with natural and synthet-ic materials. The carpentry trade has developed new skills and techniques for installing them properly.

It is not enough that you be aware of some of the new and improved products and materials used in the construction industry. You must also know how to work with them.

Special tools are often needed to cut and pre-pare the various types of materials before they can be applied. Most materials which are cut to size on the jobsite are generally cut with a power hand saw. The power hand saw is equipped with interchangeable blades, each of which is designed to effectively cut a particular material. Pneumatic nailers are used to fasten wood structural members together. Wood members are fastened to masonry, concrete, or steel beams with powder activated fasteners. An explosive shell drives a bolt or a stud through the wood member and fastens it to the wall, floor, etc. Power staplers are used to fasten sheathing and roofing.

Hand held power tools, such as the power plane, the power drill, power nailer, power stapler, power router (used with holding and template devices) are now commonly found on the job. While such power tools have eliminated much of the hard work, they often require know-how to operate them safely and efficiently.

Welding may also be sometimes required. When a building is framed, for example, spot welding may be required to assemble some of the metal parts. Welding is also used during the installation of various metal door frames.

Plywood is applied to wood structural members, using nails alone, or a combination of adhesive and nails. The adhesive is spread out over the member. Once the plywood is in place, nails are driven through the plywood into the structural members at specified intervals to increase the holding power.

Putting it simply, you must be geared for change. You must be prepared to adapt to new materials as they are developed, new procedures which are introduced, and in general be ready to solve a variety of problems which arise on the job.

TYPES OF CONSTRUCTION

Changes in the types of buildings constructed and the methods with which they are built directly influence the work and long range thinking of the carpenter. It has been the common dream of persons to own their own home on their own piece of land. This is no longer always possible or even practical. The cost of land is high, due to

some extent to the shrinking supply of vacant land in both the city and the suburbs. This has caused some builders to shift their thinking away from individual homes and devise other uses for the land still available. The idea is to find a practical yet economical solution to the land/housing problem.

One solution now offered revives and adapts the concept of row houses. This results in attractive town houses of the type shown in Fig. 1-3. Town houses are built directly next to each other. There is a common party wall between every two units, except for end units. Each resident owns the unit and the parcel of land on which it is built.

An alternative solution growing in popularity is the construction of multi-family units, such as the duplex for two families. The units in both structures are located under a single roof. The individual units are either rented or owned outright as condominiums. A condominium is an individually owned unit. The owner pays taxes on it and contributes to a general fund which is used for maintenance and upkeep of the grounds.

An increasing number of people prefer to live in highrise apartments. In addition to normally providing an excellent view of the surrounding area, the highrise eliminates the routine outdoor

Fig. 1-3. Town House units are placed side by side with a party wall between the pair of units. Using setbacks and a variety of materials helps make these units attractive. (Ponderosa Pine Woodwork Association)

maintenance chores usually associated with owning a home. Many highrise buildings are set up for individual ownership of the units and in general are operated on the same principle as condominiums.

Conventional Construction

Platform framing. Single family homes and multiple family dwellings which do not exceed three stories in height are generally built with wood framework. The type of framework ordinarily used is called *platform framing.* They are constructed with individual wood members. The members are precut to the designated length whenever this is possible. They are nailed together on the jobsite, making up the floors, the walls, and the partitions.

Industrialized Construction

Components. A number of components can be made in factories, away from the jobsite. This increases production and efficiency because a quantity of components are made and/or assembled at the same time, using machines and special holding devices called *jigs.* Weather is likewise not a factor. Packaged doors and windows and trusses are examples of such components.

Packaged doors and windows are called *packaged* because they are delivered to the jobsite complete with trim and hardware and mounted in their frames. Once delivered, they are fitted into rough openings in the framework and fastened in place. Assembled trusses are delivered by truck and swung into place to form the supporting members of the roof. See Fig. 1-4. Prebuilt cabinets and vanities provide some of the finishing touches.

Panelized construction. Some homes are built exclusively with components. This type of construction is known as *panelized construction.* An example of panelized construction is shown in Fig. 1-5. The entire house is built in a factory, in various sections. Wall sections are made in units, complete with windows, doors, and exterior and interior finish, then delivered to the jobsite. Floor, ceiling, and roof sections and trusses are also erected so they can be quickly

Fig. 1-4. Finished trusses are trucked to the jobsite and then swung into position to form the supporting members for the roof.

Fig. 1-5. Two dimensional components are assembled complete with windows, doors, and interior and exterior wall finish. This is an example of panelized construction. (Wausau Homes)

installed. This type of construction is given fuller coverage in Chapter 8, *Industrialized Building.*

Modular construction. An increasing number of homes is being built in factories in the form of *modules.* See Fig. 1-6. Modules are generally completed sections of the building or apartment. They resemble a box rather than a flat wall section.

A module can be a complete bathroom with all the piping and fixtures installed. It can be half a house, cut lengthwise, to contain several rooms. It is made with exterior finish and the roofing in place. Windows and doors are installed. The interiors are likewise finished, including the cabinets and the electrical work.

One common way to manufacture a home is to divide it lengthwise into two modules (units) so that it can be transported on the highway. The modules are delivered by truck to the jobsite where a proper sized foundation has been prepared in advance. A crane lifts the module into place, or the module are slid into position on rails. Once supporting posts are installed, the modules (units) are bolted together.

A special type of module, used for homes, apartment buildings, and motels, is called a *wet core.* A wet core consists of a complete bathroom, often with a kitchen sink and cabinets on the opposite side (outside) of one of the walls. The sink and cabinets will fit into the rest of the kitchen as the building is completed. The wet core is complete with walls, floors, and ceiling. Wall covering and finish are completed. Fixtures and piping are already installed. The only thing remaining on the jobsite is to connect them to the supply and waste lines. In some apartment

Fig. 1-6. A module is a complete section of a building. It is delivered to the jobsite and then lifted into place.

buildings, the wet cores are stacked on top of each other so that all of the plumbing for the building is located in one vertical wall which extends through several stories.

There are other factors to consider to complete the overall picture of industrialized building. Factories which produce either panelized or modular homes are set up with specialized machinery. Specially equipped trucks or tractors are then required to transport the finished product units to the jobsite. Naturally all of this special equipment represents a huge money investment. Both the production lines in the factories and the transport vehicles must be kept rolling to keep costs competitive with the costs for the conventional methods of home building. Only through high production and careful cost control can the industrialized builder remain competitive in the market.

Heavy Construction

Highrise apartment buildings are generally built with a structure of reinforced concrete. Here the job of the carpenter is a major one because it entails building the foundation forms and the formwork for the columns and the floors. Column forms are set in place, then the supporting platforms for each floor are erected.

After the concrete has set, the formwork is removed. The operations are repeated until the building is completed.

Heavy concrete construction goes into building dams, bridges, highway projects, and other types of public works projects. Carpenters must be familiar with the properties of concrete so that the forms which are erected are strong enough to contain the concrete in its semi-liquid state. Emphasis is placed on strengthening the form sheathing and providing strong internal ties and adequate bracing. Reinforced concrete construction of course involves skills which are different from those needed for conventional home and apartment building. In fact, carpenters working in the field of reinforced concrete construction need an almost completely different background, including both skills and specific job information.

KNOWLEDGE OF CONSTRUCTION METHODS

In addition to knowing materials, products, and equipment, carpenters need a strong background in building methods. They must understand and appreciate the need for quality construction, why certain jobs are best done in a certain way. Building methods often differ from one jobsite to another. Building methods and even building materials can differ considerably from one part of the country to another. Progressive carpenters learn that in some instances there are several equally effective methods of completing a job. To become competent and a real asset on the job, carpenters have to be familiar with these various methods and, where there is a choice, go with the one best suited to the particular situation.

Carpenters also have to face up to the challenges of specialization and mass production. On some sites, such as a huge building project, carpenters might be expected to do exactly the same job day after day. Such projects might require the specialization of a skilled crew to get greater efficiency while holding down costs. As already noted, technical developments, such as those in the home building industry, have resulted in mass produced components and modules.

Carpenters who have a wide background shouldn't be discouraged or feel threatened by specialization or mass production. The whole idea is to be technically and emotionally prepared for the job situation which is there. Confidence results when you are aware of yourself as a skilled person with a wide background who is able to contribute to the building industry in a variety of ways. This is particularly true when you have the ability to adapt to the particular job and grow with the changes and developments in the industry.

ENERGY CONSERVATION

The problem of conserving energy is having a significant effect on the building industry. It has moved in several directions. The first seeks to increase the efficiency of heating systems and reduce heat loss in buildings. These goals are being achieved through creating better insulation, improving house planning, and making some structural modifications. The second direction is to explore new sources of energy, seeking to minimize our dependence on the use of coal, oil, and natural gas as sources of heating.

The great diversity of heating and cooling needs across the country and the wide range in the cost of electrical energy further complicates the problem. Several alternative solutions are being pursued to solve the problems created by these differences.

Insulation is a primary topic in the conservation of heat. Experiments are being continually conducted to increase the insulation values of foundations, exterior walls, attic areas, and roofs. The efficient use of present insulation materials and the development and introduction of new products are phases of the on-going study. Changes are made in the rough framing of buildings to provide for better insulating techniques. Emphasis is placed on preventing leakage and heat loss at windows and doors. Insulating glass, consisting of two panes of glass with a sealed air space between, is frequently used. An additional pane of glass is installed to provide triple glazing in locations where the weather is severe. Doors are made of solid wood or are clad with metal or vinyl.

Special attention is also being given to tight weather stripping. Attic spaces are vented. This along with the use of fans can provide adequate ventilation to exhaust heat in summer.

Architects and builders have also focused in on designing and building dwellings which conserve energy. There has been a trend toward construction of multi-family dwellings and two story homes instead of the ranch home which is spread out over a large area. Cutting down on the number and size of windows is one consideration for colder climates. A wide overhang (the projection of the roof beyond the face of the wall) is designed to shade the building and the windows from the direct rays of the sun, thus cutting down on heat gain during the summer.

One of the structural changes being tried is the increased thickness of exterior frame walls. By using studs and other support members with larger than the usual dimension, it is possible to increase the thickness of the insulation in the walls by two inches. Roof trusses have been redesigned so that insulation over the ceiling in attic spaces can be applied up to 12 inches, all the way to the outside wall.

(Note: Insulation materials and their installation are discussed in detail in *Fundamentals of Carpentry: Tools, Materials, Practices.* Framing a house for energy conservation is discussed in Chapter 4, *Wall and Floor Framing.* Energy conserving doors and windows are discussed in Chapter 6, *Exterior Finish.*)

The problem of conserving energy is forced on us by the fact the fuels used in the past are diminishing in quantity while increasing in price. Most modern homes are heated with forced warm air, circulating hot water, or steam. The furnaces or boilers burn coal, oil, or natural gas. Electricity is used to heat an increasing percentage of homes each year. It has several advantages. It is clean, quiet, and flexible. A single room or the entire house can be heated at any given time. However much of the electric power is produced by burning coal, oil, or gas. The rates fluctuate widely throughout the country. In a limited number of areas electric power is provided by nuclear or hydroelectric generating plants.

Alternative sources of power are continually being explored. The most promising of these is solar energy. Solar energy has great potential for the supplementary heating of homes. Eventually it may be able to considerably reduce the expenditures for the fuels used presently. Solar heat-

Fig. 1-7. Solar energy collection panels are placed on the roof to collect heat from the rays of the sun. This system is used to assist the heating and cooling systems in the house. (Chamberlain Manufacturing Corp.)

ing plants are on the market and are designed to provide much of the heat needed in the average home. The amount can vary from 30% to 90% depending on various key factors. The most important factor is the outside temperature range. The solar heating plant is generally not adequate to contend with consistently low outside temperatures. The reserve heat capacity on cloudy days and during the night is again often inadequate. Therefore in many areas it is neces-sary to have a standby system that supplements the solar heating system. The cost of installing a solar heating system is several times the cost for a more conventional heating plant. However the heating costs of a house with a solar heating system spread out over a number of years will average out in the owner's favor.

A solar heating plant includes a bank of solar collector panels located on the roof to collect heat from the sun rays. See Fig. 1-7. The panels

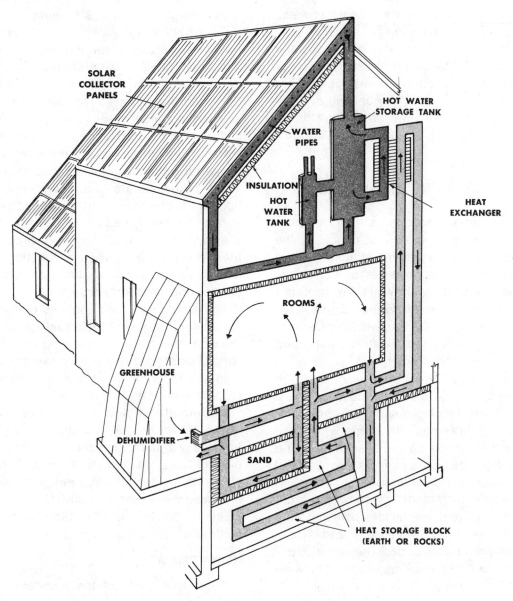

Fig. 1-8. A solar heating system. The water is heated by the solar collector panels on the roof. The heat is transferred to air ducts and circulated to the rooms. Heat from the greenhouse is circulated directly to the rooms.

are insulated and made of sheet steel or copper painted black to absorb the maximum amount of heat. The panels are protected by a glass covering. Copper tubing is arranged in serpentine fashion beneath the panels. It contains water or an antifreeze mixture which is circulated into the house. The fluid is pumped to a storage tank from which it is pumped throughout the house to the radiators.

An alternative solar heating system has the fluid pass through a heat exchanger where the heat is transferred to a stream of air-confined in ducts. See Fig. 1-8. Some of the heat circulates to convectors throughout the house. Other portions of the hot air are routed to a storage bank in the basement. The bank consists of rocks or earth confined in a concrete enclosure. After dark and on cloudy days, the heating cycle is reversed so that heat is drawn from the storage bank and circulated in the house.

READING WORKING DRAWINGS

The person who intends to become a carpenter must learn about the various tools of the trade and how to use them skillfully. This is of course basic. The carpenter must also be familiar with the materials which are routinely handled on the job and how they are best applied. (Note: Refer to *Fundamentals of Carpentry: Tools, Materials, and Practices*, published by American Technical Society.)

In addition to gaining experience and skill with tools, equipment, and materials, you must have a thorough understanding how to use the architect's working drawings. The prints which you find in this chapter are copies of actual working drawings. They are typical of prints (copies) used by home builders in all sections of the country.

These working drawings, which appear from pages 20 to 25, are drawings for a contemporary single story home. They are adaptable to regional variations in building codes. They are also designed to meet the housing needs of a wide cross section of people. This house, House Plan A, is used in several other places in the book, such as the sections on the rough framing and

the foundation, so that we can take a closer look at specific building problems.

Overall, three buildings are studied in this text to bring out the different elements of construction. House Plan A, just mentioned above and appearing in this chapter, first serves as a print reading review. Again, it is later used to increase other knowledge and skills. House Plan B appears in Chapter 2. It is tied into a discussion of site planning and foundation layout. House Plan C is found in Chapter 3-1. There it is used to describe foundation forming. In Chapter 4-1, it is used to illustrate house framing.

Construction really begins with the study of the working drawings (prints) and building specifications. Since all details cannot be shown on the drawings, a word picture of the work to be done and the materials required must be available to the builder in the form of written specifications. This information, together with the working drawings, will help the builder and the owner have the same mental picture of the finished house before the actual construction begins. Specifications can be defined as instructions to the builder. As such, they must be clear, simple, and complete. They are a supplement to the working drawings. Their function is to be explicit about all items that cannot be clearly indicated on the working drawings.

You should read the remainder of this chapter carefully. It discusses an actual set of working drawings and the accompanying specifications. Questions appear at the end of the chapter to help you determine how well you have understood the prints and specifications.

It is important that you get an idea of how this house, House Plan A, is built. As you study this text, many of the things shown on the prints will become clear to you. Note in particular how complete the details given in the specifications are. Read these over carefully. They will give you a clearer idea of your obligations and responsibilities as a carpenter.

House Plan A

The set of prints of the architect's working drawings included in this chapter is used to introduce you to prints used by builders. These

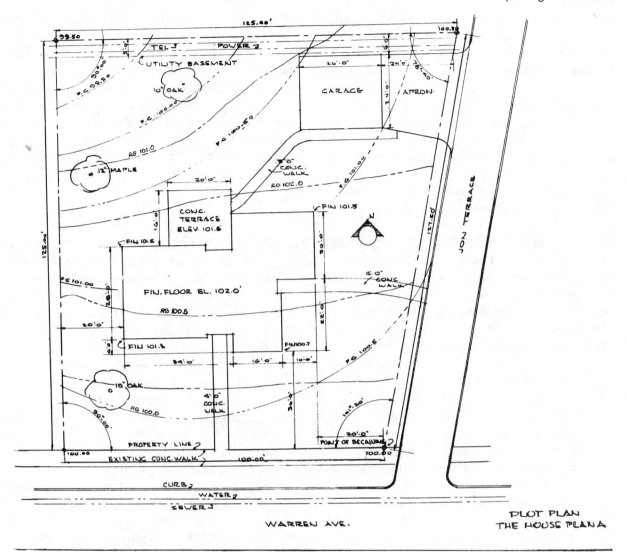

Fig. 1-9. The Plot Plan for House Plan A.

prints are later used for discussing the site work involved in laying out the house on the lot. They will also be the basis for a study of concrete forming and rough framing.

The architect who designed this house was given a copy of the Plot Plan. See Fig. 1-9. He was asked to prepare working drawings for a house which would be practical for a wide cross section of the country. The house was to be contemporary without strictly adhering to any particular architectural style. It was specified that

the foundation was to be slab at grade with a foundation wall extending below the frost line in colder climates. The same plans with a simpler foundation could be used in warmer climates. Electric heat was also specified. Ample space was to be provided for forced air conditioning equipment.

A glance at the Plot Plan shows how the house was designed to fit the lot. As you study the working drawings, you will discover how the parts have been arranged to provide for maximum view, privacy, and accessibility. You can further analyze the plans to see three distinct areas. One area is the quiet area which includes the bedrooms. A second area is the family room. The third takes in the kitchen, the dining room, and living room, which has been separated for more formal use. A terrace adjacent to the dining room provides an outdoor living space. The kitchen is central to the dining room, the utility room, and the family room. Each of these overall areas has its own roof structure which adds to the appearance and interest of the exterior of the house.

As a beginner in carpentry, you will quickly learn how important it is to be able to read prints. Once you have acquired this skill, you become an asset to the building crew, working out a series of construction problems. You must know how to take off (read) dimensions and then accurately transfer them so that all of the doors, windows, and partitions are placed exactly according to the working drawings.

Right at the outset, you must learn to recognize the different conventions used on the plans to indicate all the various types of material. You must understand the symbols used for equipment and fixtures and how to read the abbreviations and notations. (Note: Refer to *Fundamentals of Carpentry: Tools, Materials, and Practices.* See Appendix C on material symbols and the Glossary.) You should be familiar enough with the items which concern other trades so you can make allowances for their work as you build the framework for the house. For example, you should know where the heating and air conditioning ducts are placed, where the plumbing fixtures are located in the bathroom, so that when you place the wall and floor rough members you can provide for them.

You can acquire skill in reading working drawings in several ways. When you are a trainee on the job, use every opportunity to study them to see how the building is progressing step by step. Experienced carpenters are often willing to point out special details and show you how to watch for and solve construction problems. In situations where you do not have the opportunity to get firsthand information, you can study working

drawings at home, using a set such as those appearing in this chapter. The information and skill in print reading you get from a systematic, step by step study of one set of working drawings can be carried over for reading other sets of working drawings. Gradually, you will gain the confidence and knowledge to read the working drawings for more complex structures. Eventually, you will be able to read any set of working drawings required for your work as a carpenter.

Visualizing House Plan A

Looking at the set of working drawing, you might feel they are confusing and complicated because of the enormous amount of detail shown. However, an experienced builder can study them for a short while and fully understand them. Such an experienced person not only knows what the finished building will look like, but can also foresee the problem areas, which parts will be difficult or costly to build. The roof of a building, for example, frequently poses special problems. How are the rafters to be cut? How are the trusses arranged? Sometimes, in the course of framing a building, the carpenter has to make special provisions for the heating and air conditioning ducts and plumbing pipes. Occasionally in looking over the plans, the observant carpenter discovers problems or errors, such as conflicting dimensions, which could affect the whole layout of the structure. In such cases, the carpenter passes this information on to the architect or the owner. Obviously you must first learn how to read the working drawings thoroughly before you can make a serious critical study of the carpentry problems involved.

Here is a method which is useful for studying a set of working drawings. When you first look at them, disregard the many details. Instead, focus your attention on getting a good idea of the appearance and layout of the building. Figures 1-10, 1-11, and 1-12 are schematic drawings which are simplified to give you an overall picture. The Floor Plan (Fig. 1-10) shows the arrangement of the rooms. If you make an imaginary horizontal (crosswise) cut across the building about five or six feet above the floor level, you get the same view of the rooms as

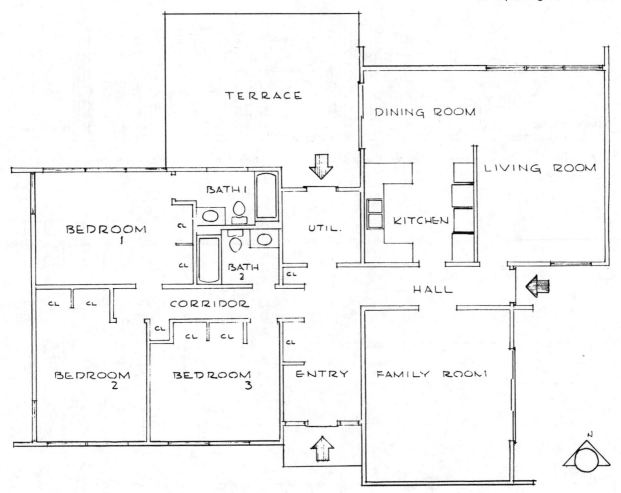

Fig. 1-10. The First Floor Plan for House Plan A. You can familiarize yourself with the layout by tracing a path from the entry through the living area then through the bedroom area.

shown in the plan view. A set of prints includes a plan view for each floor level of the building. The prints for a two story home would include a plan view for both the first (or downstairs) level and the second (or upstairs) level.

The best way to familiarize yourself with the entire layout is to study the working drawings as if you were looking at the home after it is completed. Start at the front door and then move from room to room. Once you are familiar with the floor plan, you should be able to answer various questions. How many bedrooms are there? How many exits to the home are included? Do the closets have adequate room and good location? And so on.

The next step is to look at the Front Elevation, which is shown in Fig. 1-11. In many ways, the elevation views or drawings are the easiest to understand because they give you the view you get looking directly at that part of the house. For example, the Front Elevation shows you the front of the house just as you would see it standing some distance away from it. The elevation views also indicate the type of doors and windows (which are drawn to scale), showing how they will appear on that particular part of the finished building. Each side of the house is shown on a separate view. Once you have studied this information, you can answer more questions. Does the house have a basically simple roof? Is the

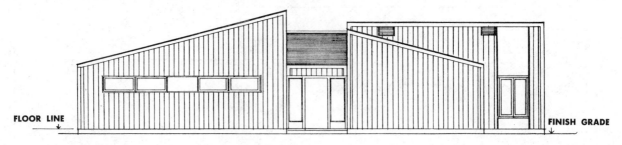

Fig. 1-11. The Front Elevation for House Plan A. Elevations, such as the one shown here, enable you to observe the roof, windows, and doors.

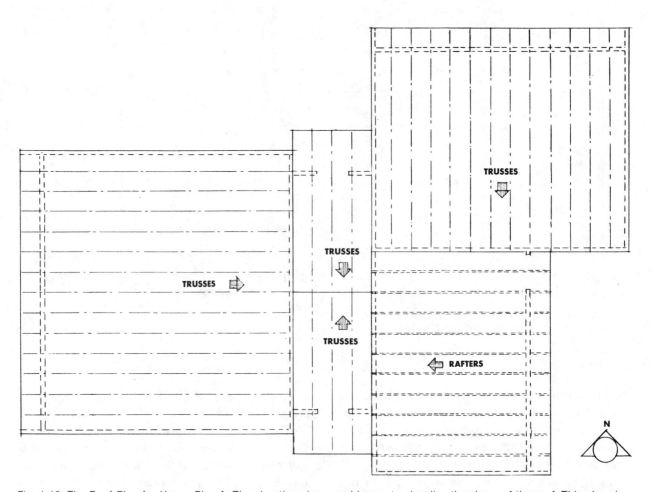

Fig. 1-12. The Roof Plan for House Plan A. The elevation views enable you to visualize the shape of the roof. This plan view explains the use of trusses and rafters. The arrows indicate the upward direction of the slope.

floor level close to the grade (ground) level or some feet above grade? What kind of windows and doors are used on that particular elevation of the building? It is essential you have a good overall idea of the house or building before you proceed to examine it in further detail.

A plan view of the roof is shown in Fig. 1-12. It shows the arrangement of the roof areas and how they intersect. Ordinarily you would get this information from studying the plan view and the elevation views.

A perspective drawing of House Plan A is given

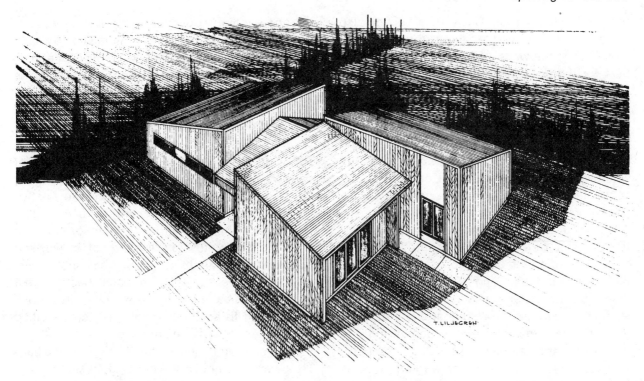

Fig. 1-13. A Perspective Drawing of House Plan A. This drawing, while not usually available to the carpenter, enables you to see how the finished house will look.

in Fig. 1-13. This type of drawing is not usually made for homes nor available to the carpenter. However, it is quite valuable for study purposes. It also gives the prospective owner a very good idea of how the finished house will look. It gives the builder a better understanding of the exterior finish, the windows and doors, and how the architect designed the roof.

Using Working Drawings

We can use such terms as *prints, plans,* and *working drawings* interchangeably. This is often done, and it should not be considered incorrect. However, the most accurate and accepted term in the building trade is *working drawings.* The term *plans* is sometimes limited to mean the floor plans. *Prints* are reproduced copies of the architect's original drawings (tracings) which are in fact the working drawings. Once you have a good overall idea of what is on the working drawings, you should analyze them in detail so that you can begin to interpret and use them intelligently.

One easy way to study the various parts of the house in detail is to follow the steps which are used for building it. This immediately takes us to a third type of view given on the working drawings, namely the *section view.* A section view is a vertical (top to bottom) cut made through the building. This type of drawing shows the interior of walls, floors, and foundations and gives information how the building is put together.

Let's look first at Section A-A, which is Sheet 5 of the working drawings. Along with other information, this sheet gives details on the foundation and the footings. Our picture of the foundation is rounded out by looking at the Foundation Plan, Sheet 2. These sheets together enable us to study the foundation in full detail, including the footings, the dimensions, and the entire foundation layout.

Since our study is following the routine construction pattern of the builder, we are next interested in the framing structure of the building. For this information we again turn to Section A-A, Sheet 5, where we have just looked at details for the foundation. This vertical section also gives much information about the structural members of the walls and the supporting members for the ceiling and how they are to be assembled.

Next are the wall sheathing and insulation, also found on Section A-A, Sheet 5. Interior wall finish is included on Sheet 6, along with the schedules for doors and windows. To study the windows and doors, you must coordinate the schedules on Sheet 6 with the window locations shown on the plan view and the elevation views.

It is important to use absolute accuracy in getting dimensions from the working drawings. This requires reading the dimensions accurately then measuring and checking the actual measurements on the material. Also remember that, although the working drawings are very detailed and complex, they do not include everything which is needed to build the house. The architect has confidence that skilled workers, such as the carpenters, are able to do many things directly related to the fine points of their crafts which are not spelled out on the drawings. For example, you as a carpenter must put in the necessary headers over openings, lay out all the rafters for the roof, position the roof trusses, and perform many other tasks which are vital to the overall strength, safety, and beauty of the house. The true challenge in your job begins after you have read the drawings and proceed with the actual construction phase.

Scaling Working Drawings

A little time should be spent on the subject of scaling a working drawing. The general rule is simple—*Don't.* Occasionally a dimension is left out, and there is a temptation to go ahead and scale the drawing (measure the unscaled item and try to figure out what dimensions it should have, using the scale shown on the working drawing). A drawing should not be scaled except as a last resort. Most of the time, the information you need is found elsewhere on the set of drawings.

The architect does not attempt to indicate every single dimension on the drawings. It would be time consuming and repetitious. When a partition is continuous and common to several rooms of the building, the dimension is only shown in one of the rooms involved. Sometimes you are expected to do a little arithmetic, such as adding together a group of dimensions and subtracting one of them from the total to get the needed information.

It is not a good practice to attempt scaling a drawing simply because the drawings are not always completely made to scale. Sometimes there are last minute changes by the prospective owner. Once such changes are decided on, it is impractical, both from the viewpoint of time and cost, to completely redraw the set of working drawing so everything is put back into scale. Usually a note is attached to each sheet where it applies, indicating that the working drawing is not drawn to scale and that only the specified dimensions have been changed.

You can avoid most errors if you make it a habit to carefully check the dimensions as they appear on the working drawings and follow them.

READING SPECIFICATIONS

A house would have to be very simple indeed for the working drawings to include every detail of information necessary to build it. Because there are so many things to be described, it is not practical to include all information on the working drawings. The working drawings for a house such as House Plan A would quickly become so cluttered with notes and details, they would be nearly impossible to read. For this reason, the architect, or designer responsible for the working drawings, usually prepares sheets of additional information and agreements which are called the *specifications.*

The specifications explain many details about the house. First of all, the specifications provide general information about the legal responsibilities connected with building the house. They spell out guarantees for work to be done, who obtains and pays for the various types of permits needed, and how the work is to be supervised. This part of the Specifications is called the *General Conditions.* They outline the broad provisions of the contract and the specific responsibilities of the architect, the contractor, and the subcontractors.

The second part of the specifications covers the technical information and is called the *Technical Specifications*. This part lists the work of each trade or subcontractor included and indicates what materials are to be used and how the work is to be carried out. It constitutes the body of the specifications and is made up of such divisions as Site Work, Concrete, Wood, etc. In many cases, the architect spells out the types of material to be used. In other cases, various choices which are both desirable and economical are given. Suppliers of lumber, roofing, and other materials study the specifications to determine the exact materials they are to provide. Equipment suppliers are asked to bid by name and model number for such specific items as bathroom fixtures, cabinets, and the heating and air conditioning units. When the contractor and the subcontractor study the plans in order to find out what quantities of materials are needed, they can estimate the overall cost with reasonable accuracy.

If the individual specification is properly written, it leaves no room for argument or guesswork. The specification for each trade likewise indicates to the subcontractor exactly how the material is to be used in the house. The subcontractor can estimate from previous experience how long the work will take. From the information given in the specifications and the current facts about labor costs, the subcontractor can present a cost estimate which is fair to the prospective owner and still shows a reasonable profit.

The specifications along with the working drawings serve as the legal basis for the general contract and subcontracts.

A paragraph, such as the one which follows here, is often used to spell out the relationship between working drawings and the specifications.

> *Specifications and the Working Drawings:* Anything mentioned in the Technical Specifications and not shown on the Working Drawings, or shown on the Drawings and not mentioned in the Technical Specifications, shall be of like affect as if shown on or mentioned in both. In case of difference between Working Drawings and Technical Specifications, the Technical Specifications shall govern. In case of any discrepancy in the Working Drawings or Technical Specifications, the matter shall be immediately submitted to the Architect without whose decision said discrepancy shall not be adjusted by the Contractor, save only at his own risk and expense. In case of differences between small and large scale drawings, the larger scale drawings shall take precedence.

The technical part of the specifications for a small building generally include the divisions given below. They can have different headings and be arranged in a different sequence.

1. Site Work (Excavating, Filling, and Grading)
2. Concrete (Footings, Foundations, Walks, and Drives)
3. .Masonry
4. Metals (Structural Steel)
5. Wood (Carpentry and Millwork)
6. Thermal and Moisture Protection (Waterproofing, Sheet Metal, Insulation, Roofing, Flashing, etc.)
7. Doors and Windows (Glazing)
8. Finishes (Drywall, Tile, Resilient Flooring, Painting)
9. Equipment (Kitchen, Laundry)
10. Mechanical (Plumbing, Heating, Air Conditioning)
11. Electrical

Wood or Carpentry Specifications

The carpenter should be interested in the entire set of specifications, even though they include the work of other trades. However, the following pages only include those parts of the specifications for House Plan A which directly apply to the work of the carpenter. (Note: At this point we can assume that the student will not be familiar with all of the language and items written in the specifications. However, the student can get an overall idea of the purpose for specifications. The student should make frequent reference to this section as the corresponding topic is covered in the text.)

In general, specifications should be studied with great care. This will avoid mistakes and help the contractor or carpenter keep within the work limits set by the contract.

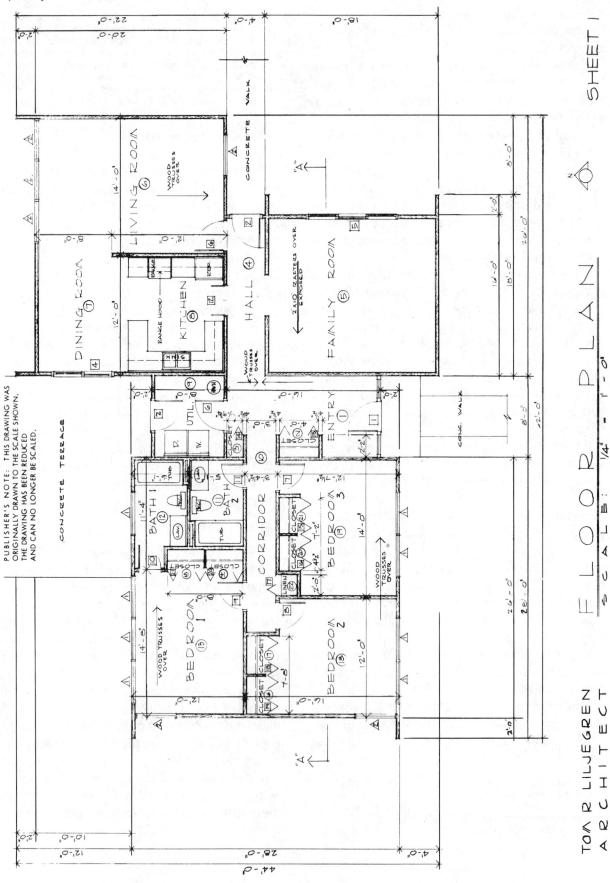

PUBLISHER'S NOTE: THIS DRAWING WAS
ORIGINALLY DRAWN TO THE SCALE SHOWN.
THE DRAWING HAS BEEN REDUCED
AND CAN NO LONGER BE SCALED.

FLOOR PLAN

SCALE: 1/4" = 1'-0"

TOM R LILJEGREN
ARCHITECT

SHEET 1

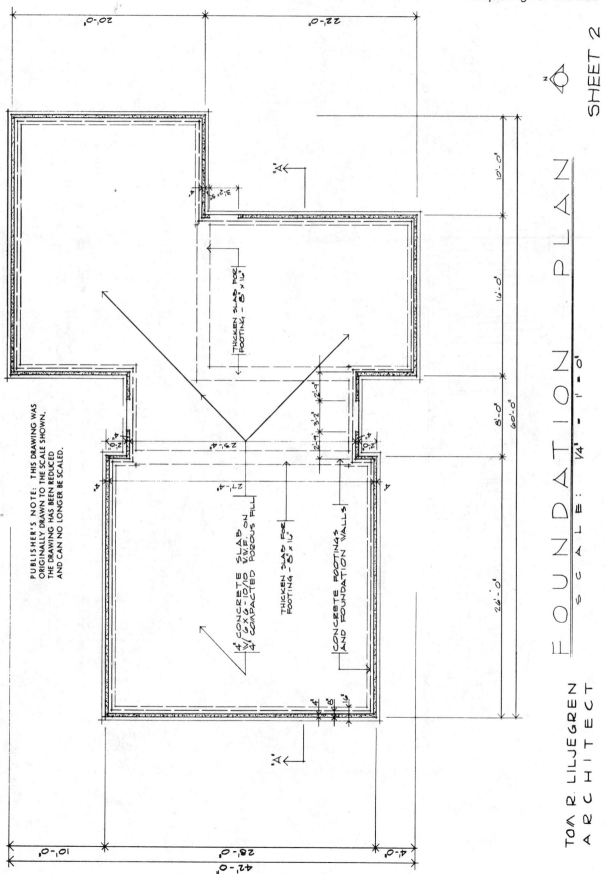

FOUNDATION PLAN
SCALE: ¼" = 1'-0"

TOM R. LILJEGREN
ARCHITECT

SHEET 2

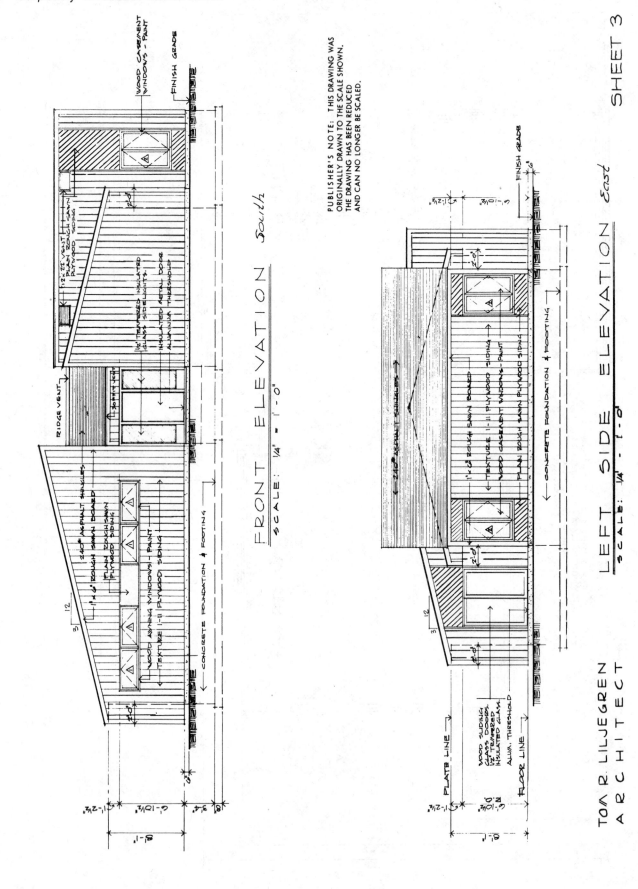

FRONT ELEVATION South
SCALE: 1/4" = 1'-0"

LEFT SIDE ELEVATION East
SCALE: 1/4" = 1'-0"

PUBLISHER'S NOTE: THIS DRAWING WAS
ORIGINALLY DRAWN TO THE SCALE SHOWN.
THE DRAWING HAS BEEN REDUCED
AND CAN NO LONGER BE SCALED.

TOM R. LILJEGREN
ARCHITECT

SHEET 3

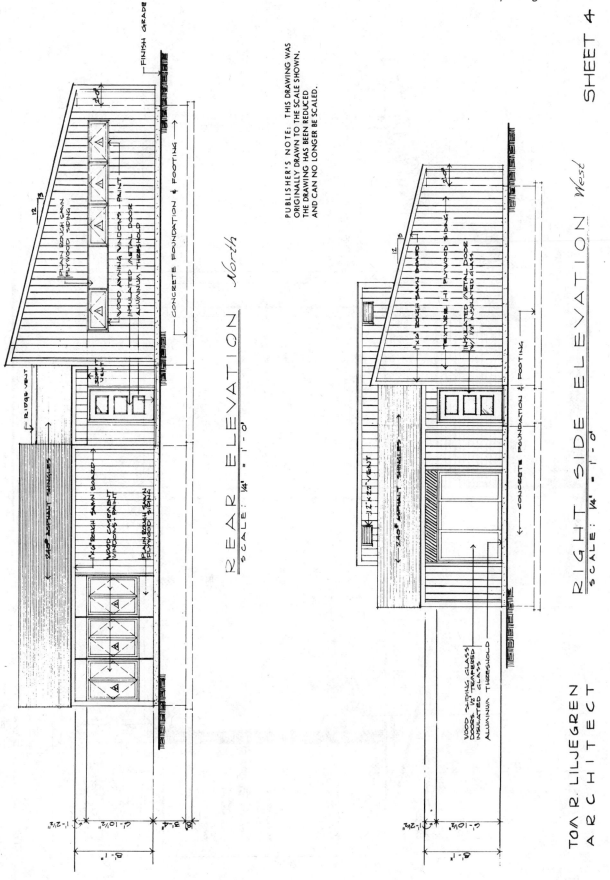

SHEET 4

PUBLISHER'S NOTE: THIS DRAWING WAS
ORIGINALLY DRAWN TO THE SCALE SHOWN.
THE DRAWING HAS BEEN REDUCED
AND CAN NO LONGER BE SCALED.

REAR ELEVATION *North*

SCALE: ¼" = 1'-0"

RIGHT SIDE ELEVATION *West*

SCALE: ¼" = 1'-0"

TOM R. LILJEGREN
ARCHITECT

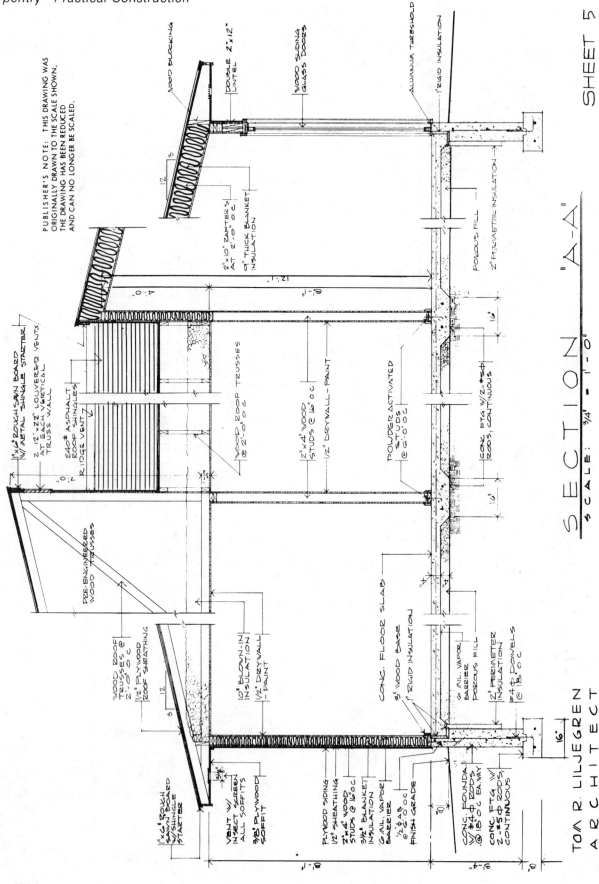

PUBLISHER'S NOTE: THIS DRAWING WAS
ORIGINALLY DRAWN TO THE SCALE SHOWN.
THE DRAWING HAS BEEN REDUCED
AND CAN NO LONGER BE SCALED.

SHEET 5

SECTION "A-A"

SCALE: 3/4" = 1'-0"

TOM R. LILJEGREN
ARCHITECT

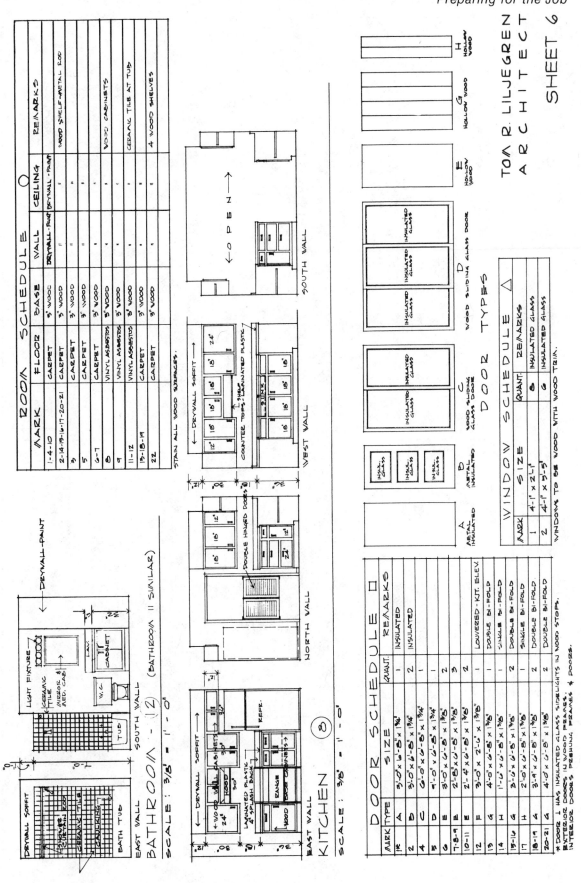

TOM R. LILJEGREN
ARCHITECT
SHEET 6

SPECIFICATIONS FOR HOUSE PLAN A

(Note: Working drawings for House Plan A are found at the end of this chapter.)

(NOTE: Only parts of the Specifications which have a direct bearing on the work of the carpenter are included. These are indicated with an asterisk. The numbering of paragraphs will not be in numerical order.)

GENERAL CONDITIONS

The latest edition of the standard form of "General Conditions of the Contract" published by the American Institute of Architects shall be understood to be a part of this specification and shall be adhered to by the Contractor (the General Contractor).

SPECIAL CONDITIONS

Sec. 1. EXAMINATION OF SITE. It is understood that the Contractor has examined the site and is familiar with all conditions which might affect the execution of this contract and has made provisions therefor in his bid.

Sec. 2. TIME FOR COMPLETION. The work shall be completed within 150 calendar days after written Notice to Proceed is issued to the Contractor.

Sec. 3. GUARANTEE. The acceptance of this contract carries with it a guarantee on the part of the Contractor to make good any defects in the work of the building arising or discovered within one year after completion and acceptance of same by the Architect, whether from shrinkage, settlement, or faults of labor or materials.

Sec. 4. RESPONSIBILITIES OF CONTRACTOR. Except as otherwise specifically stated in the Contract, the Contractor shall provide and pay for all materials, labor, tools, equipment, water, light, heat, power, transportation, temporary construction of every nature, taxes legally collected because of the work, and all other services and facilities of every nature whatsoever necessary to execute the work to be done under this contract and to bring the building to

1

completion in every respect within the specified time, all in accordance with the drawings and specifications, and applicable laws and codes. The Contractor shall carry public liability, workmen's compensation, and vehicular insurance. (The specifications include the amount of insurance to be covered in each category.) The Contractor shall coordinate all trades.

DIVISION 1 -- GENERAL REQUIREMENTS

1.01   SUMMARY OF THE WORK

1.01.1   Work under the Contract shall include all work shown on the drawings and indicated in these specifications and required by applicable laws and codes and standard trade practices. All work shall conform to local rules and ordinances. The General Contractor shall complete all work within the allotted time as indicated in the Time for Completion.

1.01.2   The carpenter shall do cutting of wood necessary for other trades and shall erect ladders inside of building. Scaffolding shall be erected, maintained, and removed by Contractor for whose work it is necessary.

1.01.3   Items provided by Owner are shown on the drawings and will be installed by Owner unless noted to be installed by the General Contractor.

1.01.4   Owner occupancy shall occur at the completion of the work. The General Contractor must complete the work within 30 days after the substantial completion. (Substantial completion date is the date when owner, architect, and contractor go over a check list of things in the contract.)

1.1   PROGRESS AND PAYMENT (Project meetings)

1.1.1   Progress of the work for payment purposes shall be determined by the Architect. The Contractor shall submit his claim for payment to the Architect for approval. The Architect shall determine that the work in place meets the quality specified and the claim for payment is for the work in place and material stored at the building site.

1.1.2   Payment to the Contractor shall be made by the Owner within ten (10) days of the Architect's approval of the claim for payment.

1.2   SUBMITTALS

1.2.1   Samples of finish materials shall be submitted to the Architect for his approval.

1.2.2   Cost Breakdown for purposes of payment shall be submitted within 30 days of the signing of the contract.

1.3   TEMPORARY FACILITIES AND CONTROLS

1.3.1   Utilities for temporary use shall be provided by the Contractor.

1.4   PROJECT CLOSEOUT

1.4.1   Cleaning up shall be the responsibility of the General Contractor (and every Subcontractor for his own debris). All rubbish shall be removed from the building and hauled to the city land fill site each week. Floors, walls, windows, and all other surfaces shall be cleaned ready for occupancy. Turn over building broom clean. The owner will wash windows and the plumbing fixtures.

2

1.4.2 Final Inspection shall be held with the Owner, Contractor, and Architect, or a representative of each, present. Within one week (7 days) the Contractor shall correct all items found to be defective.

DIVISION 2 -- SITE WORK

2.1 Trench for footings and carport foundations allowing sufficient room for form work. Place footings and foundations upon undisturbed and firm bottoms.

DIVISION 3 -- CONCRETE

3.1 CONCRETE FORMWORK

3.1.1 Forms may be job built or of prefabricated construction.

3.1.2 Forms shall conform to the shapes, lines, and dimensions called for on plans and be substantial and tight to prevent leakage of mortar. Prior to pouring, concrete forms shall be thoroughly wetted or oiled.

3.1.3 Braces and ties shall maintain forms in position and shape. Contractor shall coordinate with other trades on all inserts, sleeves, anchors, and other embedded items.

3.1.4 Remove forms without damage to concrete.

3.2 CONCRETE REINFORCEMENT

3.2.1 Provide and install reinforcing bars and welded wire fabric as indicated on drawings.

3.2.2 Provide all metal accessories required to hold steel reinforcing in position as shown on the drawing.

3.3 CAST-IN-PLACE CONCRETE

3.3.1 Portland cement shall conform to ASTM "Specifications for Portland Cement", C150.

3.3.2 Aggregates for concrete shall conform to ASTM "Specifications for Concrete Aggregates", C33. Grade coarse aggregate 3/4 inch maximum.

3.3.3 Water shall be clean and free from injurious amounts of deleterious substances.

3.3.4 Concrete shall attain a compressive strength of at least 3000 pounds per square inch at 28 days. Concrete shall be ready-mixed and shall comply with ASTM "Specifications for Ready-Mixed Concrete", C94.

3.3.5 All debris and ice shall be removed from the spaces to be occupied by the concrete. Reinforcement shall be free of ice and other coatings and shall be thoroughly cleaned.

3.3.6 Compacted fill under slab shall be approved by the Architect. A 90% compaction is required.

3

3.3.7 The 4-inch porous fill under the slab shall be composed of gravel or crushed stone of uniform-size particles, 3/4" in size, compact and level. Cover this fill with a vapor barrier polyethylene sheet 6 mils nominal thickness.

3.3.8 Concrete for slabs shall not be less than 4 inches thick. Concrete floor finish shall be true and level as called for by the drawings with maximum tolerance of 1/8 inch in 6 feet. Pitch terrace floors away from building. Trowel finish slabs.

3.3.9 Concrete paving for walks and drives shall be 4000 lbs./sq. in. compressive strength. Drives shall be 5 inches thick, and walks shall be 4 inches thick. Slope walks and drives a minimum of 1/4 in./foot.

3.3.10 Concrete shall be maintained in a moist condition for at least 7 days by water curing or membrane curing.

DIVISION 6 -- WOOD

6.1 CARPENTRY

6.1.1 Workmanship shall conform to the Uniform Building Code. Trim shall be set level and plumb, well joined. Set nails.

6.2 ROUGH CARPENTRY

6.2.1 Material. Mudsills, studs, and rafters shall be construction grade Douglas fir or No. 2 or better yellow pine. Wall and roof sheathing shall be ½ inch C D interior grade plywood.

6.2.2 Methods of Framing

6.2.2.1 Lay out carpenter work as called for by the drawings. Cut and fit for conditions encountered. All work shall be plumbed, leveled, and braced with nails, spikes, bolts, etc., to ensure rigidity. Solid bridging for trusses.

6.2.2.2 Sole framing members shall be single, 2 inch nominal thickness members for all walls and partitions.

6.2.2.3 Studs shall be 2 x 4 inch wood at 16 inch OC, doubled at openings and tripled at corners, placed to provide end nailing for sheathing and wall board. Toenail to sole with two 8d nails on each side face of each stud. One stud per four feet of exterior wall shall be fastened by means of 19 gage zinc coated metal anchor, as per manufacturer's instructions.

6.2.2.4 Top plates shall be double, 2 inch nominal thickness members for all partitions.

6.2.2.5 Plates shall be same width as studs and splice plates, nailed to studs and corner posts. Ends of soles shall be provided with splice plates, nailed to studs and corner posts. Top plates shall be nailed together with 16d nails at 24 inch OC. Two 16d nails shall be used at ends of upper members. No joint in upper member shall occur over a joint in a lower member and shall be staggered at least 4 feet. Lintels shall be placed over openings in walls and bearing partitions. Plate splices shall not occur over where plate forms part of a lintel.

6.2.2.6 All wood members shall be anchored and fastened together to ensure sound, sturdy construction.

4

6.3   WOOD TREATMENT

6.3.1   All outside wall sills in contact with concrete shall be pressure treated with preservative meeting Federal Specification TTW571.

6.4   FINISH CARPENTRY

6.4.1   Wood Siding

6.4.1.1   Vertical wood siding shall be 5/8 inch rough sawn cedar plywood-flat sheets and texture 1-11 with aluminum or galvanized finish nails.

6.4.2   Millwork and Trim

6.4.2.1   Finish lumber shall conform to American Lumber Standards and shall be clear, new, unbroken, uncracked, kiln dried, #1 Common Ponderosa pine.

6.4.2.2   Exterior millwork and trim shall be installed with tight joints, securely nailed with galvanized case nails. Interior trim and finish lumber shall be fastened in place with finishing nails, the heads set for putty and finish sanded.

6.4.2.3   Millwork shall be in long lengths with jointing where solid fastenings can be made and bedded in white lead paste. Corners shall be mitered or coped as is standard practice.

6.4.2.4   Install window trim, base, closet plywood partitions, shelves, hanging rods as shown on drawings.

6.4.2.5   Custom built oak kitchen cabinets to be built and installed by kitchen cabinet sub-contractor.

6.4.2.6   Vanities by Owner to be installed by this contractor.

DIVISION 7 -- THERMAL AND MOISTURE PROTECTION

7.2   INSULATION

7.2.1   Blanket or batt insulation shall be 3½ inch thick fiberglass stapled in place.

7.2.2   Ceiling insulation shall be 10 inch "Zonolite" vermiculite attick insulation, blown or poured in conformance with manufacturer's directions.

7.2.3   Insulation shall be installed only after conduits, pipes, etc., are in place.

7.3   ROOF SHINGLES

7.3.1   Install asphalt shingles, 235-240 lb. strips. Install as recommended by the manufacturer.

7.4   FLASHING AND SHEET METAL

7.4.1   Install flashing where roofs intersect exterior walls. 26 gage galvanized iron.

DIVISION 8 -- DOORS AND WINDOWS

8.1   WOOD DOORS

5

8.1.1   Stock. Outside entrance doors. Keyed alike. Pine frame stock design.

8.1.2   Interior doors, premium grade hollow core flush oak doors, as scheduled. Prehung.

8.2   SPECIAL DOORS

8.2.1   Wood sliding glass doors, stock type, with 5/8 inch tempered insulating glass and manufacturer's screen.

8.3   WOOD WINDOWS

8.3.1   Awning type, with insulating glass, and manufacturer's full screens. Casement window same, rototype operators.

8.3.2   Provide screened vents in soffits and louvered vents in exterior walls as shown on plans. Ridge vent over entry-utility room roof.

8.4   HARDWARE

8.4.1   All material and work in this section by carpentry contractor. Rough hardware, nails, screws, hangers, anchor bolts, and fastening as required.

8.4.2   Tracks and associated hardware for bi-fold doors. Hinges, lock and latch sets, cabinet hardware, medicine cabinets. Bedroom and bathroom doors shall have push button knob locks. Allow $000.00 Contractor's cost for same as selected by Owner. Install same.

8.4.3   Aluminum thresholds with vinyl inserts at outside doors. Weather-strip interlocking type jambs.

DIVISION 9 -- FINISHES

9.1   GYPSUM DRYWALL

9.1.1   Gypsum board. Bathroom walls, ½ inch water resisting type. On ceilings and other walls and partitions, ½ inch gypsum board for taping. Install metal beads at outside corners. Apply drywall as per manufacturer's instructions. Tape joints, putty nail heads, and corner beads to smooth finish job.

9.2   RESILIENT TILE FLOORING

9.2.1   Furnish labor and materials to complete resilient tile flooring installation.

9.2.2   All colors and patterns will be selected by the owner from the Manufacturer's current standard colors and patterns. All materials shall be packed, stored, and handled carefully so as to prevent all damage.

9.3   RESILIENT FLOORS

9.3.1   Apply materials as per manufacturer's directions. Lay out for minimum number of seams, which shall be tight. Fit tight to base, door jambs, and casings, etc. Check conditions of floors before starting work. If not satisfactory, report to general contractor for correction.

6

QUESTIONS FOR STUDY AND DISCUSSION

1. What general procedure should a builder follow when studying a set of prints to get a picture of the completed house?
2. How would you describe a plan view?
3. How would you describe an elevation view?
4. What are the three main purposes of a set of specifications?
5. How are specifications and working drawings related?

The Floor Plan, Sheet 1

6. How many rooms are there not including baths, entry, utility room and closets?
7. How many exits are provided?
8. Do all bedrooms have windows on two walls? Explain.
9. The windows in the living room face which compass directions?

The Elevations, Sheets 3 and 4

10. Is the roof flat or sloped?
11. The windows shown on the front elevations are in which rooms?
12. The glass sliding door on the left side elevation is in which room?
13. The right side elevation faces which compass direction?
14. The doorway on the rear elevation leads to which room?

The Foundation, Sheet 2

15. How thick is the floor slab?
16. What are the two overall dimensions of the building?
17. Is Section A-A taken in an East-West or North-South direction?

Section A-A, Sheet 5

18. How is the roof to the left supported?
19. How is the roof to the right supported?
20. What is the dimension from the finish grade to the bottom of the footing?
21. What material and finish are used on interior walls?
22. What insulation is used in exterior walls?
23. What insulation is used over the ceiling in the bedroom areas?
24. A door is shown in the Section View A-A to the right. What room is it in?

Schedules and Details, Sheet 6.

25. What kind of floor, base, wall and ceiling finishes are used in the Living Room?
26. *Mark 22* in the room schedules applies to which room?
27. How many windows are there altogether?
28. What is the size and type of the door marked C? In which room?
29. Where is ceramic tile used in the bathrooms?
30. The double hinged doors shown on the North wall elevation of the kitchen lead to which area?

Specifications, pp. 26 to 28

1. How long is the work guaranteed? (Special Conditions)
2. What special responsibility do the carpenters have toward mechanics of other trades? (1.01.2)
3. What kind of concrete shall be used? (3.3.4)
4. How much and what kind of porous fill shall be used? (3.3.7)
5. What kind of material shall be used for the rough framework? (6.2.1)
6. What is the spacing of studs? (6.2.2.3)
7. Which wood members shall be treated against insect infestations? (6.3.1)
8. What kind of wood is used for interior trim? (6.4.2.1)
9. What kind of nails are used for exterior millwork? (6.4.2.2)
10. What type insulation is used in walls. (7.2.1)
11. What kind of glass is used in sliding doors? (8.2.1)
12. Where is a ridge vent used? (8.3.2)
13. What does rough hardware include? (8.4.1)
14. What kind of drywall gypsum board is used in bathrooms? (9.1.1)
15. Who chooses the resilient tile for floors? (9.2.2)

CHAPTER 2
LEVELING INSTRUMENTS AND SITE WORK

The importance of accurate work in laying out the footings and the foundation of a building cannot be overemphasized. A foundation which is level, square, straight, and has accurate dimensions is the mark of excellent workmanship and makes the subsequent work of both carpenters and other mechanics easier to do. For example, the laying of floor joists and the erection of walls and partitions are simplified. It is much easier to apply trim, install cabinets, and hang the doors if the walls are plumb and true.

Architects develop plans which make good use of all space and call for prebuilt components with specified dimensions. It is important that the angles of walls and the dimensions shown on the working drawings be accurate, beginning with the foundation.

The carpenter is engaged on the construction site of a building before workers in other trades. The carpenter, for example, locates the prospective building in relation to a base point and the lot lines. He or she establishes all of the corners so that the dimensions are correct. The carpenter also has the responsibility for establishing the height of the concrete slab or foundation walls above the base point of reference. This is key to establishing the floors in the building at the correct elevation. The height (elevation) of the top of the concrete foundation wall has a direct bearing on the correct floor elevation. Level-

ing instruments, such as the builders' level or transit level, are used for this purpose. A steel tape and a leveling rod round out the basic equipment required.

The builders' level is the most common instrument used by the carpenter. See Fig. 2-1. It is used with a measuring rod, held by another worker, for such operations as those listed below. (Note: All of these operations are explained in detail in the course of this chapter.)

1. Establishing elevations in relation to a point from which measurements on a lot are made. When on or near the lot, the point is called a *point of beginning.* It can be a cross marked in the sidewalk, a point on a manhole cover in the street, or a surveyor's concrete marker. Other fixed points might also be used.

2. Laying out right angle corners and measuring horizontal angles of any number of degrees.

3. Leveling floors.

4. Providing for the desired contour of the lot by determining elevations for the finished grade at various points.

The transit level is a more versatile instrument than the builders' level because the telescope can be operated with vertical as well as horizontal motion. The transit level is particularly valuable when it is used on land that is steeply sloped. It is equipped to measure angles with greater precision than the less expensive builders' level. It can measure horizontal angles with greater accuracy, for example, when a building corner has to be made square. It can be used in all of the applications mentioned above in connection with the builders' level and in the following:

1. Laying out building lines.
2. Lining up stakes.
3. Establishing slopes for drainage tile.
4. Plumbing walls, columns, and building corners.
5. Measuring angles on a vertical plane.
6. Setting wall facing material and setting partitions.

The automatic level is found on many jobsites. It can serve all the functions listed for the builders' level. However, the time-consuming job of setting up the instrument and making it level— an operation which may be required several times in the course of laying out the building lines for a structure when you use the conventional leveling instrument—is nearly eliminated. The automatic level is equipped with a device which levels the instrument with a single simple adjustment.

A new development in leveling instruments is the laser. This device uses a concentrated beam of light emitted from the source to strike a sensor on a distant leveling rod. The laser is especially accurate over long distances. It can be set up and put into operation quickly. The beam is transmitted at close intervals of revolution in either horizontal or vertical circles. It, however, does not completely replace all other leveling instruments, since it is not designed to measure angles. It is expensive, a fact which usually limits its use to comparatively large construction companies.

THE BUILDERS' LEVEL

The builders' level, shown in Fig. 2-1, is the leveling instrument commonly used by the carpenter. It is inexpensive and quite serviceable for all the basic steps in laying out building lines and establishing elevations. The builders' dumpy level, shown in Fig. 2-2, is a more precise level. It

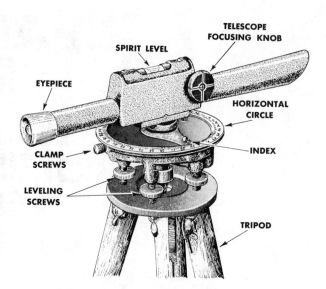

Fig. 2-1. The builders' level is an accurate instrument for determining points in a horizontal plane. (Realist, Inc.)

has a greater range because of the longer telescope and superior lens with greater magnifying power. It is also more versatile under adverse lighting conditions, such as those on an overcast day or in the interior of a building under construction before proper lighting has been installed. The principles of set up and operation are the same as for the builders' level.

Before we proceed to a fuller discussion of the builders' level and its operation, it is essential you understand the mathematics of measuring

TABLE 2-1. MEASUREMENT OF A CIRCLE.

A circle has 360° (degrees)
1° (degree) has 60′ (minutes)
1′ (minute) has 60″ (seconds)
¼ of a circle is 1 quadrant or 90°
An arc is any part of a circle

angles in degrees and parts of degrees and be able to add and subtract them. Table 2-1 shows how a circle is divided into parts.

Parts of the Builders' Level

The essential parts of the builders' level are the telescope, the spirit level, the horizontal circle, and the leveling screws. The more expensive builders' levels have a leveling base plate which is a part of the instrument itself. The inexpensive levels require the use of a special tripod, equipped with a steel ring, which serves as a base plate.

The telescope is a tubular optical instrument equipped with precision lenses which are adjusted with a screw to bring them closer or further apart for fine focusing. Two cross hairs are located in a fixed position on each end of the barrel of the telescope. When the centers of the two sets of cross hairs line up, the instrument is focused directly at the point or object being observed.

An inexpensive builders' level is usually equipped with a 12 power telescope. (The term *power* when used in connection with a telescope indicates how many times closer an object appears with the telescope than with the naked eye. For example, a 12 power telescope makes objects appear 12 times closer than if they were viewed with the naked eye.) Moderate priced instruments are 16 and 18 power. Transit levels used for big construction jobs have a higher power of magnification so that they can provide greater range. Very fine instruments are equipped with a telescope as high as 32 power.

Four leveling screws are provided so that delicate adjustments which bring the instrument into a perfectly level position can be made. See Fig. 2-3. The spirit level indicates when the adjustments are correct. The telescope frame or support is made so that the instrument can revolve on a horizontal circle marked in degrees. Some horizontal circles are divided into four quadrants of 0° to 90°, others into two half circles of 0° to 180°. See Fig. 2-5. An indicator inside of the horizontal circle serves as a pointer or index for measuring angles.

By noticing the reading at the index point and then revolving the telescope through the desired number of degrees, you can accurately determine the angles. The better made instruments are equipped with another important feature called a *vernier*. The vernier greatly increases the precision. By adjusting the lines on the vernier scale with the lines on the horizontal circle, you can measure angles to the nearest 5 minute divisions. See Fig. 2-7. (Later in this chapter we will fully discuss how to read the vernier scale.)

A plumb bob is suspended directly below the center of the horizontal circle of the tripod. See

Fig. 2-3. This instrument is leveled by adjusting opposite leveling screws.

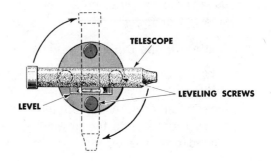

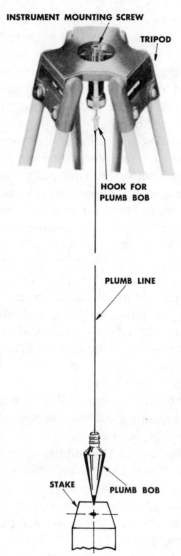

INSTRUMENT MOUNTING SCREW

TRIPOD

HOOK FOR
PLUMB BOB

PLUMB LINE

STAKE   PLUMB BOB

Fig. 2-4. The builders' level is positioned over a point on a stake using a line and a plumb bob hung from a hook on the instrument. (Realist, Inc.)

Fig. 2-4. The instrument is moved until the plumb bob is directly over the center of the measuring point on the ground. The measuring point is usually crossed lines on a stake, or a mark on a sidewalk or street.

Setting Up the Builders' Level

When you set up the builders' level, first place the tripod so that it provides a firm base. The legs should be spread about three feet apart and firmly pressed into the ground. If the tripod is placed on a paved surface, be extra careful.

Make certain the points hold securely. Most tripods have adjustable legs which enables them to be used on sloped or irregular ground.

When you remove the level from its case or container, lift it by its frame or base and not by the telescope. Loosen the clamp screw and screw the instrument on to the tripod so that the plumb bob hook hangs through the tripod head. Lightly tighten the leveling screws.

If the level is to be set up over a point on a stake, fasten the plumb bob on a cord and suspend it so that it swings freely and is almost at the level of the top of the stake. It may be necessary to relocate the instrument and tripod so that the plumb bob is not more than 1/4 inch horizontally away from the point. To get the plumb bob directly over the point, loosen two *adjacent* leveling screws and shift the instrument on the leveling base plate to the required position.

Leveling the Builders' Level

Most of the builders' levels are equipped with four leveling screws. (Some newer models have only three screws.) The accurate leveling of the builders' level is a delicate operation. Too much pressure will damage the screws or the base plate.

First loosen two *adjacent* leveling screws to free both pairs of opposite screws. Turn the telescope so that it is directly over two *opposite* leveling screws. Turn one of these opposite screws clockwise, the other screw counterclockwise, and watch the bubble in the spirit level. Continue to adjust the screws until the instrument is in a level position. Turn the telescope through a 90° arc and level the instrument in the same way by adjusting the other pair of *opposite* leveling screws. Continue swinging the telescope between the first and second position, then adjust the screws until the bubble indicates it is perfectly level. The final adjustment is made by tightening the screws more firmly. The spirit bubble should now remain centered while the telescope is revolved in a complete circle.

Sighting the Builders' Level

The next step is to sight the builders' level. With the clamp screw released, revolve the telescope, and line it up with the object by sighting along the top of the barrel. Look through the eyepiece, and adjust the telescope focusing knob until the object becomes clear. When lining up stakes, using the builders' level and a leveling rod, tighten the clamp screw so that the telescope will remain in a fixed position. When laying out or measuring angles, release the clamp screw and take readings on the horizontal circle as indicated by the index.

READING THE CIRCLE SCALE AND VERNIER

The horizontal circle, shown in Fig. 2-5, is always divided into degrees. The inexpensive levels have a pointer or index which indicates the exact reading. See Fig. 2-6. You can determine the measurement of an angle by recording the reading when the instrument is lined up on one point, then noting the reading when the

instrument is swung to line up with the second point. By subtracting the one from the other, you can find the angle.

Better quality instruments are equipped with moveable horizontal circles so that readings can begin at 0°. These instruments also have vernier scales which help measure angles with great precision. Using them, you can measure divisions as small as $\frac{1}{12}$ of a degree, or 5 minutes.

The carpenter uses the vernier to determine the right angle corners for a building. A rod is held above one point and sighted, then the instrument is swung until the scale on the circle reads 90°. But this is not accurate enough. The carpenter uses the vernier so that the angular measurement can be made to be 90° to an accuracy of $\frac{1}{12}$ of a degree. Occasionally a building is constructed so that one of the walls is parallel with a street which runs diagonally. Since in this case the building is not rectangular, all of its corners are not right angles. The carpenter should read the circle scale to determine the

Fig. 2-5. The horizontal circle of 360° is divided into quadrants (four arcs of 90° each). An angle which is less than 90° is read directly from the circle. An angle which is greater than 90° but less than 180° is determined by subtracting the reading on the circle from 180°. For an angle between 180° and 270°, the reading on the circle is added to 180°. The reading for an angle which is over 270° but less than 360° is subtracted from 360°. The example above shows an angle greater than 90°. The reading on the circle is 62°-30′. 180° − 62°-30′ = 117°-30′. Therefore the angle shown above is 117°-30′.

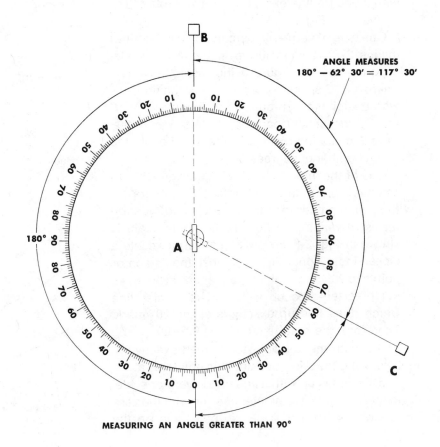

ANGLE MEASURES
180° − 62° 30′ = 117° 30′

MEASURING AN ANGLE GREATER THAN 90°

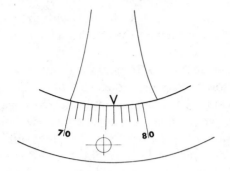

Fig. 2-6. The pointer or index indicates which degree on the circle is to be read. The pointer above indicates 76°.

angle then use the vernier to make the measurement precise.

The vernier, as indicated earlier, is a scale which lines up with the horizontal circle. (A twelve unit arc on the vernier is made equivalent to an eleven degree arc on the circle.) When an angle measures a whole degree with no fraction, the index on the vernier, which is also the pointer for the arc, will coincide with the degree line. In this case, no other vernier line will coincide with a degree line except at the two ends of the vernier. See Fig. 2-7.

One type of vernier in common use is featured in Figs. 2-7 to 2-11 inclusive. In this vernier scale, a pointer or the figure *0* indicates the index. There are 12 spaces on each side of the figure *0*, with designations representing 30 and 60 minutes. Each small space represents 5 minutes. Always make the reading in the direction of the rising number of degrees.

One of the important uses for the vernier is to establish accurate right angle (90°) corners for the building lines that are used to set foundation forms. A reading of 90° on the circle scale is shown in Fig. 2-8. By giving the vernier scale a closer look, you will notice that the indicator points to 90°, although the vernier line at 90° does not precisely line up with it. The vernier line which does line up exactly is at the 20′ mark. Therefore, the precise reading is 90° – 20′.

The instrument is adjusted to read exactly 90° by moving the telescope until the pointer is at 90° and also lines up with a line on the vernier scale. In this example, the pointer lines up with the zero line on the vernier scale. No other line on the

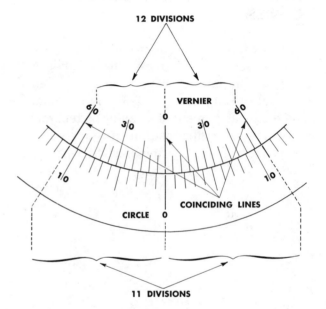

Fig. 2-7. A vernier scale is designed to find parts of a degree in 5 minute intervals. The scale above is set at 0 degrees, 0 minutes.

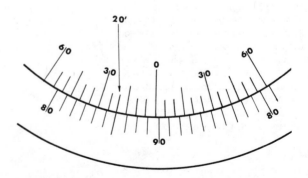

Fig. 2-8. The 20′ line on the vernier lines up with a line on the circle scale. Therefore the reading on the instrument above is 90°-20′.

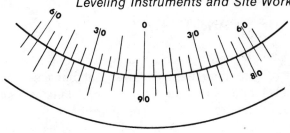

Fig. 2-9. The instrument above reads exactly 90°. The pointer (0 on the vernier scale) indicates 90°. The vernier line at 0 is also in line. No other vernier line lines up with the degree line.

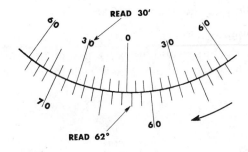

Fig. 2-10. This instrument shows a reading of 62°-30′.

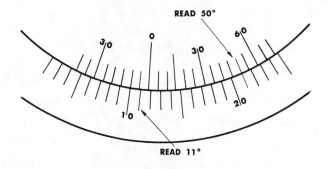

Fig. 2-11. This instrument shows a reading of 11°-50′.

vernier scale lines up with a degree line on the circle scale. See Fig. 2-9.

Figures 2-10 and 2-11 give examples of how to read typical angles on the vernier scale. The horizontal arrow shown in Fig. 2-10 shows the direction used for reading the vernier. (Note: This arrow is *not* a part of the actual circle scale. It has been added to provide this information *only*.) Notice that the index point in Fig. 2-10, represented by the figure *0,* is located between 62 and 63 degrees. To read a fraction of a degree, look at the vernier scale (upper scale) and find the line which coincides with a degree line. (Again,

you must read in the direction of the rising number of degrees, as indicated by the horizontal arrow.) In the scale shown on Fig. 2-10, the line which coincides with a degree line is the 30 minute line. Therefore, your reading is 62 degrees, 30 minutes, ordinarily written 62°–30′.

Another example is given on Fig. 2-11. However, in this case, the reading is made in the opposite direction. The index shows the angle is between 11 and 12 degrees. The 50 minute line coincides with a degree line on the vernier scale. Therefore, the reading is 11 degrees, 50 minutes, or 11°-50′.

THE TRANSIT LEVEL

The transit level is a much more versatile instrument than the builders' level because the telescope can swing in the vertical plane as well as the horizontal plane. The instrument shown in Figs. 2-12 and 2-13 has several features which increase its precision. The verniers help in measuring angles. The tangent screws permit very fine adjustments.

Setting up the Transit Level

When you set up the tripod, make certain the legs are spread and firmly planted in the ground. Carefully remove the transit level from the case by taking hold of the base plate rather than the telescope. Then mount and level the instrument in the same manner as the builders' level. The instrument is set up over the stake or point of reference using a string and a plumb bob. You must keep the telescope locking levers engaged in order to keep the telescope in a fixed horizontal position during the leveling procedure.

Sighting the Transit Level

The eyepiece may have to be adjusted and refocused for each person using the transit level. The eyepiece is focused by turning it until the cross hairs show up sharply when the main focus knob is sighted on a distant object.

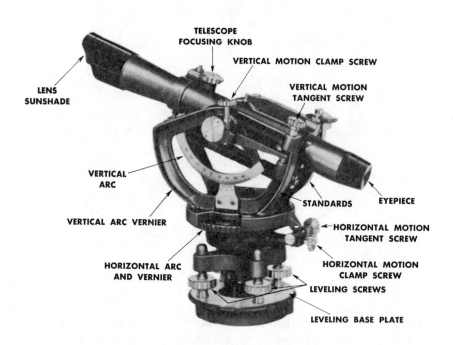

Fig. 2-12. The transit level can be used to line up stakes or to plumb walls when it is adjusted in the transit position. It can also be used to measure the angle of elevation from a horizontal plane. (Realist, Inc.)

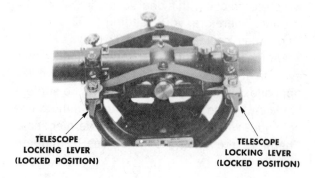

Fig. 2-13. A transit level viewed from the opposite side. The transit level is accurate for establishing grades, laying out building lines and foundations, plumbing walls, and lining up stakes. (Realist, Inc.)

When the instrument is going to be used as a level, loosen both the horizontal-motion clamp screw and the horizontal-motion tangent screw. However, maintain the telescope locking levers in a locked position. The telescope should now have free movement on the horizontal plane.

You are now ready to find the object to be sighted. First aim the telescope by looking along the top of the barrel. Then look into the eyepiece and turn the telescope focusing knob to clearly sight the object.

If the transit level is to be used in a fixed position or used to measure angles, use horizontal-motion tangent screw to bring the cross hairs into perfect bearing on the target, then tighten the horizontal-motion clamp screw.

If the instrument is to be used as a transit, as shown in Fig. 2-12, leave the horizontal-motion clamp and tangent screws in a fixed position and release the telescope locking levers. This will enable the telescope to point up and down.

The instrument is also used to read angles on a vertical plane, but it is rarely used for this by a carpenter. It will be explained later under the topic *Measuring Vertical Angles,* p. 54.

THE AUTOMATIC LEVEL

A lot of time can be spent on the jobsite adjusting the four leveling screws above the base plate and observing the bubble in the spirit level when you use a builders' level or a transit level. The automatic or self-leveling instrument, shown in Fig. 2-14, has been perfected and nearly eliminates this leveling procedure. Leveling on this instrument is accomplished by a simple quick rough adjustment of the screws. An internal compensator takes over and positions the instrument into perfect level alignment.

To use the instrument, simply place it on the tripod over the point of reference as explained earlier. Once it is in position, make rough adjustments using the three screws (as opposed to four on the conventional instrument) while observing the bull's eye (a circular vial) through the lens. You are able to see the bull's eye through the lens by means of a series of mirrors. (Some

Fig. 2-14. An automatic or self-leveling instrument. (Keuffel & Esser Co.)

instruments are designed without leveling screws. They are tilted on a ball and socket joint until the circular vial shows that the instrument is approximately in the correct position.) The compensator then operates automatically and levels the line of sight. The compensator is a complex, sensitive mechanism. However, it is insulated against shock and overall it is quite sturdy. The automatic level has a high level of accuracy because it uses high powered lenses and is not affected by a wide range of temperature change.

TIPS FOR USING LEVELING INSTRUMENTS

Any leveling instrument is a delicate precision instrument. Although it is designed and built to withstand continual use on the job, you should be aware of various safeguards, which will not only increase its useful life but assure its day to day serviceability. The following tips cover the maintenance and operation of leveling instruments.

1. It's a good policy to read the instruction sheets which come with any instrument. These sheets tell you how to operate the device, point out special features, and spell out what routine maintenance is required.

2. The instrument should be kept clean and lubricated and stored in a dry place. If it is ex-

posed to dust and grime on the job, the instrument should be cleaned before it is put away. It is important to store it in its carrying case.

3. A waterproof cover should always be available in case of rain or other inclement weather.

4. Set the tripod on firm ground. The leveling base should be as level as possible. The tripod legs should be positioned about 3 to 3½ feet apart, and even more if it is windy. The legs must be seated firmly.

5. If you use the instrument on a smooth floor, make certain to provide a base so that the tripod legs will not slide.

6. Carefully remove the instrument from its carrying case. Lift it out by the leveling base plate, not by the telescope.

7. The mounting screws should be carefully engaged to prevent cross-threading.

8. The leveling screws should be set snugly, firmly, without being too tight. (Setting the screws too tight can spring the instrument.)

9. Avoid touching the tripod legs when you are using the instrument. Also you should not straddle a leg if this can be avoided, because you are more apt to jiggle the leg.

10. The instrument should be allowed to reach air temperature before it is used. It should be checked every 10–15 minutes to verify that it is still level, since changes in temperature affect the setting. It should also be checked to avoid errors caused by any shifting of the tripod.

11. The spirit level should be checked before readings are taken. The bubble is inaccurate when the vial is unevenly heated. Keep the instrument in the shade if possible.

12. The lens should be cleaned occasionally. Use a camel's hair brush or lens paper.

13. When you are passing through doorways, under trees, or scaffolds, carry the instrument in front of you and not over your shoulder. (Ordinarily there is no problem with carrying the instrument over your shoulder.)

THE LASER

The laser is a beam of light emitted from a closely controlled source. The beam travels across space and is intercepted by a light sensi-

tive target. The laser as used in construction work consists of helium-neon sealed in a tube. In a general way, it is similar to the neon tubes used for electrical advertising signs. The beam itself is a form of radiation in which the light rays are synchronized, of the same color (red), and fortified enormously into concentrated energy. When the beam leaves the instrument barrel, it is itself approximately ⅜ inch in diameter. It travels toward the target with virtually no spread.

The equipment includes a portable power supply. Ordinarily, a 12 volt storage battery is used to activate the laser. The laser unit itself consists of the tube which contains the laser element and a device, shown in Fig. 2-15, which

Fig. 2-15. This laser leveling instrument is mounted on a tripod and operates from a portable power supply. (Laser Alignment)

accurately emits the beam. The unit mounts on a tripod or a universal mount. It can be revolved mechanically to sweep in a circle, with speeds from 60 to 180 revolutions per minute (rpms). The instrument is designed so it can be leveled quickly. Most laser leveling devices are placed in position simply by adjusting two screws.

The laser sensor or beam detector is mounted on a leveling rod. It is then moved up or down to intersect with the laser beam. See Fig. 2-16A and B. The reading of the elevation on most sensors is made on the side of the rod. See Fig. 2-16A. Some of the sensors are equipped with a handheld readout device which provides a digital display of the elevation to the nearest .01 of a foot.

The laser principle has been adapted for many broad applications since its invention in 1960. Today its applications include cutting, drilling, and welding of a variety of materials. One important application of the laser in the construction field is for boring tunnels. The great accuracy of the laser provides perfect alignment of the tunnel as the work progresses. It has been adapted, with excellent results, for laying pipe for sewer and drainage systems. The laser is mounted so it fits into the end of a pipe and transmits a horizontal beam. The sensor is then introduced into the end of each succeeding section of pipe as it is laid so that it is kept perfectly in alignment.

Millwrights and other persons involved in setting up machinery likewise find the laser very useful. The machines to be set up are roughly placed in line. Then a readout box, in conjunc-

Fig. 2-16. The laser sensor rod (A) receives the laser light and indicates the correct height with the use of a green light. Before the correct height is achieved, red lights indicate whether the sensor should be raised or lowered on the rod. (B) shows the laser and the sensor rod in field application. (Laser Alignment)

tion with the laser, is brought into play. Horizontal and vertical measurements (readings) indicate how much the machines should be moved to place them in perfect position.

The laser has shown its value in the heavy construction industry. Prior to the adaptation of the laser to the building industry, all accurate layout of field work on construction projects was done with leveling instruments, involving telescopes mounted so they could be focused to sight distant objects. The laser has proven itself to be both a time-saving and labor-saving device. The equipment can be set up quickly. Some operations are done by one person instead of two.

Laser leveling equipment can be used for many operations on the jobsite in place of the builders' level or the transit level, provided the air is not turbulent. This equipment is used for site work, in setting footings, and establishing grades. One person takes elevation readings wherever they are required, while the unattended laser continues to transmit a beam in a constant circle. At a range of 500 feet, the laser is accurate to within .01 of a foot. This assumes that air turbulence is minimal. The ultimate accurate range is up to 1500 feet, depending on the atmospheric or working conditions. The laser is also versatile. When it is mounted in a horizontal position, it can swing on a vertical plane and make a perfect circle at right angles to its axis. In the vertical position, the laser is used to align reinforcing steel, line up exterior wall facing, plumb corners, set partitions and curtain walls, and to line up light fixtures.

Laser Safety Practices

Although there are various safety practices which must be observed when the laser is used, the safety record for lasers used on construction sites is excellent. In fact, the laser can even help eliminate some of the problems and potential hazards resulting from improper use of conventional leveling devices.

Lasers can be highly dangerous to the eye sight. However, the modern low-powered laser is designed so that the potential danger to those working in the area is minimal while it is emitting a beam in a continuous circle. The red helium-neon lasers used on construction sites emit a very low-powered beam, usually less than 5 milliwatts (0.005 watts).

The Occupational Safety and Health Act (OSHA) of 1970 has listed various regulations which are to be observed when a laser is in use. Manufacturers likewise provide various regulations and warnings to prevent hazardous situations.

1. Prominent warning signs must be posted in the area where a laser is being used. An example of a warning sign is shown in Fig. 2-17A. Be alert to warning signs.

2. Avoid going into marked-off areas unless it is necessary because of your job.

3. Only those employees who have been trained are permitted to set up or operate the laser or make any adjustments on it.

4. Qualified employees must carry their Operator's Card at all times when they are operating laser equipment. Fig. 2-17B shows a typical Operator's Card.

5. Read and follow the instructions for the particular instrument supplied by the manufacturer.

6. Avoid looking directly at the beam. Also avoid looking at any surfaces, such as polished metal or a mirror, which can cause the beam to be reflected into your eyes.

7. Special hazards are present when the laser is used in a fixed position, such as when it is used for tunnel or sewer work. Never look at the concentrated beam. Never look along the axis from the laser toward the point you are sighting because the beam can be reflected from another surface.

8. Wear safety goggles when the laser has a power output of 5 milliwatts (0.005 watts) or greater. The laser equipment must have a label indicating the maximum power output. See Fig. 2-17C.

9. Never point a laser beam at anyone. The laser should be set up well above or below the heads of workers.

10. The laser beam should be turned off, or shuttered or capped, when it is not in use.

11. If you suspect any problem caused by the

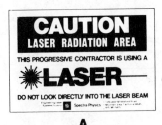

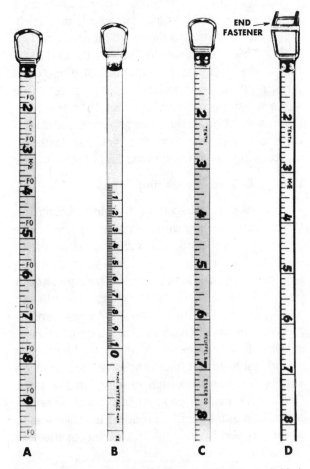

Fig. 2-18. The measuring tape shown in (A) above is divided into inches and 8ths of an inch. (B) is a metric tape divided into metric units. (C) is an engineer's or surveyor's tape divided into inches and 10ths of an inch. (D) is also divided into inches and 10ths of an inch and has an end fastener. (Keuffel & Esser Co.)

Fig. 2-17. Laser safety requires the use of various visual warnings and identifications. The warning sign shown in (A) must be posted in areas where the laser is in use. (B) shows a typical laser operator's card. The laser instrument must have a label, such as the one shown in (C), which indicates its maximum output. (Spectra-Physics)

laser, such as a persistent after image, report for immediate medical attention.

Laser Care and Maintenance

The laser is expensive, sophisticated equipment. It should be handled and maintained with care. The laser and all its components should be properly stored in the carrying case when not in use. The laser should be kept reasonably clean. However, the laser optics should not be cleaned with harsh solvents. Use Windex or cleaning material designed for photographic lenses. The power cables and connectors should be periodically checked to make sure they are in good repair. When the laser is stored for a long period of time, it should be kept in a cool dry place. It should be operated overnight once a month.

MEASURING TAPES

The most common measuring tapes (builders' tapes) used in construction work are steel or fiberglass tapes, 50 or 100 feet long, which are divided into feet, inches, and eighths of an inch.

Fig. 2-19. A tape with an end fastener enables a worker to make measurements as shown.

See Fig. 2-18A. Most tapes of this type shown in Fig. 2-18A have a ring included in the first inch. The tape shown in Fig. 2-18B is divided into metric units. The engineer's or surveyor's tape is divided into inches and tenths of inches. See Fig. 2-18C. The tape shown in Figs. 2-18D and 2-19. has an end fastener and is typical of those used by carpenters. The end fastener is a convenience when a person is alone and has to make measurements. It does not have the all around usefulness of a tape with a ring. The end fastener, like the ring, is included in the first inch.

Using the Steel Measuring Tape

Whenever you begin to use the measuring tape, place the ring of the tape over a nail on a stake or batterboard. The nail should be offset enough to allow the end of the ring to coincide with the point of measurement. The tape should be held flat and taut, and the measurement read accurately at the other end. Some carpenters prefer to set up the tape so that the point of measurement coincides with the one-foot mark on the tape. In such instances, a foot must be subtracted from the final reading on the tape. Carpenters who engage in this practice feel it provides a more accurate reading and eliminates errors caused by damage to the ring or the end of the tape.

Caring for the Measuring Tape

Since the measuring tape is basically a very thin band of steel or fiberglass and often gets used over rough ground, it should be given special care to prolong its life and usefulness. Avoid kinking and twisting the tape. Whenever the tape becomes looped, do not attempt to straighten it out by pulling on it. When you are winding the tape after taking a measurement, do not allow the tape to drag on the ground more than is absolutely necessary. Breakage usually occurs in the first few feet of the tape simply because this is the part of the tape which gets the greatest wear. To avoid excess wear, get in the habit of using a pocket tape for short measurements. Occasionally clean the tape with a rag. Steel tapes should be lightly oiled with household or sewing machine type oil.

Metric Measuring Tapes

Tapes graduated in metric units are available. These tapes are divided into meters, decimeters, and centimeters. The centimeters are further divided into millimeters. A metric measuring tape is shown in Fig. 2-18B.

The basic unit for measuring land and laying out the buildings is presently the foot. Once metric dimensioning is adopted for buildings and other types of construction, metric measurements will also be used for laying out and locating the structures on the land site. The carpenter will find it necessary to use metric tapes and rules and sometimes convert from English (the foot-pound system) units to metric measure.

Actually reading a metric tape is quite simple. The units used are all multiples of the number 10. One meter, for example, is made up of any of the following: 10 decimeters, 100 centimeters, or

1000 milimeters. Whenever all of the dimensions are given in metric units, it is quite simple to add or subtract quantities because all units are expressed as decimals.

1 meter = 39.97 inches
1 decimeter = 3.93 inches
1 centimeter = .39 inches
1 millimeter = .039 inches

Conversion to metric units. Chapter 9 discusses metric measure as it applies to the construction industry. It further discusses the procedure for converting from English to metric units.

LEVELING RODS AND TARGETS

When sighting long distances (more than 150 feet), you must use a leveling rod. See Fig. 2-20. The carpenter, however, generally uses a ripping and a rule for shorter distances. See Fig. 2-21. The carpenter who is doing the sighting can turn the instrument to bear on the four corners of the proposed building. The ripping can be held in the same way at each of these corners. One method is to hold a measuring tape or rule against the ripping in such a way that the inches can be read through the instrument. It is important to hold the rod or ripping in a vertical position. For very precise work, hold a hand level against the rod or stick.

Leveling rods are used to measure the difference in elevation between the point of beginning, where the level is located, and the place where the rod is held. Some rods are called *self-reading.* The measurement is read by the person operating the level. Other rods have targets that allow the person holding the rod to read the measurement. Rods are graduated in two ways. Some rods are divided into feet, inches, and eighths of an inch. Other rods are divided into feet, tenths, and hundredths of a foot. Obviously, when you use a rod, you must know how it is divided.

Signals

Hand signals A and B, shown in Fig. 2-22, can be used when the persons at the instrument and

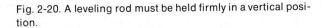

Fig. 2-20. A leveling rod must be held firmly in a vertical position.

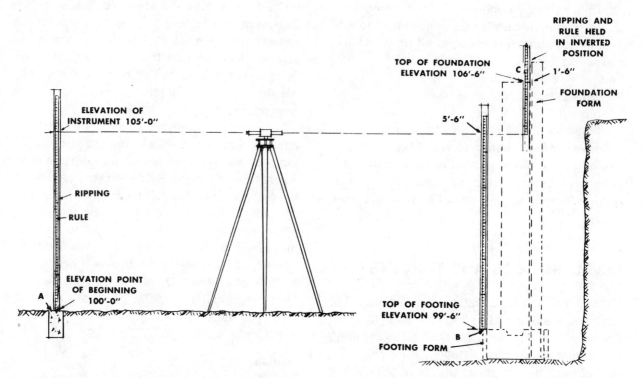

Fig. 2-21. A rule, held against a ripping, and a leveling instrument are used to transfer elevations from the point of beginning (A) to footing forms (B) or foundation wall forms (C). The marks on the rule are read at the point of beginning. The instrument is then revolved to bear on a rule at the desired location. The top of the foundation wall is determined by inverting the rule, sighting the lower end, and marking the height on the inside of the form with a nail. The elevations are shown on the working drawings. The elevations shown in this illustration are typical of those used.

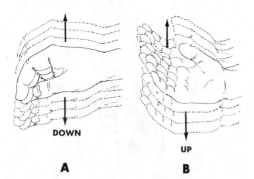

Fig. 2-22. The level operator uses hand signals to tell the helper to raise or lower the rod. The hand is moved up and down from the wrist as shown. Palm down signifies *Down*, palm up signifies *Up*.

the rod (the carpenter and a helper) are comparatively close together. Signals A and B are made from the wrist, with a rapid up and down motion. The motion becomes slower as the target approaches the horizontal hairline when sighted in the telescope. Hand straight out indicates the target is *on grade*. See Fig. 2-23.

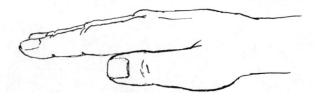

Fig. 2-23. When the target is on grade, the level operator signals by holding the hand in a level position.

MOVE TARGET LEFT

MOVE TARGET RIGHT

Fig. 2-24. The signal to lower the target. The photograph shows the level operator at the instrument. (Realist, Inc.)

TARGET PLUMB

Fig. 2-25. The signal to raise the target.

Fig. 2-26. Signals used to tell the helper to get the target on the proper plumb line. To indicate that the target should move sideways (right or left), an arm is extended and the other is held at the side. The rod is moved in the direction that the arm points. Both arms extended straight up indicates that the target is plumb.

Arm signals are necessary when the distance is over 150 feet. Figures 2-24 and 2-25 show how the arm is moved closer to the horizontal as the target approaches the instrument hairline. Figure 2-26 shows the arm motions for proper rod placement. One arm is extended sideways, and the other arm hangs at the side. Both arms extended horizontally indicate *on grade.* See Fig. 2-27.

Fig. 2-27. Both arms extended horizontally (level) signals that the target is on grade.

USE OF THE LEVEL OR TRANSIT LEVEL

Either a builders' level or a transit level can be used for most simple operations. However, the transit level is more versatile. With it, you can not only perform a wider range of operations. You can also do some of the more basic operations with greater accuracy and speed.

Leveling

Finding the difference in grade between several points and the transfer of the same elevation from one point to another is called *leveling.* Either instrument can be used to transfer the elevation established at the point of beginning (a reference point on a sidewalk or a stake) to the point where it is used to determine the elevation of the first floor, the top of the footings,

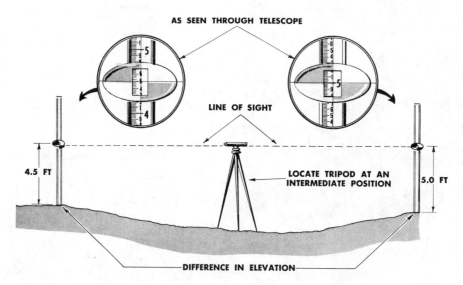

Fig. 2-28. When transferring an elevation from one point to another, place the instrument at an intermediate location. The elevation of the ground to the left is ½ foot higher than the ground to the right. (**Note:** The enlargements at the top show that the rod is divided into tenths of a foot.)

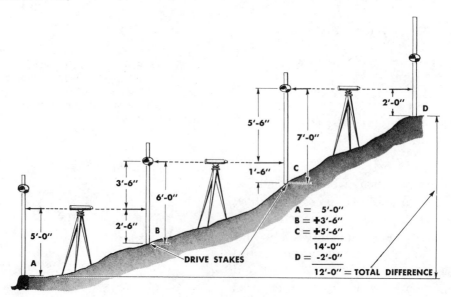

Fig. 2-29. The instrument may have to be moved several times when the rise from one elevation to another cannot be sighted from a single position.

levels of batterboards, etc. To transfer an elevation from one point to another, set up instrument at some intermediate point and adjust it as outlined in the previous sections. A rod is held at one of the points, then a reading is made through the telescope. A rod is held at the other point, and another reading is made. The difference between the two readings is the difference in elevation. See Fig. 2-28.

If you cannot sight the distance from the highest to the lowest point from one position, then you have to relocate the tripod as often as necessary. Each time you move the tripod, set it up carefully and level the instrument. The successive readings must be added or subtracted accurately. See Fig. 2-29. You should note that the rod is only moved after the tripod is set up in the new position and an accurate reading is taken.

Determining Elevations

Elevations in most local urban areas are established in relation to the height of certain designated markers. The urban building commission is responsible for setting up this system of reference. The basic marker or markers are called *bench marks*. The surveyor uses these points to locate the position and the elevation for a *point of beginning* on or near a particular piece of property. The point of beginning is an important reference point for determining floor elevations, the bottom of foundation walls, the drainage of water after grading, etc. A common procedure is to arbitrarily designate the elevation of the established point of beginning as 100′–0″. Therefore, an elevation of 110′–0″ would be 10 feet above the point of beginning, and an elevation of 95′–0″ would be 5 feet below the point of beginning.

You may have to make two or more position settings with the instrument when you encounter steep hillside lots. Use the same principle as shown in Fig. 2-29.

When other stakes or batterboards are about to be erected, set up the tripod as near the middle of the building site as possible. See Fig. 2-30. This minimizes errors in leveling and can somewhat simplify the focusing. The need for focusing is reduced if A,B,C,D,E, and F, shown on the illustration, are as equally distant as possible. Once the instrument has been leveled and the telescope locked into a horizontal position, the elevation established at any one corner can be easily transferred to all the other corners. Simply revolve the telescope and sight each corner.

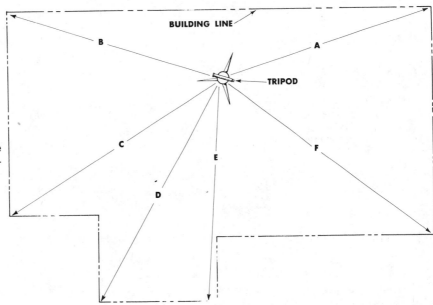

Fig. 2-30. The instrument should be located at some central position within the excavation.

Laying Out Building Lines

The transit level is a fast and accurate means for laying out building lines. It is used as a level to establish the elevation at each corner. It is used as a transit to determine right angle corners or other angles, as called for on the working drawings.

As an example, let's assume a building 100 feet by 190 feet is to be laid out on a particular lot. This example is illustrated in Fig. 2-31. First, you must establish the first corner stake at point *A*. Measure the required distance of 20 feet on one side and 30 feet on the other side from the lot lines. Next set up and level the transit level over the point of beginning. Now, with the telescope locking levers engaged, sight a pole held on the top of corner stake *A*. Drive the stake into the ground until you establish the required elevation. If the lot is flat, it may not be necessary to establish the elevations at this time, provided you are careful all the other stakes are driven to approximately the same elevation. Again measure the distance from the lot lines and mark a cross on the top of stake *A* so that it indicates the exact point.

Set up the instrument over the point on corner stake *A*, as shown in Fig. 2-32. The helper then drives a temporary stake far down the lot at *B-1*, 20 feet from the lot line, and makes a mark on the stake. While you are holding the end of a steel tape on the mark on stake *A*, have the helper measure 100 feet and drive in a stake which is in line with both *B-1* and corner stake *A*. Drive the new stake, *B*, into the ground so that the elevation is equal to the elevation at corner stake *A*. With the telescope locking lever in a fixed position, sight a rod, or a line and plumb bob which is held over the mark on stake *B-1* by the helper. Tighten the horizontal-motion clamp screw and release the telescope locking levers. Depress the telescope so that the mark on stake *B-1* is seen. Depress it further to bear on stake *B*. See the detail shown in Fig. 2-32. The helper is directed by hand signals and marks a line on stake *B*. After again checking the measurement of 100 feet from the point on corner stake *A*, the helper marks a cross on stake *B*. This locates a corner of the house.

Now you must measure a right angle to establish stake *C*. Keep the telescope set and leveled over point *A* and engage the locking levers. Now sight a line and plumb bob which are held over the point at stake *B*. The sighting here should be the same as the previous sighting. Then adjust the graduated horizontal circle so that the index

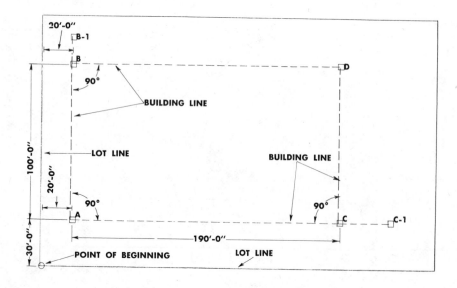

Fig. 2-31. The problem of laying out building lines, such as those shown to the left, can be solved with a transit level.

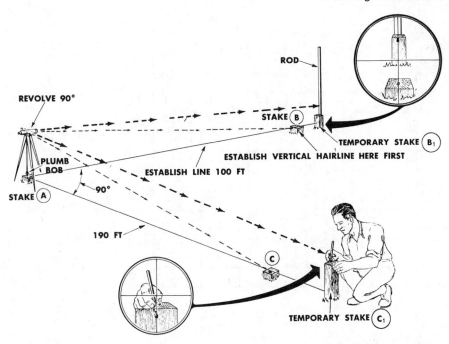

Fig. 2-32. The transit level provides real advantages when you are laying out the corners of a lot or a foundation.

reads 0°. Release the horizontal-motion clamp screw and turn the telescope clockwise until it reads exactly 90°. The helper holds a rod beyond the required distance so that he can locate the position of a temporary stake at *C-1* and drive it into the earth. Using a line and a plumb bob and observing your hand signals, the helper establishes a point on stake *C-1,* as shown in Fig. 2-32.

Use the same procedure used for stake *B* to establish the position and elevation of corner stake *C*. The helper marks the point on corner stake C which indicates the location of another corner of the building.

The last corner stake to be located is stake *D*. See Fig. 2-31. Set up the instrument directly over the mark on corner stake *C*. While the telescope locking levers are engaged, sight a line and a plumb bob held over the mark on corner stake *A*. Tighten the horizontal-motion clamp screw and adjust the horizontal circle to read 0°. Now release the horizontal-motion clamp screw and move the telescope clockwise until it reads exactly 90°. Tighten the horizontal-motion clamp screw again. Hold a tape on the mark on corner stake *C* while the helper locates a point 100 feet

away and drives in the corner stake at *D*. This corner stake should be driven to an elevation which corresponds to corner stakes *A, B,* and *C*.

The helper measures a distance of 100 feet from corner stake *B* to make a cross mark on corner stake *D*. Now release the telescope locking levers and depress the telescope to bear on the corner stake. The helper, following your hand signals, marks the exact line at right angles to line *AB*. The helper then draws the measuring line taut again and drives a nail at point *D* on the line.

Take one end of the steel tape, while the helper holds the other end, and check sides *AC* and *BD* to verify that they are exactly 190 feet. Now check sides *AB* and *CD* and verify they are 100 feet long. Then measure the diagonals *AD* and *BC* to make sure they are equal.

The 6-8-10 method. The 6-8-10 method can be used for laying out small buildings when a leveling instrument is not available. It can be used in many different applications, during the construction of a house, to determine right angle corners. It is based on the geometrical fact that a triangle with sides of 3, 4, and 5, or multiples of

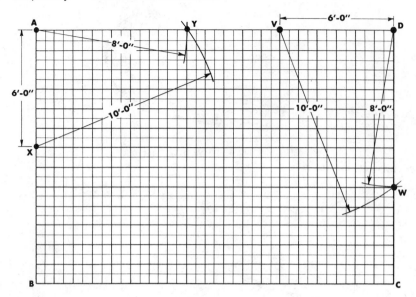

Fig. 2-33. The 6-8-10 method can be used to lay out the right angles of a building.

these numbers, contains a right angle. However, this method is only as accurate as the precision with which the sides of the triangle are measured. It cannot replace the leveling instruments.

Use the largest multiple you can in each application. For example, when checking the corners of layout shown in Fig. 2-31, use 36, 48, and 60 feet, or some other large proportional measurements.

Assume a building with dimensions *AB* and *AD,* shown in Fig. 2-33, is to be laid out. Establish corners *A* and *B* by measurement and stretch a line between them. Measure a distance of 6 feet from point *A* and drive a stake to locate point *X*. With a radius of 10 feet from point *X* and 8 feet from point *A*, locate point *Y*. Drive a stake at point *Y*. Extend line *AY* and measure distance *AD*. Drive a stake at point *D*.

Angle *XAY* is a right triangle. Points *A, B,* and *D* have been located. With point *D* as the apex of a right triangle, located and mark point *V* on line *AD,* 6 feet from point *D*. Using two tapes, with a measurement of 10 feet from point *V*, and 8 feet from point *D,* locate point *W*. Extend a line from point *D* through point *W*. Measure a distance on this line equal to *AB*. Drive a stake and locate point *C*. Line *BC* should equal line *AD* if you have measured accurately.

When you check the layout you will find that line *AB* equals line *CD,* line *AD* equals line *BC,* and the diagonals, dimensions *AC* and *BD,* are equal. If the building were larger, 12-16-20, or 30-40-50, could be used instead of 6-8-10.

Irregular shaped buildings. When buildings have an irregular shape, as shown in Fig. 2-34, first you lay out lines *A, B, C,* and *D,* then complete offsets and wings. Chapter 3-1 explains

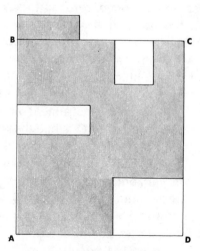

Fig. 2-34. A building with an irregular shape should be laid out from a large rectangle. The offsets and wings are laid out later.

how House Plan A, which is irregular in shape, is laid out.

Running lines.   The transit level is in a transit position when the telescope is locked for horizontal motion. However, it is free to move on a vertical plane. The instrument is therefore useful both for setting a series of stakes in a straight line or for plumbing walls and columns.

Set up the transit level directly over a reference stake which is the base point for the line of stakes. After the instrument has been leveled, release the telescope and tilt it until you see a stake set in the line of sight in the far distance. See Fig. 2-35. Tighten the horizontal clamp screw so that the telescope cannot move except in a vertical plane. By depressing or elevating the telescope, you can find any number of intermediate points. All of them will be exactly on the line of sight. This operation cannot be done in the same manner with a builders' level.

Plumbing walls or columns.   When a building wall or other vertical line is to be checked, set up and level the transit level at some convenient distance but approximately in line with one of the walls. See Fig. 2-36. Release the locking levers and swing the telescope so that it bears on the line or point. Tighten the horizontal clamp screw and make fine adjustments with the horizontal tangent screw so that the vertical cross hair lines up exactly with the point. As the telescope is rotated in a vertical plane, all of the points which constitute the corner of the build-

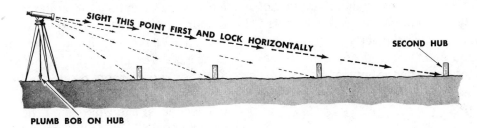

Fig. 2-35. A row of stakes can be set in line with a transit level.

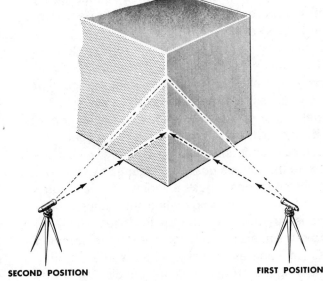

Fig. 2-36. A vertical line can be plumbed when the instrument is used in a transit position.

ing or vertical member are sighted on a plumb line.

Set up the instrument in a second position about 90° from the first position and about as far away. Repeat the operation to determine that the line is truly vertical. See Fig. 2-36.

Measuring vertical angles. On rare occasions, a carpenter may have to measure an angle on a vertical plane. Other mechanics, particularly those who deal with pipe and use angle fittings, find considerable use for this particular feature of the transit level. The operation is identical to that for measuring angles on a horizontal plane. The instrument is brought to a perfectly level position and sighted on a base point. The horizontal line of sight is one side of the angle to be measured. The telescope locking levers are released, and the telescope elevated or depressed to bear on the line or point on the other side of the angle. By tightening the vertical-motion clamp screw and using the vertical-motion tangent screw, you can make fine adjustments before reading the angle on the vertical arc.

SAFETY TIPS

Frequently, a discussion of leveling instruments puts a lot of emphasis on caring for the instruments themselves, while little or nothing is said about the safety of the persons using such instruments. Personal accident prevention is most important on the jobsite. Various safety tips can provide the guidelines for an accident free work environment.

Sighting work is often done on rough terrain. The jobsite is frequently wet, muddy, icy, slippery, and may have a steep slope. An accident can occur if you are not careful where you are walking or where you stand. Whether you operate the instrument or serve as the helper, you can be in danger of slipping or falling while moving to the various positions required by your job. Such situations require a variety of precautions, all of which cannot be spelled out because situations vary considerably from job to job. The thing to remember is that a lot of safety is common sense, such as watching where you are going, wearing the proper equipment, including sturdy

work shoes, a plastic hat, and protective clothing. Shoes have been designed for a variety of conditions. They should have reinforced toes which can prevent injury from falling objects. The soles should be designed to prevent slipping on oily or other slippery surfaces and strong enough to prevent punctures from nails or other sharp objects.

Reference points are sometimes located in relatively dangerous places, such as on a street or road or a busy walk. In instances of this sort, steps have to be taken to reroute or otherwise control the traffic flow until the work is completed. The level or transit operator is especially exposed to danger because of the concentration the job requires. The helper holding the rod is also vulnerable because of the need to maintain the position of the point. When traffic is a factor, a flagperson can help direct the traffic. The flagperson can also help the persons with the instrument and the rod by giving them verbal instructions in case of any danger.

It should be a general rule on the job that excavating and grading are temporarily discontinued whenever site work is in progress. Excavating machinery and trucks are potential sources of danger. Deep holes made during excavations cannot always be protected with fail-safe barriers. The helpers, working with the level and the rod, must be alert in order to avoid serious falls.

Whenever a laser beam instrument is used, you must take special care to avoid damaging the eyes. Various safety tips regarding the use of laser beams have already been discussed earlier in this chapter. See pp. 42–43.

THE SURVEYOR'S PLAT

One of the responsibilities of the carpenter in most residential work is to lay out the building, footing, and foundation lines. The necessary information can be found on those sheets among the working drawings which show the plot plan, the basement plan, the wall section, and the elevations. Occasionally the carpenter must consult the *surveyor's plat* in order to locate reference points, the lot corner stakes, and to obtain other information.

The surveyor's plat is a drawing which gives the legal visual description of the lot. It includes accurate dimensions of the boundaries of the property, the angles at each corner, the exact location and width of streets, alleys, and public walks, the existing trees, utility poles, grade elevations, location of sewers, and the grade elevation at the bottom of sewers, and so on.

It is extremely important that all the data about the lot be verified by a licensed surveyor on the site. This data must be checked by the surveyor so that it complies with the official plat of the area on file with the local building or zoning authority. The surveyor may have to determine the information by accurately surveying from a bench mark or other intermediate reference point located some distance away. In most communities, the surveyor's plat is required before a building permit is issued, or the loan is negotiated.

THE ARCHITECT'S PLOT PLAN

The plot plan is part of the set of working drawings. It serves several purposes. The local building authority studies the plan to make certain all zoning ordinances are complied with. The carpenter studies the plot plan to locate the dimensions he needs to establish and run the lines for the excavation and the concrete form work. Cement finishers or concrete workers use it to obtain information about walks and driveways. Electricians need to determine the location of utility poles or underground cables. Plumbers find the location of city water and sewer pipes. Excavating and grading workers find varioius information about preparing and finishing the ground.

Reading Plot Plans

The plot plan shown in Fig. 2-37 is the plot plan for House Plan A already studied in print form in Chapter 1. It is typical of many plot plans. The following items are some of the things you should study whenever you examine a plot plan.

The scale. Plot plans must be drawn at a smaller scale than other working drawings because they include the entire lot. Some of the scales frequently used are $\frac{1}{8}'' = 1'0''$ (each $\frac{1}{8}$ of an inch on the drawing is equivalent to one foot on the actual building site), $1'' = 20'$, or $1'' = 50'$.

Compass direction arrow. The compass direction arrow, as shown here, is always shown pointing North. If the craft worker keeps this direction in mind while reading the working drawings, he or she can better understand the relationship between the plot plan, the elevation views, and the setting of the house on the lot.

Dimensions of the lot. The property lines, which are the boundaries of ownership, are sometimes shown with strong dot-dash lines. Dimensions are given along each lot line, and the angles for each corner are shown. These dimensions are taken from the surveyor's plat and are given in feet and tenths and hundredths of a foot, or in feet and inches.

The point of beginning. When the lot is surveyed, the surveyor makes certain the corners are correctly located in relation to the fixed point of reference for that neighborhood established by the local zoning authority having jurisdiction, e.g. engineering, public works, or zoning departments. The surveyor again makes certain each corner of the lot is identified by a permanent marker such as a cross cut in the sidewalk, a stake, a pipe, or a concrete monument. One of these points, which might be a corner or some other marker on or adjacent to the lot, is often designated as the *point of beginning,* a term we have already used quite frequently. The carpenter uses the point of beginning to establish elevations and to measure distances for the layout of the building lines.

Streets, sidewalks, and trees, etc. Existing streets, sidewalks, and other special features, such as retaining walls, outcropping of rock, boulders, and so are, are shown on the plot plan. Present buildings or structures, which are not slated for removal, are likewise indicated. The plot plan also includes such detail as the approximate location and size of trees and shrubs.

Utilities. The plot plan gives the location of gas, water, and sewer pipes in streets or parkways. Power and telephone poles are drawn in their respective locations. An easement is required when utility cables are buried under-

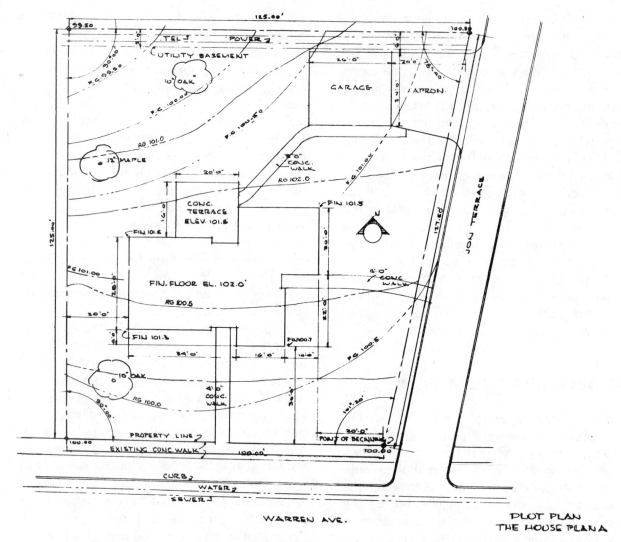

Fig. 2-37. A typical plot plan provides a variety of data including the location of building elevations at several points.

ground or cross over the lot. The easement is simply a strip of land set aside by mutual agreement of the owner and the utility company so that utility workers will have easy access to power lines.

(All of the above information is generally shown on the surveyor's plat and then transferred to the architect's plot plan.)

Dimensions for the location of the building. The location of the building must comply with the local zoning and building codes. The local zoning and building commission or authority review a set of the working drawings before they

issue a building permit. The plot plan is examined to verify that the building is setback from the front property line as required, and the side and rear yard limitations are observed. The ratio between the area of the building and the overall area of the lot is calculated to likewise verify that this ratio falls within the limits set up by the local ordinance.

The architect and owner agree on the best location for the house or building. They take into consideration such factors as the view, the best use of sunlight and the prevailing winds, provisions for outdoor living, the location of existing and projected trees, and the overall contour of the lot, and so on. The architect then locates the building on the plot plan and includes all necessary dimensions to property lines and the point of beginning.

Elevations. The plot plan not only shows the horizontal dimensions which locate the building, but it also includes the various elevations. (The height of any point on the lot is its *elevation*.) This elevation is measured in relation (above or below) some point of reference, such as the point of beginning. The elevation of the first floor is extremely important because the depth of the excavation and concrete formwork has to be calculated so that the finished first floor is exactly at the stipulated height.

Contour lines. Contour lines are imaginary lines which are drawn to pass through points of the same elevation. The surveyor can use contour lines on his plat to indicate the rough (undeveloped) grade, especially if the lot is located on a hill or has some other unusual slope. The architect often uses contour lines on the plot plan to show the existing and finished grades (i.e., *R.G.* means rough grade, and *F.G.* means finish grade). The shape of the finished lot must be designed to provide adequate runoff of surface water while still enhancing the overall beauty of the property. The contour lines are drawn so that each level represents an increment of the same amount of elevation. For example, an increment of ½ foot is commonly provided for relatively flat lots. On steep lots, however, the increment between adjacent contour lines may be as much as 2 feet or more.

New walks and drives. The plot plan shows all the new walks and drives which are to be installed by the concrete contractor. This plan often shows the precise dimensions for the walks and drives because the dimensions are not given on any of the other working drawings.

STAKING OUT A HOUSE

The plot plan for House Plan B is used as an example for staking out a house. A small house was chosen for this study of laying out the building lines in preparation for the excavation and the erection of the concrete forms. This house is rectangular in shape, with an offset. Since there is a crawl space rather than a basement under the house, the foundation wall is comparatively low, and the excavation is shallow. See Figs. 2-38, 2-39, and 2-40.

Staking Out House Plan B

The lot for House Plan B is 75 feet by 125 feet, and fairly level. The garage and storage area will be laid out and poured later. The corner stakes for the lot have already been located by the surveyor. The Southeast intersection of the sidewalks, is used as the point of beginning from which measurements and elevations are taken. See Fig. 2-40.

Procedure for staking out house plan B. Locate a point 30 feet from the south lot line and 12 feet from the east lot line and drive corner stake *A* at this point. It is important that the lines are held approximately level when the horizontal measurements are made. Drive a nail into the top of the stake to indicate the exact location of the point. Ideal stakes can be made by cutting a 2 × 2 inch stick into pieces 2 feet in length and sharpening one end. Set corner stake *B* 48 feet to the west of corner stake *A* and 30 feet from the south lot line.

The manner of making a right angle corner at corner stake *A* and establishing corner stake *C* can be done using a transit level or a builders' level.

If a transit level is available, set it up over corner stake *A*, using a plumb bob to get the exact location over the stake. After the instrument has been leveled, release the lock-release lever so that it can move in a vertical plane. Depress the telescope and make a sight on corner stake *B*. Turn the telescope through 90°, measure 28 feet from corner stake *A*, drive corner stake *C*, and locate point *C* on the stake with a nail.

If a builders' level is available, follow the same procedure except that the telescope cannot be depressed to sight the stakes. A plumb bob attached to a string can be held above the nail on the stake. Now swing the telescope through a 90° arc by reading the horizontal circle. Using the rod for sighting, set corner stake *C* and mark

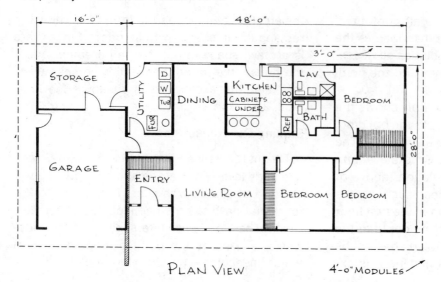

PLAN VIEW

4'-0" MODULES

Fig. 2-38. The Plan View for House Plan B.

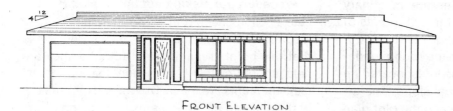

FRONT ELEVATION

Fig. 2-39. The Front Elevation for House Plan B.

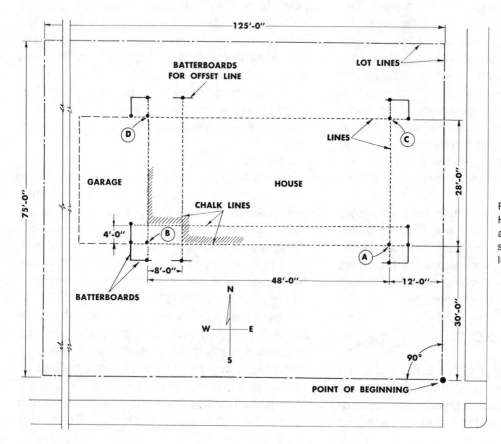

Fig. 2-40. The Plot Plan for House Plan B. Batterboards are erected and lines stretched to indicate the location of building lines.

the point with a nail.

Move the transit level or the builders' level to a position directly over corner stake *B* and follow the procedure indicated in the previous step to locate corner stake *D*. Mark point *D* with a nail.

Check all of the building lines to make sure that they have the right measurements. Check the diagonals of the rectangle to see that they are equal. Measure the distance to the lot line to check that the building is correctly located.

(Note: In the parts of the country where there is no frost problem, the floor slab can be laid directly on the ground. No deep excavation is necessary. A trench which is large enough to provide for the footing formwork and the low foundation wall is dug. Stakes which have been accurately laid out are sometimes considered sufficient. Batterboards are dispensed with in these instances.)

The next operation is the erection of batterboards. Select three pieces of 2 × 4 inch material for each corner of the building and sharpen the ends. Make them long enough so that when they are driven firmly in the ground, they will extend to a convenient height above grade. Drive the stakes to be parallel to, and about 4 feet away from, the building lines. See Fig. 2-41. Verify

with the excavating contractor the clearance needed for the machinery to operate, then set the batterboard stakes farther out if needed.

Nail a 1 × 6 inch ledger board to the stakes at corner *A* at the same height as the top of the foundation wall. (Ledgers are horizontal boards.) You can do this by leveling with the instrument from the point of beginning, as shown in Fig. 2-42. Carpenters often set the batterboards a uniform dimension above the foundation grade at each corner so that the lines, when strung, do not interfere with the form building operations.

Set the transit level at a central location in the building area and sight the ledger height on the other batterboard stakes and mark this elevation on each of the stakes.

Nail ledger boards in place at the marks on the stakes. Use a hand level to check that the ledger boards are perfectly level.

Using lines and a plumb bob, stretch a line so that it passes over the points marked on corner stakes *A* and *B*. Mark the top of the ledger where the lines cross. Make a shallow saw kerf in the ledger, pass the line through the kerf, and fasten the line to a nail. (A *kerf* is a saw cut.) Some carpenters, however, drive a nail into the top edge of the ledger instead because the nail can be

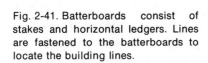
Fig. 2-41. Batterboards consist of stakes and horizontal ledgers. Lines are fastened to the batterboards to locate the building lines.

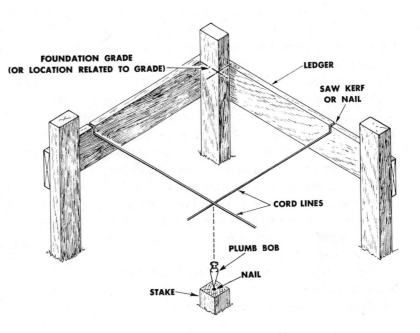

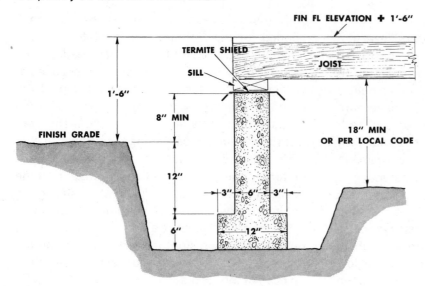

FIN FL ELEVATION **+ 1'-6"**

TERMITE SHIELD

SILL

JOIST

1'-6"

8" MIN

FINISH GRADE

18" MIN
OR PER LOCAL CODE

12"

3" 6" 3"

6"

12"

Fig. 2-42. This section view taken from the working drawings shows the relation between the grade and both the top of the foundation and the bottom of the footing.

easily moved if there is an error. Repeat this operation for the other three building lines.

(Note: When an offset occurs, such as at corner *B,* extend the batterboards or erect new ones and stretch additional lines. See Fig. 2-40.)

Alternative Method of Staking Out House Plan B

There is a fundamental difference between the first method and the alternative method of staking out the house. With the alternative method, you can eliminate some of the work of setting the corner points of the building, while maintaining great accuracy and precision. Basically, it is done by determining the location of lines on batterboards by measuring from property lines rather than plumbing to points on the stakes.

Procedure. Locate a point 30 feet from the south lot line and 12 feet from the east lot line and drive in corner stake *A* at this point. Set corner stake *B* 48 feet to the west of corner stake *A* and 30 feet from the south lot line. Set corner stake *C* 28 feet north of corner stake *A* and 12 feet from the east lot line. Set corner stake *D* 48 feet west of corner stake *C* and 28 feet north of corner stake *B.* Check the measurements and the diagonals.

The next operation is the erection of batterboards. Make stakes of 2 × 4 inch pieces, long enough to extend at least 6 inches above the

foundation level. Drive the stakes so that the ledgers are parallel to, and about 4 feet away from, the building lines. Verify with the excavating contractor the clearance required to permit the excavating machinery to operate. Move the batterboard stakes back if more clearance is necessary.

Set the transit level at a central point and sight the exact foundation grade, marking it on the batterboard stakes at each corner. Check the section view of the working drawings to establish the grade elevation. See Fig. 2-42.

The marks on the batterboards which indicate the placement of the ledgers are usually made 2 inches or higher above the actual point so that the lines, when strung, will not interfere with the formwork. Nail 1 × 6 inch ledgers on to the stakes. The top edge is nailed on the grade marks, or the grade level is adjusted as described. Make sure that the ledgers are level.

With the use of a tape and a plumb bob, tape the exact dimension from the lot lines and place marks on the top of the batterboards at corner *A,* shown in Fig. 2-40. The method is shown in Fig. 2-43. Make measurements at corner *B* in the same manner.

Using a transit level or a builders' level, establish points on the batterboards at corner *C* and corner *D,* as shown in Fig. 2-40. After all of the

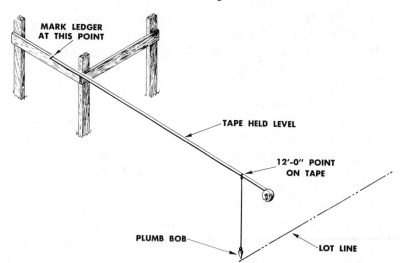

Fig. 2-43. The building lines are marked on the batterboards by a direct measurement from the lot lines.

ledgers have been marked, drive nails at each point on the top edges of them. Then string lines between the nails. Drop plumb lines from the intersecting lines to locate the building corners precisely on the stakes at corners *A,B,C,* and *D.*

Check the dimensions between the points and the lot lines. Check to make sure that sides *A-B* and *C-D* are both 48′—0″. Verify that sides *A-C* and *B-D* are both 28′—0″. Likewise check that diagonals *A-D* and *B-C* are equal. See Fig. 2-40.

QUESTIONS FOR STUDY AND DISCUSSION

1. What is the basic difference between a builders' level and a transit level?
2. How does a dumpy level differ from a light builders' level?
3. What is a laser beam?
4. How is a plumb bob used with a leveling instrument?
5. What steps are followed when you level a builders' level?
6. How do you use a builders' level to lay out a line at right angles to another line?
7. How is a vernier used?
8. What is the smallest division you can measure with the vernier as discussed in this chapter?
9. How is a transit level used as a transit?
10. What is the feature of the automatic level which gives it final perfect leveling?
11. What important directions must be followed when you set up a tripod?
12. What is the safe method for carrying a mounted instrument around trees and scaffolding?
13. How is the laser instrument used to establish elevations?
14. How is the laser read at the sensor end of the beam?
15. What important safety practices should be followed when laser equipment is used?
16. What divisions (units) are used on each foot of a builders' tape?

17. What divisions (units) are used on each foot of an engineer's or surveyor's tape?
18. What is distinctive about the first inch of a builders' steel tape?
19. What is the purpose of the target on a leveling rod?
20. How do you use a leveling instrument for leveling from the point of beginning to other points?
21. What is a bench mark?
22. What is a point of beginning?
23. What final check do you make to verify that the foundation lines are square?
24. Describe the 6-8-10 method.
25. How are stakes set in line with a transit level?
26. Why must the transit level be transferred to a second point when you plumb a column?
27. What is a surveyor's plat?
28. How does the carpenter use an architect's plot plan?
29. The plot plan is checked to make sure it complies with what zoning restrictions?
30. What is the function of contour lines?

CHAPTER 3
FOUNDATIONS FORMWORK

Concrete formwork is a phase of construction which is distinctly the work of the carpenter. Some carpenters specialize exclusively in preparing the job site and erecting the forms used for the foundations of small homes. For other carpenters, concrete formwork is simply one part of their overall skills and knowledge. They work in all areas of house construction, from formwork to roofing.

Reinforced concrete construction is a division of primary importance in the industry today. In fact, an entire sector of the carpenter's trade is devoted to the erection of buildings which use reinforced concrete construction. Carpenters are responsible for erecting all the formwork for very tall buildings made from reinforced concrete. This includes buildings fifty stories or higher. They likewise build the formwork for bridges, dams, and public works, such as sewage disposal plants, water filtration plants, and highway projects. They are essential in the building of nuclear and electrical power plants. Architects and engineers frequently design strikingly imaginative structures made from reinforced concrete. It is the task of the carpenters to create forms which are strong enough to hold the concrete in the desired shapes.

Concrete formwork for residential construction requires a certain amount of skill and know-how. For example, the carpenter must know how to read plans, understand the properties of concrete as it is placed, and be familiar with the different reactions of concrete to hot and cold temperatures. The carpenter who is doing formwork must also understand the basic engineering principles which go into building and bracing the forms. The materials used must provide safe and strong structures which are capable of retaining the concrete until it sets.

It is nearly impossible to overemphasize the care with which the footings and foundations for homes must be constructed. Footings must always be laid on firm soil to prevent cracks from developing in the foundation wall. (Cracks in the foundation wall allow water seepage.) Even though forms are cumbersome and awkward to handle and fasten in place, the carpenter must make certain the form walls are erected so that the corners are square and plumb and that each side has the required dimensions and is in perfect alignment. The top of the concrete in the forms must be made level at the designated height so that the finished wall will meet grade requirements. When these procedures are not carried out accurately, it is necessary to make adjustments in the framework of the entire structure as it is erected. This of course creates great problems, not to mention increasing cost.

REGIONAL VARIATIONS

Since formwork problems vary from region to region, the formwork which results is often quite different. Carpenters who work exclusively in the South and West will find that some sections of the material covered in this chapter are unfamiliar. Likewise, carpenters who work exclusively in northern areas will notice coverage of types of formwork used primarily in other geographic regions. However, it is valuable for all carpenters to become somewhat familiar with at least some of these different methods.

In the South and West, light wood frame construction is more widely used than masonry. There is no need for deep footings and foundation walls since frost is not a problem for most warmer climate areas. The footings and the foundation are often constructed in one piece in the shape of an inverted *T*. When the soil composition permits, the footing is formed by directly cutting a trench in the ground. The foundation wall is low, generally just high enough to prevent termites from infesting the wood members and yet provide access under the floor.

One type of foundation used here is called the *slab at grade.* In this application, a simple foundation or footing is placed around the perimeter, and a slab is placed over the ground and serves as the floor. Both the foundation and the slab can be placed in one piece.

A second type of foundation used is a combination of a footing and a low foundation wall. The wall is placed high enough to allow room for a crawl space under the floor joists.

In northern regions, the foundation and footing must be protected against frost damages. Almost all types of soil hold moisture from rain and snow. This moisture freezes during colder weather. Frozen soil expands and tends to rise above its normal elevation. The level recedes again once the soil thaws. In order to prevent damage caused by the expansion of the frozen soil, the footings and foundation walls are extended below the frost line. The exact depth required is specified locally. Figure 3-1 shows the depth of frost penetration across the continental United States. Table 3-1 gives the depth for footings as required in various U.S. cities.

Many residential buildings in northern climates have a basement as well as a second level or story. This arrangement provides a minimum amount of heat loss in relation to the overall living area in the dwelling. Brick masonry is used in some areas. When it is used, the footings and the foundation walls are specifically designed to

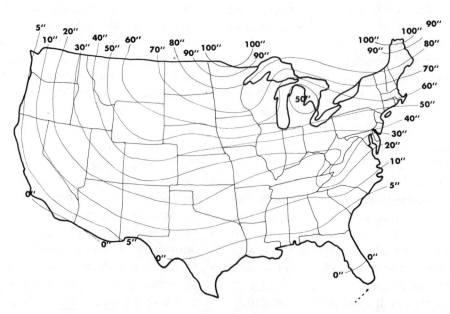

Fig. 3-1. Map showing maximum frost penetration in the continental United States. (U.S. Weather Bureau)

TABLE 3-1. FOOTING DEPTHS AS REQUIRED BY
REPRESENTATIVE CITY CODES.

City		City	
Milwaukee, Wisconsin	4'-6"	Baltimore, Maryland	3'-0"
Chicago, Illinois	3'-6"	Kansas City, Missouri	3'-0"
St. Paul, Minnesota	3'-6"	Philadelphia, Pa.	3'-0"
Boston, Massachusetts	4'-0"	Louisville, Kentucky	2'-6"
Halifax, Nova Scotia	4'-0"	St. Louis, Missouri	2'-6"
New York City, N.Y.	4'-0"	Denver, Colorado	1'-6"
Detroit, Michigan	3'-6"	Seattle, Washington	1'-6"
		Jacksonville, Florida	1'-0"

carry heavier loads. The foundation walls in houses with basements are usually seven or eight feet high. They support the beams and the joists and provide adequate headroom in the basement.

Some houses in the North are built with slab at grade construction. This type of construction is illustrated in a later section. See Fig. 3-34 on p. 91. The foundation and the footing are placed in such a manner that they extend below the frost line. Edge insulation and moisture protection must be provided for the floor slab since it rests on the ground. Houses in the North are also built with crawl spaces. The footings and foundations are laid deep enough to prevent frost damage.

The particular design of the foundation depends to a great extent on the soil composition. Parts of the country have a sand base. Other areas have loam, clay, or a combination of rock and other substances. Each type of soil has a different bearing ability. Some permit trenching for footings without requiring forms for the footings. The nature of the soil is a factor in determining the size of the footing needed to support the load of the building.

The forming systems used by builders vary from region to region. Job-built 4 × 8 foot panels are common in some areas. Many patented forming systems are also popular. Some are used nationwide, others are only used in local areas. These forming systems generally consist of special panels complete with the hardware (fastening devices). They can be rented or purchased directly from the manufacturer. Some of them use steel frames, with plywood inserts for the panel faces. Such forms are very durable and can be used over again many times. Several of the more common forming systems are described later in this chapter.

After the general discussion of concrete forming, we will look at two particular problems in greater detail in Chapter 3-1, *Practice in Concrete Forming.* We will study the forming required for the foundation and floor of House Plan A. We will refer back to the working drawings in Chapter 1. This particular house requires a slab at grade foundation which is adaptable to either warm or cold climates. In addition, we will also study the forming for the footings and foundation of House Plan C. Here we will refer to the working drawings at the end of Chapter 3-1. This particular house has two stories and a full basement. Unit panel forms are used.

CONCRETE AS A MATERIAL

It has already been stressed that the carpenter needs an understanding of concrete as a material because the forms must be constructed strong enough to hold the concrete in place until it sets. Also, there are several types of concrete for different purposes. If the forms fail to do the job, the time spent erecting them, plus a lot of costly material, is completely wasted. Labor and forming material can likewise be wasted if the forms are made overly strong, beyond what is required to do the job safely.

In most areas, concrete is delivered by truck as *ready mix*. You might be tempted to assume that the material has been properly proportioned and mixed and therefore meets the requirements for the particular job. You should not make such assumptions. Each particular job requires specific ingredients and proportions. Each batch must be prepared at the plant according to specifications. The contractor, particularly on big jobs which require large amounts of concrete, inspects samples and further tests the concrete in the structure during the period of curing. (*Curing* refers to the hardening of concrete over a period of time during which it gradually gains strength. Most concrete reaches its maximum strength after 28 days. The concrete must be kept wet and protected from freezing temperatures in order to cure properly. The name for the chemical change which takes place in the concrete is *hydration*.)

The concrete used for most construction jobs is ready mix and delivered to the site in specially designed trucks. There are situations, however, in which ready mix is difficult or impractical to get because of distance, a need for only a small batch, etc. In these cases, the concrete is prepared on the job. The concrete must be prepared with strict adherence to good procedures. These include carefully choosing the ingredients and properly proportioning and mixing them.

The carpenter must be familiar with the various types of concrete which have been developed to serve different functions. The specific type of concrete depends on the type of cement used, the type and size of the aggregates, and the amount of water used in the mixture.

The primary objective of the carpenter is to build forms which can retain the mixture (in the fluid state) in the shape desired. The forms must have sufficient strength to withstand the pressure. They must also be made so that the faces of the resulting structure achieve the desired smoothness.

As already stated, concrete is a mixture of cement, aggregates, and water. Two types of aggregates are used. One is a fine substance, usually sand. The other is coarse, usually consisting of some form of stone. The cement used is called *portland cement*. Portland cement is not a trade name. The process for manufacturing it was invented over 150 years ago. The name *portland* is derived from the Island of Portland, which is a part of England. Since walls made with this cement resembled stone quarried on the island of Portland, the cement was called *portland*.

The cement is made out of some form of lime plus silica, alumina, and iron oxide. These substances are heated to a high temperature in a rotary kiln so that they form clinkers. The clinkers are ground to a fine powder. A small amount of gypsum is added to control the setting process. Other additives can be incorporated to give the resulting concrete special properties. When the cement is mixed with water, it sets (becomes firm) and then hardens. The hardening process (hydration) is a chemical reaction which takes place over a number of days. The sand, gravel, or crushed stone make up the filler material for the concrete. (Note: The terms *cement* and *concrete* are not interchangeable. You cannot substitute one word for the other. For example, it is incorrect to refer to a walk as a "cement walk". Cement is technically an ingredient of concrete.)

Portland Cement

Concrete which is made with regular (Type I) portland cement is a general all-purpose product. It is capable of satisfying the needs for residential foundations and large industrial and commerical structures using reinforced concrete.

Other types of concrete have been developed with special admixtures. These are designed to solve various construction problems. When concrete sets, the chemical reaction of hydration produces heat. Some admixtures called *retardants* minimize this heat rise. They are used when concrete is placed in warm weather. Minimizing the heat factor is of course a considerable problem in building large structures such as a dam or bridge abutment. Because of the large mass of concrete, the heat cannot be dissapated easily. A retarding cement is used in the concrete, and pipes used to circulate water are often embedded in the structure.

There are other problems. Concrete which is exposed to the alkaline soil found in various parts of the country is subject to sulfate action. This action attacks the concrete and causes it to deteriorate. Special *sulfate resistant cement* has been developed to combat this situation.

Air-entrained portland cement concrete is highly resistant to severe frost. It is also highly resistant to the action of salt and other chemicals often used in cold climates to melt snow and ice on streets and sidewalks. This type has millions of air bubbles trapped inside each cubic foot of concrete.

The all-purpose concrete does not completely cure and reach its ultimate compressive strength until 28 days after it is placed. However, it is frequently important that the concrete set and reach a high degree of strength more quickly. The concrete developed for this purpose is called *high early strength.* High early strength concrete has various advantages. Among these are the lessened amount of time the newly placed concrete has to be protected against freezing temperatures. This reduces the cost of heating and protecting the concrete while it cures. It speeds up the overall construction time by allowing the builders to go ahead with the construction of other sections of the structure or project. In the case of reinforced concrete structures, the builders can also continue on to the upper floors with less delay. Since high early strength concrete needs less setting time, the formwork can be disassembled and used over again more quickly.

Additional Ingredients for Concrete

The inert material, which makes up from 66 to 78 percent of the volume of concrete, falls into two classes: fine particles and coarse aggregate. Clean sand serves as the fine particles. The coarse aggregate is principally stone or crushed rock. The size of the stone varies with the requirements of the job. Ordinarily concrete, which uses stone for the coarse aggregate, weighs about 145 pounds per cubic foot. When cinders are used instead of stone, the concrete weighs about 110 pounds per cubic foot. Lightweight concrete is made with trade name products, such as Haydite, Waylite, Perlite, and Zonolite, as the aggregate. These aggregates are made from expanded blast furnace slag, volcanic glass, or mica. The resulting concrete weighs as little as 50 pounds per cubic foot.

Water which is potable (fit to drink) is considered suitable for mixing concrete. No silt or organic matter should be visible to the naked eye. If such water must be used, it should be left in settling basins before it is used.

Proportioning Ingredients

Concrete is a mass of fine and coarse aggregates held together with portland cement paste. If the paste is strong, and the aggregates are hard, the resulting concrete will be strong. The proportion of the water and the cement in the paste is the most critical factor. If excess water is used, the paste will be diluted, and the concrete will be weak once it hardens. The following is the Water-Cement Ratio Strength Law:

For given materials and conditions of handling, the strength of concrete is determined primarily by the ratio of the volume of mixing water to the volume of cement, as long as the mixture is plastic and workable.

Table 3-2 gives the proportional ingredients for concrete as developed through scientific testing by the Portland Cement Association. When you use this table to mix batches of concrete, follow the proportions carefully. Give particular emphasis to the quantity of water. This particular table is based on proportioning by volume. This is the practice commonly used for small batches. It is based on the fact that a bag of cement is equal to one cubic foot. The example which appears on the following page, after Table 3-2, shows you how to use the table effectively to achieve the proper proportion of ingredients you need to mix various amounts of concrete.

TABLE 3-2. PROPORTIONS BY VOLUME.

Maximum-size Coarse Aggregate, In.	Air-entrained Concrete				Concrete Without Air			
	Cement	Sand	Coarse Aggregate	Water	Cement	Sand	Coarse Aggregate	Water
$3/8$	1	$2\frac{1}{4}$	$1\frac{1}{2}$	$\frac{1}{2}$	1	$2\frac{1}{2}$	$1\frac{1}{2}$	$\frac{1}{2}$
$1/2$	1	$2\frac{1}{4}$	2	$\frac{1}{2}$	1	$2\frac{1}{2}$	2	$\frac{1}{2}$
$3/4$	1	$2\frac{1}{4}$	$2\frac{1}{2}$	$\frac{1}{2}$	1	$2\frac{1}{2}$	$2\frac{1}{2}$	$\frac{1}{2}$
1	1	$2\frac{1}{4}$	$2\frac{3}{4}$	$\frac{1}{2}$	1	$2\frac{1}{2}$	$2\frac{3}{4}$	$\frac{1}{2}$
$1\frac{1}{2}$	1	$2\frac{1}{4}$	3	$\frac{1}{2}$	1	$2\frac{1}{2}$	3	$\frac{1}{2}$

For air-entrained concrete with ¾ inch maximum size coarse aggregate:

1 bag cement	1 cu. ft.
2¼ parts sand	2¼ cu. ft.
2½ parts coarse aggregate	2½ cu. ft.
½ part water	3½ gallons

(7 gallons water = 1 cu. ft.)

FACTORS IN FORM DESIGN

When building formwork, the carpenter is primarily concerned with providing the shape and dimensions specified on the working drawings. This is relatively simple when the building is rectangular in shape. However, sometimes the carpenter has to provide forming for offsets, pilasters (rectangular masonry columns attached to the walls), fireplaces, windows, doorways, and concrete stairs. Forms must be built tight enough so that none of the liquid concrete leaks between the form panels. The finished walls must have the smoothness required, especially when they are exposed to view.

It is very important of course that the formwork be constructed so that it can withstand the strain placed on it. The lateral (side) pressure of the liquid concrete must be counteracted with strong internal ties spaced to absorb the full load. Horizontal external members (called *walers*) are used primarily to line up the form panels. Additional vertical members (called *stiffbacks*) are used on heavy construction to firm up the entire form assembly. Diagonal braces keep the forms in position when they are being assembled and while the concrete is being placed. Occasionally a foundation wall is placed so that the outer surface is against a vertical earth bank. Only an inner form would be required in such a case. Strong diagonal and horizontal bracing is used to hold the inner form in position to take the thrust of the pressure caused by the liquid concrete.

Generally, when the rate of pour is slow, the concrete loses part or all of its fluid pressure at the bottom as it begins to set. This reduces the pressure at higher levels. In other applications, the rate of pour may be so rapid the forms must withstand great pressure throughout their depth. A pour of from one to two feet (in depth) of concrete per hour is considered to be a slow rate. The rapid delivery of ready mix concrete provides a fast pour rate for filling the forms and may average three to seven feet per hour. The forms, however, have to be constructed more strongly to withstand the added pressure. This is usually done by increasing the number of ties per unit area, particularly at the lower part of the forms. As a result, the walers are also placed closer together in the lower part of the form.

Temperature is another important factor to be considered when the forms are being designed. The setting of concrete is delayed to some degree at lower atmospheric temperatures. Since the concrete remains liquid for a longer time, the formwork again must be made to absorb the additional pressure. Ordinarily concrete should be maintained at some temperature between 50 to 70 degrees Fahrenheit (10 to 21 degrees Celsius) when it is placed in the forms.

Vibrators also affect form design. Immersion vibrators are used to compact freshly placed concrete and to eliminate air pockets. See Fig.

Fig. 3-2. A vibrator is used to consolidate the concrete in the form. (Adolphi Studio)

3-2. The vibrators tend to keep the concrete in a fluid state throughout the forms. The concrete therefore exerts greater pressure on the forms when vibrators are used.

MATERIALS USED IN FORMWORK

The materials used in making and assembling formwork consist mainly of plywood, wood structural members, and a wide variety of metal devices used to fasten the forms together.

Panel Material

A type of plywood called *plyform* is manufactured specifically to serve as form facing material. Two classes of plyform are supplied. They are distinguished by the species of wood and various structural characteristics. Plyform is made with smooth finished solid outer plies which are coated with oil or other sealer. The edges of the panels are also sealed. The sealer keeps the plywood faces of the forms from absorbing water from the concrete. Some panels are made with a *high density overlay*. They have a plastic surface that is waterproof, resistant to abrasions, and easy to clean. The smooth surface overcomes

the wood grain effect left on the finished wall by ordinary plyform panels. Tempered hardboard faces serve a similar purpose. The hardboard must be kept wet for some time before the concrete is placed in the forms. Form liners made from polymer alloy are used to achieve unusual finish effects, such as ribbed, simulated random boards or brick.

Lumber and Steel Framing Members

The lumber which is used should have good structural qualities, nail holding ability, and ruggedness so that it can be reused many times. The members are generally construction grade Douglas fir, west coast hemlock, or long leaf southern yellow pine. Two by four inch material is used for most applications for panel frames, studs, walers, bracing, and stakes. Wide boards, 1 or 2 inches thick, are often used for the forming for footings and low walls. Plywood or plyform can also be used for this purpose.

Some framing systems use panels which have steel angle frames and plywood inserts to provide the actual panel surface. The frames are almost indestructible under ordinary use. They can be reused two hundred times and more. When the plywood inserts deteriorate or are damaged, they are removed and replaced by new ones.

Miscellaneous Equipment

Various types of metal devices are used to hold forms together. Most of those used in residential construction involve some type of metal tie with a wedge or cam lock feature which drives it tight. See Fig. 3-3. Other tying devices use rods with threaded ends and nuts or threaded washers to hold the forms together. Plastic cones are used on the inside ends of the form ties. These provide a uniform, easily repaired hole once the tie end is broken off. See Fig. 3-3.

Steel Reinforcing

Concrete has great compressive strength but does not have great resistance to bending or tension stresses. Steel reinforcing compensates for this problem. Reinforcing bars (also called *rods*) are used in footings and foundation walls

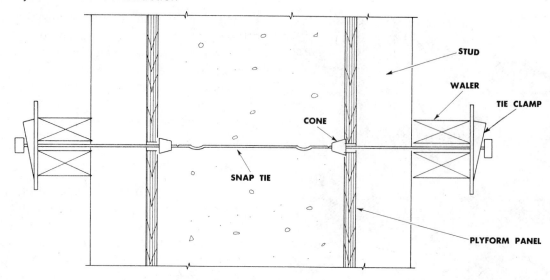

Fig. 3-3. The snap tie holds the form assembly together. The tie clamps draw it tight. The holes left in the concrete by the cones are easy to repair.

whenever the soil conditions are unstable, or other factors such as the danger of seismic disturbances (earthquakes) requires it. Reinforcing bars are also used when parts of the foundation are to be added later, or when walks or drives are to join the foundations. Reinforcing bars are described by number. Each number represents 1/8 inch in diameter. For example, a #4 reinforcing bar is 4/8 (1/2) inch in diameter, and a #5 bar is 5/8 inch in diameter.

Driveways and slabs are often reinforced by welded wire fabric (mesh). On working drawings, this fabric is designated by the spacing between the wires and the gage of each wire. A designation WWF 6 × 6 × 10/10 is read "Welded wire fabric with a 6 × 6 inch spacing using 10 gage wire in both directions".

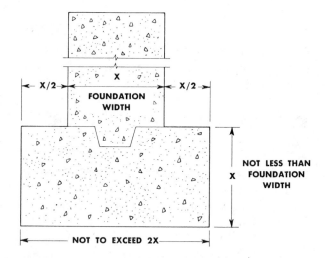

Fig. 3-4. Footings are designed to transfer the load of the building to the earth.

BASIC FORMWORK

Forming for Footings

Footings are required under foundation walls to transfer the load of the building to the ground. The sectional view through an outside wall, which is a part of the working drawings, usually gives the information about footing and foundation dimensions. The basement plan gives information about the column footings.

The major factors in determining the size of the footing are the weight of the house on each lineal foot, the bearing value of the soil, and its stability. Generally, the horizontal dimensions of the footing is made about twice the thickness of the foundation wall. The depth of the footing is at least made equal to the thickness of the foundation wall. See Fig. 3-4.

When frost is not a problem, and the soil is stable, a trench is dug the width and depth of the footing. The concrete is carefully placed directly into the trench so that it does not cause the banks on either side to cave in. This method can also be used for forming footings when there is a frost problem. The excavation for the basement or the foundation wall is dug to the proper frost line depth. A trench is then dug for the footing. The soil bottom for the footing should be undis-

turbed earth (not dug up then filled in) so that it provides a firm bearing for the footing.

Procedure for erecting footings. The procedure for erecting footings is given below. Also see Fig. 3-5.

Fasten lines to the batterboards, running them through the saw kerfs which have been made previously. See Fig. 3-6. (Nails can be used instead of saw kerfs.)

Drop a line and plumb bob at the intersection of the lines. This locates the position of the corner stake. Cut 2 x 2 inch stakes about 16 inches

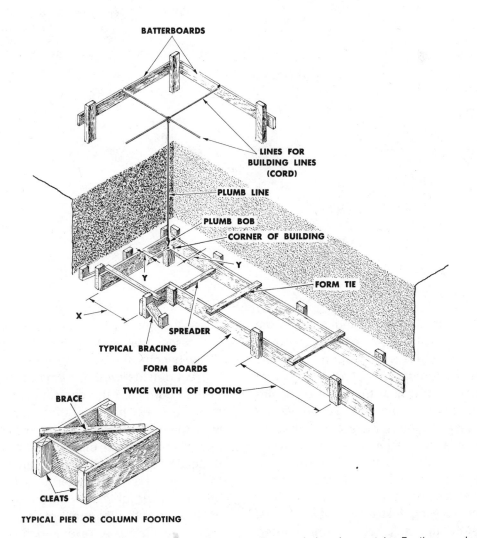

Fig. 3-5. A point is dropped from the intersection of the building lines and placed on a stake. Footings are located from this point. Form boards are spaced by spreaders. Stakes and bracing are installed. Form ties hold the form boards in perfect alignment.

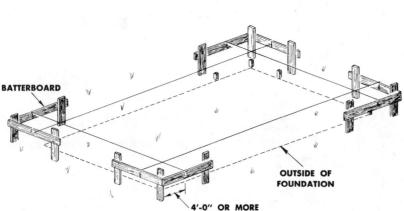

FOUNDATION GRADE
(OR LOCATION RELATED TO GRADE)

LEDGER

SAW KERF
OR NAIL

CORD LINES

PLUMB BOB

NAIL

STAKE

BATTERBOARD

OUTSIDE OF
FOUNDATION

4'-0" OR MORE

Fig. 3-6. The batterboards, as shown in the top view above, consist of stakes and horizontal ledgers. Lines (heavy cord) are fastened to the batterboards to locate the building lines. The bottom view shows that the lines are run between kerfs (or nails) on the batterboards.

long. Drive a stake at the point indicated by the plumb bob. Use a nail to indicate the exact point.

Drive the stake to a level corresponding to the elevation for the top of the footing. This height should be checked against the point of beginning. Use a leveling instrument and a rod. Varifying the footing elevation is important so that the house will set at the required elevation when completed. Drop a line with a plumb bob attached where the lines intersect at other corners. Drive stakes at each point.

Set up the leveling instrument at a central point and sight the stake at the first corner. Transfer the elevation to the other three stakes, driving them to the same level and placing a nail

at each of the building corner points. Connect the corner stakes with lines tied to the nails. These become the building lines.

Erect outside footing forms so that the inside of the boards are the correct distance from the building lines. The top of the forms must be level with the top of each corner stake. If 1 inch boards are used, the stakes holding them should be set 2 or 3 feet apart. Stakes can be spaced farther apart if 2 inch material is used for form boards. Wood or metal stakes are used. Brace each stake with diagonal bracing.

Erect inside forms, reversing the procedure for erecting the outside forms. The spreaders should be as long as the footing is wide. These help line up the inside forms so that the form boards and stakes are placed correctly. Brace the inside footing forms at the stakes. The spreaders are removed as the concrete fills the forms. Place the form ties at 4'-0'' intervals, or as required. Check the footing forms in relation to the building lines. Check the level of the form boards all around so that the footing is level when the concrete is placed. (In some parts of the country, the footings and foundation walls particularly for low walls are placed monolithically.)

The next operation is the placing of the concrete. Be careful not to disturb the corner stakes because they will later be used to set the foundation wall forms.

When the concrete begins to set, press a piece of wood into the center. This forms a *keyway*. The keyway provides the form for a *key*, which is a projection of the foundation wall after the concrete is placed. The key serves to resist the lateral pressure of the backfilled soil against the foundation wall. To some extent it also helps to prevent the seepage of water. See Fig. 3-7. Make forms from 2 x 4 inch material. Place a slight chamfer on the edges which allows them to be pulled out of the concrete.

Remove the form boards after the concrete of the footing has set. Drain tile is frequently needed, depending on the nature of the soil, whether or not excessive water is present in the soil. Lay the tile in a bed of gravel around the outside of

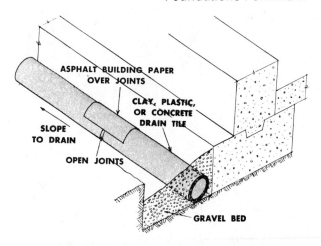

Fig. 3-8. Drain tile is laid along the footing to divert water.

the footing as shown in Fig. 3-8. Lay it with open joints (a space between each section). Place strips of asphalt building paper over the top of the joints to prevent gravel or soil from falling into the pipes. Slope the tile as you lay each section so that the water drains toward a sump pump in the building, or a drywell, as required by the local building code. After laying the tile, cover it with gravel at least to the top of the footing.

Regional footing construction procedures. In warm climates where frost is not a factor, the footings and foundation are placed in a single monolithic (one piece) unit which resembles an inverted T. This type of footing foundation will be covered later in this chapter under *Low Wall Forming.*

When a house is built on severely sloped ground, stepped footings are used. Stepped footings avoid the necessity of a very deep excavation and high foundation walls. If the steps are not too great, the footings are placed in one piece. The vertical boards retain the concrete at each step. See Fig. 3-9.

In areas which are prone to earthquakes, or where the soil is otherwise unstable, reinforcing bars are placed in the footings. The architect or engineer specifies the number and size of the bars and shows their approximate location on the section view of the working drawings. A typical notation appears in the detail shown on Fig.

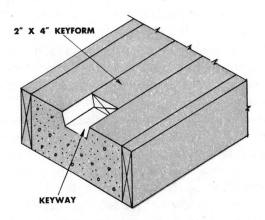

Fig. 3-7. A keyway is made in the footing by pressing a piece of wood into the concrete before it sets.

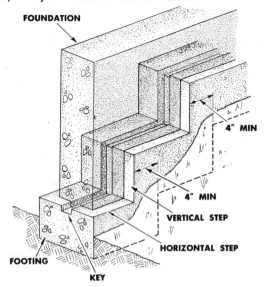

FOUNDATION

4" MIN

4" MIN

VERTICAL STEP

HORIZONTAL STEP

FOOTING

KEY

Fig. 3-9. Stepped footings are designed to give horizontal support when the ground is sloped or the house levels are not on the same plane.

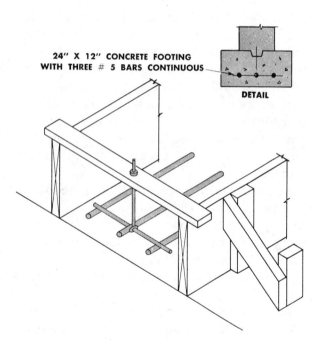

24" X 12" CONCRETE FOOTING
WITH THREE # 5 BARS CONTINUOUS

DETAIL

Fig. 3-10. Reinforcing steel bars are suspended or supported in the footing to provide additional strength.

3-10. Three reinforcing bars, $5/8$ inch in diameter, are required.

Basic Foundation Forming Types

Four basic foundation forming types, along with some of their variations, are discussed in this section. The next section broadens the scope of this subject by analyzing specific types of manufactured forming systems.

The following are the basic types of forming:

Conventional wall forming. Wall forming, used for foundations which range four or more feet in height above the footing, requires a wall of paneling placed on two sides. The entire assembly is free standing except for the bracing required for temporary support and alignment. See Fig. 3-11A for this type of wall.

Low wall forming. Low wall forming is used primarily for buildings which have a crawl space. The Western T type is one type of footing commonly used. The footings and foundation are laid monolithically in the shape of an inverted T, as shown in Fig. 3-11B. Other types are constructed in a way similar to the double wall forms except simpler forming techniques are used.

Slab at grade. The slab at grade foundations are used when the house is constructed directly on the ground. The forming is usually quite simple, often consisting of a trench in the soil. The foundation used in some warm, dry climates requires that the edge of the slab be thickened to support the building. See Fig. 3-11C. A low perimeter wall is also used for this purpose. Here, a floating slab can be used to form the floor. (The slab is independent of the foundation.) See Fig. 3-11D. Buildings in other warm climates, where excessive water is present in the soil, require

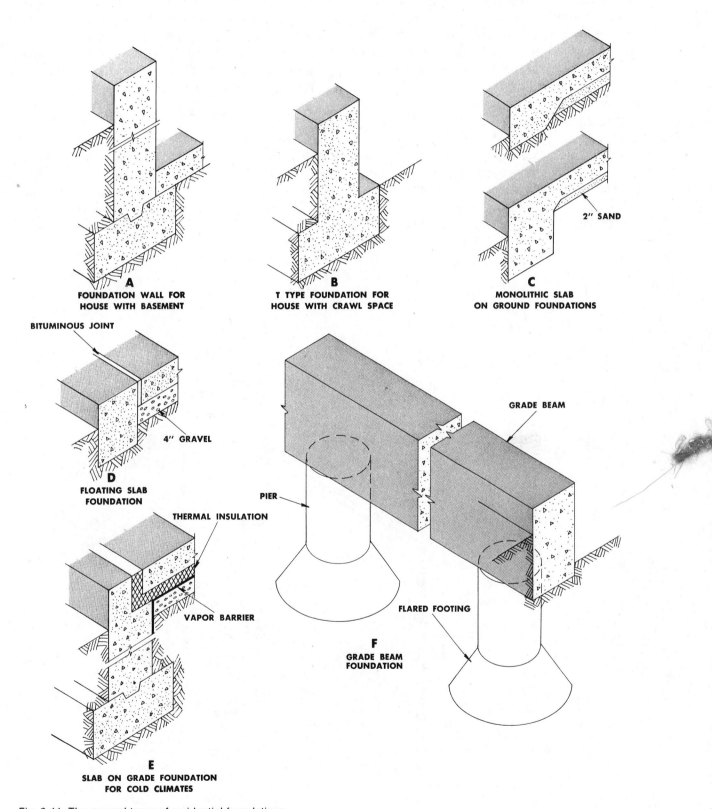

A
FOUNDATION WALL FOR
HOUSE WITH BASEMENT

B
T TYPE FOUNDATION FOR
HOUSE WITH CRAWL SPACE

C
MONOLITHIC SLAB
ON GROUND FOUNDATIONS

2" SAND

BITUMINOUS JOINT

D
FLOATING SLAB
FOUNDATION

4" GRAVEL

THERMAL INSULATION

VAPOR BARRIER

E
SLAB ON GRADE FOUNDATION
FOR COLD CLIMATES

GRADE BEAM

PIER

FLARED FOOTING

F
GRADE BEAM
FOUNDATION

Fig. 3-11. The general types of residential foundations.

that careful waterproofing procedures be followed. A vapor barrier is used to prevent water vapor from entering the building under the foundation or the slab. The foundation wall for structures in colder climates is extended below the frost line which may eliminate the moisture problem. Slab edge insulation, which prevents frost damage and heat loss, is required. See Fig. 3-11E.

Grade beam foundations.   Grade beam foundations are primarily used when the building is erected on unstable soil. Unstable soil conditions can be caused by quicksand, underground water veins, sloped ground which develops into mud slides during the rainy seasons, etc. Piers or columns are spaced around the perimeter of the building. These support reinforced concrete beams on which the house is to rest. The main forming problem centers around the forms for the beams. See Fig. 3-11F.

Conventional Wall Forming

The basic manner of preparing formwork for residential buildings which have foundations exceeding four feet is shown in Fig. 3-12. Plyform panels and vertical studs which are lined up by horizontal walers are used. The plyform panels are drilled in a pattern which accomodates

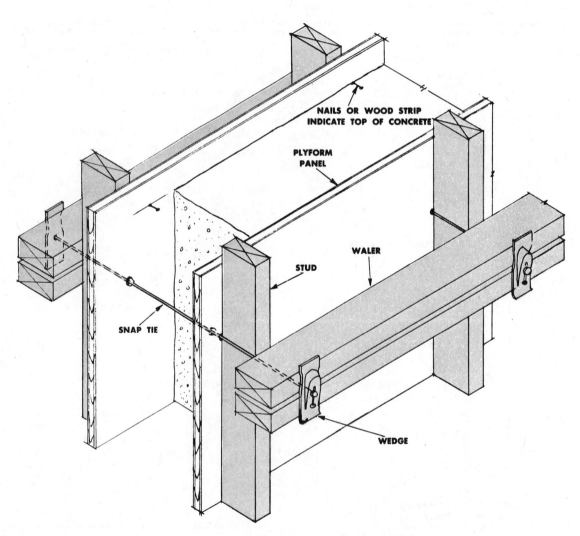

Fig. 3-12. Conventional stud wall forming uses studs aligned by walers. Wedges are used to tighten snap ties.

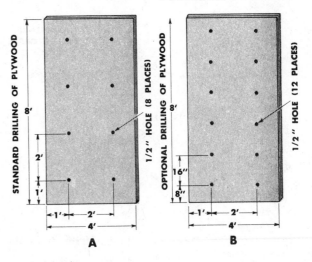

Fig. 3-13. Several sheets of plyform are stacked and drilled at the same time. The panels shown in (A) provide holes for snap ties so that the walers are spaced 2 feet (24 inches) apart. The holes for the snap ties shown in (B) are spaced 16 inches apart.

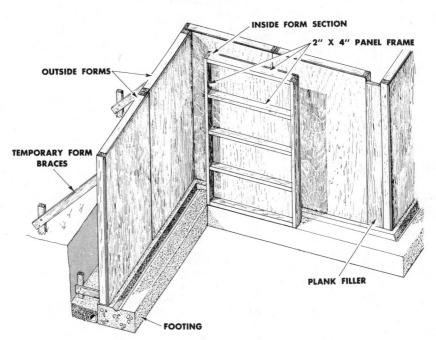

Fig. 3-14. Wood panels with fillers of various sizes can be adapted to most forming problems. They can be re-used many times. Forms of this type are generally made by the carpenters. Holes for ties are drilled as needed.

the number of walers needed in the particular application. See Fig. 3-13. One half inch holes are drilled for the snap ties to pass through.

There are several variations of the stud wall forming principle. Some of these adaptations use framed panels. Here the studs are part of the panel. Units built by carpenters, such as those shown in Fig. 3-14, are made so that the panels can be assembled into a wall with more flexibility. Holes can be drilled in the outside forms before they are set. However, new holes are generally made in the inside panel because they rarely line up if pre-drilled.

Snap ties. The snap tie is the most common and popular type of form tie. (Other types of form ties are discussed later.) A snap tie is a steel wire or a rod. It is long enough to pass through the width (thickness) of the foundation wall and provide a space equivalent to the thicknesses of the stud and waler. The snap tie also has a projection on each end to engage the form clamps. See Fig. 3-15.

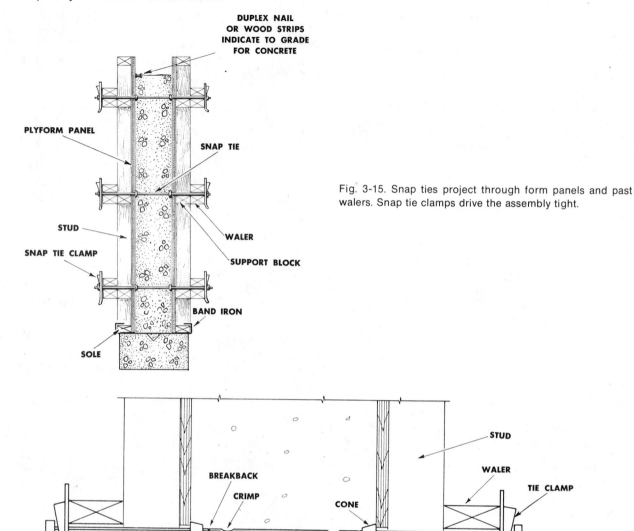

DUPLEX NAIL
OR WOOD STRIPS
INDICATE TO GRADE
FOR CONCRETE

PLYFORM PANEL

SNAP TIE

STUD

SNAP TIE CLAMP

WALER

SUPPORT BLOCK

BAND IRON

SOLE

Fig. 3-15. Snap ties project through form panels and past walers. Snap tie clamps drive the assembly tight.

STUD

WALER

TIE CLAMP

BUTTON
SNAP TIE
END

BREAKBACK

CRIMP

CONE

SNAP TIE

PLYFORM PANEL

Fig. 3-16. The snap tie is provided with a notch called a *breakback* where it can be broken off in the wall. It is crimped (or flattened) to prevent it from turning when the ends are twisted off.

Snap ties are available in a range of sizes for various foundation wall thicknesses and with varying safe working loads.

Snap ties have various features as shown in Fig. 3-16. Two spacer washers are welded to the tie. These serve to keep the forms at the distance required to accommodate the thickness of the wall. Notches are made in the tie a short distance inside of the spacer washers. These are called *breakbacks*. When the forms are stripped, the tie extends from the wall about 8 inches on either side. The tie ends are twisted and broken off. The

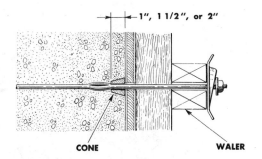

CONE   WALER

Fig. 3-17. Some snap ties are made with plastic cones which when removed from the wall leave a smooth conical hole for patching.

tie breaks at the weakened breakback point in the wall.

Some snap ties are made with plastic cones positioned against the plyform surface. See Fig. 3-17. A smooth conical hole remains once the panels are stripped away. The hole can be neatly repaired with grout or a plastic plug.

Two flattened or crimped portions of the snap tie are located closer to the center of the tie. Af-

ter the concrete sets, the flattened or crimped section prevents the entire rod from turning when the carpenter breaks off the tie end. A button welded to each end of the snap tie enables it to bear against the clamp. The clamp is driven tight with a few taps of a hammer.

Procedure for foundation formwork. The procedures to follow when you are building conventional foundation formwork using plyform

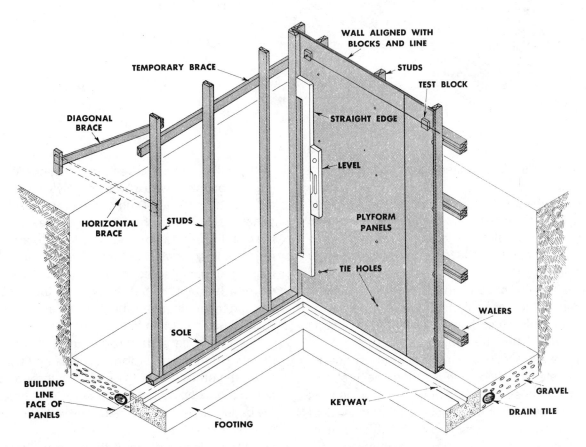

WALL ALIGNED WITH BLOCKS AND LINE

TEMPORARY BRACE

STUDS

TEST BLOCK

DIAGONAL BRACE

STRAIGHT EDGE

LEVEL

HORIZONTAL BRACE

STUDS

PLYFORM PANELS

TIE HOLES

WALERS

SOLE

BUILDING LINE FACE OF PANELS

KEYWAY

GRAVEL

DRAIN TILE

FOOTING

Fig. 3-18. Double wall forming is begun with the fastening of a plate to the green concrete. The studs are erected and braced, plyform sheets are applied, and the walls are aligned and plumbed.

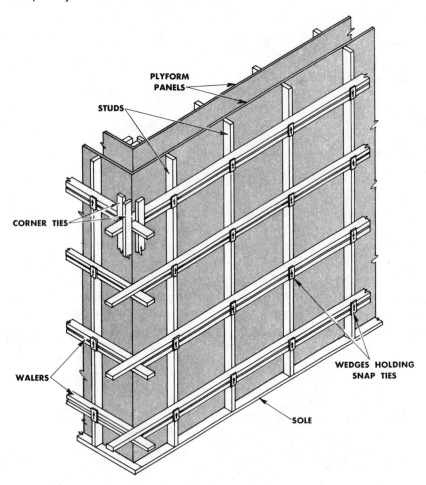

PLYFORM
PANELS

STUDS

CORNER TIES

WALERS

WEDGES HOLDING
SNAP TIES

SOLE

Fig. 3-19. Double wall foundation forming using studs, walers, and snap ties. The panels are predrilled so that holes line up between inner and outer walls. Several methods are used to brace the corners.

panels and studs are listed below. Also see Figs. 3-18 and 3-19.

String the building lines through the saw kerfs on the batterboards. Drop a line with a plumb bob attached at each corner intersection to check the marks on the footing which locate the building corners.

Holding the ends of the line on the building corner points, snap a chalk line on top of the footing. This chalk line represents the outside face of the foundation wall.

Nail a 2 x 4 inch sole into the green concrete. (Concrete is called *green* during the first few days after it is placed and before it hardens.) Set the sole back from the chalk line the thickness of the plyform panel. Use duplex (double headed) nails on all framing and bracing members whenever they are to be pulled to strip the forms.

You will provide the best arrangement for using full panels if you place the plyform panels around the outside of the excavation. The short panels needed to fill the odd spaces are cut once the exact space can be measured. Try to arrange the panels so that short pieces which are less than a half panel do not occur adjacent to a corner or offset. Set the panels in place, temporarily bracing them diagonally back to the earth if necessary, then nail them to the sole.

Insert snap tie ends through the holes in the panels. Bring the inner form panels into line and insert the other end of the snap ties through the holes. It may be necessary to make new holes if the panels do not line up perfectly with the holes already made.

Place studs alongside of the snap tie ends on both sides of the form. Place walers in position.

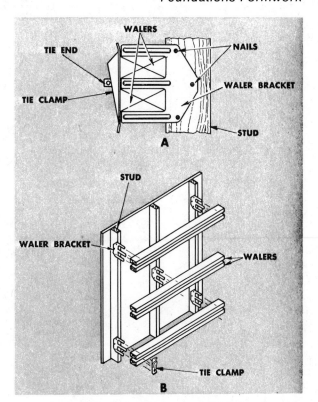

Fig. 3-20. The waler brackets shown at (A) are fastened to the studs of the form to hold the walers in position until the ties are inserted and clamped in place. (B) shows how the form is assembled.

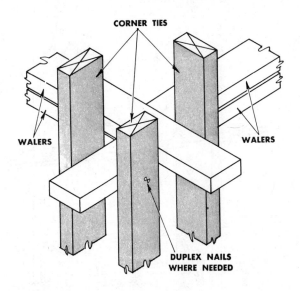

Fig. 3-21. Walers are reinforced at the corners by corner ties.

Use waler brackets, as shown in Fig. 3-20, or blocks to temporarily support them. Extend walers at corners as shown in Fig. 3-21. Nail two or three vertical backing boards to hold the walers in place. Insert wedges over the button ends of the snap ties.

Check the alignment of the entire assembly. For each wall use testblocks and a line stretched from one corner to the next corner. See Fig. 3-22. Make sure the walls are plumb by checking with a level and a straight edge at several places along the wall. Drive the wedges tight.

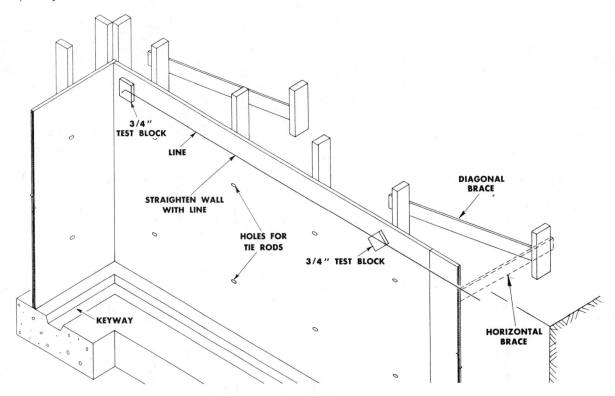

3/4"
TEST BLOCK

LINE

STRAIGHTEN WALL
WITH LINE

HOLES FOR
TIE RODS

3/4" TEST BLOCK

DIAGONAL
BRACE

HORIZONTAL
BRACE

KEYWAY

Fig. 3-22. Test blocks are used to line up forms before nailing braces to stakes.

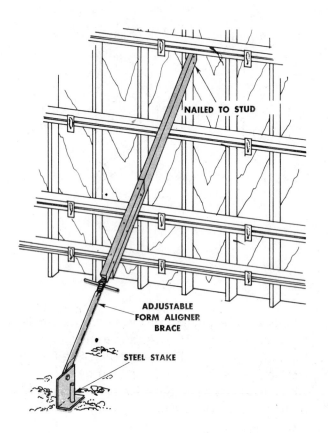

NAILED TO STUD

ADJUSTABLE
FORM ALIGNER
BRACE

STEEL STAKE

Fig. 3-23. Aligner braces serve to make fine adjustments in alignment and hold forms in place during assembly. (Burke Concrete Accessories, Inc.)

Brace the formwork with 2 × 4 inch braces which extend down to stakes in the ground. Form aligner braces can be used for this purpose. See Fig. 3-23.

Sectional forms. Much of concrete forming is done with sectional forms made by the carpenter. See Fig. 3-24. Frames are made the same size as the plyform sheets. Cross members are spaced to stiffen the sections. Some builders have them made on the jobsite or constructed in a shop or yard and transported from job to job. They are generally durable and can be reused a number of times. The most common size for sectional forms is in 4 × 8 foot sections. Other popular sizes are 2 × 8 and 2 × 6 feet. Ordinary snap ties are used with walers. Holes for the snap ties are drilled as needed. Fillers are made in one foot multiples in the same manner as the larger sections. Planks used to fill small spaces are ripped to the exact width required.

One system for erecting the sectional forms is shown in Figs. 3-25 and 3-26. In this system, the carpenters put up the forms so that the studs on either side of the wall are opposite each other. Figure 3-25 shows the manner in which the full

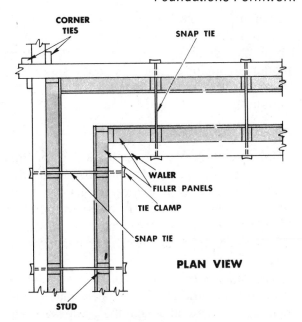

Fig. 3-25. With the use of fillers at the inside corner, low wall sectional forms are arranged so that the snap ties pass between the panels. The view shown above is a Plan View, looking down into the formwork from above.

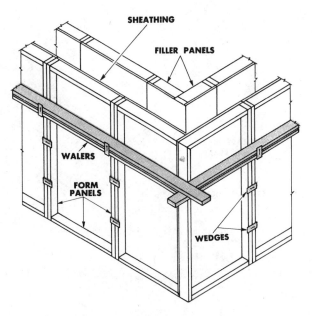

Fig. 3-26. Low wall sectional forms are kept in line by walers. Snap ties bear against the walers or studs.

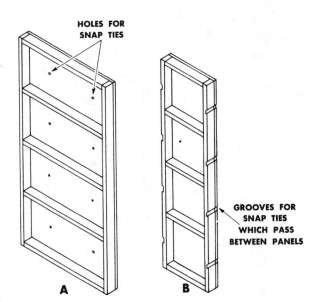

Fig. 3-24. Two types of sectional forms, serving different forming systems, are built by the carpenter. The form panel in (A) is used when ties pass through the panel face. The form panel in (B) is used when ties pass between panels.

panels are arranged. With this system, the arrangement is begun at the outside corner. A filler panel of the specified width is placed on the inside of one of the walls. All subsequent panels in both directions fall into alignment. Two different lengths of snap ties are used. One is long enough to include walers on both sides. The other is used to provide wedging action over the studs.

Low Wall Forming

One advantage of using sectional forms is that they can be adapted for forming low foundation walls. Generally, 4 × 8 foot panels are used. When the required height is less than four feet, 4 × 8 foot panels are laid on their side. However, panels can be made to any specific height as the job dictates.

Monolithic forms. In warm climates which have no frost problem, footings are laid so that their top surface is at or near the grade level. The

forming can be erected so that the footings and the foundation wall make one unit. In areas where the earth is compact and firm, it is a common practice to cut a trench the size of the footing, thus eliminating the need for footing formwork.

T type monolithic forms. T type monolithic forms with trench footings are shown in Fig. 3-27. The procedure for erecting these follows:

(Note: We can assume that the footing trench has already been dug in the proper location to the specified depth and that the lines have been run from various points on the batterboards to provide the building lines for the foundation.)

Drive stakes at outside foundation wall corners. Set back the thickness of the plyform sheathing material. Check the points by dropping a line and plumb bob from the building lines

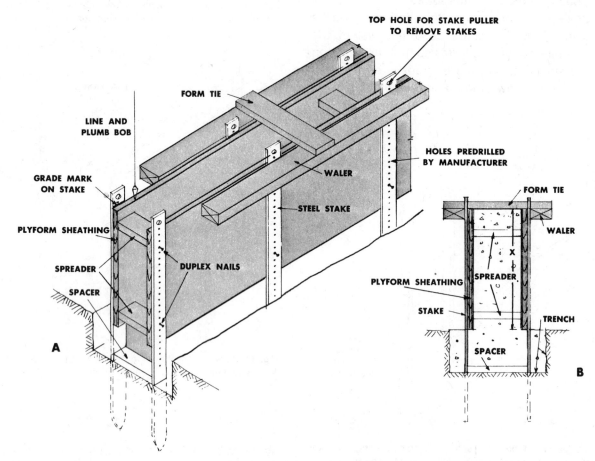

Fig. 3-27. T type monolithic foundation is supported over a trench. The view shown at (A) illustrates how members of the formwork are arranged. The view at (B) shows a section through the formwork with the concrete in place.

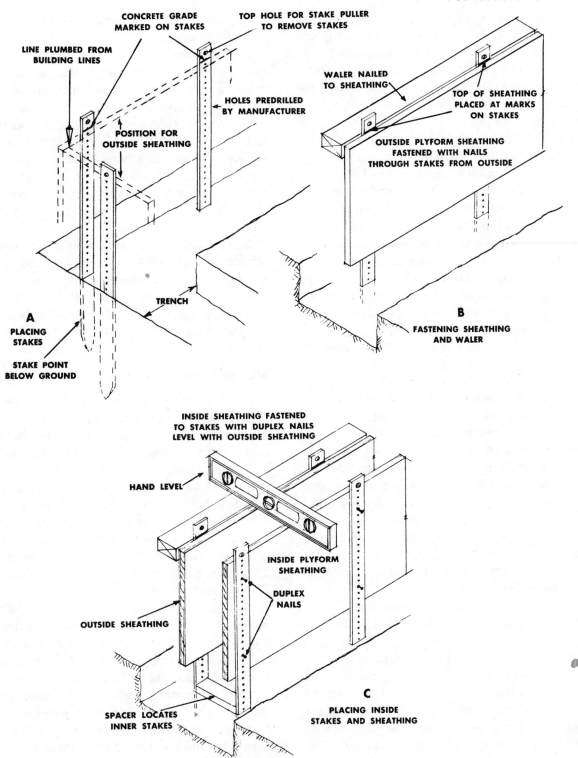

Fig. 3-28. Steps in erecting T type foundation walls. (A). Stakes are placed at uniform intervals (2 feet or 3 feet) in accordance with lines dropped from building lines. An allowance is made for the thickness of the sheathing. The concrete grade mark is placed on each stake. (B). Plyform sheathing cut to the required width is positioned to the marks on the stakes then nailed in place. (C). Spaces cut to provide the correct foundation width are placed on the ground at each stake. Inside stakes are driven opposite outside stakes. Inside plyform sheets are positioned with a hand level and nailed in place.

strung between batterboards. See Fig. 3-28A. Drive the stakes so that they project above the grade level of the top of the foundation wall high enough to allow the actual grade to be marked on their sides. (Wood or steel stakes can be used. Steel stakes, available in either round or flat shapes, are preferred because they can be reused indefinitely. When the forms are stripped, the stakes can be loosened from the green (uncured) concrete with a few blows of a hammer or pulled out with a stake pulling device. Holes are provided at one inch intervals to permit any combination of nailing.) Drive additional stakes at intervals (3 feet is suggested) around the foundation, in line with the stakes at the corners.

Set up the leveling instrument near the center of the foundation. Establish the grade point for the top of the foundation wall on a corner stakes. Mark this elevation on all of the stakes. (Information on this procedure is covered in Chapter 2, *Leveling Instruments and Site Work.*)

Cut the pieces of plyform sheathing material into strips that are equal in width to the height of the foundation wall which extends above the footing. See Dimension *X*, Fig. 3-27B.

Nail the plyform sheathing in place to line up with the grade marks on the stakes. See Fig. 3-28B. Use two nails, one near the top and one near the bottom of the sheathing. Use any holes in the stakes which are convenient. Three nails are used if the plyform sheathing is more than 18 inches wide. Use duplex (double headed) nails when assembling forms so that they can be stripped quickly.

Install walers outside the stakes. See Fig. 3-28B. Line up each outside form wall by stretching a line between corner stakes, or by plumbing from lines stretched between batterboards.

Cut spacers that are equal in length to the thickness of the foundation wall plus twice the thickness of the sheathing material. Locate the position of the stakes for the inner form wall by using the spacers at each stake position. See Fig. 3-28C. Drive an inside stake opposite each outside stake.

Apply the inside plyform sheathing in the same manner as the outside sheathing except, using a hand level, put it in position level with the outside sheathing. See Fig. 3-28C. Place walers on the inside forms in the same manner as you did for outside forms.

Cut spreaders the exact thickness of the foundation wall. Place them near the top and the bottom of wall forms at about 4 foot intervals. If the formwork is constructed properly, the spreaders will fit snugly. (Some carpenters cut a spreader to check the foundation width. It is used to test the work at several places. If the width checks out OK, the spreaders are then omitted.) Once the concrete is placed, knock out the spreaders as the liquid concrete reaches their respective level.

Tie the top of the forms with form ties as shown in Fig. 3-27A. If necessary, brace the stakes to keep the formwork in alignment.

Strip the forms shortly after the concrete has reached its set. Remove the stakes immediately without delay. Fill the holes left by the stakes with grout.

Slab at Grade Foundations

The floor for most dwellings in warmer climates is a concrete slab placed directly on the ground over a bed of gravel and a plastic membrane. This is also becoming an increasingly accepted practice in colder climates. This method eliminates the cost of excavating basements, building high foundation walls, and providing the support framework for floors. The need for basements has decreased to a large extent with the development of compact heating and cooling units. Utility rooms provide the space for home laundry facilities.

Heating, cooling, plumbing, and electrical work is done in several ways. Ducts for heating and cooling with forced air are placed in the gravel bed and covered with the concrete slab. The same is done with the plumbing piping. The heating and cooling units are often placed over the ceiling. Ducts are dropped down in walls and partitions. The wide acceptance of electric heat eliminates the need for ducts. Baseboard units are located in each room, or wires are embedded in the ceilings. Hot water heat usually requires some piping under or in the slab. Electrical supply to outlets is placed in the walls and partitions, with overhead connections located in the attic space.

From an architectural standpoint, the slab at grade foundation gives the house a low profile. The ranch and other contemporary styles of home having this design are very much in demand. Adjacent outdoor living space, such as a patio or a porch, is generally provided at the same level or approximately the same level as the slab for the building.

Types of slab at grade foundations. Supporting exterior walls and providing a satisfactory concrete slab floor involve relatively simple procedures in warm climate construction. In very dry areas, the weight of a building with light construction is transferred to the ground, using a monolithic footing and a low wall. The slab is laid inside the wall but separated from it by a bi-

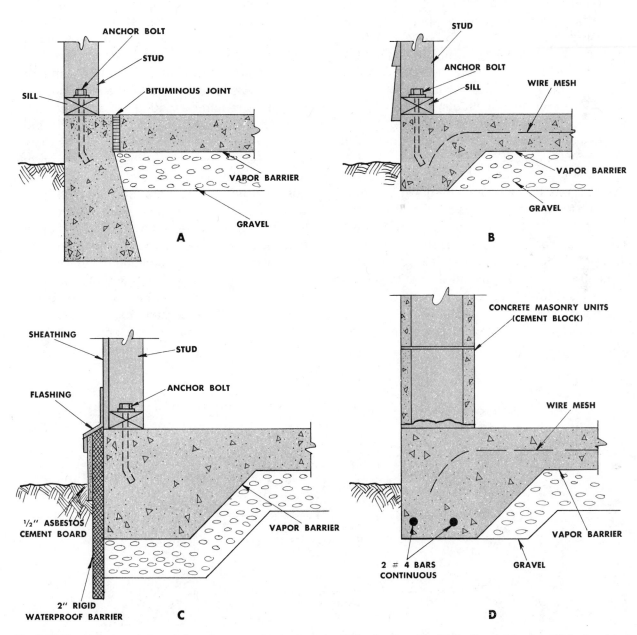

Fig. 3-29. The slab at grade foundations for warm climates include: (A) the floating slab, (B) the flared slab edge, (C) the insulated slab edge, and (D) the slab for the masonry wall.

tuminous expansion joint. The slab and wall are indepedent. Thus the slab can rise or fall without causing cracks at the edges of the wall and is therefore called a *floating slab*. See Fig. 3-29A.

A vapor barrier of polyethylene film is applied over the ground fill wherever there is a problem with moisture. The vapor barrier serves to retain the fine parts of the concrete so that they do not sink down into the fill. It also prevents moisture from rising from the soil under the slab.

Another commonly accepted procedure is to lay the slab with a perimeter support consisting of a flared thickening of the slab edge. Reinforcing mesh is installed in the concrete and continued into the footing to add strength. See Fig. 3-29B. A 2-inch rigid insulation board can be applied to the outside of the footing. It is extended below the gravel fill, giving a form of edge thermal insulation and providing an additional vapor barrier. See Fig. 3-29C.

A version of this procedure modified for use with concrete masonry units is shown in Fig. 3-29D. The foundation consists of a flared footing.

Slabs, foundation walls, and partition foundations are placed in one piece as shown in Fig. 3-30. Reinforcing bars are placed at the base of the foundations. Anchor bolts are used at the points where wood partitions are to be erected. The vinyl vapor barrier is covered with two inches of sand. This forms a cushion for the concrete.

Some warm areas have a problem with insects or high humidity. Here the builders use concrete masonry units (concrete block) for the exterior walls. See Fig. 3-31. The floor slab is extended into the wall to make a tight seal. Bearing partitions are supported on footings. The exterior wall footings and partition footings require steel reinforcing bars. The slab is strengthened with wire mesh. The slab inside the walls is finished with terrazzo. (*Terrazzo* is a type of concrete using marble chips for aggregate. It is applied as a topping bonded to the floor slab. After it has set, it is ground and polished to achieve a smooth surface.)

Slab at grade foundation, when used in moderate climates where there can be some possibility of frost, requires a footing and a foundation of some depth. A slab at grade foundation built with two sizes of concrete masonry unit is shown in Fig. 3-32A. The edge of the slab rests on the wall. The footing is reinforced with two continuous steel reinforcing bars. Steel dowels (rods) which are ½ inch in diameter are placed on 48 inch centers to tie the footing and the foundation together. The voids in the masonry units are filled with concrete, embedding the dowels. Figure 3-32B shows how the foundation and slab are modified when heating ducts are placed around the inside perimeter of the house.

In colder climates, the builders must make

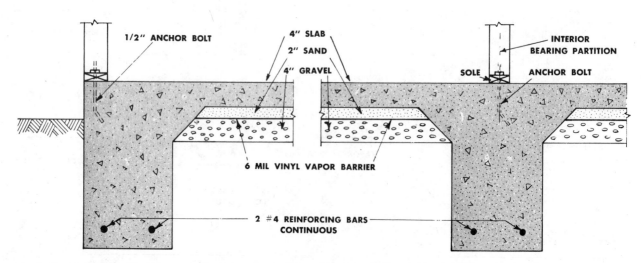

Fig. 3-30. A monolithic slab at grade foundation requires a wide base and steel reinforcing.

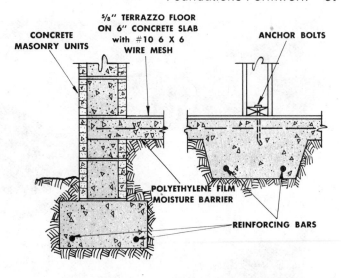

Fig. 3-31. Detail shows method used in warm climates for houses with terazzo floors and exterior walls made with concrete masonary units (concrete block).

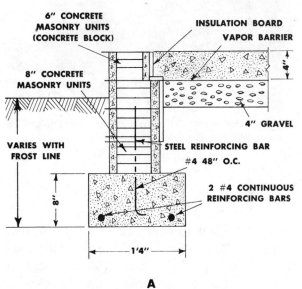

A

Fig. 3-32. A slab at grade foundation for a moderate climate (A above) requires minimum insulation. Detail shown in (B) makes provisions for heating and air conditioning ducts.

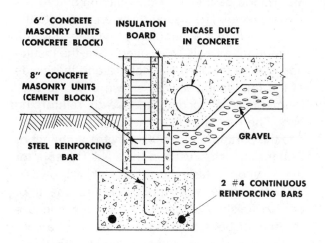

B

Fig. 3-33. Perimeter cellular glass insulation and polyethylene film is used to prevent heat loss and moisture penetration in slab houses. (National Gypsum Co.)

provisions for the thermal insulation of the slab. See Fig. 3-33 and Fig. 3-34A. The foundation and the footings extend to a point below the frost line. Two feet of rigid insulation are installed around the perimeter of the building before the slab is placed. The insulation extends up into the notch made in the foundation to the top level of the foundation wall. As an alternative, the insulation can be installed vertically, along the inside of the foundation wall as shown in Fig. 3-34B. Reinforcing is accomplished with reinforcing bars. The bars are set in the footing of the wall and the partition as specified by the architect or designer. Mesh is used in the slab. The sole for the partition is fastened to the slab by powder-activated studs. With this procedure, the floor can be finished as one plane surface because the anchor bolts do not project out of the concrete. (Some local codes require anchor bolts instead of powder-activated studs.)

Grade Beam Construction

Grade beam construction uses concrete piers to support concrete grade beams. See Fig. 3-35. This is a practical solution for the foundation when the ground cannot support the footings at normal depth. It is used with both slab at grade and crawl space construction. Pier holes are drilled at specified intervals. The diameter and depth of the piers are both determined by local codes. A common specification for single story houses is 12 inch round holes, 4 feet deep, and 5 feet on centers. The size and spacing vary according to soil conditions, building design, and other factors. Some building codes require that the pier base be flared.

Governmental agencies, such as HUD (Housing and Urban Development) and FHA (Federal Housing Administration), have set up requirements for sound building practices to be fol-

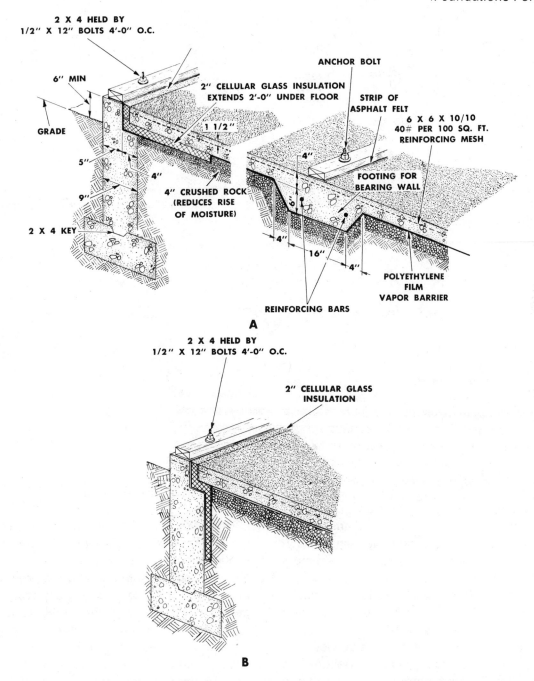

2 X 4 HELD BY
1/2" X 12" BOLTS 4'-0" O.C.

6" MIN

2" CELLULAR GLASS INSULATION
EXTENDS 2'-0" UNDER FLOOR

ANCHOR BOLT

STRIP OF
ASPHALT FELT

6 X 6 X 10/10
40# PER 100 SQ. FT.
REINFORCING MESH

GRADE

1 1/2"

5"

4"

9"

4" CRUSHED ROCK
(REDUCES RISE
OF MOISTURE)

FOOTING FOR
BEARING WALL

4"

2 X 4 KEY

6"

4"

16"

4"

POLYETHYLENE
FILM
VAPOR BARRIER

REINFORCING BARS

A

2 X 4 HELD BY
1/2" X 12" BOLTS 4'-0" O.C.

2" CELLULAR GLASS
INSULATION

B

Fig. 3-34. Slab at grade construction for cold climates requires edge insulation and a vapor barrier. (A) shows insulation applied horizontally around the perimeter of the outside walls. (B) shows an alternate method for applying insulation, placing it vertically inside the foundation wall. Powder activated studs may be used in place of anchor bolts to fasten plates for interior partitions.

lowed when using grade beam construction. The materials on piers and grade beams below has been adapted from FHA No. 300, *Minimum Property Standards For One and Two Living Units.*

Piers. Piers shall be spaced not more than on 8 foot centers. They shall have a minimum diameter of 10 inches and be increased to cover at least 2 square feet at their base. A dowel (a rein-

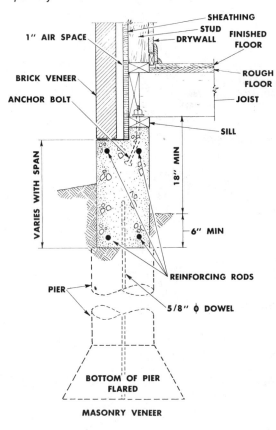

SHEATHING
STUD
DRYWALL FINISHED FLOOR
1" AIR SPACE
BRICK VENEER
ANCHOR BOLT
ROUGH FLOOR
JOIST
SILL
18" MIN
VARIES WITH SPAN
6" MIN
REINFORCING RODS
PIER
5/8" φ DOWEL
BOTTOM OF PIER FLARED
MASONRY VENEER

Fig. 3-35. A grade beam is supported on concrete piers as shown in the view at left.

forcing bar, ⅝ inch in diameter) shall extend the full length of the pier and into the beam.

Grade beams. The minimum dimensions for grade beams for frame buildings shall be 6 inches wide by 14 inches deep. The minimum dimension for grade beams for masonry or masonry veneer buildings shall be 8 inches wide by 14 inches deep. When the house has a crawl space, the grade beam shall be deep enough to provide an 18 inch minimum space between the earth and the wood floor members. Reinforcing shall consist of 2 #4 reinforcing bars at both top and bottom of the beam for houses with frame construction. The reinforcing bars are increased to #5 for houses with masonry or masonry veneer construction.

Typical details of grade beam construction for slab at grade dwellings are shown in Fig. 3-36A, 3-36B, and 3-36C. Typical construction for a house having a crawl space is shown in Fig. 3-36D.

CONCRETE FORMING SYSTEMS

There are a number of quite distinctive forming systems available. These go beyond the conventional stud and double waler system we have already discussed. Manufacturers produce various types of form ties, holding devices, and form panels, which together make very practical forming systems. The various materials and components can be rented or purchased. Whether they are purchased or rented depends on the type and number of construction jobs the builder routinely handles. When the systems are rented, all but the expendable items, such as the form ties, which are used up with each forming application, are returned to the lending firm.

When choosing among the available systems, the builder selects the one best suited to meet overall needs then stays with it so the building crew can become proficient using it. Since the various manufacturers are in competition, they

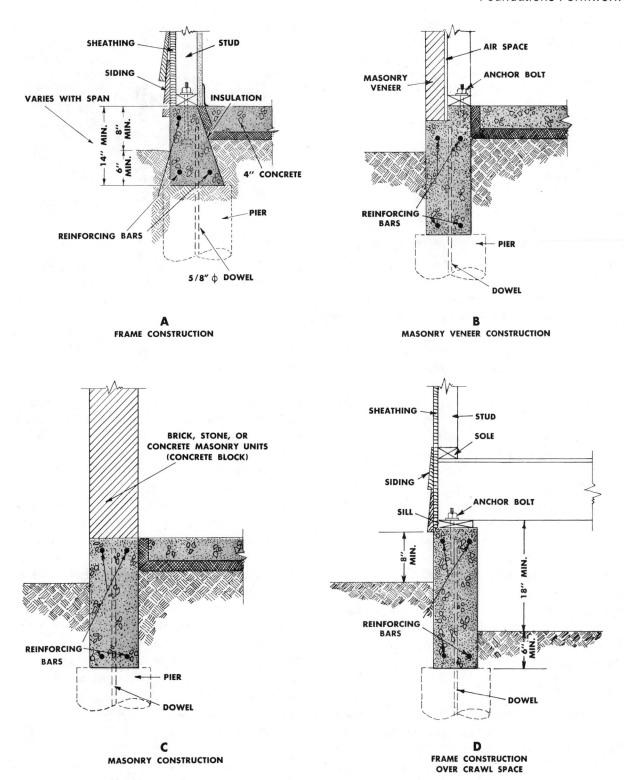

Fig. 3-36. Grade beams are used to support exterior walls in all types of small building construction. (A) shows how a grade beam supports a frame wall in slab at grade construction. (B) illustrates how a grade beam supports both the masonry and the frame parts of a masonry veneer building. A solid masonry wall is supported as shown in (C). (D) illustrates how a frame wall and joists over a crawl space are supported. Reinforcing steel is generally required in all grade beams.

naturally look to provide systems that can be set up and stripped efficiently, are relatively easy to handle, and are able to be reused again and again. These systems are designed to minimize the use of lumber for studs, walers, and braces. Many manufacturers offer the services of their engineering staff to assist the contractor in planning and executing the job. Some systems are more versatile than others and are capable of solving unusual forming problems. Other systems are designed for heavy concrete construction. Of course economy is a very important factor. When choosing a system, the builder must consider the initial cost or rental charges, the number of times the forms can be reused, and the day to day cost of expendable ties. Form ties are arranged in various ways among the various systems. One of the following three methods is ordinarily used.

The first method requires that holes be drilled in the panels on one side of the wall. After the forms are in place, holes are drilled through the opposite panel to receive the form ties. The holes in some panels have to be plugged after each use. Although it provides maximum flexibility, this method is time consuming.

With the second method, all of the panels are pre-drilled according to a particular set pattern. The panels for the two walls are arranged so that they are directly opposite each other. The holes are then lined up so that the form ties can be inserted quickly through both sides.

A third method does not require holes in the panels. The panels are arranged directly opposite each other. However, the form ties pass *between* adjacent panels rather than *through* them. This type includes panels equipped with steel frames.

Forming with Snap Ties

Double waler construction. Double waler construction uses either plyform panels (sheets) with vertical studs placed alongside the snap tie holes or framed panels. The snap tie ends pass between two walers. The walers strengthen the wall and keep it in alignment. The entire assembly is clamped together with wedges. You may want to review the discussion of this method

Fig. 3-37. Some manufactured panels have steel frames and plywood faces. Patented corners and form ties are used to speed up assembly. (Universal Form Clamp Co.)

which appeared on pages 78 to 83. (Also see Figs. 3-15, 3-18, and 3-19.)

Several companies produce steel framed panels with either plyform or plastic coated face inserts. These inserts can be easily replaced when worn or damaged. Steel framed panels are shown in Fig. 3-37. The forms are usually 2 feet wide and vary in height from 3 to 8 feet, in one foot increments. The weight of the steel frames makes these panels more difficult to handle than other form panels. However, they can be reused almost indefinitely. Many builders find this an important factor.

Foundation wall corners are formed with steel angles and bent steel shapes. The edges of panel frames and corner pieces are drilled or slotted to receive the form ties. Filler panels, shown in Fig. 3-38, are easily made by cutting pieces of plyform to fit the open area. Steel angles drilled with holes which match the edges of panels are fastened to each side of the filler plyform piece to make the filler panel. Generally the forms for low walls require no walers. Single or double walers, however, can be used to assure perfect alignment. Stiffbacks (vertical members placed at 4 feet or wider intervals to provide added strength) are used when forms are stacked to make high walls.

The major difference among the various form-

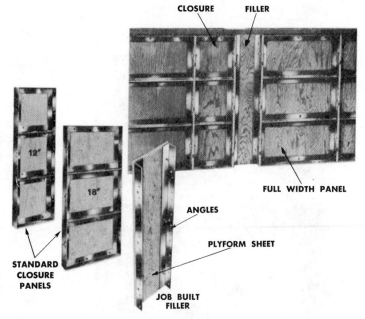

Fig. 3-38. Closure panels are available in smaller than standard widths. Very small job-built fillers are made by using a piece of plyform and two steel angles. (Universal Form Clamp Co.)

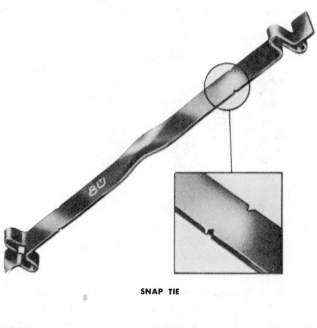

Fig. 3-39. A typical flat snap tie and tie wedge. (Universal Form Clamp Co.)

ing systems is in the form ties. One type of form tie is a flat bar notched to provide a breakback. It is likewise formed at each end to receive tapered pins. See Fig. 3-39. The tie is placed between form panels with two loops on each end projecting through slots in the panel frames. Tapered pins are dropped into place on both sides. See Fig. 3-40. Waler clamps (bent wires with hooks)

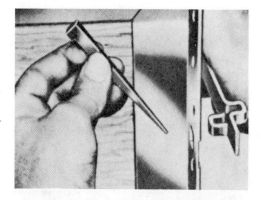

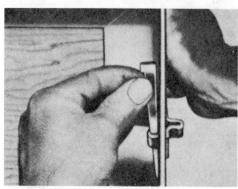

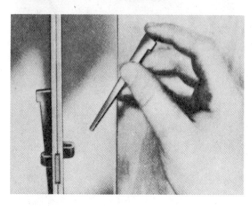

Fig. 3-40. The flat snap tie passes between the form panels and is held by two tie wedges. (Universal Form Clamp Co.)

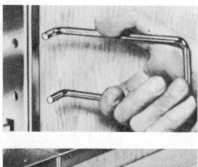

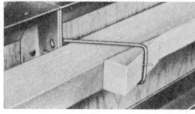

Fig. 3-41. The waler is held in place by a waler clamp which is hooked to fit into holes in the panel frame. A wedge is inserted between the liner and the waler and held in place with a nail. (Universal Form Clamp Co.)

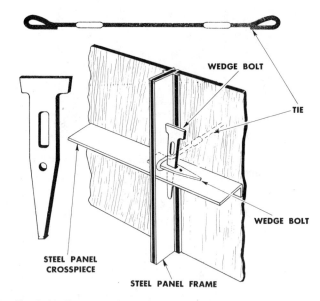

Fig. 3-42. One type of steel framed panels requires a wire tie held in place with two wedge bolts. (Symons Corp.)

fit into holes in the panel frames. The waler is inserted into the clamp and held in place with a wedge and nail. See Fig. 3-41.

Another type of form tie consists of a typical wire tie which passes between form panels and is held in place with a wedge bolt. The wedge bolt passes through slots in the two panels. See Fig. 3-42. A second wedge bolt locks it in place. Double walers are fastened with a hook gripped to one of the holes in a form angle. A plate and a wedge further hold the waler firmly in place. See Fig. 3-43.

Single waler construction. Pre-drilled panels using snap ties and brackets designed to hold over single walers provide a quick system for

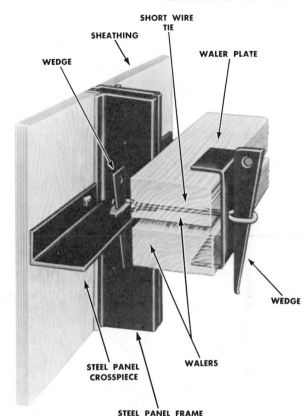

SHORT WIRE TIE

SHEATHING

WEDGE

WALER PLATE

WEDGE

STEEL PANEL CROSSPIECE

WALERS

STEEL PANEL FRAME

Fig. 3-43. Double walers are held in place with a wedge assembly and a short tie. (Symons Corp.)

forming light construction concrete foundation walls. Since the walers bear across several panels, their main purpose is to align the panels. They also add to the stiffness of the assembly. Some systems include the use of stiffbacks. Ties are either standard snap ties, which are shorter because of the absence of studs, or formed wire ties. The wire ties are made of heavy gage wire with loops at both ends to receive the wedges or clamping devices. They come in several sizes to withstand specific loads. Cones are provided to serve several functions. They leave clean holes when the tie ends are broken off and the forms are stripped. They also serve as spacers. The primary difference between the various systems is in the ties, the snap brackets or wedging devices, and the liner clamps.

A type of snap bracket is shown in Figs. 3-44A and Fig. 3-45. The bracket has a foot, which presses against the plyform panel, and a wedge to drive the tie tight. Vertical liners are held in place with liner clamps. They are also tightened

with wedge action. See Fig. 3-44B. The liners are fastened to walers with duplex nails. The form aligner braces are 2 x 4 wood members to which adjustable screw-type metal parts are added. They extend from the walers to stakes in the ground. They serve to align the forms accurately and hold them firmly in place.

A type of construction which uses patented brackets and single walers is shown in Figs. 3-46 and 3-47. The snap ties are similar to conventional ties except that they have wood spacers. The spacers fill the holes in the panels to prevent the escape of grout from the concrete. Steel spreader washers hold the faces of the form whatever distance apart is required. The 2 x 4 inch walers serve to line up the walls and put pressure on the snap ties. Braces are only placed on one side at widely spaced intervals. See Fig. 3-48.

A variation of the bracket tie clamp, shown in Fig. 3-49, uses wire ties. The ties are pulled tight by a hammer blow on the hooked part of the

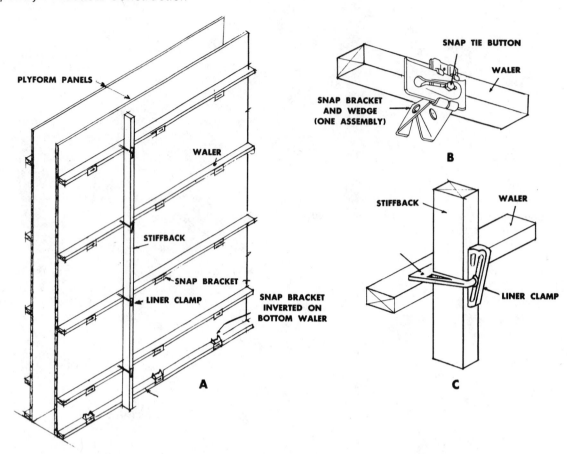

Fig. 3-44. The single waler system uses snap brackets which press against the form panel surface. The wedge is driven over the snap tie button to draw the formwork tight. (A) shows the entire formwork assembly, including the stiffbacks. (B) shows the snap bracket assembly with the wedge driven tightly over the snap tie button. (C) shows the liner clamp which is driven tightly to hold the stiffback against the waler. (Burke Concrete Accessories, Inc.)

Fig. 3-45. Single waler forms are held by ties and snap brackets. Liner clamps hold the vertical liners. (Burke Concrete Accessories, Inc.)

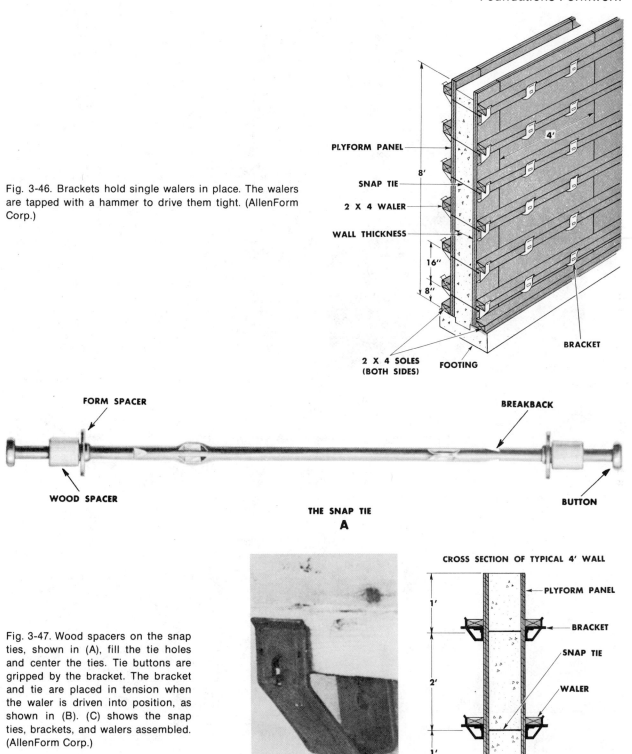

Fig. 3-46. Brackets hold single walers in place. The walers are tapped with a hammer to drive them tight. (AllenForm Corp.)

PLYFORM PANEL

SNAP TIE

2 X 4 WALER

WALL THICKNESS

4'

8'

16"

8"

2 X 4 SOLES (BOTH SIDES)

FOOTING

BRACKET

FORM SPACER

BREAKBACK

WOOD SPACER

BUTTON

THE SNAP TIE
A

Fig. 3-47. Wood spacers on the snap ties, shown in (A), fill the tie holes and center the ties. Tie buttons are gripped by the bracket. The bracket and tie are placed in tension when the waler is driven into position, as shown in (B). (C) shows the snap ties, brackets, and walers assembled. (AllenForm Corp.)

CROSS SECTION OF TYPICAL 4' WALL

PLYFORM PANEL

BRACKET

SNAP TIE

WALER

2 X 4 SOLES (BOTH SIDES)

FOOTING

1'

2'

1'

C

Fig. 3-48. Single walers hold the plyform panels in line. A minimum amount of bracing is required to keep forms plumb and aligned. (AllenForm Corp.)

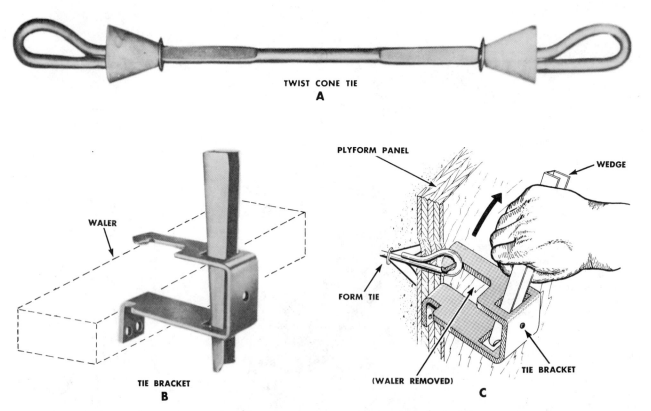

Fig. 3-49. (A) shows a wire tie. The tie bracket which has a hook to engage the tie loop is shown in (B). The tie bracket also holds the waler in place. A metal wedge is driven down to hold the waler tight and at the same time to pull the hook on the tie end tight. (C) shows how the tie end is snapped off by a twist of the tie bracket. A wedge is used for leverage. (Symons Corp.)

bracket and then driven tight by blows on the end of the wedge. The ties are broken off with a twist on the loosened wedge before the forms are stripped. Stiffback brackets are attached to the tie ends and enclose the double 2 x 4 inch stiffback members.

The foundation wall forming system, shown in Figs. 3-50 and 3-51, uses pre-drilled panels. The studs and walers are held in position by a single holder. Plyform panels are positioned horizontally in the wall. The panels can be stacked to make a wall 8 or 12 feet high or higher. Studs are placed so they line up alongside each row of tie holes. They are also placed over the vertical intersection of adjacent sheet of plyform panels. The snap ties on one side of the wall are placed through the forms. The button is passed through the hole in the holder which fastens the snap ties in place. The studs are placed, and the walers are slipped into place. The form on the other side of the wall is placed so that the tie ends pass

through the tie holes. The tie holder is placed over the button on the snap tie, and the studs and walers are put in position. The whole assembly is then driven tight with a few hammer blows on the waler.

Another type of formwork which uses $3/4$ inch plyform panels depends on special wire ties and $5/8$ inch steel rods (instead of walers) to provide walls which are strong enough to retain the concrete. See Figs. 3-52 and 3-53. The ties are made of twisted wire in such a way that they do not revolve when broken off, and with bent ends which serve as spacers between the forms. Breakbacks are provided in the tie loops. See Fig. 3-52A. Panels are cut with slots at designated intervals to line up and receive the tie ends. The hardware attached to the corner panels consists of loops arranged to work like a hinge. See Fig. 3-52B. A vertical $5/8$ inch steel bar (bent like the handle end of a cane) is passed down through the loops and serves as a pin. See Fig. 3-

SNAP TIES

A

Fig. 3-50. Snap ties and brackets (A) are used to hold walers against studs (B). (AllenForm Corp.)

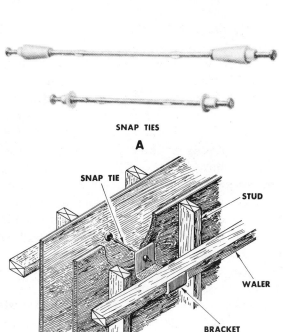

B

Fig. 3-51. Ties and brackets are used to form a high wall. Note the work platform supported by brackets. (AllenForm Corp.)

52C. The first two outside panels shown in Fig. 3-52D are assembled, then a 2 x 4 inch waler is nailed across the top to line up the next panels. When the panels are all in place, the steel waler rod is inserted through the tie loops and extends over several panels. The inside wall of the form is erected in the same manner except that no 2 x 4 inch member is required across the top. The tie ends are placed to extend through the form face and are held in place by horizontal steel waler rods. Channel top ties are placed around the top of the wall on 2 foot centers to give it added strength and provide the designated wall thickness. The entire assembly is shown in Fig. 3-53.

Unframed form panels without walers. Another type of forming system depends on 1⅛ inch high density overlaid plywood panels with 2¼ inch backing bars attached to each panel. This provides a system which requires no walers.

See Figs. 3-54 and 3-55. The backing bars extend beyond the edges of the panels to help line up adjacent panels. The panels are placed opposite each other to accommodate the ties.

The ties are made out of steel rod with a ¼ inch breakback. See Fig. 3-55A. The locking lever locks down over the tie end and draws the forms snuggly together. See Fig. 3-55B. The type of panel lever shown in Fig. 3-54 is reversible. The panel can thus be used upside down with levers appearing on the opposite side of the panel. See Fig. 3-55C.

Inside and outside corners are formed with the use of steel angles which have slots to fit the hooks on the panel levers. Fillers of the correct width are necessary at outside corners so that panels fall in line opposite each other. Make-up fillers are placed at the far end of the wall once the exact width of the space is known. A steel

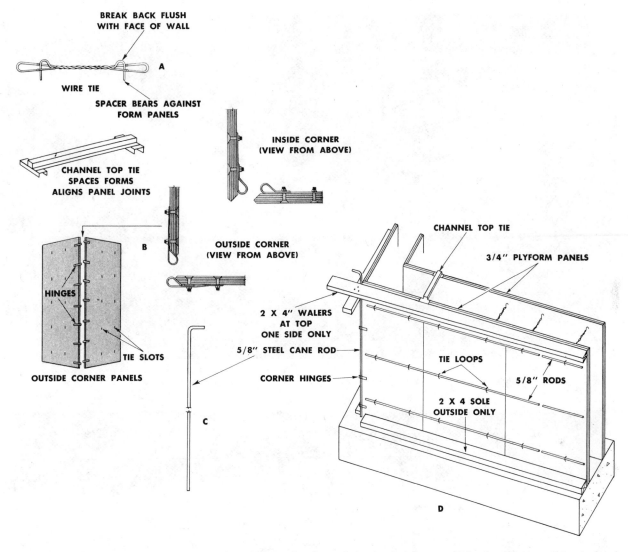

Fig. 3-52. A type of concrete formwork involves the use of unframed sheets of plyform and wire ties (A) secured to steel rods instead of walers. Corner panels have hinges (B) which are held together with a steel rod (C). Corner panels are shown in (D). (Gates and Sons)

line-up rail, made to fit over the top edge of several adjacent panels, helps hold them in line. A minimum amount of bracing is required. See Fig. 3-54.

A forming system, shown in Figs. 3-56 and 3-57, is designed to provide low walls, ranging from 36 to 66 inches high. An integral flared footing is placed directly on the ground. The panel frames are made of steel with plywood faces. Flat ties are placed under the forms to space the footing. Holes on the ties (top) and pins on the form panel (bottom) hold the assembly together at the designated dimension for the wall thickness. Flat ties are used at the top of the forms. They have two holes at each end to fit the pins on the adjacent forms. They hold the forms together and space the form faces for the wall thickness. See Fig. 3-57. Combination ties are used to provide different wall thicknesses. The forms are quickly assembled. They can be readied for stripping by prying off the top ties. The bottom ties are removed from under the wall with a tool designed for the job. All of the material, including the ties, can be used again.

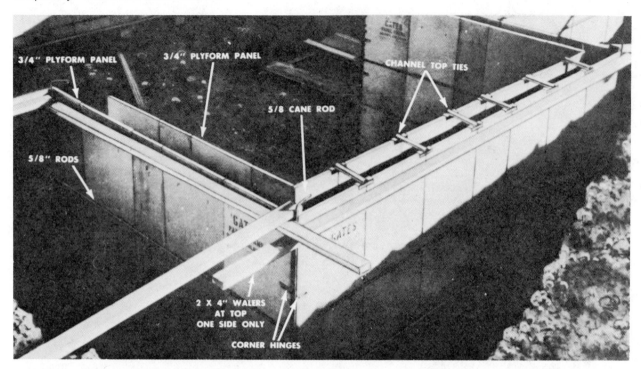

Fig. 3-53. The assembled plyform panel system described in Fig. 3-52. (Gates and Sons)

Fig. 3-54. Form panels are made of 1⅛ inch high density overlaid plywood. Walers are not used. Steel backing bars are attached to each panel. (Simplex Forms)

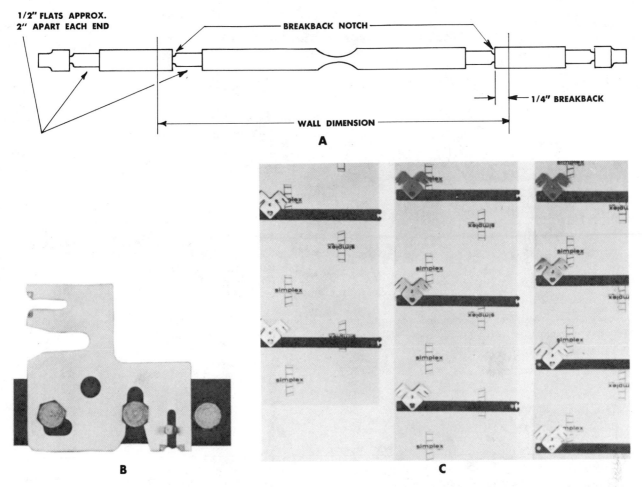

Fig. 3-55. The unframed form panel system shown in Fig. 3-54 uses a special tie (A). The lever (B) fastens over the tie end to draw the forms tight. The complete form panels (C) with levers and reinforcing steel backing bars are reversible. (Simplex Forms)

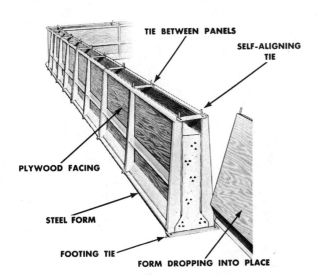

Fig. 3-56. A one piece footing and foundation form for low walls has a steel frame and reusable ties. (Proctor Products Co., Inc.)

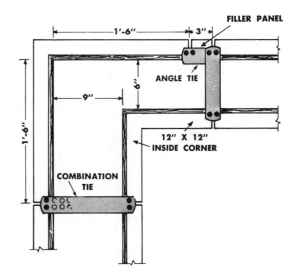

Fig. 3-57. Combination ties provide for thicker walls. Forms are held together with pins which pass through the ties. (Proctor Products Co., Inc.)

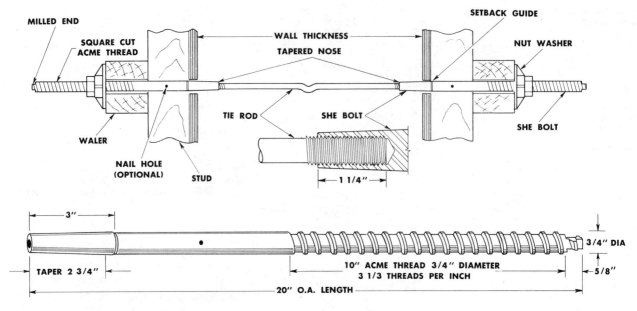

Fig. 3-58. The she bolt (stud rod) and threaded tie rod system is used for heavy construction. The tie rod remains in the wall after the forms are stripped. (The Dayton Sure-Grip & Shore Company)

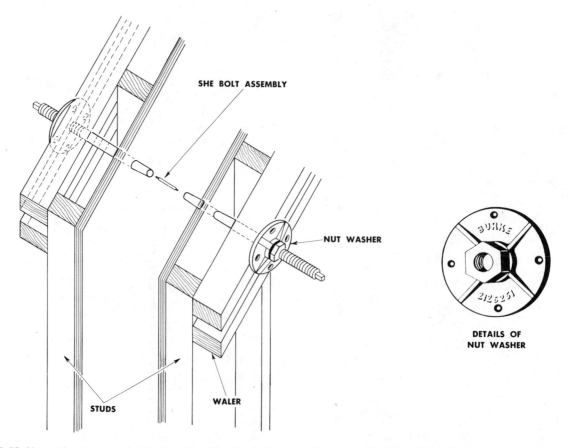

Fig. 3-59. Nutwashers are used at both ends of the she bolt assembly to draw the forms tight. (Burke Concrete Accessories, Inc.)

Forming with Threaded Rod Ties

Threaded tie rods and coil ties are two types of ties used extensively in the heavy construction field. They develop great strength and are versatile in the forming of complex structures.

She bolt tie system (Stud Rods). The assembly, shown in Fig. 3-58, includes an internal tie, ½ inch in diameter, which is threaded at both ends. (Larger bolts are made for heavy duty use.) It remains in the wall after the forms are removed. A she bolt is screwed on each end of the tie rod so that it protrudes through the form and provides a means for holding the tie rod in place. The she bolt has internal threads on one end which fit the tie rod. See Fig. 3-59. It has external threads on the opposite end to fit the nut washer. (The nut washer is also called a *cathead*.) The

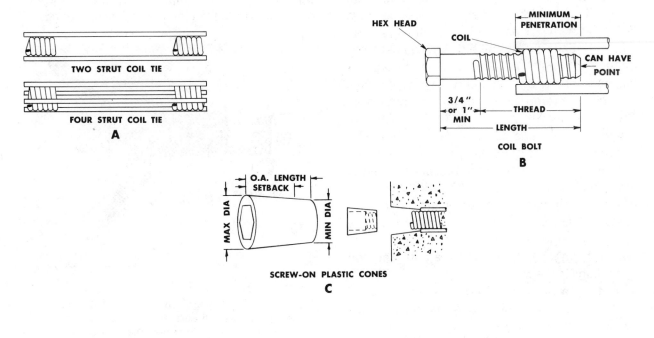

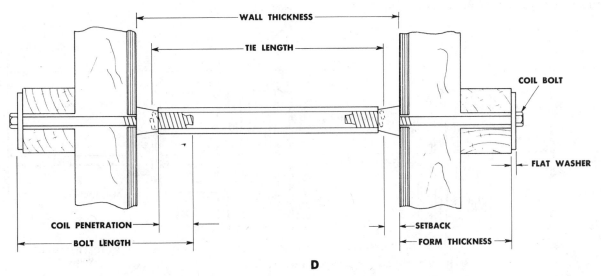

Fig. 3-60. Coil ties (A) are used for light and heavy concrete forming. A coil bolt, such as the one shown in (B), threads into the coil. Cones (C) provide an easily patched hole. The entire assembly is shown in (D). (The Dayton Sure-Grip & Shore Co.)

end with the longer threaded part extends through the form panel then between two walers. A nut washer is screwed on to the end. The nut washers on both ends of the assembly are turned until tight. Some of the forms are spaced the desired distance apart by using cones which are threaded on to the tie rods. The conical hole in the concrete left by the cones can be easily repaired.

Coil ties. Coil ties are made in several sizes to receive ½ to 1¼ inch bolts. See Fig. 3-60. They are made with 2 or 4 struts, which satisfies various safe working load requirements. The coarse threads of the coil bolt turn quickly into the coil to bring the assembly up tight against the flat washers. Several different types of cones are manufactured to serve various functions. The cone shown in Fig. 3-60, for example, screws on to the outside of the coil tie. The cones on the ends of the form tie serve to determine the wall thickness. The conical holes left in the concrete can be easily repaired. When the forms are stripped, the coil bolts, the walers, and the form panels are removed in that order. The cones are removed with a twist of a tool which fits the hexagonal opening in the end. The coil tie remains in the wall.

Forming with clamps and rods. The use of ¼, ⅜, or ½ inch steel rods in the place of ties can solve many of the forming problems connected with heavy construction. See Fig. 3-61. In the case of ¼ inch rod, pencil rod is furnished in 100 pound coils. Each coil averages 600 linear feet which is cut to size as needed. The rod is installed in the same manner as form ties. However, it must project enough beyond the walers so that there is room for the form clamps and the tightening tool. A form clamp is tightened firmly on the rod on the side of the form. A second form clamp is slipped on the other end of the rod. A tightening wrench, shown in Fig. 3-62, is used against the form clamp to drive the forms into line and to hold the clamp until it is tightened.

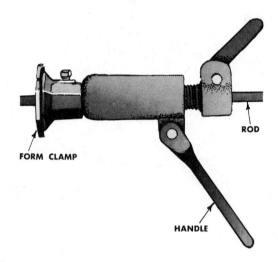

Fig. 3-62. The tightening wrench is used to force the form clamp tight against the waler so that it can be made secure. The handle is revolved to tighten the rod. (Universal Form Clamp Co.)

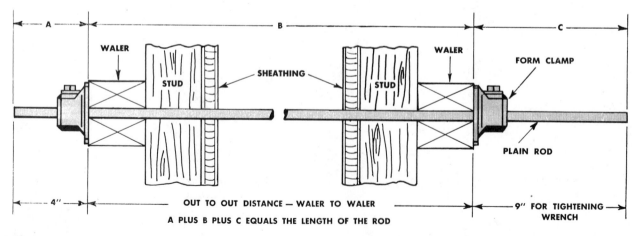

Fig. 3-61. Form clamps and rods are used for heavy construction. When the rods are to be reused, they are covered with a paper or plastic tube. (Universal Form Clamp Co.)

The rods are often left in the wall and cut off flush with the wall surface. If the rods are to be removed from the wall, they are enclosed in a paraffin impregnated paper tube at the time they are installed. A rod puller is used to extract the rod from the wall. The rod slides out easily, leaving the tube in the wall. Grout is forced in the hole to fill it.

ERECTING FORMWORK

There are some differences in how formwork is erected. This is directly related to the various types of form panels and tying devices available. However, the procedure which follows generally applies.

First bring and lay the outside panels around the perimeter of the excavation. They can be temporarily leaned against the bank in the approximate location they will occupy during the forming process. The inside wall forms should next be piled inside the excavation within easy reach of their final position. The outside forms are lined up to a chalk line which has been snapped on the footing. The chalk line is directly below the building line strung between the batterboards. Some systems require that a 2 × 4 inch member be nailed to the green concrete to serve as a sole for the form framework. The member is set back the thickness of the plyform panel. The next step is to erect and temporarily brace an exterior corner. Some panels are predrilled and require that the holes in the outside panel and the holes in the inside panel be directly in line. In other systems, the panels are arranged so that the ties pass between the edges of two outside panels and then between the edges of two inside panels. In all of these systems, the panels must be placed in position so that they are directly opposite each other, and the ties line up properly. Adjustments must be made at corners by introducing smaller panels or fillers. See Fig. 3-63A.

When large carpenter-built panels are used, the corners are the starting points for both the inner and the outer form walls. See Fig. 3-63B. The outer panels are drilled for tie ends in the specified pattern. Holes are drilled in the inner panels opposite each hole, using a long drill. A filler panel is usually required to make the wall come out even to the required length. The job of measuring and cutting the filler panel is usually done last so that it accurately completes the wall, as previously explained.

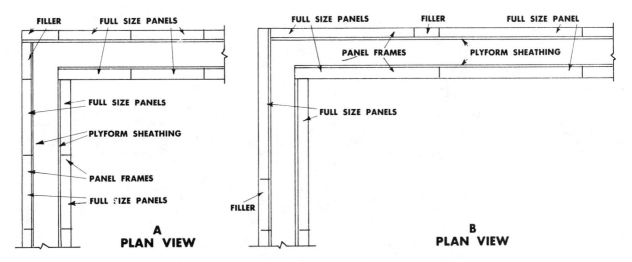

Fig. 3-63. The placement of filler panels and the corner construction depends on the type of panels and snap ties used. In (A), panels are arranged opposite each other to accomodate the ties or tie holes. Fillers occur at the corners. In (B), the panels are set beginning with the corner. Fillers are placed away from the corners. Holes for the ties are drilled in the side panels where they are required.

Openings in Walls

Allowances for any openings in the foundation walls are made before the inner form wall is erected. Various methods can be used. One example is shown in Fig. 3-64.

In some instances, the frames for windows and doors which will be located in the foundation are installed before the concrete is placed. The frames for some metal or plastic basement windows, for example, are put in place before the concrete is placed. In other applications, wood forms are used to make the sill and sides of the window opening. A strip is nailed to the form.

Fig. 3-64. When a concrete wall is placed, a buck must be used to provide for any opening which is to appear in the finished wall. (Symons Manufacturing Co.)

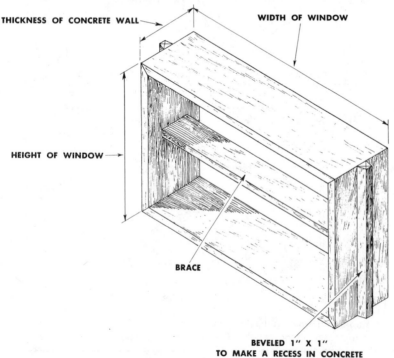

THICKNESS OF CONCRETE WALL

WIDTH OF WINDOW

HEIGHT OF WINDOW

BRACE

BEVELED 1" X 1" TO MAKE A RECESS IN CONCRETE FOR FLANGE OF STEEL SASH

Fig. 3-65. The rough form for steel basement windows is removed after the wall is placed, leaving a recess for the window frame.

**SIDE LOCKING
HOPPER WINDOW**

**SLIDE WINDOW
CLOSED**

Fig. 3-66. Metal basement windows eliminate the need for special window forms when a foundation wall is placed. (Inland Steel Products Co.)

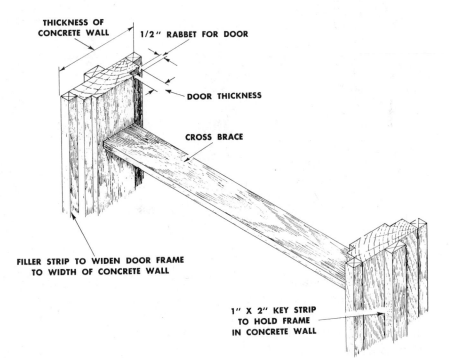

THICKNESS OF
CONCRETE WALL

1/2" RABBET FOR DOOR

DOOR THICKNESS

CROSS BRACE

FILLER STRIP TO WIDEN DOOR FRAME
TO WIDTH OF CONCRETE WALL

1" X 2" KEY STRIP
TO HOLD FRAME
IN CONCRETE WALL

Fig. 3-67. The door frame is often installed before the wall is placed. The key strip holds the frame in place.

The strip will cause a recess in the wall when the concrete is placed. After the form for the window opening is removed, the metal sash is dropped into the exposed recess and sealed with grout. Figure 3-65 shows an example of a rough form used for steel basement windows.

Other metal basement windows come completely packaged (including frames, sash, and glass). The frames of these windows, shown in Fig. 3-66, fit exactly between the form faces and remain in the finished wall.

Finished door frames are installed between the form faces so that they remain in place when the forms are stripped. They are located exactly as indicated on the working drawings. Blocks are tacked in place so that the width of the frame is equal to the thickness of the concrete wall. See Fig. 3-67. A key strip is firmly nailed to the door frame. When the concrete is placed, the key strip is embedded in the concrete so that the door frame is permanently fixed in place.

The traditional method of providing for door frames and other openings is to make a *buck*. A buck is a temporary framework which is fastened inside of the form to provide whatever size opening is required. The buck is made so that it can

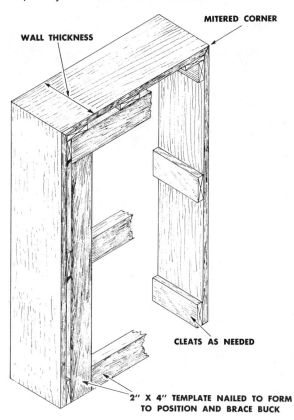

WALL THICKNESS

MITERED CORNER

CLEATS AS NEEDED

2" X 4" TEMPLATE NAILED TO FORM
TO POSITION AND BRACE BUCK

Fig. 3-68. The illustration to the left shows an example of making a rough buck. The template is fastened to the form sheathing on one side. It is made part of the buck on the other side.

be easily removed after the forms are stripped. Figure 3-68 shows a door buck. The first operation is to make a template which will be used to position the buck in the proper location. The template is nailed to the inside face of the outside panel (through the panel) with duplex nails. The sides and top of the buck are then put in place with internal cross braces as they are required. After the concrete has set and the forms are ready to be removed, the duplex nails holding the template are drawn. The forms are stripped, the template and bracing are removed, and the buck is taken apart. Mitered corners permit the sides and top to be pulled out. The buck and template are ready to be used again. The finished door frame is installed in the opening and fastened in place with powder-activated fasteners.

Additional Formwork Problems

The concrete foundation walls may require modifications other than those for window or door openings. An example would be a shelf needed for the ends of a girder. A strip of lumber, which is the right size, or a box is fastened to the inside of the form to provide for the change.

After the special problems are taken care of, the inner wall is erected by passing ties through the form panels, then the studs and walers are placed. Braces which extend to the ground inside the foundation are used to plumb and align the wall. The tie clamps are driven fast to lock up the whole assembly. Anchor bolts, usually $1/2$ inch diameter threaded bolts, are then installed. They are bent to resist turning and to hold more effectively. They are placed in the top of the concrete wall at 4 foot intervals. This dimension can vary. These bolts are hung from strips of wood fastened to the top of the form or from cleats within the form, as shown in Fig. 3-69.

One of the final steps is to establish the top line for the concrete in the form. Duplex nails are driven at 10 foot intervals around the inside of the form wall to indicate this elevation. Strips of

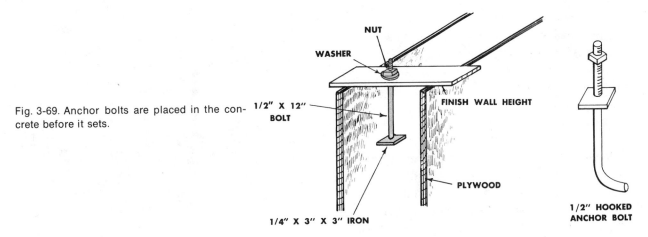

Fig. 3-69. Anchor bolts are placed in the concrete before it sets.

wood nailed inside the form at the right elevations serve the same purpose. In the case of low forms, the top of the form sheathing is made to coincide with the top of the concrete. The carpenter studies the plot plan and section view of the working drawings to determine the elevation of the finished floor and the dimensions of the floor thickness, the floor joists, and the sill. By subtracting these dimensions from the elevation of the finished floor, you can find the elevation of the top of the concrete. After setting up a leveling instrument over the point of beginning (or some central point of reference), you can sight a rod held at each position where nails are to be driven.

PLACING CONCRETE

Concrete is usually delivered to the job by truck as ready mix and placed in the forms using a chute. The chute can be extended as required. The concrete should be deposited near where it is to be placed so it doesn't have to be moved any distance inside the form. When the formwork is inaccessable to the truck chute, the carpenter must build a runway so that the concrete can be wheeled to its location. The runway is a temporary strong framework. One side of it rests on posts extending to the ground, and the other is fastened to the formwork frame. The concrete should not be allowed to drop over 3 or 4 feet. Dropping concrete beyond this distance can

cause the coarse aggregates in the concrete to separate (segregation). Drop chutes should be used when the walls are high. The cement should be placed in lifts (layers) which do not exceed 18 inches. It should then be spaded and puddied with puddling sticks which are long enough to reach the bottom of the form. Stone pockets and air bubbles are eliminated as the concrete is worked. The agitation distributes the aggregate so that a smooth wall surface will result when the forms are stripped. The concrete should be worked around corners and angles and under and around window bucks. Vibrators can be used to increase the agitation. Excessive vibration has the tendency to separate the cement and the aggregate.

STRIPPING THE FORMWORK

It is important to know when to strip the forms. Builders wish to proceed with the construction without delay or they may wish to use the form components on another job. Generally, wall formwork can be removed three or four days after the concrete has been placed, and the building process can continue. However, the concrete does not reach its design strength until approximately 28 days after it is placed. The curing process continues throughout this period. Keeping the wall wet by leaving the forms in place for a longer period of time, or covering it with wet burlap or canvas will give the finished wall greater strength and make it more waterproof. In cold weather, the temperature of newly placed concrete must be kept at a minimum of 50 degrees Fahrenheit (10 degrees Celsius) for at least five days after it is placed. The concrete is covered to retain moisture and prevent freezing.

Forms are generally stripped in reverse fashion to the way they are set up. The bracing is removed first. Form tie clamps are removed and placed in a container so you can keep track of them. Walers are taken off. Duplex nails holding bucks are withdrawn. The panels are then carefully pulled from the wall, leaving the tie ends protruding from the concrete. It is not recommended that you use pinch bars or other metal tools to remove the forms when they fail to respond to light pressure. Otherwise you can mar the surface or edges of the wall. Instead you can use wood wedges on stubborn forms.

When all of the panels have been removed and stacked, the ties are broken off, and the wall surface is repaired. Wire ties are removed by bending or twisting them with a tool. Flat ties are broken off with a vertical hammer blow. Threaded she bolts and coil tie bolts are screwed out of the wall. The hole which is left is patched with grout. When ties with cones have been used, the hole can be filled with grout or with a plastic plug which is cemented into place.

The forms are cleaned with a hose and scraped. They are readied for the next job by oiling or coating them with an acrylic preparation. This preparation serves a dual purpose. It prevents the plyform panel from absorbing water from the concrete and permits easy separation of the forms from the concrete wall. Some form panels are made of high density overlaid plywood panels which require no surface preparation other than cleaning.

Before the earth is backfilled around the foundation, one or two coats of bituminous material can be sprayed on the outside of the wall for moisture protection. When the area has a problem with water concentration and requires more adequate waterproofing, several alternative materials can be used. A membrane is fixed to the wall with the use of mopped asphaltum. It can consist of 2 plies of hot mopped felt (asbestos or other fibers which are supplied in long

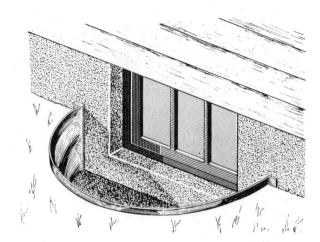

Fig. 3-70. A basement window protected by a steel areaway.

rolls and saturated with tar or asphaltum), polyvinyl film, 6 mils thick, or roll roofing. All edges must be sealed. A gravel bed and drain tile are placed around the perimeter of the footing.

Areaways are installed at basement windows which are below grade. The type usually used is made of sheet metal and is fastened to the wall so that it projects above the finished grade and below the bottom of the window. See Fig. 3-70.

CONCRETE STAIRS

The forming for a set of stairs is shown in Fig. 3-71A. These stairs will be open at the sides once the form panels and riser form boards are removed. One of the first steps in constructing the formwork is to place the side form panels and then anchor and brace them. Mark the profile of the treads and risers on the two form panels. Cut riser form boards the exact width of the stairs. Nail blocks to the ends. Position these assembled pieces so that they line up with the marks drawn on the form boards. Nail the blocks to the form panels with duplex nails which enable the blocks to be readily removed. Place a 2 × 4 inch piece across the center of the riser form boards, extending down to a stake in the ground. Nail the piece at each riser. Use additional blocks to hold the risers in position and to stiffen the whole assembly. An alternate method with plank sides is shown in Fig. 3-71B.

A

Fig. 3-71. (A). Forming for a set of concrete stairs. (B). In this alternative method, the stairs are formed with a minimum amount of lumber. Planks are used to form the sides. (Portland Cement Association)

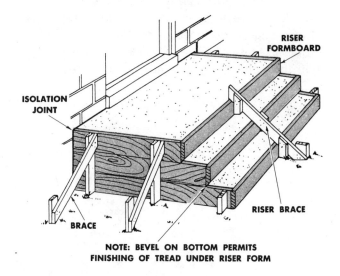

One method of finishing the stairs is to remove the riser form boards after the concrete has set sufficiently, which might be within a few hours. The cement finisher then trowels the surfaces. When this method is used, the bottom edge of the riser form boards is beveled so that the cement finisher can finish the treads all the way to the intersection of treads and risers shortly after the concrete is placed.

(Note: The method for determining the exact riser height and tread width for any individual situation is discussed in detail in Chapter 7, *Interior Finish*. As a general rule, the riser height of stairs should not be more than 8 inches. The tread width must not be less than 9 inches. The Uniform Building Code specifies various riser and tread dimensions depending on the building's use and occupancy load. All the risers on any stair must have the same height, and all the treads must have the same width.)

WALKS AND DRIVES

Either the carpenter or the cement mason may have the job of preparing the formwork for walks and drives. Walks and drives are installed late in the building schedule. They are laid at a time when the traffic flow of mechanics from other trades going in and out of the building is minimal. The earth where walks and drives are to be laid must be thoroughly compacted.

If the grade has to be built up, the preparation of the bed is generally given priority over the building of the formwork. A 4 to 6 inch bed of granular material, sand, stone, or crushed rock is recommended if the site has excessive surface water or a potential problem with frost. Under normal conditions, the concrete is placed directly on the earth, after it has been dug out to the required depth. All vegetation must be removed.

Precise information about the location of the walk or drive in relation to building and property lines can be obtained from the plot plan. This plan likewise indicates the elevation in relation to the point of beginning, discussed earlier in Chapter 1.

Once you have all the necessary information, drive stakes at the corners of the walk. String lines to indicate one side of the walk. Additional stakes are placed along one side, at three foot intervals, as shown in Fig. 3-72. A leveling instrument is set up over the point of beginning. Using a leveling instrument, transfer the correct elevation of the top of the walk to each stake and mark it. Then adjust the line according to these marks.

Form boards, 2 × 4 inches, are nailed to the stakes so the top of the boards are in line with the marks on the stakes. A spreader is cut from the same material as the form boards to a length equal to the width of the walk. The spreader is used to locate the form boards on the other side of the walk and to temporarily retain the concrete while it is being placed in one area at a time. Form boards are then placed on the opposite side of the walk, spaced by the spreader, and nailed to stakes in a position which is level (except for a slight slope) with the form boards on the initial side. A hand level and a straight edge are used to adjust the height. A slope of 1/8 of an inch per foot is generally made for drainage purposes. The end boards are then cut to the

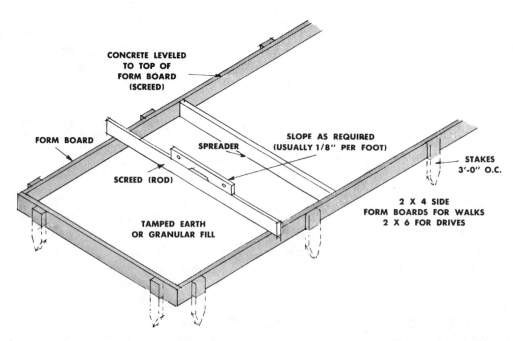

Fig. 3-72. Forms for walks and drives consist of side and end form boards held firmly by stakes. The form boards serve as screeds to determine the top level of the concrete.

Fig. 3-73. A vibrating straightedge is used for the strike off operation. (Portland Cement Association)

Fig. 3-75. A control joint provides a weakened depression in the concrete. Cracks develop at the control joint in the event the slab settles or contracts. (Portland Cement Association)

Fig. 3-74. A power vibrated mechanical screed is used to level (screed) large concrete areas. Note the use of wire mesh. (Stow Manufacturing Co.)

proper length and nailed to the side form boards. The end boards are also reinforced with stakes. The form boards, spreader, and end boards serve as guides to determine the top level of the

concrete. They are often called *screeds* when used in this capacity. Another board, also called a *screed,* is used to level off the concrete. It is laid on top the form boards and moved back and forth with a saw-like motion. A vibrating straightedge simplifies the problem of striking off the concrete for a walk. See Fig. 3-73. A mechanical screed, such as the one shown in Fig. 3-74, increases speed and efficiency.

Isolation joints, made of asphalt-impregnated material, are placed wherever the walk or drive butts on the building foundation, or other places where slabs should be separated. This allows the walk to float freely in event of any settlement. Otherwise, settlement could cause severe cracking.

Control joints are partial cuts made in the freshly placed concrete following finishing. See Fig. 3-75. These joints provide a weakened place for the concrete to crack should the concrete shrink during the curing process or after. This generally prevents random cracking on the walk or drive. The joints are placed at 4 or 5 foot intervals on walks and 10 foot intervals on driveways.

The above discussion also applies to driveways. They are formed in much the same manner as walks. However, the form boards used for driveways are made 2 × 6 inches if a thicker slab is specified. Welded wire fabric is supported so that it is about 2 inches above the bottom of the slab. This is accomplished by placing half bricks at random over the fill in the area of the drive to support the welded wire fabric.

SAFETY IN CONCRETE FORMING

An attitude of safety is essential in the construction of concrete formwork. The problem of retaining the concrete as it is placed demands adequate safeguards. Carpenter-made or patented panels and hardware are relatively safe for residential foundation work provided they are assembled and then erected according to good practices. Unusual forming problems and heavy construction require greater precautions. For example, the failure of shoring can present great hazards to life. (Shores are the supports which hold up the formwork for floors above.)

When you work on any type of formwork, from residential work to tall buildings or public projects, you must be alert to your surroundings. As you perform your job, you should inspect the materials and discard those which are faulty. Ties and other holding devices should be properly spaced and tightly fastened. Adequate bracing should always be used.

Concrete work also has other hazards. Concrete is worked in a fluid state. Skin burns can result from contact with fresh concrete because of the heat caused by hydration and the chemical reaction between the cement and the water. Gloves should be worn. The skin can also be protected with hand cream or petroleum jelly. Other areas of the body should likewise be protected. Long sleeved shirts and jackets are required. Eye protection is mandatory. The same chemical action which can harm the skin can also harm the eyes. There is also danger from flying particles of sand and other debris. Dry cement dust is particularly irritating to the eyes.

Other hazardous situations can be caused by sharp objects protruding from the forms such as duplex nails and form tie ends. You must be careful to avoid injuring your skin or tearing your clothes. Reinforcing bars sometimes project from the walls where additional concrete parts will be tied in later. These exposed bars are particularly dangerous. They should be properly covered to prevent people from tripping over them or becoming impaled on them.

A variety of safety equipment is available which helps minimize such on-job hazards. Safety footwear with a range of safety features, such as steel tips and puncture-proof steel midsoles, are especially important to the carpenter. These can help prevent injury when heavy objects are dropped. They also help prevent the foot from becoming impaled on nails and other sharp objects, a common serious hazard on the construction site.

Sufficient barricades should be erected around excavations. Although excavations for smaller structures are often deep enough to cause severe injury from falls, they are rarely adequately barricaded The worker on the job must constantly be alert to this danger. The potential for danger of course increases when excavating machinery and other heavy duty equipment are in operation.

When formwork is under construction, the worker can sustain strains, sprains, and back injuries because the work often requires heavy lifting. Heavy and awkward forms must be lifted, pulled, and pushed into place. It is essential that the worker know how to lift properly, using leg and arm muscles rather than straining the back. The potential for such injuries can sometimes be reduced if carpenters work in pairs.

Formwork is often erected at heights which are dangerous. A fall from such heights can be fatal. Proper scaffolding, runways, and bracket platforms are essential in such job situations.

QUESTIONS FOR STUDY AND DISCUSSION

1. What ingredients go into making ordinary concrete?
2. Explain the difference between concrete and cement?
3. What are admixtures used for?
4. How long does it usually require for ordinary concrete to reach full compressive strength?
5. Why is the quantity of water used in making the concrete important?
6. What is the function of walers?
7. Describe the function of stiffbacks.
8. Why are rate of pour and the atmospheric temperature two very important factors to consider when you are building the forms?
9. What is the purpose of vibrating the concrete?
10. How do plyform panels differ from ordinary plywood sheets?
11. Describe the steel reinforcement used in concrete construction.
12. Explain the function of snap ties in regard to: a) spacing the width of the wall, b) holding the wall together, c) providing breakbacks at the tie ends.
13. What is the purpose of a vapor barrier in slab at grade foundations?
14. What method of thermal insulation is used for slab at grade foundations?
15. What is a grade beam?
16. Describe the advantages and disadvantages of steel framed panels.
17. Explain the method for holding snap ties in a single waler type of forming.
18. What is the function of threaded tie rods and nut washers in foundation formwork?
19. Describe the function of a coil tie.
20. Explain the function of filler panels.
21. Explain how basement windows are installed in a concrete wall.
22. What is a door buck?
23. Explain the method of installing anchor bolts.
24. What is the method of determining the top of the concrete elevation in the form?
25. What care should be given to the forms once they have been stripped?
26. Describe a method for forming concrete stairs.
27. Explain the method of making forms for walks and drives.
28. What is an isolation joint? A control joint?
29. Describe some of the potential hazards of mixing and placing concrete.
30. Describe some of the potential hazards of erecting concrete formwork.

CHAPTER 3-1
PRACTICE IN SITE WORK AND FOUNDATION FORMWORK

HOUSE PLAN A

House Plan A is used to illustrate how footings and foundation walls are built in a specific application. House Plan A also gives us an example of how to locate the building and prepare the site for a slab at grade building. The working drawings for House Plan A appeared at the end of Chapter 1, *Preparing for the Job.* The Plot Plan (Fig. 3-76), the Foundation Plan (Fig. 3-77), and a portion of the Section View A-A (Fig. 3-78) are included here for convenient reference.

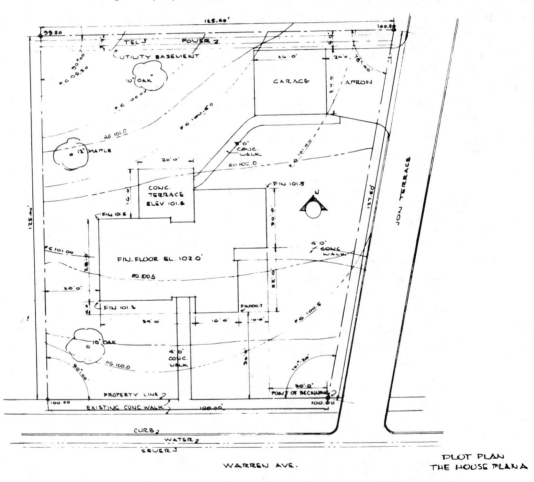

Fig. 3-76. The Plot Plan for House Plan A shows the location of the building in relation to lot lines.

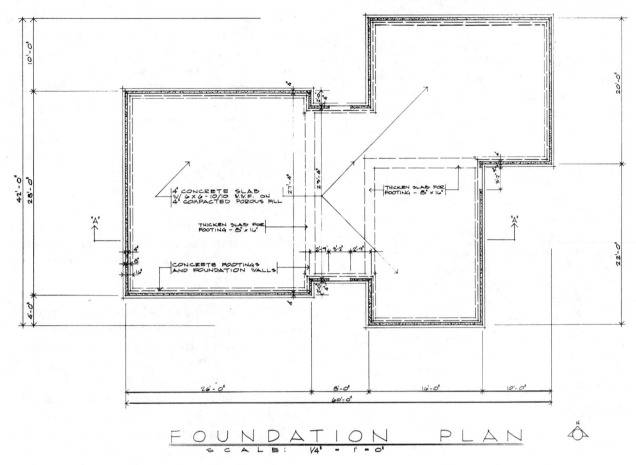

FOUNDATION PLAN
SCALE: 1/4" = 1'-0"

Fig. 3-77. The Foundation Plan for House Plan A.

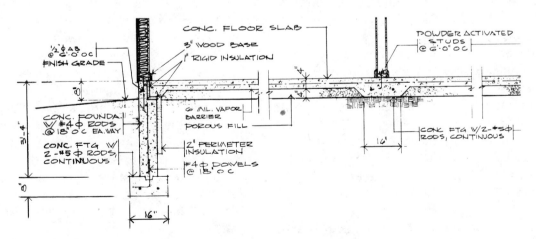

Fig. 3-78. This section view is taken through the foundation and slab and shows the details of construction.

Physical Characteristics of the Lot

The physical characteristics of the lot on which House Plan A is built are shown on the Plot Plan. See Fig. 3-76. Such characteristics are always important. This plan tells the carpenter where the building is to be located and at what height above the point of beginning it is to be placed. Other important information is shown. If the house is to be built on level ground, the ideal plot is one which is already relatively level and requires little or no modifications. The building should be set high enough above grade so that when the grading is finished, the surface water flows away from the foundation. When excessive moisture is a problem, drain tile should be placed along the outside of the footing to divert any seepage of water.

If you look closely at the Plot Plan, you will notice that the building is placed at the highest point on the lot. There is a gentle slope downward toward the northwest and southeast corners. A minimum of grading and fill is required for the slab.

Staking Out The Building

The method used for staking out a house and erecting batterboards generally follows the procedures laid down earlier in Chapter 2 for House Plan B. The first job is to locate the four basic corner stakes, numbered *1, 2, 3,* and *4* on Fig. 3-79. To locate point *A*, measure 20 feet from the Point of Beginning toward the west, along the edge of the sidewalk. (The edge of the sidewalk here is the property line.) The leveling instrument is set up over point *A*.

Sight along the property line back to the Point of Beginning, then swing the telescope 90 degrees in order to locate building line 1-2. (Line 1-2 represents the eastern building line and is a base line for other measurements.) With the instrument fixed in this position, measure 30 feet from point *A*. Sight a rod at that distance and drive a stake at Point 1. Mark this exact spot on the stake with a nail. This corner stake is designated stake *1*.

With the instrument still fixed over point *A* and the telescope bearing in the direction of Point 1, locate Point 2 by measuring 42 feet toward the north from Point *1*. Drive a stake at this point and mark the exact point on the stake, which is designated corner stake 2.

The next step is to set up the instrument over point *B*. Remove the instrument from over point *A*. Point *B* is located by measuring 60 feet west along the edge of the sidewalk. (Again, the sidewalk is the property line.) Set up the instrument over point *B*.

Sight along the property line back toward the Point of Beginning, then swing the telescope 90 degrees in order to locate Points 3 and 4. Fix the telescope in this position. Sight a rod held 30 feet away from point *B*. Drive a stake and mark the exact point with a nail. This corner stake is designated stake *4*.

With the telescope fixed in the same position over point *B*, sight a rod held 42 feet toward the north from the point on stake *4* to locate corner stake *3*. Drive a stake and mark the exact point on the stake with a nail as before. This stake is designated stake *3*. Line 3-4 represents the western building line and is a base line for other measurements.

Stretch a line from the nail on corner stake *1* to the nail on stake *2*. Continue on with the line to the nail on stake *3* then to stake *4* and finally back to stake *1*. You have now established the building lines for the largest dimensions of the building, as shown in Fig. 3-79. Check lines 1-2 and 3-4 to verify they measure exactly 42 feet. Also measure lines 1-4 and 2-3 so that they are exactly 60 feet. Measure diagonally from Point *1* to Point *3* and from Point *2* to Point *4* to make certain these two measurements are equal. If this procedure has been followed correctly and checks out, the area enclosed by the building lines is a perfect rectangle with square corners.

Several types of instruments, such as the transit level and the automatic level, can be used in place of the builders' level for the above operations. When a transit level is used, it has the advantage that the telescope is depressed to bear directly on the stakes. A rod is not required. This and other procedures for establishing building lines have been outlined earlier in Chapter 2.

Corner stakes *1, 2, 3,* and *4* are removed during the excavating operation. Their location

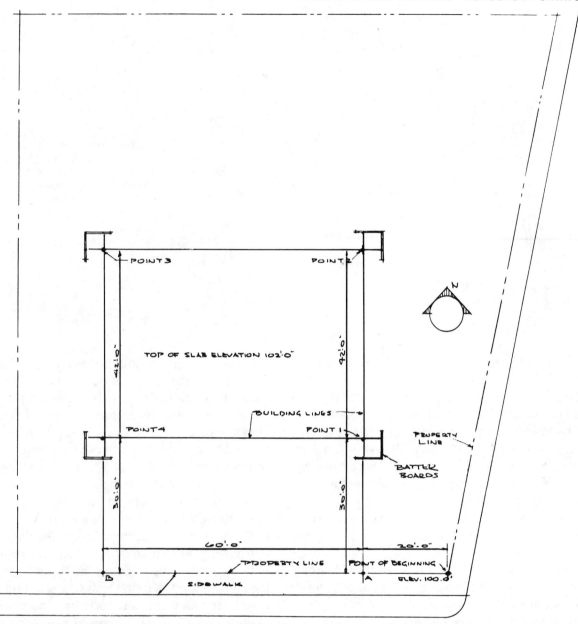

Fig. 3-79. Building lines are established in relation to the point of beginning. Batterboards are erected so that lines are available for later reference.

would be lost unless batterboards, as shown in Figs. 3-79 and 3-80, are erected far enough outside the excavation so that they will not be disturbed. The purpose of the batterboards is twofold. They provide a means for stringing lines to locate the foundation corners after the excavating work is completed. They also establish the

elevation of the top of the slab, and other elevations related to it.

Erect the batterboards as follows. Drive long stakes into the earth so that they project a minimum of 3 feet above the earth. Set the stakes back about 4 feet from the building lines which have been strung between points *1, 2, 3,* and *4.*

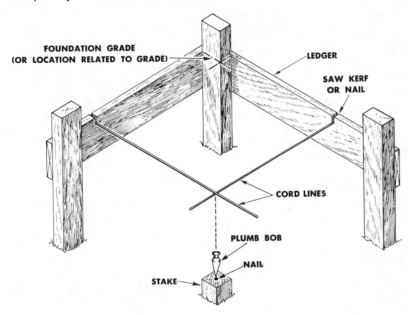

FOUNDATION GRADE
(OR LOCATION RELATED TO GRADE)

LEDGER

SAW KERF
OR NAIL

CORD LINES

PLUMB BOB

NAIL

STAKE

Fig. 3-80. Batterboards consist of stakes and horizontal ledgers. Lines are fastened to the batterboards to locate the building lines.

Horizontal boards called *ledgers* are nailed to the stakes according to the instructions which follow. A leveling instrument is required to establish the location of the horizontal boards.

The Point of Beginning, as shown in Fig. 3-79, is at an elevation of 100 feet. The elevation for the top of the slab is at 102 feet. Therefore the tops of the ledgers must be placed 2 feet above the Point of Beginning. To place the ledgers correctly, set the instrument over the Point of Beginning. Level the instrument and adjust it so that it can revolve in a horizontal plane. Place a ripping alongside of the instrument, with one end resting on the sidewalk. Measure 2 feet above the sidewalk and draw a line on the ripping at that height. Mark a second line opposite the center of the telescope.

Hold the ripping against the stake and adjust it while a helper sights it from the instrument. When the *upper* line on the ripping is adjusted to the height of the telescope, mark the stake with a line even with the *lower* line on the ripping. Lines marked on the stakes at this elevation coincide with the height for the top of the slab. Nail the ledgers in place at these marks. (It may be advisable to place the marks on the stakes one foot above the elevation for the slab so that lines will be above and out of the way of the foundation forms.)

The next step is to locate the saw kerfs which are used for stringing the building lines. (Some carpenters use nails instead because nails can be easily moved to make final adjustments.) These lines when strung will pass directly over the building lines which have been strung between corner stakes *1, 2, 3,* and *4,* as shown in Fig. 3-79. Tie lines to the ledgers and adjust them until a plumb line dropped from the point where they intersect coincides with the location of the nails on each of the corner stakes. See Fig. 3-80. Use a string and a plumb bob. Make saw kerfs (or use nails) in the top edges of the ledgers at each location. Once the batterboards are strong and firmly fixed, and the saw kerfs have been accurately located, you can remove the building lines and the corner stakes. Then the excavating machinery can begin operation. The saw kerfs made in the ledgers (or the nails) enable the building lines to be restrung at any time. The building lines for each wall must be available when needed for the setting of the formwork.

The building is irregular in shape. Therefore it requires more points than those at corners *1, 2, 3,* and *4.* Some carpenters erect batterboards at each secondary location. For a building such as House Plan A, stakes are satisfactory and can be quickly installed. See Fig. 3-81. (On a large struc-

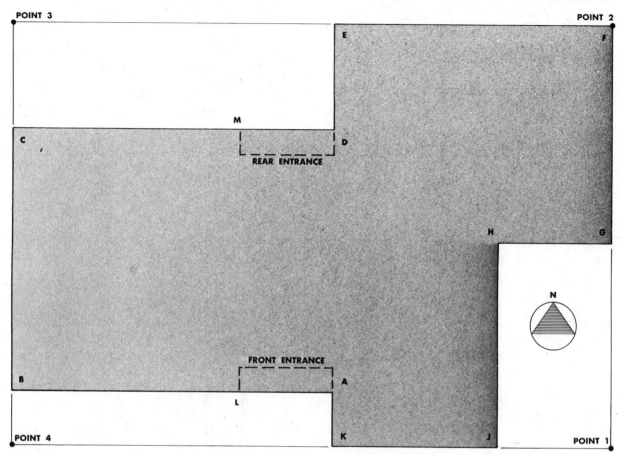

Fig. 3-81. The irregular shape of the building requires that offset walls AB, CD, and GH be located by stakes which are in line and parallel to the East and West walls. The offset walls EK and HJ must be located by stakes which are in line and parallel to the North and South walls. An additional line is required for locating points L and M. The dotted portions at the entrances do not require special lines since they are measured from other lines.

ture, batterboards would be used to provide accurate walls for each offset.) The stakes are driven into the ground in line with the batterboards. Since they are placed in line with the batterboards, as shown in Fig. 3-81, they will be at least 4 feet away from the building line and therefore out of the way of the excavating machinery. The stakes are driven so that they project approximately 18 inches above the ground. This measurement does not have to be precise. The important thing is that the lines, when strung later, are above the formwork.

String lines from batterboard to batterboard (rather than between batterboards over the building lines), using the kerfs already made on the ledgers to hold the lines. See Fig. 3-82. These lines will be parallel to the building lines established earlier (1-2, 2-3, 3-4, and 4-1). In order to pick up all of the offsets shown on Fig. 3-82 for the building, you must make the following measurements. Start with point *X* on the batterboards near Point *1*. See Fig. 3-82. Along the line parallel to building line 1-2, measure dimensions of 4'-0'', 18'-0'', 10'-0'', and 10'-0'' respectively. The fourth measurement on the line which is parallel to building line 1-2, the measurement of 10'-0'', returns to the kerf marked *Y* on the batterboards near Point *2*. Drive stakes at each of the three points as shown. Using a plumb bob, mark the exact points with a nail. These points as shown on Fig. 3-82 are the points designated as *5, 6,* and *7.*

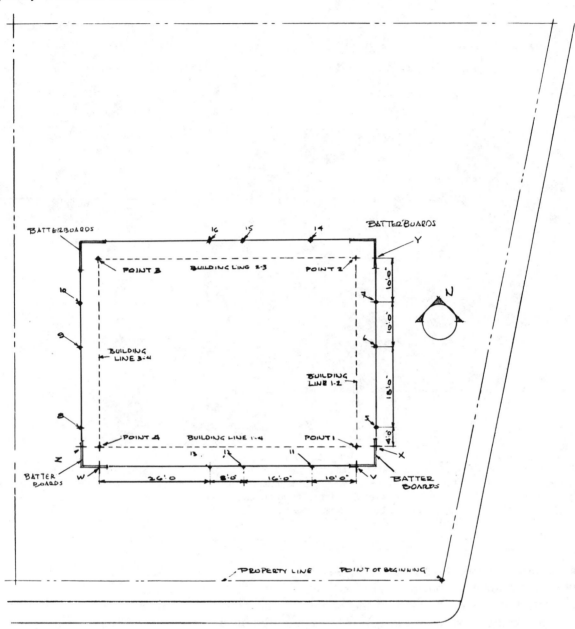

Fig. 3-82. Secondary stakes are needed to locate the forms for all of the walls of the buildings. Lines parallel to the building lines are drawn between the batterboards. Measurements are taken from the kerf points on the batterboards to locate the position of the stakes.

Use the same procedure along the line parallel to building line 3-4. Start from the saw kerf marked *Z* near Point *4* on Fig. 3-82 and locate the stakes and points designated *8, 9,* and *10.*

The Foundation Plan, Fig. 3-77, indicates that measurements, and therefore stakes are required in an east-west direction also. The measurements shown on the Foundation Plan correspond with those shown on the botton of Fig. 3-82, parallel to building line 1-4. Begin at the saw kerf marked *V* on the batterboards near Point *1.* Measure distances of 10'-0", 16'-0", 8'-0", and 26'-0". The measurement of 26 feet extends to the kerf marked *W* on the batter-

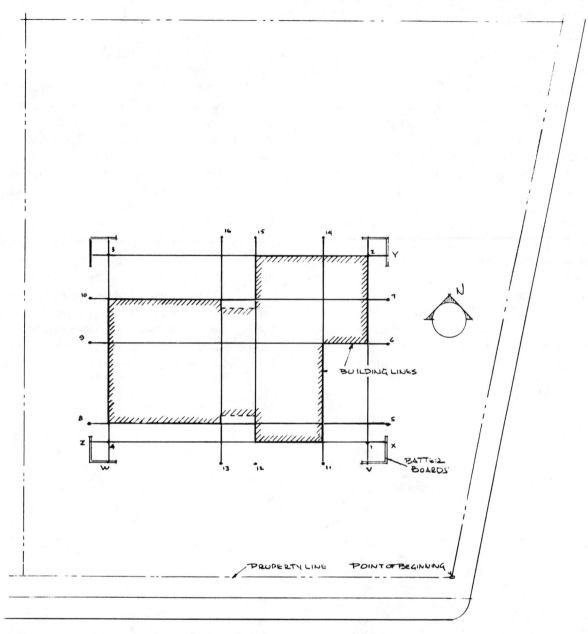

Fig. 3-83. After the grading for the slab is finished and the trenches for the footings and the foundations walls are dug, the building lines are strung to locate the foundation wall formwork.

boards near Point *4.* Drive stakes at each of the three points shown on Fig. 3-82. Using a plumb bob as before, mark the exact point on the stake with a nail. These points are designated Points *11, 12,* and *13.*

Repeat the procedures for the north side, establishing Points *14, 15,* and *16* shown on Fig.

3-82. Once all the points are established, check the measurements to make sure they are correct, then remove the lines. The ground is graded in preparation for the slab, and the excavations for all of the foundation wall are dug. The lines are strung between the batterboards and the stakes across the excavations. See Fig. 3-83.

The Excavation

The excavator removes the debris and top soil in the area where the building is to be located. The soil is leveled and firmly tamped until it measures exactly 8 inches below the elevation for the finished slab. (4 inches for the slab and 4 inches for the porous fill.) This work should be completed before the excavation for footings and the foundation wall begins. Additional work with excavating equipment should be kept at a minimum once the foundation walls are in place.

The excavator needs to know where the earth is to be removed for the footings and the foundation walls, and to what depth. The depth must be exact so that the footings can be placed on firm earth. The bottom of the footing, as shown in Section A-A (Fig. 3-78), is 4'-0" below the top of the slab. The ledgers of the batterboards are located at an elevation equal to the top of the slab. The excavator determines the bottom of the footing by measuring down 4'-0" from the top of the ledgers. The excavation for the footings and the foundation walls must be dug wide enough so that the carpenters have room to work as they assemble the formwork. A minimum width of 4'-0" is considered adequate for a low wall. The plumber usually brings in the water and sewer lines, which pass beneath the footings, before the footings are placed. It is much easier to cut trenches for the pipes before the footings are placed than to dig under the footings afterwards.

Erecting Footings

Once the excavating machinery is removed from the area, string the building lines between the batterboards and the secondary stakes. See Fig. 3-83. Check the dimensions from point to point where the lines intersect to verify that the batterboards and the stakes have not been disturbed during the excavating operation. Check the basic dimensions shown in Fig. 3-82. Section

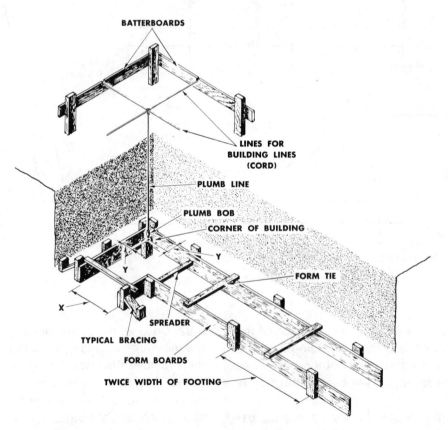

Fig. 3-84. Footings are constructed to provide a broad base for the foundation walls.

A-A, Fig. 3-78, indicates that the footing is to be 16 inches wide and 8 inches high. Erect the forms for the footing, following the procedures discussed earlier in this chapter. See Fig. 3-84.

After the forms for the footings are in place, suspend the steel reinforcing bars in the designated locations and secure them to supports so that they do not move while the concrete is being placed. Suspend them approximately 3 inches above the bottom of the footing and 8 inches apart.

The reinforcing bars as described in Fig. 3-78 are two #5 bars ($^5/_8$ inch) continuous. They are ordered in the length needed so that they extend the length of each footing. The word *continuous* means that the bar must extend continuously the

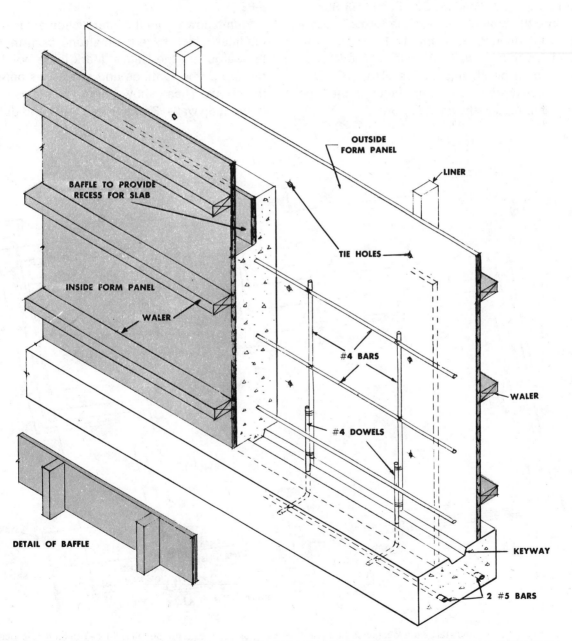

Fig. 3-85. A cutaway view of the formwork shows the reinforcing rod and provision for forming the recess for the slab.

length of the footing. If two bars are used to make up the dimension, they must overlap each other and be wired together. The reinforcing bars are lapped and bent around corners as specified by local building codes.

Rest the #4 reinforcing dowels on the footing reinforcing bars while the concrete is being placed in the forms so that the vertical part of the dowel projects upward at the center of the foundation wall. The dowels are placed all around the footings spaced 18 inches O.C. See Fig. 3-85.

Before the concrete sets in the forms, press a piece of 2 × 4 inch stock into the top of the concrete to form a keyway. The sides of the keyway form should be chamfered (beveled) so that it can be easily withdrawn from the concrete. The finished footing is shown in Fig. 3-86.

The Foundation

The type of forming used for the foundation walls of House Plan A is shown in Fig. 3-85 and 3-87. The sheathing consists of 4 × 8 foot sheets of plyform. Alignment is provided by 2 × 4 inch horizontal walers. Vertical liners (2 × 4 inch) add stiffness to the assembly. Snap ties pass through holes drilled in the panels and are driven tight by snap brackets bearing on the walers. See Fig. 3-88.

Plumb down from the intersection of the building lines which have been strung from the batterboards and the stakes. Mark these points on the top of the footings and snap lines between the points. These show the outside lines of the foundation walls. Prepare the inside and outside

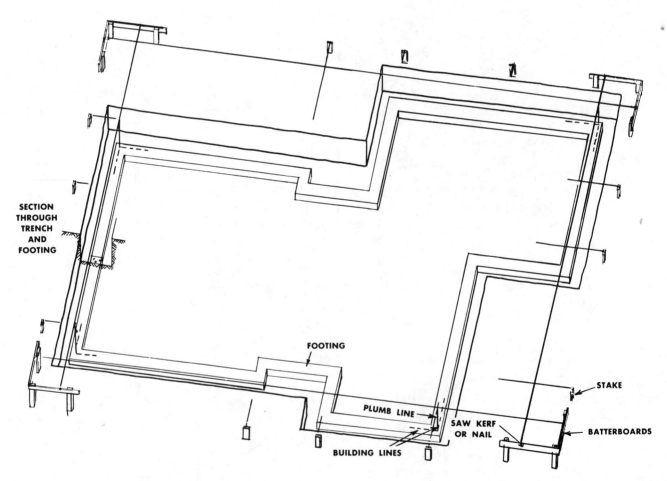

SECTION THROUGH TRENCH AND FOOTING

FOOTING

PLUMB LINE

SAW KERF OR NAIL

BUILDING LINES

STAKE

BATTERBOARDS

Fig. 3-86. Batterboards and stakes are located so that lines can be strung to locate the building lines. The footing is located 4 inches outside of the building line.

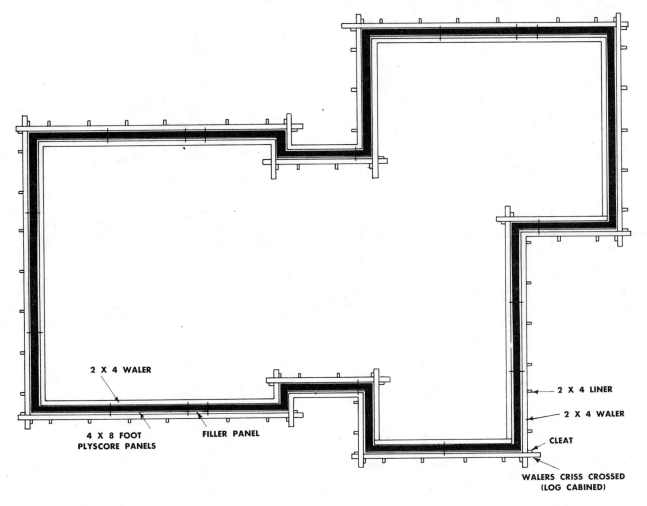

Fig. 3-87. Formwork layout using predrilled plywood sheets and the forming system shown earlier in Fig. 3-7 and 3-8.

panels by drilling holes which match the tie spacing. The foundation wall is recessed at the top to allow for the end of the slab. A 4 × 8 inch baffle is used. It is fastened to the inside of the panels, as shown in Fig. 3-84. This illustration also provides a detail of the baffle. The baffles are made in advance so they are ready to nail in place once the forms are erected. Start erecting the outside form sheathing at one of the corners.

Place ties in the pre-drilled holes, as you install the sheathing panels. Place snap brackets over the tie ends. Install walers which are long enough to extend beyond corners and bridge the joints between panels. Brace the assembly with temporary diagonal members tacked to the walers and stakes in the ground.

The foundation wall requires reinforcing. The reinforcing must be placed and tied before the inside sheathing is applied. See Figs. 3-78 and 3-85. Wire vertical bars to the stubs which project from the footings. They are spaced 18 inches O.C. Wire the horizontal bars to the vertical bars. Place horizontal bars approximately 3 inches above the footing, 3 inches below the recess for the slab, and at the mid point between the upper and lower bars, as shown in Fig. 3-85. Tie the whole assembly to the form ties after all of the formwork is in place.

Put the inside panels in place, and insert the ties through the holes. The ties are manufactured with spacing washers which automatically provide for an 8 inch thick wall. Place the walers

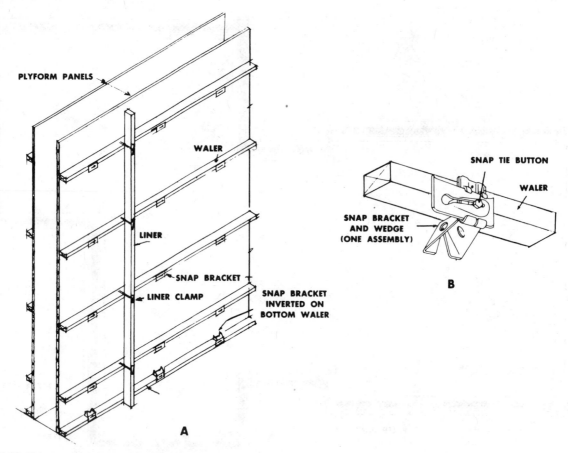

Fig. 3-88. This forming system requires single walers and stiffbacks.

against the sheathing. Insert the snap brackets, shown in Fig. 3-88 in the detail, over the tie ends and drive the clamps tight. Returning to the outside, adjust the alignment of panels as needed by adjusting and renailing the braces. Drive the clamps on the snap brackets tight. Install liners on the outside at 4 foot intervals. Nail the baffles, shown in Fig. 3-85, inside the forms. The final job before placing the concrete is setting the anchor bolts. Hang the bolts from wood pieces which you nail in place at the top of the form on 6'-0" centers.

Waterproofing and Insulation

When the wall formwork has been removed, a bituminous waterproof coating is spread on the outside of the foundation wall. Insulation is ap-

plied to the inside of the walls with an adhesive. (These operations are done by mechanics other than carpenters).

Protection from heat loss is essential in colder climates since there is a wide temperature variation between the inside and the outside of the house. Engineers have found that great heat loss in a slab at grade house occurs around the outer edge of the slab. The Section View, Fig. 3-78, shows insulation placed at the edge of the slab and along the inside of the foundation wall. It consists of 1 inch insulation board at the point where the slab and foundation wall intersect and 2 inch insulation board along the inside of the foundation wall, down 2 feet below the notch in the wall. The material is cellular glass or rigid urethane slabs. It is adhered to the wall with mastic.

Preparing for the Slab

Once the insulation is in place, the excavator returns to backfill the earth around the outside of the foundation and to level the earth inside the foundation area. The earth is brought to the required level and compacted. The level is to be made 8 inches below the elevation of the top of the finished slab. (The top edge of the batter-board ledgers have been placed at this elevation.)

The plumber completes the job of running supply and waste lines to various locations in the house, including the bathrooms, utility room, and the kitchen. The pipe stubs are made long enough so that they extend up through the slab when it is placed. Figure 3-89 shows a typical example for a warm climate dwelling.

Gravel to the depth of 4 inches is placed, leveled and compacted. A trench is left in the ground bed wherever bearing partitions are to be located. The Foundation Plan shown in Fig. 3-77 indicates their locations. A cross section view is shown in Fig. 3-78. When the slab is later

Fig. 3-89. The plumbing extends above the slab.

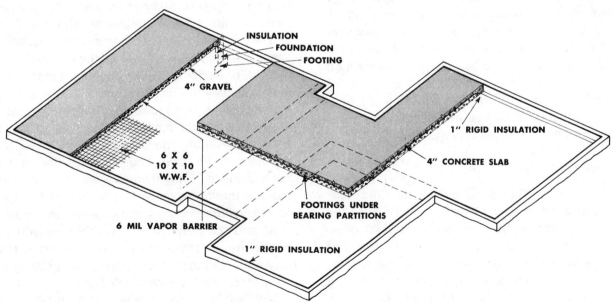

Fig. 3-90. Steps in preparing for the placing of the slab include placing insulation, spreading gravel, spreading the vapor barrier, placing the wire mesh, and providing for footings under the bearing partitions.

placed, the trench is filled with concrete. This forms a footing 8 inches thick with a base of 16 inches and flared sides. See Fig. 3-90.

A vapor barrier of polyethylene film, 6 mil (.006 inch) thickness is placed over the gravel fill. (A mil is one thousandth of an inch.) The barrier must be strong enough to resist puncture from the stone below it while retaining the concrete above. A 6 inch lap is required at joints between strips. The vapor barrier continues up to the top of the foundation walls and makes a waterproof seal. The vapor barrier has several functions. Water or water vapor is prevented from rising up from below. The fine particles are retained in the concrete above. Water in the concrete is prevented from being drawn into the porous fill.

Reinforcing the Slab

The next step is to put the welded wire fabric (mesh), shown in Fig. 3-90, in place. The fabric is supported on reinforced concrete bar supports called *chairs* or on half brick placed at intervals around the area. Chairs are bent wire devices which hold the reinforcing bars at the proper height above the supporting surfaces. The Foundation Plan, Fig. 3-77, describes the fabric as 6 × 6 × 10/10 WWF. This designation means that the fabric has 6 inch squares and is made of 10 gage wire. The mesh is available in wide rolls. It is placed from wall to wall and lapped where edges join. This provides maximum reinforcement. Workers should be careful not to walk on the mesh once it is in place, so that it will not bend or be displaced. Two #5 (⅝ inch) reinforcing bars are placed in each bearing partition footing supported on chairs.

Placing the Slab

The concrete is placed to the 4 inch thickness and finished perfectly level. The surface of the concrete should be kept wet for several days. This helps the curing process. Large sheets of polyethylene film are used to cover the surface of the concrete if there is a likelihood of freezing temperatures, or if the temperature is so high that evaporation can occur too rapidly. (Note: The job of placing the reinforcing steel and placing and finishing the concrete is generally not the work of the carpenter.)

HOUSE PLAN C

The house shown as House Plan C is used to illustrate how to prepare footings and formwork for foundations when the house has a basement. It is used to give additional site work practice in staking out a house and determining elevations. (The working drawings for House Plan C appear at the end of this chapter, pp. 143 to 148.) The Plot Plan is shown in Fig. 3-91. The method for staking out the house and erecting batterboards follows the same general procedure outlined for House Plan A.

Staking Out The House

Study the Plot Plan Fig. 3-91. The particular things to note are the lay of the land, the manner of locating the building on the lot, and the elevations of the floors in relation to the point of beginning. You will notice that the contour of the lot is shown by lines running diagonally across the lot. Two numbers are shown on each line. The first number is the present or rough grade. This information is supplied by the surveyor. The second number is the finished grade decided on by the architect. The quantities are shown in feet and decimals of a foot. To convert these quantities to feet and inches multiply the decimal part by 12. (For example, .7 foot equals 8.4 inches or approximately 8½ inches since .7 × 12 = 8.4)

The excavator is required to do very little grading. Notice that the highest point of the lot is the 102 feet elevation. The lot is sloping down to 101 feet near the Southwest corner and 97 feet at the back of the lot. The finished grade requires that most of the lot be made nearly level at the 101 feet elevation, sloping slightly toward the southwest and the north. It is important for you to notice that batterboards can be erected in the conventional manner and before the grading is done since the lot is relatively level.

The next thing to consider is the location of the building. The building proper (not including the garage) is located 40'-6" inches from the

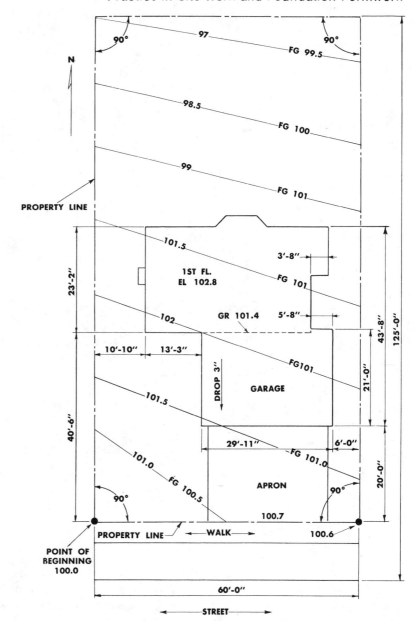

Fig. 3-91. The Plot Plan for House Plan C shows the contour of the lot and the grade elevations for the floor of the garage and the floor of the house. The location of the house is given to the lot lines.

south building line, measured from the point of beginning. The southwest corner of the building is located 10'-10" from the west property line.

The elevation for the first floor is shown as 102.8 feet or (102'-9½"). The first floor is therefore 2'-9½" inches above the point of beginning. There is a drop of 1.4 feet (approximately 1'-5") where the house and garage meet. The garage floor slopes 3 inches from the wall, where

the house and garage meet, toward the front of the garage.

Fasten lines between stakes set up at lot corners *A, B, C,* and *D* (Fig. 3-92). Locate a point 10'-10" from the west lot line and 40'-6" from the south lot line and drive stake *1* at this point. Drive a nail in the stake at the exact point.

Set stake *2* a distance of 37'-6" to the east of stake *1,* and 40'-6" from the south lot line. Mea-

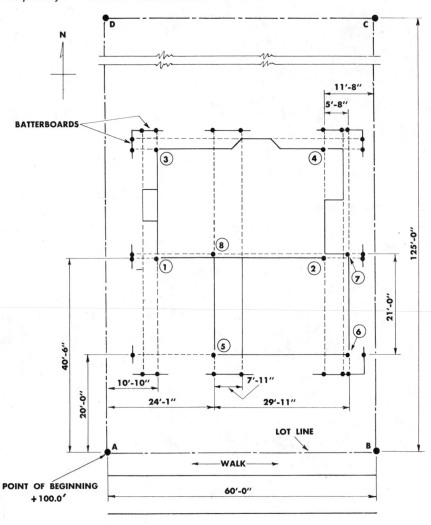

Fig. 3-92. After the corner stakes are located. the batterboards are erected and the lines are stretched to indicate the building lines.

sure a 23'-2" from point *1* and 10'-10" from the west property line to locate point *3*. Measure 37'-6" from point *3* and 23'-2" from point *2* to locate point *4*. See Fig. 3-92.

Check the location of points *1, 2, 3,* and *4* as follows: Measure 40'-6" from points *1* and *2* back to line A-B. Measure 10'-10" from points *1 and 3* to line A-D. Measure sides 1-3 and 2-4 to see that they measure 23'-2". Measure sides 1-2 and 3-4 to see that they measure 37'-6". Measure the diagonals 1-4 and 2-3 to see that they are equal. (Some of these operations are done using a builders' level or transit level, as discussed earlier in Chapter 2.)

Stake out the garage and entry passage area, using the following procedures. Locate point *5* a

distance of 24'-1" from the west lot line and 20'-0" from the south lot line. Locate point *6* a distance of 29'-11" from point *5* and 20'-0" from the south lot line. Locate point *7* by measuring 21'-0" from point *6* and 5'-8" from the line stretched between points *2* and *4*. (This point is 6 inches north of line 1-2 extended). Locate a stake at point *8* by measuring 21'-0" from point *5* and 24'-1" from the west lot line. Check the measurements and diagonals for points *5, 6, 7,* and *8*. The main points of reference for the house and garage area have now been located. The location of the bay, stairway to basement, and the chimney can be determined after the batterboards are erected.

Erect batterboards in each corner. Select

pieces of 2 × 4 inch material long enough to extend firmly into the ground and still project 2 to 3 feet above the ground. Place the batterboards approximately 4 feet back from the building lines so that they are not disturbed when the excavation is dug. Place stakes so that they are able to serve two or more lines where necessary once the ledgers are nailed in place. See Fig. 3-92.

Set the builders' level over the point of beginning. Sight and mark each stake so that the ledgers are level and are related to the elevation of the first floor. The elevation of the first floor is to be at 102.8 feet (102'-9½"), or 2'-9½" above the point of beginning. Mark lines on all batterboards at the 3'-9½" level. (One foot is added to bring the lines above the rough grade, which in some places is at 102 feet.) Nail ledgers so that they coincide with the marks on the stakes. Using a line and a plumb bob, stretch the line so that it passes directly over points *1* and *2*. Mark the top of the ledger where the lines cross. Make a shallow saw kerf in the ledger and pass the line through the kerf. After driving a nail into the ledger, tie the line to the nail. (Some carpenters use a nail on the top edge of the ledger as mentioned earlier.)

Measure two points 6 inches north of the line just established and make a second saw kerf on each batterboard. A line through these kerfs locates the short wall at the back of the garage passing over point *7*.

Establish line 3-4 at the rear of the building, using the same procedure of plumbing down to points *3* and *4* on the stakes and fastening lines to the batterboards. The bay projects 2 feet. See the First Floor Plan of the working drawings, p. 144. Measure 2'-0" on the batterboard from line 3-4 and establish a line which indicates the face of the bay. Stretch a line to pass over points *5* and *6*. Then stretch a line which passes over points *1* and *3*.

The chimney projects 2'-1". Measure 2'-1" on the batterboards from line 1-3 to establish a line which indicates the face of the chimney. Stretch a line to pass over points *5* and *8*.

The west wall of the garage is 7'-11" east of line 5-8. Measure this distance on the batterboards and stretch a line over points *2* and *4*. Then stretch a line over points *6* and *7*.

The wall for the exterior stairs to the basement is located 3'-8" from the building. Measure this distance on the batterboards from line 2-4 and stretch a line. Check all measurements so that they agree with the dimensions shown on the Basement Plan, Sheet 1, p. 143.

Excavating The Basement

The stakes marked *1* through *8* on Fig. 3-92 are eliminated as the work of excavating proceeds. The excavator must make a hole large enough for the carpenters to work between the forms and the earth bank but not so large that it disturbs the batterboards. The batterboards have been placed 4 feet from the building lines. The excavator should stay 2 feet from the batterboards so that a 2 foot space is provided for the carpenter to work on the forms.

The next step is to study the Plot Plan (Fig. 3-91) and Section View A-A on Sheet 6 of the working drawings to determine the depth for the excavation. The working drawings for House Plan C appear at the end of this chapter.

The Plot Plan shows that the top of the first floor is at an elevation of 102.8' (102'-9½"). Section A-A, shown on Sheet 6 of the Working Drawings, indicates the bottom of the concrete floor is 8'-4" below the elevation of the finished first floor. The bottom of the footing is 9'-2" below the elevation of the finished first floor. See Fig. 3-93.

The excavator digs the hole for the foundations to a depth of 8'-10" below the elevation for the first floor. This allows for 6 inches of ground fill. The depth is 9'-10" below the top of the batterboard ledgers. The bottom of trenches is 10'-2" below the top of the ledgers for footings for the house. (Note: The ledgers have been placed one foot above the elevation for the first floor as a matter of convenience to keep the lines above the foundation formwork.)

The garage presents several problems. It has a sloped floor. There is a stepped footing where the garage foundation joins the house foundation. A study of Sheets 4 and 5 of the working drawings (pp. 146 and 147) shows details of the garage foundation walls. The top of the foundation walls for both the garage and the building are both at

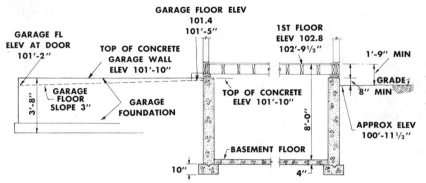

Fig. 3-93. Determining the bottom of the excavation and the top of the concrete foundation wall and slabs is a very crucial phase of the operation.

the same level. The bottom of the footing is 3'-8" below this elevation. The excavator digs the trenches for the foundation walls for the garage and the entry passage to this depth. Earth is removed to provide for the sloped floor and the sand fill.

The stepped footing is considered necessary because the footings of the house and garage are at different elevations. A trench 2 feet outside the house foundation has been made to allow the carpenter to work on the forms. After the trench is filled in, the earth is not compact enough to provide adequate support for the garage footings and foundation where they join the house. Therefore a stepped footing is used to overcome this problem.

Forms for Footings

After the excavating is completed, again string lines between batterboards so that you can drop other lines to locate points previously spotted. Notice that these points are building line points and not points locating footings. To find the location of footings you must measure 4 inches outside the line to find the outer line of the footings. See Fig. 3-94. To lay out the footing at the bay and at the fireplace follow this same procedure shown in Fig. 3-95. The dimensions you must follow for the foundation walls are on the Foundation Plan, Sheet 1. The footings are constructed as discussed earlier in Chapter 3 under *Forming For Footings*, p. 70.

There is no footing either under the platform at the kitchen door or for the wall enclosing the stairs to the basement. Unless the soil at the

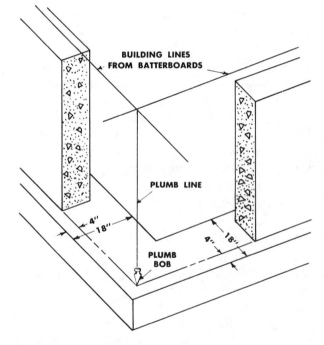

Fig. 3-94. The footings are located 4 inches outside of points dropped from the building lines.

building site is unstable, the foundation walls at these places are wide enough to transfer the load to the soil.

The footings for the garage are not placed until the foundation for the house is complete and the forms stripped. The soil is filled and compacted along the foundation wall close to the building before you place the forms for the stepped footing. These forms are located as shown on Sheet 1 of the working drawings, p.143. The forms are shown in profile on the West

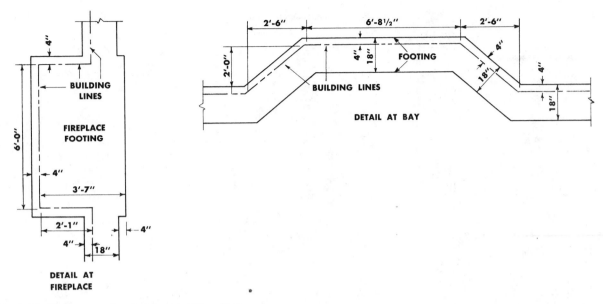

Fig. 3-95. Footings are located from building lines.

Elevation, Sheet 5. The bottom and the two intermediate steps are formed with wood boxes which do not have tops or bottoms.

Drive stakes to mark the center location of the two columns which support the girders. Build boxes 2'-6" square and 1 foot high. Locate them as indicated on Sheet 1 of the working drawings so that the column is in the center of the footings.

Foundation Formwork

The type of formwork chosen for the foundation of House Plan C uses 4 × 8 foot carpenter-built panels. These panels consist of frames that are made from 2 × 4 inch members with plyform faces. See Fig. 3-96. They can be reused many times. The forms are drilled to accomodate snap ties at intervals so that the ties bear against the walers. See Fig. 3-97. Three sets of double walers keep the forms in line and add stiffness to the assembly. Wherever it is convenient, place the panels opposite each other so that the need for drilling holes is kept at a minimum.

Figure 3-98 and the discussion which follows are intended to establish a method of operation. This does not imply that this is the so-called best arrangement. Carpenters familiar with the setting of forms generally dispense with such a layout. However, it is useful when the building presents unusual forming problems and/or when a list is made to indicate the number of forms of various sizes, quantities of form ties, and other supplies needed for a particular job.

Beginning at one corner of the building, set the first outside section. See Fig. 3-96. This section should be carefully plumbed, braced, and anchored at both the top and the bottom. It is also essential that it be carefully lined up with the building lines. The next step is to bring the second section into position and fasten it to the first section with duplex (double headed) nails. Drive the nails through the frame of the section. Sections are added until the outside wall is enclosed. Use temporary bracing to keep the wall from falling.

Place fillers where they are needed. The forms are best lined up if the fillers are placed away from the corners. See Fig. 3-98. Builders who do a lot of concrete work have common sizes of filler, such as 1 foot or 2 foot, readily available. If such fillers are not kept on hand, they are built as needed. Builders can improvise for a given situation by using pieces of plyform paneling cut to the required width with 2 × 4 inch studs at each side. Narrow strips, 2 × 4 inch members, and planks are used to fill small spaces.

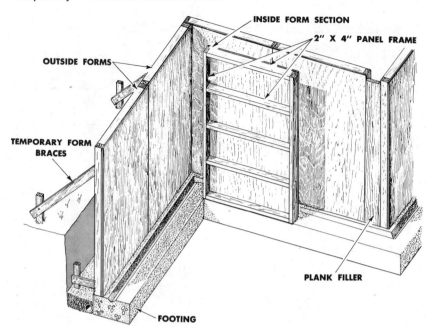

INSIDE FORM SECTION

2" X 4" PANEL FRAME

OUTSIDE FORMS

TEMPORARY FORM BRACES

PLANK FILLER

FOOTING

Fig. 3-96. Panels with fillers of various sizes can be adapted to most forming problems. They can be reused many times. Forms of this type are generally made by the carpenter.

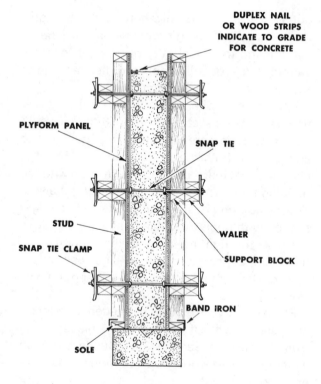

DUPLEX NAIL OR WOOD STRIPS INDICATE TO GRADE FOR CONCRETE

PLYFORM PANEL

SNAP TIE

STUD

WALER

SNAP TIE CLAMP

SUPPORT BLOCK

BAND IRON

SOLE

Fig. 3-97. The form panels are held together by snap ties and wedges.

Install the bucks for windows and doors, using the procedure discussed earlier in Chapter 3. See *Openings in Walls* on p. 110.

Erect the inside panels in the same manner as the outside panels. Insert snap ties as the forms go into place. Whenever holes are needed, drill them in the panels with an extension bit. Install double 2 × 4 inch walers and snap tie clamps in place. See Fig. 3-97.

Choose long enough material for the 2 × 4 inch

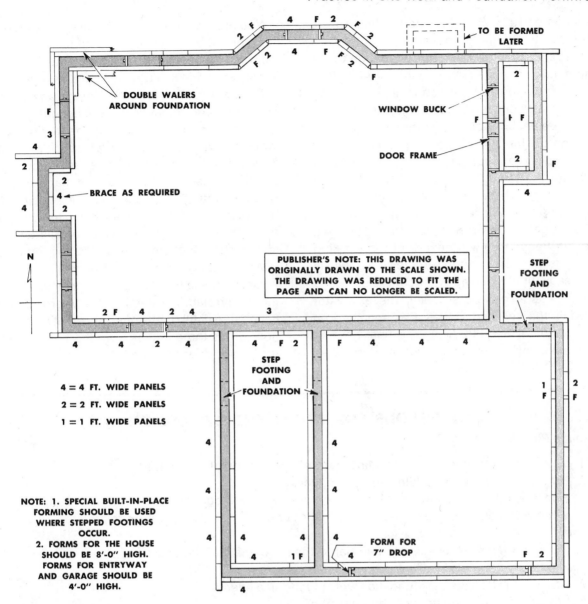

PUBLISHER'S NOTE: THIS DRAWING WAS ORIGINALLY DRAWN TO THE SCALE SHOWN. THE DRAWING WAS REDUCED TO FIT THE PAGE AND CAN NO LONGER BE SCALED.

4 = 4 FT. WIDE PANELS

2 = 2 FT. WIDE PANELS

1 = 1 FT. WIDE PANELS

NOTE: 1. SPECIAL BUILT-IN-PLACE FORMING SHOULD BE USED WHERE STEPPED FOOTINGS OCCUR.
2. FORMS FOR THE HOUSE SHOULD BE 8'-0" HIGH. FORMS FOR ENTRYWAY AND GARAGE SHOULD BE 4'-0" HIGH.

Fig. 3-98. A panel layout for a complex foundation helps the carpenter save time when erecting the forms.

walers so that it can bridge over several panels to extend the pieces beyond the corners. The snap ties bearing on the clamps hold the walers in place. The walers are nailed to each other with cleats or fastened with a band iron where they extend beyond the corners and interlace the walers from the adjoining wall.

The grade level for the top of the concrete in the forms is marked with a series of nails around the inside of the forms at approximately 10 foot intervals. The method for determining this eleva-

tion can be found by examining Fig. 3-93 and Section A-A on Sheet 6 of the working drawings, p. 148. Instead of using nails you can use a strip of wood nailed at the correct elevation. The first floor elevation is 102'-9½". By adding the plywood flooring, the joist depth, and the sill thickness, you obtain a dimension of 11½ inches. (102'-9½" − 11½" = 101'-10".)

Set up a builders' level over the point of beginning. Hold a ripping alongside it. Mark a line 1'-10" above the sidewalk. (The sidewalk is at an elevation of 100'-0".) Mark another line on the ripping at the height of the telescope. Place the ripping at the places inside the foundation forms where nails are to be driven. Sight the top line on the ripping and mark the bottom line. This indicates the elevation of the top of the concrete.

Prepare anchor bolts and hangers. Mark their position on the top of the form panels. Insert them in place as soon as the concrete is placed in the forms. Bore holes in the forms for the reinforcing bars. These tie the foundation for the platform (at the kitchen door) to the main foundation. See the Basement Plan, Sheet 1, p. 143, and the Section View, Sheet 6, of the working drawings, p. 148. The form work for the house is now ready for the placing of the concrete.

After the forms are stripped, the excavator backfills the soil and grades it as specified on the plot plan. See Fig. 3-91. The final grading and backfilling around the garage are done later.

You have yet to erect the forms for the foundation for the platform near the kitchen door. The information needed for this procedure is found on Sheets 1 and 6 of the working drawings.

Build the footing forms for the garage and the entry passage, including the stepped footings. After the concrete has set and the footing formwork is removed, erect the forms for the walls. This is done in the same manner as the formwork for the house. Use 2 × 8 foot panels at the stepped footings. You can build the remainder of the walls using forms that are 4 or 6 foot high. An alternative is to turn 4 × 8 foot panels on edge so that the 8 foot side lies on the footings.

While backfilling at the garage walls, the excavator can make the finishing touches for preparing for the slab. The bed is made to provide for a 6 inch gravel fill. It is sloped so that the finished floor drops three inches from the house toward the garage door opening.

QUESTIONS FOR STUDY AND DISCUSSION

House Plan A

1. What steps are followed for locating corner stakes at points *1,2,3, and 4?*
2. How do you verify the building lines are square?
3. How do you establish the elevation of the finished first floor from the point of beginning?
4. What is the elevation at the bottom of the footing? How is this elevation related to batterboard ledgers?
5. How do you make the recess for the slab in the top of the foundation wall?
6. How are ties, walers, and liners assembled?
7. How do you insulate the foundation wall and the slab edge?
8. What functions does the vapor barrier serve?
9. What method is used for reinforcing the slab?

House Plan C

10. Why are batterboards used?
11. Why is a stepped footing required where the garage joins the house?
12. What type of form panels are used?
13. What are fillers and what is their function?
14. How do you mark the grade level for the top of the concrete when it is in the forms?
15. Identify the following elevations with dimensions:
A. Point of beginning.
B. Top of first floor.
C. Top of basement floor.
D. Top of concrete in the foundation wall.

PUBLISHER'S NOTE: THIS DRAWING WAS
ORIGINALLY DRAWN TO THE SCALE SHOWN.
THE DRAWING WAS REDUCED TO FIT THE
PAGE AND CAN NO LONGER BE SCALED.

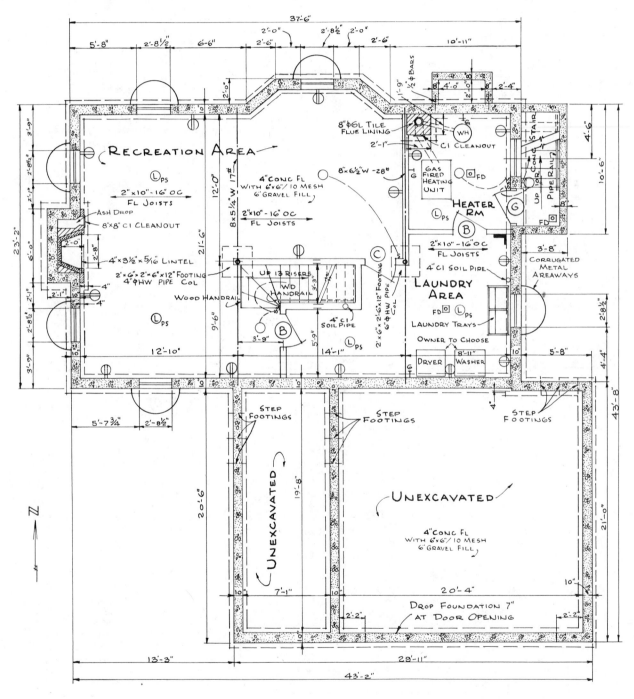

Ekroth, Martorano and Ekroth
- Architects - Chicago, Ill.   BASEMENT PLAN   SCALE ¼" = 1'-0"   SHEET 1

NOTE : ALL DIMENSIONS ARE TO
FACE OF STUDS

PUBLISHER'S NOTE: THIS DRAWING WAS
ORIGINALLY DRAWN TO THE SCALE SHOWN.
THE DRAWING WAS REDUCED TO FIT THE
PAGE AND CAN NO LONGER BE SCALED.

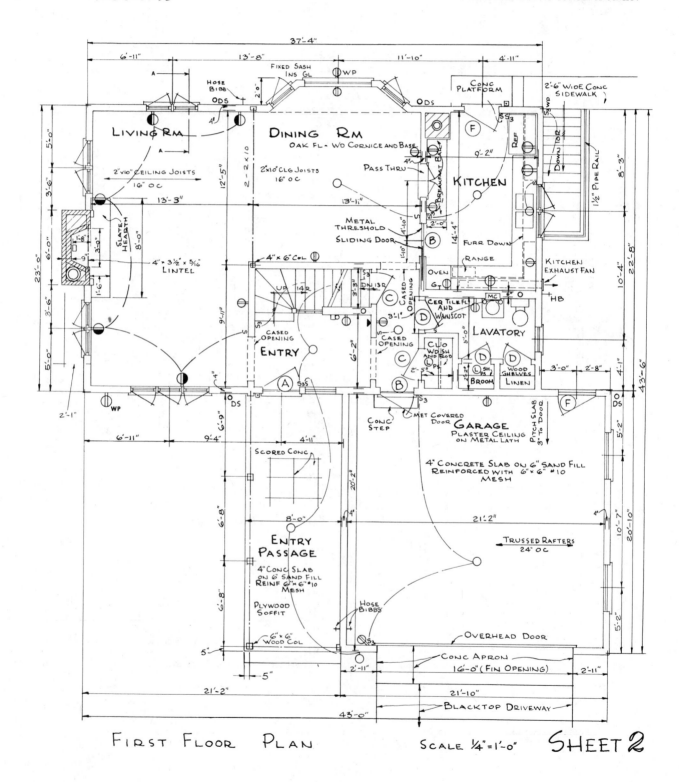

FIRST FLOOR PLAN   SCALE ¼"=1'-0"   SHEET 2

NOTE: ALL EXTERIOR WALL DIMENSIONS
ARE TO OUTSIDE FACE OF STUDS

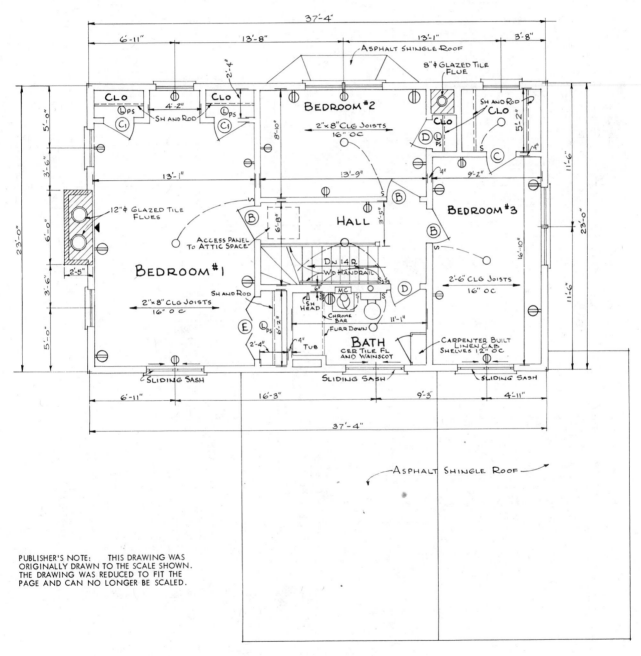

SECOND FLOOR PLAN   SCALE ¼"=1'-0"   SHEET 3

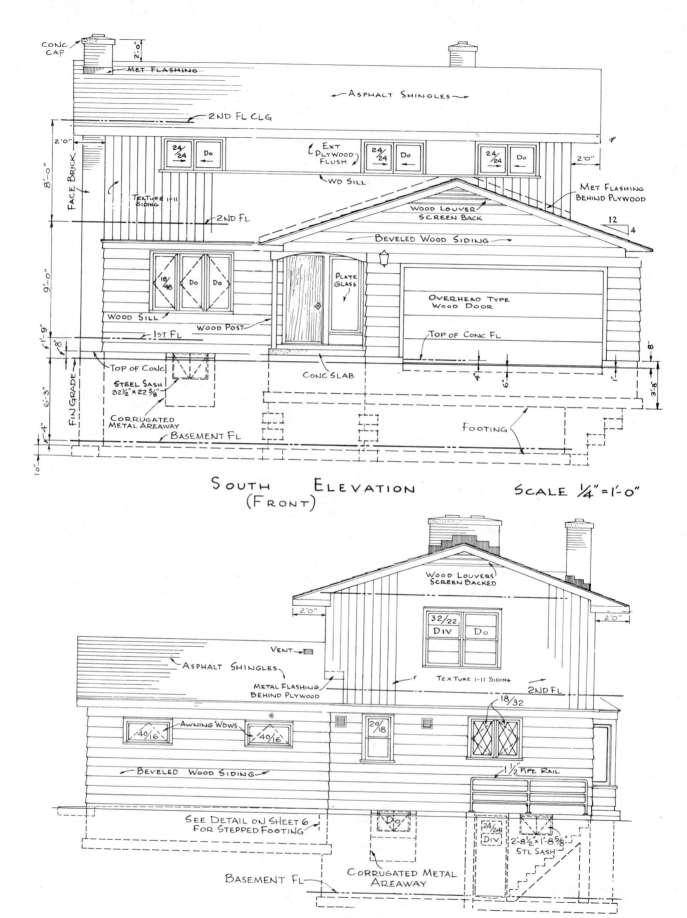

CONC CAP

2'-0"

MET FLASHING

ASPHALT SHINGLES

2ND FL CLG

2'-0"

8'-0"

FACE BRICK

24/24   Do

Ext Plywood Flush

24/24   Do

24/24   Do

2'-0"

WD SILL

MET FLASHING BEHIND PLYWOOD

TEXTURE 1-11 SIDING

2ND FL

Wood Louver Screen Back

Beveled Wood Siding

12   4

9'-0"

18/48   Do   Do

Plate Glass

OVERHEAD TYPE Wood Door

Top of Conc Fl

8"

Wood Sill

Wood Post

1ST FL

1'-9"

8"

Top of Conc

Conc Slab

4'

6'

1'

3'-8"

FIN GRADE

6'-3"

Steel Sash 32½" x 22⅝"

Corrugated Metal Areaway

BASEMENT FL

Footing

4"

1'-0"

SOUTH ELEVATION
(FRONT)

SCALE ¼"=1'-0"

Wood Louvers Screen Backed

2'-0"

32/22   Div   Do

2'-0"

Vent

ASPHALT SHINGLES

Metal Flashing Behind Plywood

Texture 1-11 Siding

2ND FL

18/32

40/16   Awning Wdws   40/16

20/18

Beveled Wood Siding

1½" Pipe Rail

See Detail on Sheet 6 for Stepped Footing

Do

24/24 Div

2'-8½" x 1'-8⅝" Stl Sash

Basement Fl

Corrugated Metal Areaway

EAST SIDE ELEVATION

SCALE ¼"=1'-0"

SHEET 4

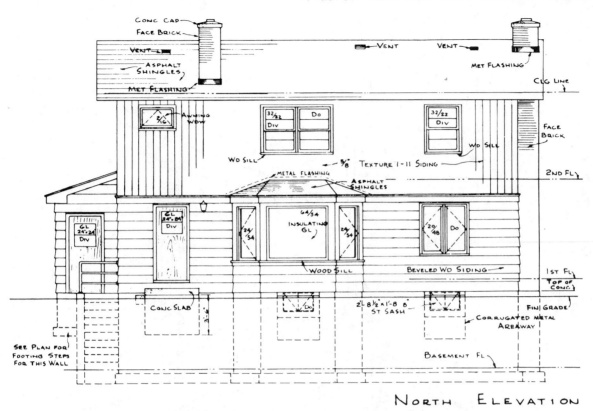

NORTH ELEVATION

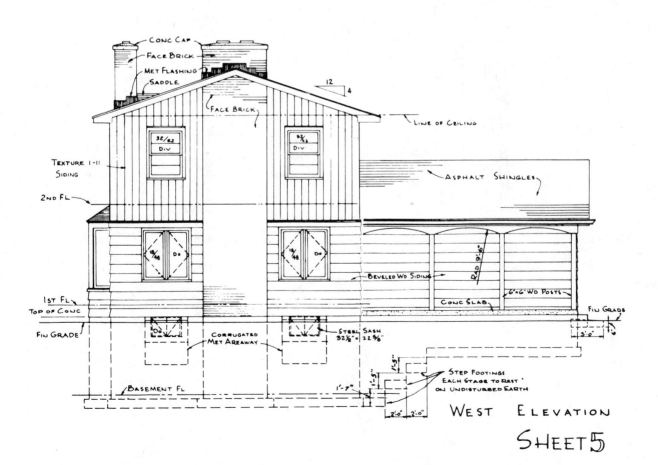

WEST ELEVATION

SHEET 5

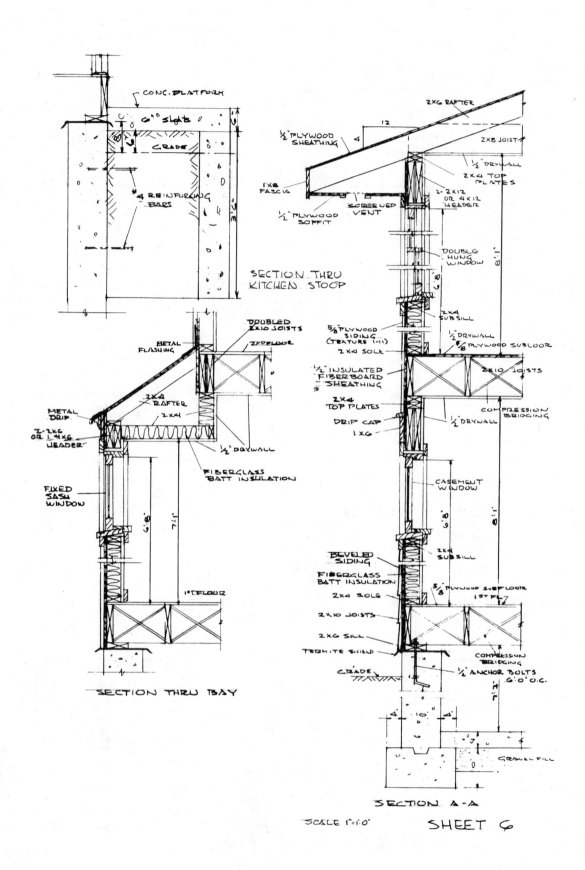

CONC. PLATFORM

6" SLAB

GRADE

#4 REINFORCING BARS

SECTION THRU KITCHEN STOOP

DOUBLED 2X10 JOISTS

2ND FLOOR

METAL FLASHING

2X4 RAFTER

2X4

1/2 DRYWALL

FIBERGLASS BATT INSULATION

METAL DRIP

2-2X6 OR 4X6 HEADER

FIXED SASH WINDOW

1ST FLOOR

SECTION THRU BAY

2X6 RAFTER

1/2" PLYWOOD SHEATHING

12

4

2X8 JOIST

1/2" DRYWALL

2X4 TOP PLATES

1X8 FASCIA

SCREENED VENT

2-2X12 OR 4X12 HEADER

1/2" PLYWOOD SOFFIT

DOUBLE HUNG WINDOW

5/8" PLYWOOD SIDING (TEXTURE 1-11)

2X4 SUBSILL

1/2" DRYWALL

5/8" PLYWOOD SUBFLOOR

2X4 SOLE

1/2" INSULATED FIBERBOARD SHEATHING

2X10 JOISTS

2X4 TOP PLATES

COMPRESSION BRIDGING

DRIP CAP

1X6

1/2" DRYWALL

CASEMENT WINDOW

BEVELED SIDING

2X4 SUBSILL

FIBERGLASS BATT INSULATION

2X4 SOLE

5/8" PLYWOOD SUBFLOOR 1ST FL.

2X10 JOISTS

2X6 SILL

TERMITE SHIELD

COMPRESSION BRIDGING

GRADE

1/2" ANCHOR BOLTS 6'-0" O.C.

GRAVEL FILL

SECTION A-A

SCALE 1"=1'-0"

SHEET 6

CHAPTER 4
WALL AND FLOOR FRAMING

Each step in building a house is important. A person who examines a finished building would need quite a background in the construction industry to be able to understand the basic structure of the building because most of the structural work is concealed by the finished work. Much of the responsibility for quality framing rests with the carpenter. The carpenter must make a great number of decisions during the course of construction on how the various members should be put together.

The primary objectives of good construction are to erect a building which is structurally sound and capable of withstanding such forces as wind and floor loads. In some instances, stresses caused by the shrinkage of lumber are also a factor. The framing must be constructed so that it provides for the heating, air conditioning, plumbing, and electricity.

The following are some of the important things you have to consider when framing a house:

You must have a thorough knowledge of the basic framing systems and adapt their features so that you can solve individual problems which arise in the constructions of a particular house.

You must know how to use building materials to their best structural advantage. You have to take into consideration such basic things as framing the floor systems so that they provide adequate support at partitions, particularly around stairwells and other openings and building strong exterior walls and interior partitions with properly framed openings. You should have some knowledge of the various stresses placed on the structural members of a building. This chapter helps you understand, for example, why a joist or beam must be a certain size.

As a carpenter, you may occasionally choose among a variety of new building products, new tools, and new techniques. You must ask yourself several questions before you make your choices. Will the materials stand up under the test of time? Are the materials economical to use? For example, systems have been developed that use metal studs as framing partitions. Plywood and fiberboard have replaced wood boards in many applications. Power tools for nailing, stapling, sawing, and planing have become standard equipment. New tools are likewise added from time to time. In short, you must constantly be alert to changes made in the industry.

You must consider methods which provide both speed and economy and yet maintain quality. Despite high costs of land, material, and labor, it is essential that the overall cost of a building be maintained at the lowest possible level. There are various ways to achieve this, such as employing labor efficiently, using labor-saving methods, and choosing materials wisely.

Because their work is so varied, some builders follow time tested building practices. It would not be practical for them to build parts of buildings away from the jobsite because each part is generally different. Other builders do not have the volume of work needed to make it economical for them to set up a shop for prefabricating parts. Still other builders, particularly those who put up a number of similar buildings in the same area, build various parts (such as wall sections) at a central location and then move them to the building site. Prefabricated units called *components* are made in factories and transported to the job. Examples include roof trusses and packaged windows and doors (prehung units ready to fit into the wall). These components have been adopted for general use throughout the industry.

A large segment of the construction industry today is devoted to manufacturing complete houses in factories, using assembly line procedures. Some houses are delivered in wall and floor sections which are assembled on the job. Other houses are delivered in the form of units (two or four units) which are then moved into place on the foundation and bolted together. (See Chapter 8, *Industrialized Building.*)

You must always give safety prime consideration. Workers must be protected from falling and other hazards while they are erecting framework and applying exterior finish. Correct procedures for erecting scaffolds and other supporting devices are considered as a part of rough framing.

(Note: This chapter provides general information on your job as a carpenter in framing buildings. The illustrations and text discuss how the various members of a building are assembled. This information can help you solve similar problems which occur in frame houses. Chapter 4-1, *Practice in Wall and Floor Framing,* which follows this chapter, provides a detailed analysis of the framing procedures for House Plan A and House Plan C. The construction of these houses is considered step-by-step. This gives practical application to the basic approach provided here in Chapter 4.)

ROUGH FRAMEWORK

The wall and floor framing, together with the roof supporting structure, form the skeleton of a building. This chapter deals with all of the framing procedures except the roof, which will be studied in detail in Chapter 5. The sheathing, outside finish, inside wall finish, trim, and rough and finished floor are fastened to the rough framework. These areas will be covered in Chapters 6 and 7. Since the strength and rigidity of a building depend on good structural procedures and workmanship, it is advisable that you give these two factors special consideration. When you are erecting the framework for a building, you have a choice of methods you can use, depending on design, conditions in the locality, materials available, and the experience and preference of the builder. Before we discuss framing in detail, we will consider some of these methods.

Types of Wall Framing

During the early history of our country, timber was abundant, but the means for making it into lumber were primitive. There were no power sawmills. Timbers and planks were cut by hand saws or were rough hewn. Because of the necessary handwork, it was cheaper to use large-sized members when framing a building than to cut the lumber into smaller sizes. Because nails were comparatively scarce and expensive and labor was cheap, it was common practice to use mortise and tenon joints where various pieces of the framework came together. Heavy timbers were used until about the middle of the nineteenth century.

However, during the subsequent years modern power sawmills have made it possible to produce lumber of small dimensions at a relatively low cost. Framing was designed with members spaced closer together. The structure was made strong enough to carry the required loads and yet not waste wood.

Two methods of framing that have been developed are called *platform framing* (also called *western framing*) and *balloon framing.* Material with a nominal 2 inch thickness is generally used throughout the structure in both systems. Occasionally a column is made of 4 × 4 or 6 × 6 inch stock. 4 × 12 inch members or wide boards are commonly used to support the walls above window and door openings. Steel girders support floor joists. In some instances, large dimension lumber or 2 or 3 members spiked together are used. The framing members are fastened together with nails driven by power nailers. Metal joist hangers and framing anchors add strength and permit rapid assembly of the parts.

A return to large dimension lumber has occurred in one sector of the industry with *plank* and *beam* (also called *post* and *beam*) *construction*. Posts (4 × 4 inches square or larger) support roof beams which are 4 × 6 inches or larger. The posts, floor beams, and roof beams are spaced at wide intervals compared to conventional framing. The beams support 1½ or 2½ inch planks which serve as flooring or roof sheathing. Because of the great distance between supports, the planking serves a structural purpose.

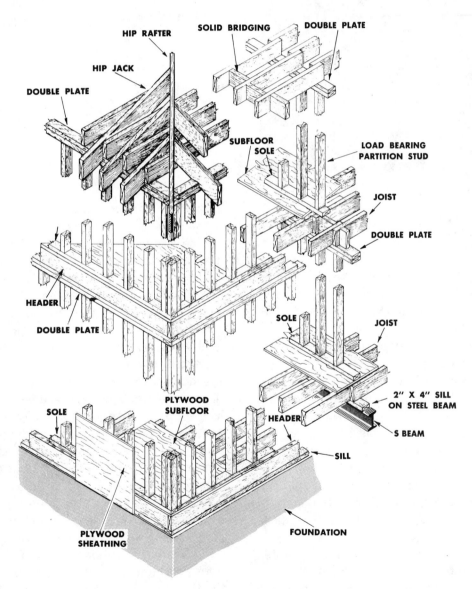

Fig. 4-1. Platform (Western) framing is the most common type of framing used for residences. It provides a platform to work on at each floor level. Walls and partitions are quickly assembled by the tilt-up method and can be made using prefabricated methods. The framework itself provides firestops in the exterior walls.

Fig. 4-2. The rough members which make up the walls and partitions for platform framing are assembled on the floor and tilted up into place. (Western Wood Products Association)

Platform or western framing. The characteristics of platform framing, shown in Fig. 4-1, include several important items. A floor system of joists is erected which is then covered with a continuous subfloor of plywood sheets. The subfloor extends to the extremities of the building and includes openings for stairwells and (if any) chimneys. The exterior walls and partitions are assembled on the floor and tilted up into place, as shown in Fig. 4-2.

There are several advantages to using platform framing. The platform on each floor provides a safe working surface as you construct the wall and partitions assemblies. Building the walls and partitions in sections is done conveniently involving a minimum of time and effort. Each stud space is blocked off by the wall and partition framing members so that the danger of fire is minimized. In other types of framing, these spaces are open and would serve as flues in event of fire. Platform framing can be adapted for modern prefabricating techniques. Wall and partition assemblies can be made away from the job.

Balloon framing. Balloon framing was one of the common basic framing systems used during the past one hundred years. Today, however, it is only used occasionally, for special purposes. The basic idea of balloon framing is to use long continuous studs between the sill, at the foundation, and the top plate at the top of the wall in order to minimize the effects of shrinkage on the lumber. See Fig. 4-3.

Second floor joists rest on a ribbon cut into the studs, and are then nailed to the studs. (The ribbon is a narrow horizontal board which serves as a support for the joists as they are put in place and helps transfer the load of the joists to the studs.) Bearing partitions are also made with the long studs. In some buildings, studs are made to extend from a girder, which supports the first floor joists, all the way to a top plate at the top of a second floor partition. In other buildings, the studs extend to the top of a top plate for the first floor bearing partitions. Studs for the second floor bearing partitions extend down to rest on the top of the same top plate. See Fig. 4-3.

The open stud spaces in balloon framing have

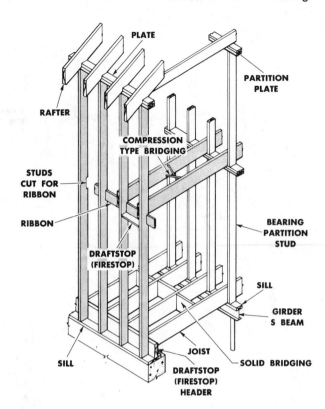

Fig. 4-3. Balloon framing includes such features as long studs reaching down to the sill, second floor joists supported by a ribbon, and draftstops (firestops) in the floors and walls.

an advantage and a disadvantage. They provide an access for introducing wiring and heating or ventilating ducts in the wall. However, the open stud spaces must be shut off with draftstops to prevent them from acting like a flue. Draftstops are pieces of board cut to fit the spaces at (or near) each floor and fastened securely in place.

There are two major advantages to using balloon framing. This type of framing provides a rigid construction. It also provides a means for reducing the effect of shrinkage in the framing members. The maximum shrinkage of lumber occurs across the grain (width) of a board. The shrinkage across the length is almost negligible. See Fig. 4-4A. In balloon framing, the joists at the first floor and at the second floor are supported in a manner designed to take advantages of these fact. See Fig. 4-4B. In contrast, the design in platform framing places the joists in a position where they are subject to shrinkage. See Fig. 4-4C. Since drywall has almost completely replaced lath and plaster for interior wall surfaces, concern about shrinkages in house framing is

less important. Plastered walls develop cracks when the wood framework shrinks. The support of the joists for an intermediate floor of a tri-level house, such as the one shown in Figs. 4-5 and 4-6, is an example of the application of the balloon framing principle.

Plank and beam framing. Plank and beam framing has become popular because, with heavy exposed roof beams and plank ceilings, it provides a novel effect. It is used for architectural effect or esthetic reasons. Some labor is saved because less members and less nailing is needed in the building. The characteristic features of plank and beam framing (also known as *post and beam framing*) are found in the floor, wall, and roof support aspects of the construction. See Fig. 4-7.

Floor Construction. Large dimension floor beams are spaced far apart. They are supported on the foundation walls and piers or posts. Planks which are 1½ or 2½ inches thick span the distance between the beams to form the floor. Heavy plank material is necessary because

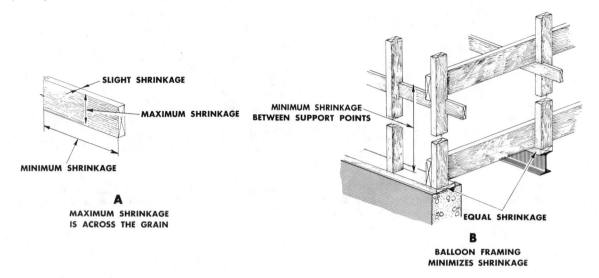

SLIGHT SHRINKAGE

MAXIMUM SHRINKAGE

MINIMUM SHRINKAGE

A

**MAXIMUM SHRINKAGE
IS ACROSS THE GRAIN**

MINIMUM SHRINKAGE
BETWEEN SUPPORT POINTS

EQUAL SHRINKAGE

B

**BALLOON FRAMING
MINIMIZES SHRINKAGE**

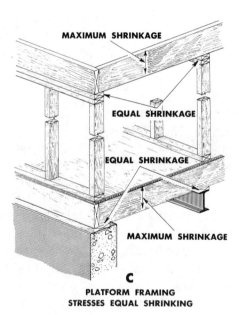

MAXIMUM SHRINKAGE

EQUAL SHRINKAGE

EQUAL SHRINKAGE

MAXIMUM SHRINKAGE

C

**PLATFORM FRAMING
STRESSES EQUAL SHRINKING**

Fig. 4-4. Balloon framing minimizes the effects of lumber shrinkage.

it must span greater distances than ordinary subflooring. There are no joists to provide intermediate support.

Wall Construction. The structural part of the exterior walls consist of 4 × 4 inch posts placed at the same interval as the floor beams. They are tied together at the top with large dimension headers, or 2 × 4 inch top plates. Posts are also arranged at the same interval along the center of

the building. They are tied together at the top by a ridge beam.

Roof Support Structure. Roof beams are supported over the posts. Planks which are 2 or 3 inches (nominal dimensions) thick span the distance between roof beams. The planks support the roofing. The underside of the planks is usually left exposed to form the ceiling in the rooms. See Fig. 4-8. This method of framing, however,

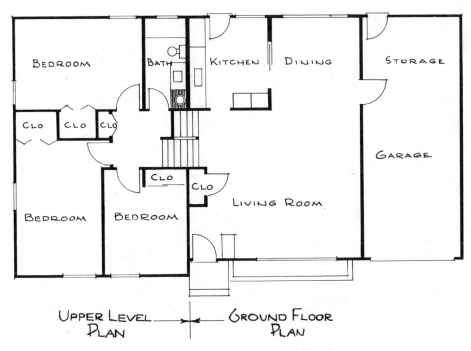

UPPER LEVEL PLAN ——— GROUND FLOOR PLAN

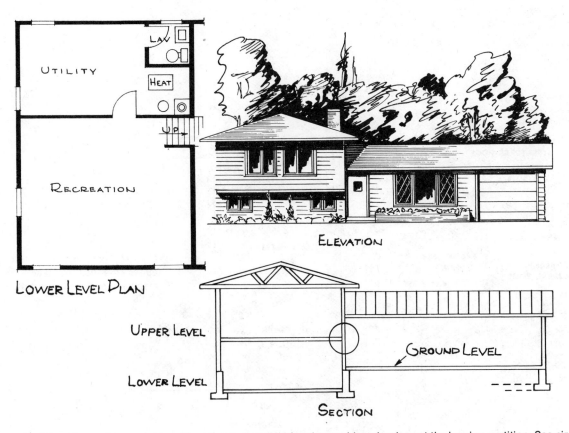

LOWER LEVEL PLAN

ELEVATION

SECTION

Fig. 4-5. A tri-level house is arranged on three levels. A special framing problem develops at the bearing partition. See circle in the section view above.

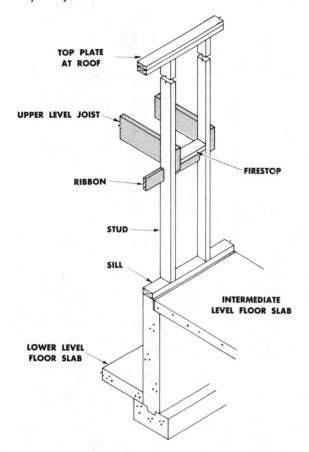

TOP PLATE
AT ROOF

UPPER LEVEL JOIST

RIBBON

FIRESTOP

STUD

SILL

INTERMEDIATE
LEVEL FLOOR SLAB

LOWER LEVEL
FLOOR SLAB

Fig. 4-6. The bearing partition in a tri-level house can use some of the features of balloon framing.

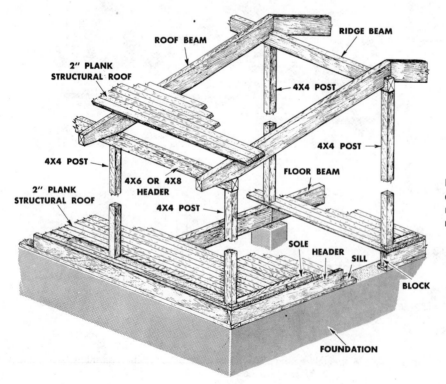

ROOF BEAM

RIDGE BEAM

2" PLANK
STRUCTURAL ROOF

4X4 POST

4X4 POST

4X4 POST

2" PLANK
STRUCTURAL ROOF

4X6 OR 4X8
HEADER

FLOOR BEAM

4X4 POST

SOLE

HEADER

SILL

BLOCK

FOUNDATION

Fig. 4-7. Plank and beam framing requires the use of heavy structural members and planks for floors and roofs.

Fig. 4-8. Plank and beam framing gives an effect of exposed heavy beams and high sloped wood ceiling. The supporting posts are uniformly spaced in the walls. (Weyerhaeuser Co.)

has distinct limitation. The posts are equally spaced in the wall. This does no allow for much flexibility when the windows and the doors are placed. The arrangement of partitions is limited in order to conceal the posts in the partitions. A building which uses this type of framing must be designed carefully because the posts supporting the structure require a fixed location.

FRAME CONSTRUCTION DETAILS

Frame construction includes many items. The basic essentials of frame construction concern building sills, erecting girders and beams and columns and posts, framing joists, applying rough flooring and bridging, building stud walls and partitions, building rough framing for openings, and sheathing walls. A discussion of these various details follows below.

Sill Construction

The sill (sometimes called a *mudsill*) is the first part of the frame to be set in place. It rests on the foundation wall or slab and extends all around the building. It is usually made of one 2 × 4, 2 × 6, or 2 × 8 inch member.

The sill is usually embedded in mortar or placed on a strip of insulating material. Washers and nuts are placed on the anchor bolts which protrude through the sills. The best way to level the sills is to use a builders' level or a transit level. Shims are used when needed to make the sills level.

Once you determine that the sills are level, tighten the nuts which keep the sill in position. The anchor bolts serve to fix the exact position of the exterior walls and keep the building from shifting or raising under the pressure of high winds. The size, length, and spacing of the anchor bolts are usually specified by the local building code and are shown on the working drawings.

A typical designation for anchor bolts would call for ½ inch diameter bolts, 16 inches long, threaded on one end and bent on the other (to prevent their turning in the concrete). The bolts are placed 4 feet O.C. along the top of the foundation wall.

The best construction to use with platform framing is the box sill. See Fig. 4-9. With the box sill, the ends of the joists are given adequate bearing. Draft stopping is automatically provided

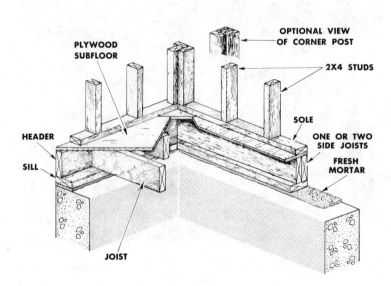

Fig. 4-9. The box sill provides adequate bearing for joists for platform framing. No firestops are needed.

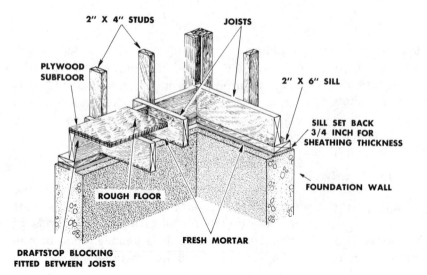

Fig. 4-10. A sill construction for a balloon frame provides draftstops. The studs extend to the sill.

by the flooring and the sole. A header of the same size as the floor joists is placed at each end of the building. It provides end nailing into the joists and keeps the joists in line to prevent them from twisting. The last joist at each side of the building is placed flush with the outside of the sill. Many builders double the last joists to provide added strength and nailing for wall soles. The headers and the side joists complete the enclosure of the perimeter of the building framework that is immediately above the foundation.

The type of sheating and exterior wall covering used determines whether or not the sill is made flush with the foundation wall. When plywood sheets 1/2 inch or more in thickness are used for siding, eliminating the need for sheathing, the sill is brought out so that it is flush with the foundation wall. When other types of siding are used over the sheathing, the sill is set back from the face of the foundation a space equal to the thickness of the sheathing, as shown in Fig. 4-10.

The type of sill illustrated in Fig. 4-10 is used with balloon framing. The studs extend down to the sill. The stud spaces are cut off from the basement by the last joist parallel to the outside wall. The draft stops are cut to fit between the joists. This provision prevents the spread of fire through the walls.

Protection Against Termites

The wood-devouring termite or white ant is one of the enemies of wood construction. Of the many different species of termites, the two most common in the United States are the subterranean and the non-subterranean (or dry-wood) types. The subterranean termite is very active and is found in almost every part of the country. However, this termite is most prevalent in the southern and western parts of the country. These destructive insects live underground and come out to feed on wood. After feeding, they must return to the ground for moisture. If they are shut off from moisture, they die. These termites can burrow through weak mortar joints or poor concrete to reach the wooden superstructure of a building. Sometimes they build earthlike shelter tubes (See Fig. 4-11.) over materials through which they cannot burrow. They then

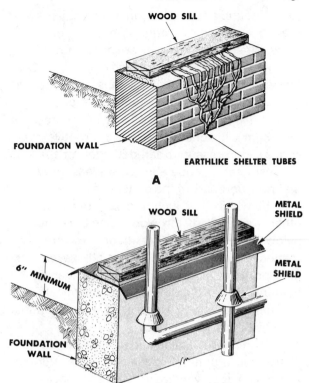

Fig. 4-11. Termites attack wood members by traveling through shelter tubes, shown in (A). Termite shields, (B), are placed on the wall and on pipes.

use these tunnels as travel routes to move back and forth between the ground and the wood on which they feed. Termite shields, shown in Fig. 4-11, can normally disable the effectiveness of this tunnelling.

In localities where the attack of the subterranean termite is inevitable, the builder must take preventive measures at the time of construction. The minimum recommended space between the joists and the soil in the crawl space is 18 inches. The minimum recommended space between the girders and the soil is 12 inches. The lowest wood member of the exterior of a house should be placed 6 inches or more above the grade. Metal termite shields which project on each side of the masonry wall should be provided. They should be 24 gage galvanized iron. Concrete should be compacted as it is placed in the forms so that rock pockets and honeycombing are eliminated.

Good footing and foundation design can prevent the development of cracks in the foundation walls. When masonry units are used, they should be laid in mortar that has a rich cement content. In areas where termite damage is prevalent, the hollow blocks are filled in solid with grout.

At the mill, the lumber is given a pressure treatment with chemicals as a preventive measure. Using pressure treated lumber for the sills is considered adequate in some areas, provided that no wood part of the building extends below a point which is 6 inches above the grade. In other areas, all of the wood members located below the finished first floor are pressure treated. When there is a potentially severe termite problem, all of the framing lumber in the building is treated on the job with chemicals.

Other precautions which can be used against termite infestation include:

1. Poison the soil inside and outside the foundation wall around the perimeter of the building.
2. Provide good ventilation for crawl spaces.
3. Suspend pipes from the floor above in such a way that they do not come into contact with the earth.
4. Periodically inspect the foundations for signs of termites.

Girders and Beams

The joists rest on sills placed on top of the foundation walls around the outside of the build-ing. However, the distance between the foundation walls (the span) is usually so great that additional support is needed between the walls. Such additional support can be in the form of a bearing wall, a wood girder, or a steel beam. The girders or beams are supported at the ends where they rest in girder pockets in the wall or on a pilaster of the foundation. See Figs 4-12 and 4-13. The girder pockets are made large enough to permit air circulation in order to prevent decay of the ends of wood girders. The ends are supported on steel pads or blocks of wood treated with wood preservative. They can be coated on the job with a chemical such as penta (pentachlorophenol).

Columns and Posts

Additional support is usually provided by one or more posts or columns placed at intervals under the girder. Footings must be constructed for these additional supports. When wood posts are used, it is advisable to construct a concrete base with a 6 inch pier. The pier brings the bottom of the post safely above the moisture of the basement floor, as shown in Fig. 4-14. A 3/4 inch iron dowel in the concrete pier will hold the post in place.

Instead of a concrete pier which extends through the floor, and must be formed along with the column footing, you can use a metal post base anchored by an anchor bolt. (The concrete forming for the footing is simplified.) The

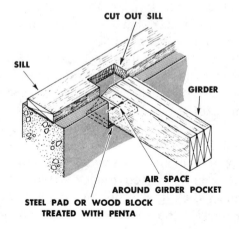

CUT OUT SILL

SILL

GIRDER

AIR SPACE AROUND GIRDER POCKET

STEEL PAD OR WOOD BLOCK TREATED WITH PENTA

Fig. 4-12. A girder is supported by the foundation in a girder pocket.

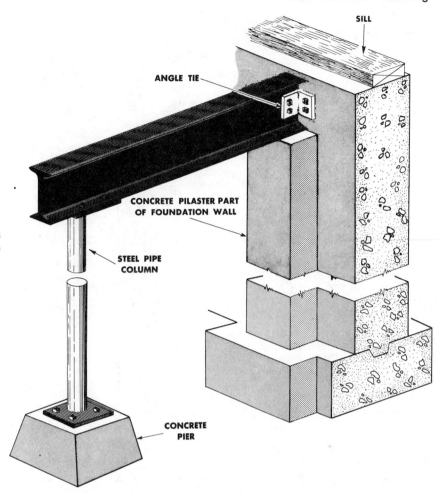

Fig. 4-13. A steel beam (girder) is supported by a pilaster and pipe column.

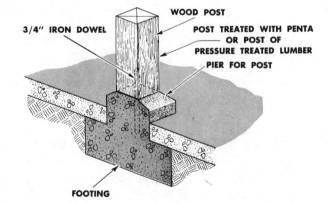

Fig. 4-14. A pier is provided to keep the post above the basement floor.

metal post base holds the post firmly in place and elevates it above the floor to keep it dry. The post base and the post cap (top tie) which hold the girder in place are shown in Fig. 4-15.

Steel beams which serve as girders are supported by pipe columns (lally columns). These are iron pipes filled with concrete with bolting plates welded to both ends. The columns are

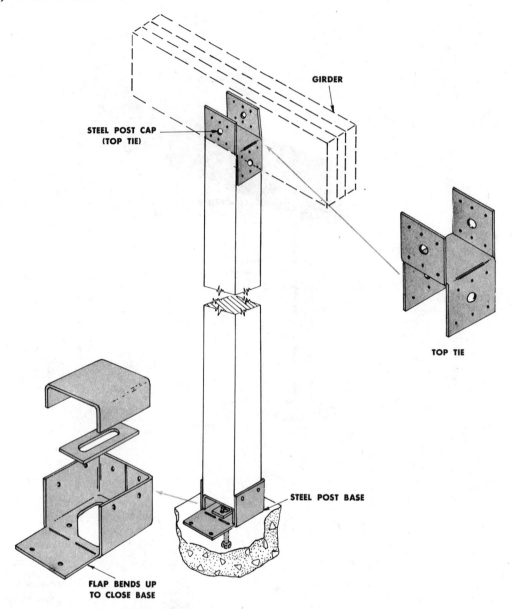

GIRDER

STEEL POST CAP
(TOP TIE)

TOP TIE

STEEL POST BASE

FLAP BENDS UP
TO CLOSE BASE

Fig. 4-15. Metal accessories hold wood posts at the top and at the bottom. (Simpson Co.)

anchored with bolts into the column footings. The tops are bolted to the steel girders. See Fig. 4-13. Adjustable steel posts are available and help make the leveling of girders easier.

Framing Joists at Girders or Beams

Various methods of framing joists on foundation walls have been explained earlier, in connection with the discussion on sill construction. See p. 159. Several different methods can be used for framing joists at the girders. The simplest method is to rest the joists on top of the girder, as shown in Fig. 4-16. When flush ceilings are required underneath the joists to provide greater clearance (headroom), the ends of joists are supported by joist hangers. Two available joist hangers are shown in Figs. 4-17 and 4-18.

Three different types of wood girders can be used. The girder can be a solid timber of the type shown in Fig. 4-19A. It can also be a solid mem-

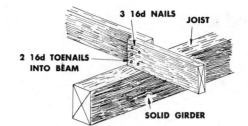

Fig. 4-16. Joists are lapped and rest on the girder.

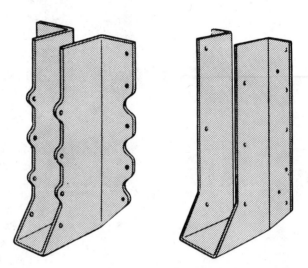

Fig. 4-17. Joist hangers support joists at the girders. (Simpson Co.)

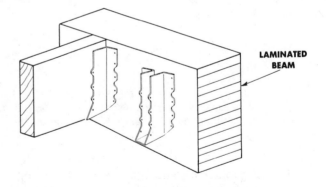

Fig. 4-18. When joists are hung using joist hangers, the maximum headroom is obtained below the girder. (Simpson Co.)

ber made in a controlled factory process in which several planks are laminated together by being glued under pressure. This type of girder (laminated beam) is shown in Fig. 4-19B. Common wood girders, however, are made by bolting two or three members side by side, as shown in Fig. 4-19C. The size and the spacing of the bolts are determined by the local building code, or by an engineer. A typical specification calls for 5/8

inch bolts spaced on 20 inch centers. When the planks are not long enough to extend the required distance, they can be spliced. The splices should occur over posts, and one or two of the planks should continue beyond each post. See Fig. 4-19C. Built-up girders have the advantage of not developing splits as easily as solid wood girders. They are also less likely to contain decayed wood.

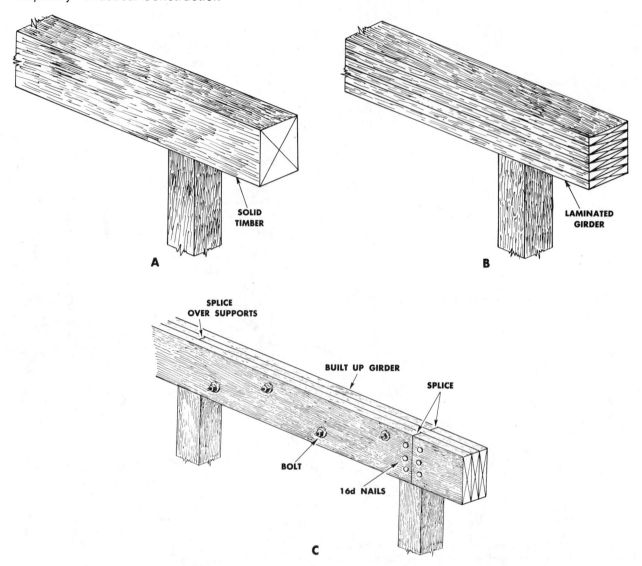

Fig. 4-19. Various types of girders. A solid girder is shown in (A) above. A laminated girder (B) is made of several planks glued together. A built-up girder (C) is made of three planks. They should be spliced over supports and spiked or bolted together.

Framing joists at steel beams presents slightly different problems than those involved in framing joists at wood girders. A common procedure, which can be used if headroom in the basement is not important, is shown in Fig. 4-20. A 2 × 4 inch sill is bolted to the top of the beam. The joists are placed in position on the sill, lapped a minimum of 4 inches, and spiked together. The joists are then nailed to this member.

Joist hangers of the type shown in Fig. 4-21 can also be used. The hangers are bolted or welded to the steel beams. The joists are hung using the same procedures followed when a wood girder is used. The hangers are made (or the joists are shimmed up) so that the top surface of the joist is about ³⁄₈ths of an inch above the top of the steel beam. If these two surfaces are made flush, and the joists shrink, the rough floor tends to buckle.

Alternative methods of supporting joists at steel beams are shown in Figs. 4-22, 4-23, and 4-24. A method in which 2 × 4 inch ledgers are bolted to the web of the beam is shown in Fig. 4-22. The joist ends are cut out to receive the top

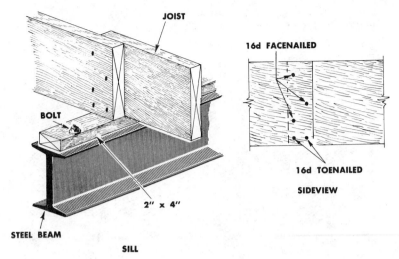

Fig. 4-20. Wood members are fastened above a steel beam when headroom is not important.

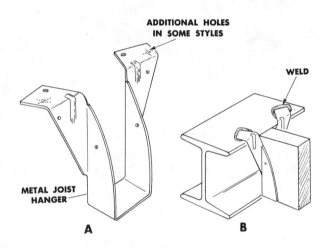

Fig. 4-21. Metal hangers are used to support joists from steel beams. They can be bolted (A) or welded (B) to the beams. (Simpson Co.)

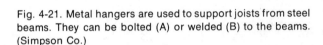

flange of the steel beam. The joists are placed so that they rest on the ledgers. A metal tie is driven into the top of the joists to hold them in position. A combination beam and plate used to support the joists is shown in Fig. 4-23. With the W beam (wide flange) shown in Fig. 4-24, the joists are cut out to fit the contour of the beam. The joists are then placed to rest on the lower flange of the beam. When joists are cut out to be fitted against steel beams, a small clearance space is allowed above the beam. This makes allowance for the joists to shrink without splitting. The method shown in Fig. 4-24 is only used with W beams. The flanges of S beams are too narrow. (Note: Beams formerly designated *I beams* are now

called *S beams*. Beams formerly designated *WF beams* are now called *W beams*.)

Doubling Joists

When floor joists are framed, you must cut the joists to make openings in the floor where stairwells occur or to frame around chimneys and fireplaces. The strength lost when joists are cut must be compensated for. The usual procedure is to frame them against double headers, as shown in Fig. 4-25. These headers, in turn, are supported by double joists called *trimmers*. A single header can be used, up to a maximum of 4'-0''. When headers longer than 6 feet must be used, you should fasten them to their supporting

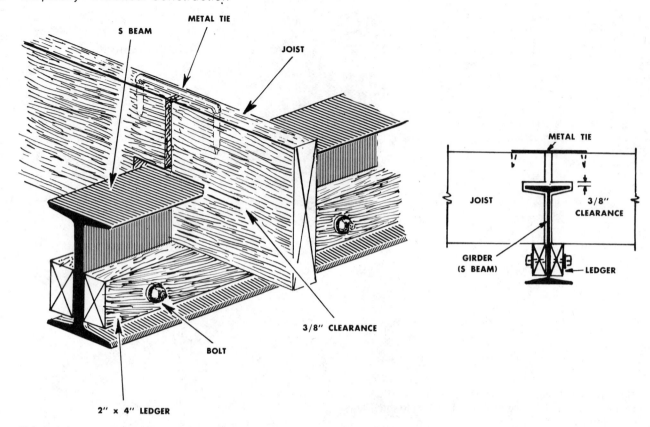

Fig. 4-22. Two 2 × 4 inch ledgers are bolted to the steel beam to provide support for joists. A metal tie is used to hold the opposite joists in position. The view to the right above provides a sectional view of the beam.

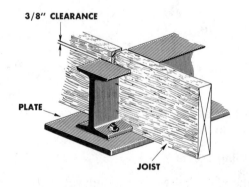

Fig. 4-23. An S beam and plate are used to support the joists so that the ceiling below can be made flush.

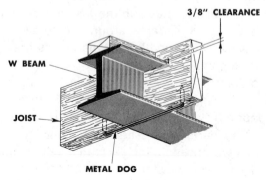

Fig. 4-24. Joists are cut to fit the shape of a wide flange beam.

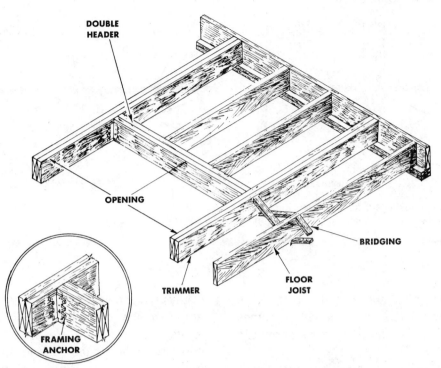

Fig. 4-25. An opening in the floor is framed by double headers and double trimmers. The detail to the left above shows a framing anchor.

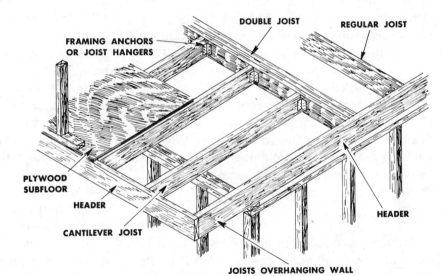

Fig. 4-26. Cantilever joists rest on the wall plate.

joists by means of framing anchors shown in Fig. 4-25.

Sometimes a building is constructed so that the second story projects beyond the wall of the first story. This is called *cantilevered construc-*tion. If the second floor joists are parallel to the joists overhanging the wall of the first story, the framing is comparatively easy. It can be done simply by using longer joists for the second floor than you use for the first floor. However, if

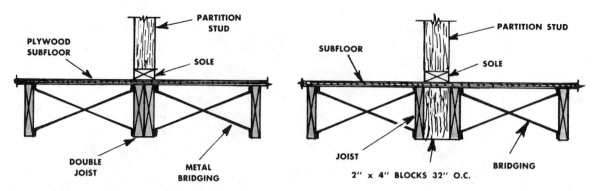

Fig. 4-27. Joists are doubled (left) or placed on either side of blocks (right) under partitions which run parallel to them. Pipes, conduit, and ducts can pass between the two joists when they are separated by blocks.

the second floor joists are at right angles to the joists overhanging the wall, then it becomes necessary to use cantilever joists as shown in Fig. 4-26.

The rooms of an upper story are not always planned so that every partition comes directly above a partition of the story below. When partitions do not line up, or when a partition of an upper story runs parallel with the floor joists, the joists supporting the partition must be doubled to carry the additional load placed on them. Frequently pipes or heating ducts are run in such partitions. When you double the joists for this purpose, place a joist on each side of the partition to leave room for ducts. The procedure for this type of framing is shown in Fig. 4-27. When a partition runs crosswise to a floor joist, you may be required to provide additional support by doubling up some of the joists.

Rough Flooring

Material for rough flooring can be either plywood or boards. Plywood has been found to be very satisfactory and economical because of the large sheets (4 × 8 feet or larger) which can be rapidly nailed or stapled in place. See Fig. 4-28. The thicknesses required are ½, ⅝, ¾, or 1⅛ of an inch, depending on the joist spacing and the floor load requirement. The 1⅛ inch tongue and groove panel has been developed for a floor system which uses 4 foot spacing between supports. See Figs. 4-29 and 4-30.

Lay plywood sheets with the face grain at right angles to the joists. Lay the sheets so that the joints between them are placed over joists or supports. Stagger the second row so that joints fall on other joists. When an underlayment for tile or carpeting is used, stagger the panels of underlayment so that the joints do not coincide with the joints of the plywood subfloor. See Fig. 4-28. The underlayment is usually ¼ or ⅜ inch plywood or particleboard. A combined subfloor underlayment is available using ½, ⅝, or ¾ inch panels with tongue and groove edges. When using this material, you must be careful to use seasoned lumber for framing the joists. If square edge panels are used in this application, place blocking between the joists so that the blocking supports the joints between the sheets of combination subfloor underlayment.

A glued floor system has been developed which increases the stiffness of the floor and therefore minimizes the squeaks caused by nail popping. This system is particularly adaptable for resilient flooring, such as linoleum, vinyl, or carpeting, because it serves as both the underlayment and the subfloor. Underlayment grade tongue and groove plywood ⅝ or ¾ inch thick is used. A bead of glue is placed along the top edge of the joists. A thin bead of glue is spread in the groove at the edges of the plywood sheets. (The gluing should be paced so that not more than one or two panels are ready for laying at any one time.) The panels are put in place with their long

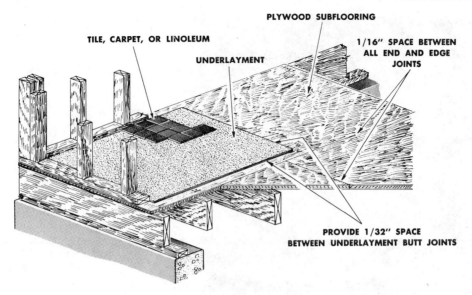

Fig. 4-28. Plywood subfloors are installed quickly because of the large panels which are used. (American Plywood Association)

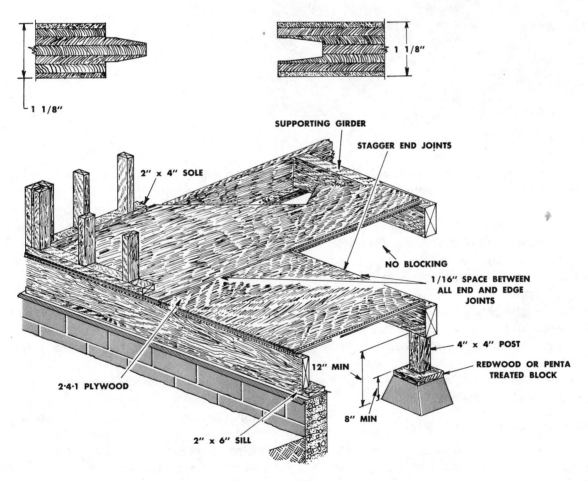

Fig. 4-29. Plywood tongue and groove panels 1⅛ inch thick serve in a wide span support system. Detail of the panel in section view appears at the top. (American Plywood Association)

Fig. 4-30. 2-4-1 plywood panels are supported 4 feet O.C. Tongue and groove edges need no support. Square ends rest on 4 inch beams. (American Plywood Association)

dimension at right angles to the joists. They are arranged so that the joints for the whole floor are staggered. A space of $\frac{1}{16}$ inch is allowed between the edges and the ends of the panels. Nails are used to supplement the glue in holding the panels. 6d deformed shank or 8d common nails are driven through the panels into the joists, 12 inches O.C.

Board rough flooring is rarely used today. However, when boards are used for this purpose, either square edge or shiplap material is chosen. It is laid so that it is at right angles to the joists. The ends of the boards are placed so that they fall over joists whenever they join.

Bridging

The term *bridging* is used to refer to a system for bracing floor joists or studs. Solid bridging is shown in Fig. 4-31, cross bridging is illustrated in Fig. 4-32, and the compression type of metal

bridging is found in Fig. 4-33. The advantage of bridging lies in its effectiveness in helping to distribute a concentrated floor load from a heavy appliance or piece of furniture, over a large area of floor space. Bridging also distributes the force and vibration of live loads such as occur when people are walking on the floor. Bridging stiffens the joists, helps to hold them in alignment, and also helps to prevent them from warping or twisting.

Solid bridging consists of single pieces of boards (or blocks) set at right angles to the joists and fitted between them. See Fig. 4-31. Cross bridging consists of transverse rows of small diagonal braces, set in pairs, which cross each other between joists. See Fig. 4-32. A row of bridging is required whenever the joist span between supports exceeds 8 feet. Exceptionally long joist spans should be bridged at 6 foot intervals.

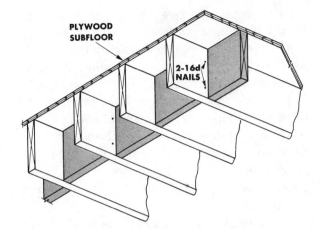

Fig. 4-31. Solid bridging provides maximum rigidity to the joists. The bridging is offset to permit end nailing.

Fig. 4-32. Wood cross bridging is cut to fit diagonally between joists. The lower end of each piece of bridging is fastened after the floor joists have adjusted themselves.

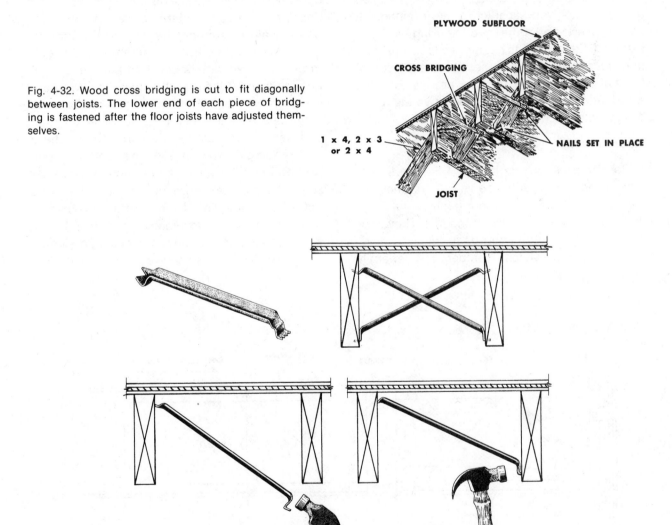

Fig. 4-33. Compression type metal bridging is installed with a few hammer blows. (Cleveland Steel Specialty Co.)

To be most effective, the rows of bridging should be in straight lines which continuously extend the entire length of the floor. (They should be offset enough to permit nailing for solid bridging). When wood bridging struts are placed in position, they are at first only nailed to the joists at the top. The lower end of each piece is left free until after the rough flooring has been nailed to the joists. See Fig. 4-32. When the floor joists have adjusted themselves to the rough flooring, the lower end of each bridging strut is nailed in place.

Compression type metal bridging is used in residential structures because it can be installed quickly and gives satisfactory performance. It can be inserted after the rough flooring has been installed. Each piece is positioned in its approximate location. A few hammer blows are given to the flat end to drive the top prongs into the wood. Additional hammer blows drive the lower prongs into place. See Fig. 4-33.

Floor Trusses

Floor trusses, shown in Fig. 4-34, are sometimes used in residential construction. They are prefabricated in a shop under quality control so that they meet engineering standards. They consist of top and bottom chords fastened to diagonals with truss plates. Members are generally 2 × 4 inch material. Assembly is done on a jig bench, then the trusses are passed between large heavy rollers which drive the prongs of the truss plates into the wood. Trusses are usually spaced on 2 foot centers and have a depth of from 12 to 18 inches, depending on the load and the span.

There are several advantages to be derived from using floor trusses instead of the conventional construction with joist and girders. Trusses are installed quickly and with relative ease. This reduces the overall cost. Trusses are also continuous, from wall to wall, with no intermediate support. This adds flexibility for room layout. There is no need for intermediate piers or foundation walls to support the first floor structure. Partitions for upper floors need not be load bearing. The trusses are lightweight and meet engineering standards for each application. Mechanics in other trades find the open spaces between diagonals excellent for passing pipe, conduit, and heating and air conditioning ducts. The 2 foot spacing between the trusses and the open web design helps the plumber and the carpenter solve the problem of framing at the toilet bend under the floor. The 4 inch (actually 3½ inch) face of the top chord member provides a wide surface for nailing the floor and ceiling material.

Typical flat trusses used for floor or roof sup-

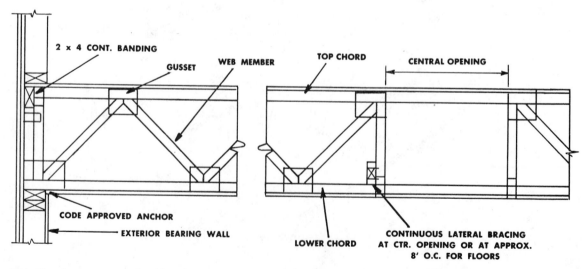

Fig. 4-34. Floor trusses are made of 2 × 4 inch material and designed to provide stiff strong floors. (Automated Building Components, Inc.)

Fig. 4-35. Floor trusses can also be used for roof trusses and are adaptable to light commercial construction. The trusses are supported by the top chord at the wall. This system permits passage of ducts and pipes. (Automated Building Components, Inc.)

Fig. 4-36. Flat trusses are used to support floors.

port are shown in Fig. 4-34. Pipes and wiring can easily pass through the open spaces. When trusses are used with masonry walls, they are hung by the top chord. See Fig. 4-35.

The arrangement of truss members is shown in Fig. 4-35. In frame construction, the truss is made an integral part of the exterior wall.

The arrangement of truss members in the construction of a small residence is shown in Fig. 4-36. The end trusses are made an integral part of the exterior wall. The second floor projects over the first floor and is supported by the cantilevered trusses.

Corner Construction

Two types of corners, shown in Figs. 4-37 and 4-38, are used in residential construction. Both types are quickly assembled, provide good nailing for both interior and exterior finish, and are structurally sound. All corners are required to be made from a minimum of three studs.

The studs for all corners should be made from the best straight and clear stock available then nailed together firmly, using 16d nails spaced not more than 24 inches O.C.

The type of corner shown in Fig. 4-37 requires that 2 × 4 inch blocks be nailed to stud 1. The blocks should be long enough so that they do not split when they are nailed and placed at the top, middle, and bottom of the stud length. Stud 2 is then nailed to the blocks. Stud 3 is framed to the adjoining wall and is nailed to studs 1 and 2 after both walls which meet at the corner have been tilted up.

The corner arrangement of studs, shown in Fig. 4-38, requires the least amount of time and material to build. This arrangement provides excellent backing for interior finish but less surface for exterior trim attachment. Stud 3 is framed to the adjoining wall and nailed to studs 1 and 2 after both walls have been tilted up.

Partition Corners

Partition corners, and the intersection of partitions with exterior walls must be firmly tied together. Backing must also be provided for the interior finish material. Place studs on 16 inch centers (24 inches in some cases) so that the 4

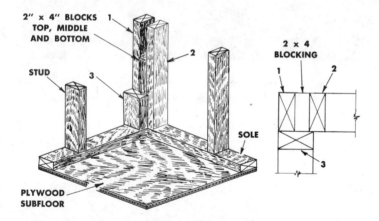

Fig. 4-37. A corner construction consisting of three studs and spacer blocks is sturdy and provides good nailing for interior and exterior finish. Stud 3 is nailed to studs 1 and 2 after the adjoining wall is tilted up into position. The detail to the right above provides a plan view of this corner construction.

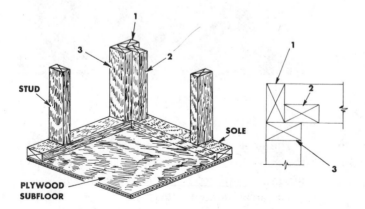

Fig. 4-38. Another type of corner construction. The corner studs are arranged to provide quick assembly. The detail to the right shows a plan view of this construction.

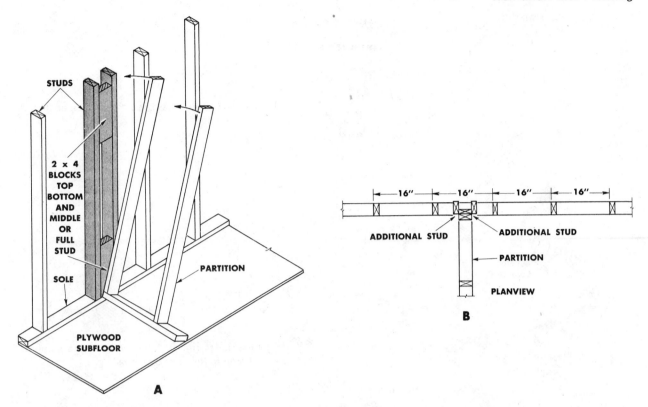

Fig. 4-39. The construction at the intersection of an exterior wall and a partition is called a *channel.* The channel is made by adding studs and strengthening them with nailing blocks. The partition is tilted into place and fastened with nails through the blocks.

foot wide sheets of exterior sheathing are provided with a nailing base. The arrangement of partitions in a building is rarely such that they fall on 16 (or 24) inch spaces where intersections occur. In most instances, the carpenter builds the exterior walls, holding strictly to the 16 (or 24) inch stud spacing from one corner. New studs and blocking are introduced when partitions join the exterior wall.

The preferred method for intersecting walls is shown in Fig. 4-39A. This type, called a *channel,* is especially useful for platform framing because it allows the partition and wall parts assembled on the floor to be tilted up into place for easy fastening. The order for introducing additional studs to maintain the 16 inch spacing of studs in the wall is shown in Fig. 4-39B.

Wall and Partition Plates

Continuous wood members placed on top of a wall as supporters for joists and rafters are called *top plates.* The top plates have three primary functions:

1. They tie the studding together at the top and thus insure stud alignment.
2. They provide support for the structural members above the plates, e.g. for attic joists and roof rafters.
3. They provide a means to tie intersecting walls and partitions together.

Top plates should be doubled at the top of walls and partitions. When walls have the same plate height, tie the outside corners and partitions together by lapping the top plates. See Fig. 4-40. Splices should not be placed closer than 4 feet for a double top plate.

The framing of exterior walls and the framing of interior partitions is done in the same manner. Each uses a bottom member (called a *sole*), vertical studs, and a double top plate. Some building codes permit the use of a single top plate for nonbearing partitions. Stud spacing is usually 16

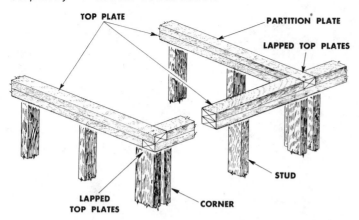

Fig. 4-40. Top plates should be lapped at corners and partitions.

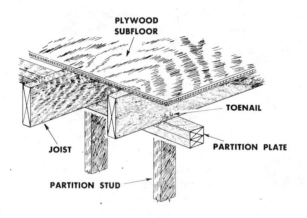

Fig. 4-41. When joists run at right angles to the partition, they are anchored to the plate by toenailing.

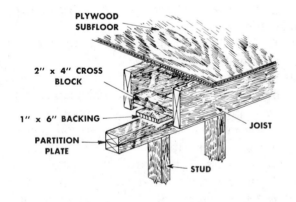

Fig. 4-42. When joists are parallel to the partition, cross blocks are used to hold them in position. The 1 × 6 inch backing provides nailing for the ceiling finish.

inches O.C. Various building systems require 24 inch O.C. spacing. This spacing is permitted by some building codes for certain applications.

Proper framing of the top of partitions is as important as providing solid anchorage for wall corners. You must firmly tie the partitions to the joists and install suitable backing. When partitions run across the joists (at right angles), toe-nail each joist to the top plates. See Fig. 4-41. When partitions run parallel with the joists and come between two adjacent joists, as shown in Fig. 4-42, the top plates are backed with 1 inch material. This backing provides a nailing base for interior ceiling finish. The walls are anchored to the joists with 2 × 4 inch cross blocks at 3 foot intervals.

Framing Rough Openings for Windows and Doors

The first step in laying out wall sections, including windows and doors, is to study the plan views of the working drawings to find their exact horizontal location from the corners of the building. The section view or the elevation views give the dimensions of the windows above the finished floor elevation. The elevation views or a window schedule give information regarding the size and type of windows required. Windows are not generally standardized. Sizes vary with the manufacturer and the type of window. Before laying out the rough opening and determining the length of the header, the carpenter requires certain information. If the windows are delivered to the job, the carpenter can take a measurement of the frame and make allowances for a fitting. If

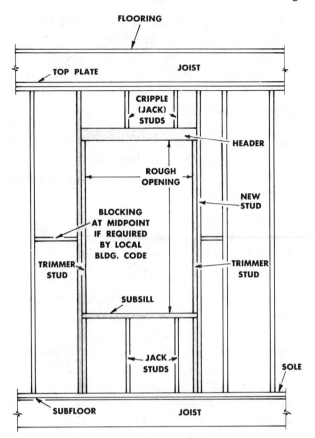

Fig. 4-44. A header supports the structure above the window opening.

windows are to be delivered after the rough framing is completed, the carpenter can obtain the allowance needed for fitting the frame into the opening from tables in the manufacturer's catalog or by consulting the dealer.

The wall sections for a building with platform framing are laid out and assembled on the floor. Lay out the sole and top plate with markings so that the studs fall on 16 inch centers (24 inches in some instances) from the corner of the building. Instead of moving the studs to accommodate the window and and door openings, place additional studs at the sides of the rough openings. See Fig. 4-43 and Fig. 4-44. The basic concepts in placing the structural members is to provide a strong horizontal member called a *header* across the opening to support the structure above, without having it rest on the window or door frame, and to transfer the load to vertical

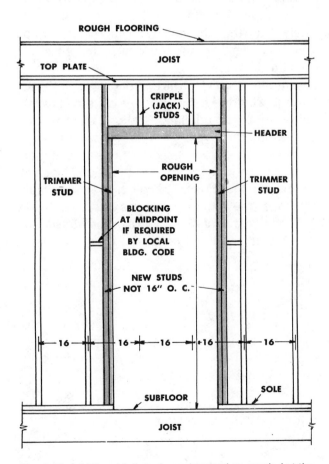

Fig. 4-43. Additional full studs are inserted as needed at the side of rough opening for a door frame.

members called *trimmers.* The trimmer studs for windows and doors rest on the sole.

Header sizes. Header size and construction depends on the load to be supported. Headers are placed on edge in order to develop the maximum strength. The actual size of the header is an engineering problem and is usually given in the building code. The information which follows on header size is from the Uniform Building Code and applies to one story buildings with normal roof loads.

Header Size	Application
4 × 4 inch	Up to 4'-0" span
4 × 6 inch	From 4'-0" to 6'-0"
4 × 8 inch	From 6'-0" to 8'-0" span

Wider spans require wider material. The exact size is determined by an engineer.

One common practice is to use 1½ inch planks with ½ inch spacer blocks between them to bring the overall thickness to 3½ inches. Thus the assembled header is the same thickness as the stud wall. Another practice, which is used to save labor, is to increase the width of the header so that it serves as the rough opening for the window or door frame and in addition fills all of the space up to the top plate. See Fig. 4-45. This method eliminates the short jack (crippled) studs over the header, shown on Fig. 4-44. A solid member such as a 4 × 12 (3½ × 11½) inch timber is used rather than a built-up member to further save on labor costs. The equivalent built-up member would be made of two 2 × 12 inch planks with ½ inch spacer blocks.

Sizes of rough openings for windows. (Note: A study of all types of windows is made in Chapter 6, *Exterior Finish.* The common types of windows used in residential work are shown in Fig. 4-46.)

It is important that you know how to determine the rough opening for each window frame to be installed in a building. Proper allowance must be made for fitting and adjusting of window frame in the opening. This allowance is necessary so that the window frame can be slipped into the opening. Once in place, the window frame must be free so that it can be moved and made plumb before it is nailed in place. The size of the opening must not be too large because there must be sufficient space for nailing the trim into the side studs on both the inside and the outside.

Locating the window in the wall is also important. The window must be placed in the wall at the location indicated on the working drawings. The rough opening must be made at the right height so that the finished window sill or top window trim is at the designated height above the finished floor.

Literature which gives the rough opening sizes for the windows to be used in a building is available from the suppliers. Sections of charts for three different types of windows are shown in Figs. 4-47, 4-48, and 4-49. The two sets of dimensions important for you are the rough opening and unit dimensions. The rough opening dimensions are the actual dimensions you use between the two side studs and between the subsill and the header. The unit dimensions are the overall

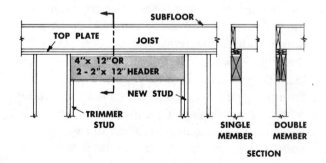

Fig. 4-45. A wide header can be used in place of a smaller member and cripple (jack) studs.

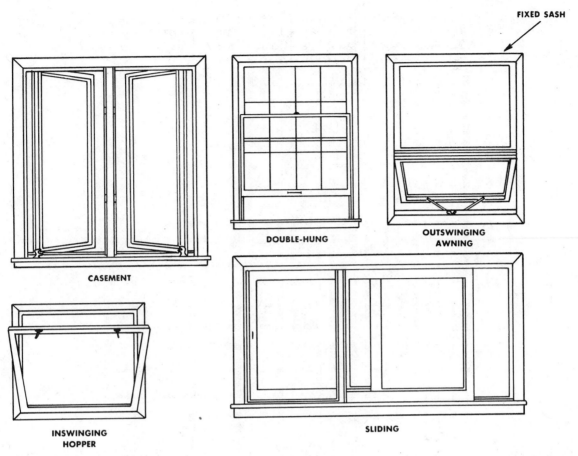

Fig. 4-46. The windows shown above are the basic types used for residential construction. These various types of windows are also available with metal frames and sash. (Andersen Corp.)

dimensions of the window assembly. See Fig. 4-50. The horizontal dimension is the dimension between the outside edges of the side casings. The vertical dimension is from the bottom of the sill to the top of the top casing. The bottom of the sill and the side casings are of the same height. With a metal window, the measurements are made or given to the outside of nailing flanges.

You may have to interpret schedules and notations on the working drawings in order to determine the rough opening sizes. These are shown on Fig. 4-51A, B, and C. Information about windows is generally shown in a window schedule. This information includes the number of windows of each type, the size of each window, and the manufacterer's catalog number or some other description. When a window schedule is not provided on the working drawings, notations

are shown on the elevation drawings. These notations can be shown in one of two ways. The description, as shown in Fig. 4-51A, includes a dimension for the unit size or frame size. If the frame size is given, it is usually correct to add ½ inch on each side and the top to obtain the rough opening dimensions. If only the unit size is given, you must determine the rough opening by examining the windows or obtain the information from the supplier. Notice that the horizontal dimension is always given first. This is standard architectural practice.

Windows are also dimensioned by giving the light (glass) size. See 4-51C. A typical double hung window can be designated 32/22. This indicates that each light is 32 inches wide by 22 inches high. The horizontal dimension includes the following information:

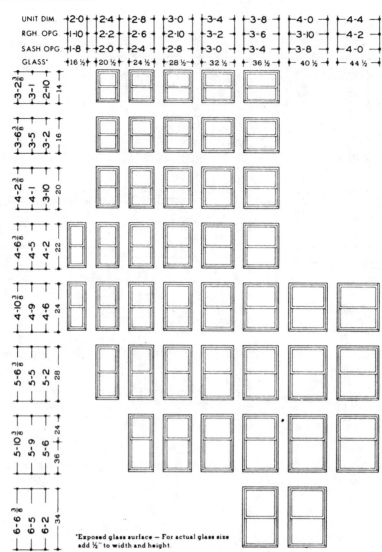

Fig. 4-47. The chart above shows the different sizes of double hung windows available. The rough opening required for each size is likewise indicated. (Andersen Corp.)

Glass	32	inches
2 Stiles (vertical side pieces of the sash), 2″ each	4	inches
2 Jambs (side pieces of the frame) ¾″ each	1½	inches
½ inch Fitting space each side	1	inch
rough opening	38½	inches

The vertical dimensions include the following information:

2 lights, 22″ each	44	inches
Top rail (top horizontal member of the sash)	2	inches
Meeting rail	1	inch
Bottom rail	3	inches
Top jamb	¾	inch
Sill	2	inches
Fitting space (top only)	½	inch
	53¼	inches

The example above, used to calculate the

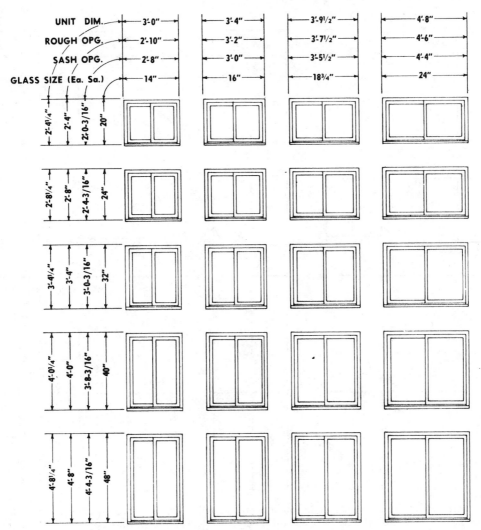

Fig. 4-48. The portion of a window chart appearing above shows sizes of wood casement windows.

rough opening, may vary somewhat in actual practice. The windows may have narrower stiles and rails than those indicated in the example. The sill may have a different type of construction. Sash adjusting devices, such as sash balances, may require an additional allowance.

Rough openings for horizontal sliding windows, casement windows, and awning windows can be calculated in a similar manner. This consists of adding the thicknesses of the sash parts, the frame parts, and the fitting spaces to the light size.

Some sash are divided into small lights by using wood dividing bars called *muntins*. The working drawing may show a light size designated as 32/22 Div. The muntins are to be ignored in calculating the rough opening because the dimensions (32/22) describe the sash as though there were no muntins.

Sizes of rough openings for doors. In order to calculate the size of a rough opening for a door, you must know the size of the door, the thickness of the jambs, and the space required to adjust the door in the opening. One half inch on each side of the door and over the head jamb is usually considered sufficient. A door which is

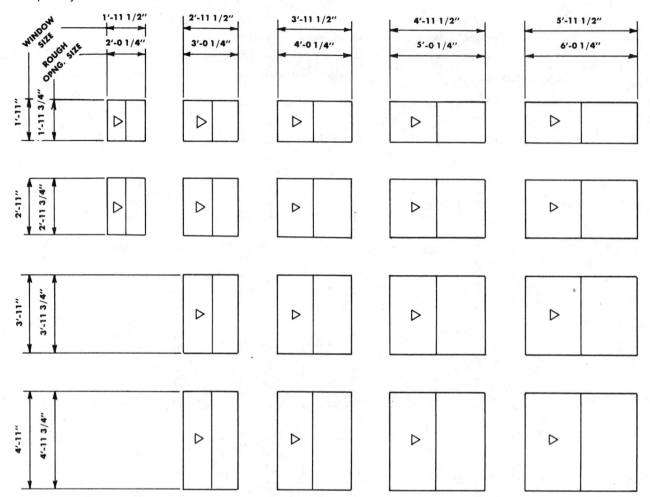

Fig. 4-49. The chart above gives the window and rough opening sizes for metal horizontal sliding windows. The arrowhead (or triangle) indicates that one sash is movable while the other is fixed. (Reynolds Metals Co.)

3'-0" × 6'-10", with 1 inch jambs would require a rough opening of 3'-3" × 6'-11½". The placing of the header above the rough floor is important and must be considered carefully. You must know the thickness of the finished floor and the allowance for the threshold and the carpeting before the rough framing members are cut and fastened into place.

Wall Sheathing

After completing the framework of a new building, fasten a covering, known as *sheathing,* to the frame. In addition to serving the purpose of covering the frame, sheathing also furnishes a base for exterior trim, such as siding stucco or masonry veneer. The sheathing also helps to stiffen the building, which makes the structure more resistant to wind pressure. It provides a small amount of insulation against extreme weather conditions.

Three distinct types of material are generally used for sheathing. They are plywood, fiberboard, and gypsum board. They each present different installation problems and vary in structural quality, effectiveness of resistance against air and moisture, and cost.

Board siding is rarely used because other materials are easier to apply and better serve the

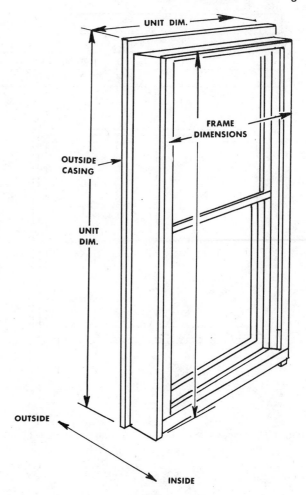

Fig. 4-50. Unit dimensions are the maximum dimensions measured to include the outside casings and sill.

purpose. With board sheathing, diagonal bracing and an application of asphalt saturated felt paper are required which add operations.

Fiberboard (insulated fiberboard) is made from several different substances, such as sugar cane, corn stalks, and wood pulp. These materials are processed and pressed into boards, generally ½ or 25/32 inch thick. They are asphalt impregnated to make them water resistant. They are cut to 2 × 8, 4 × 8, or 4 × 9 foot panels. The 2 × 8 foot panels are made with a V type tongue and groove edge on the long sides and with square cut edges on the short sides. They are installed in a horizontal position. See Fig. 4-52. Roofing nails (1½ inch) are spaced 4 inches O.C. on studs for ½ inch sheathing. 1¾ inch nails are used for 25/32 inch sheathing.

Gypsum board is similar to fiberboard except that it is made of two layers of water repellent paper with a core of gypsum reinforced with fiberglass. The common sizes are 2 × 8, 4 × 8, and 4 × 9 feet by ½ inch thick. The 4 foot wide panels do not need corner bracing. The nailing procedure is similar to that of fiberboard.

Neither fiberboard nor gypsum board should be used as nailing base for siding. Nails must penetrate the sheathing into the studs unless some form of furring is used. Both of these types of sheathing are easily cut and installed, are relatively inexpensive, and have excellent weather resistant properties.

Plywood has proved to be the strongest sheathing. The thicknesses commonly used are 5/16, 3/8, ½, or 5/8 inch. The common sizes are 4 × 8, 4 ×

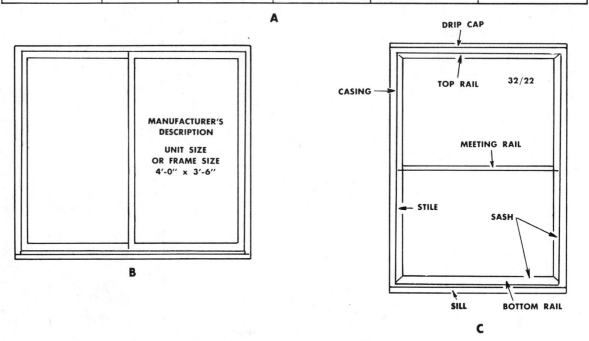

WINDOW SCHEDULE					
DESIGNATION	NUMBER	TYPE	UNIT SIZE	ROUGH OPENING	REMARKS
A	7	DOUBLE HUNG	3'-4" x 4'-6 1/2"	3'-2" x 4'-5"	UNIVERSAL #1074
B	2	CASEMENT	4'-8" x 4'-1"	4'-4" x 4'-0"	ROTO OPERATOR

A

MANUFACTURER'S DESCRIPTION

UNIT SIZE OR FRAME SIZE 4'-0" x 3'-6"

B

C

Fig. 4-51. Windows are shown on working drawings with unit size or light (glass) size.

Fig. 4-52. Insulated fiberboard is asphalt impregnated and easily installed. In this figure, is installed in a horizontal position. The 2 × 8 foot pieces are placed vertically. (Armstrong Cork Co.)

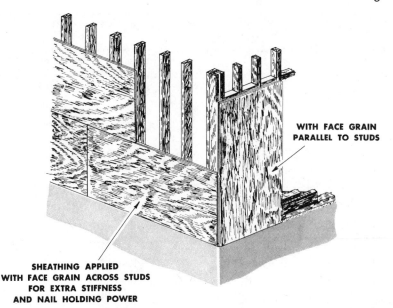

Fig. 4-53. Plywood sheathing can be applied both vertically and horizontally. (American Plywood Assoc.)

WITH FACE GRAIN PARALLEL TO STUDS

SHEATHING APPLIED WITH FACE GRAIN ACROSS STUDS FOR EXTRA STIFFNESS AND NAIL HOLDING POWER

9, 4 × 10, or 4 × 12 foot panels. The sheets are applied with common annular, spiral threaded, or *T* nails. (Nail size and spacing are as given in the Uniform Building Code.) Staples can be used if the spacing is closer together.

When installing sheathing, the carpenter must follow the manufacturer's suggestions or consult the local building code and/or other applicable building codes to determine the type of nail or staple to use with each type of material and material thickness. These sources will also indicate the proper spacing to use. The sheathing must be weathertight where sheets meet or terminate so that air cannot penetrate it. Sheathing also serves to stiffen the structure in most applications. Plywood sheets can be installed either vertically or horizontally, as shown in Fig. 4-53, depending on the strength and appearance required for the particular application.

Bracing

Diagonal bracing, shown in Fig. 4-54, is required for horizontal board sheathing, gypsum board sheathing, and, in certain cases, fiberboard sheathing. The bracing material is generally 1 × 4 inch boards let into the face of the studs. The wall is made square and plumb, then the diagonal member is laid across the studs and

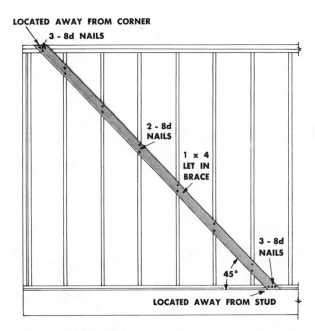

LOCATED AWAY FROM CORNER

3 - 8d NAILS

2 - 8d NAILS

1 x 4 LET IN BRACE

3 - 8d NAILS

45°

LOCATED AWAY FROM STUD

Fig. 4-54. Braces are let into the face of the studs at a 45° angle unless rough openings interfere.

marks are made along the edges. Cuts ¾ inch deep are made on the marks. The material between the saw cuts is removed. Braces should be placed so that they extend through the sole and top plate. However, the braces should be set

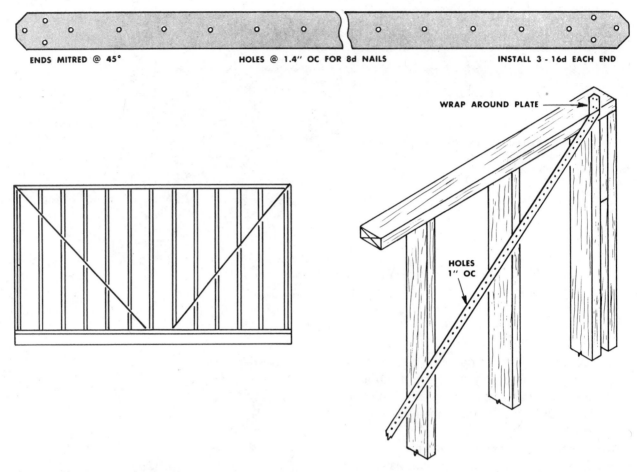

ENDS MITRED @ 45°   HOLES @ 1.4" OC FOR 8d NAILS   INSTALL 3 - 16d EACH END

WRAP AROUND PLATE

HOLES 1" OC

Fig. 4-55. Two types of steel strap bracing are shown above. Steel strap bracing is nailed to the face of the sole, the studs, and the plate through pre-punched holes. (United Steel Products Co.)

away from the studs at top and bottom so that the cuts do not run into nails. The braces should be placed as close to a 45° angle as possible.

Another type of diagonal bracing is achieved with the use of steel strap bracing. Metal straps of 16 gage steel (.059 thick) are pre-punched with holes for nails spaced at close intervals. Two lengths are provided. One makes a 45° angle, across the studs. The other makes a lesser angle which is used when space is limited. The strap metal is thin enough so that it does not interfere with the sheathing or exterior finish when it is nailed to the outside of the studs, the sole, and the top plate. It is approved for use by several national building codes and rated equivalent to 1 × 4 inch let in bracing. See Fig. 4-55.

Plywood can be used in combination with fiberboard or gypsum board sheathing, as is shown in Fig. 4-56. The plywood serves to stiffen the corners and may make bracing unnecessary, depending on the requirements of the particular application. In this case, fiberboard or gypsum board is used for the remainder of the wall to save cost. Some plywood is patterned with a grooved surface on one side so that the single sheet serves as both the sheating and the exterior wall finish. This application is discussed in Chapter 6, *Exterior Finish.*

Metal Framing Devices

Various types of metal framing hangers and straps have already been discussed and illustrat-

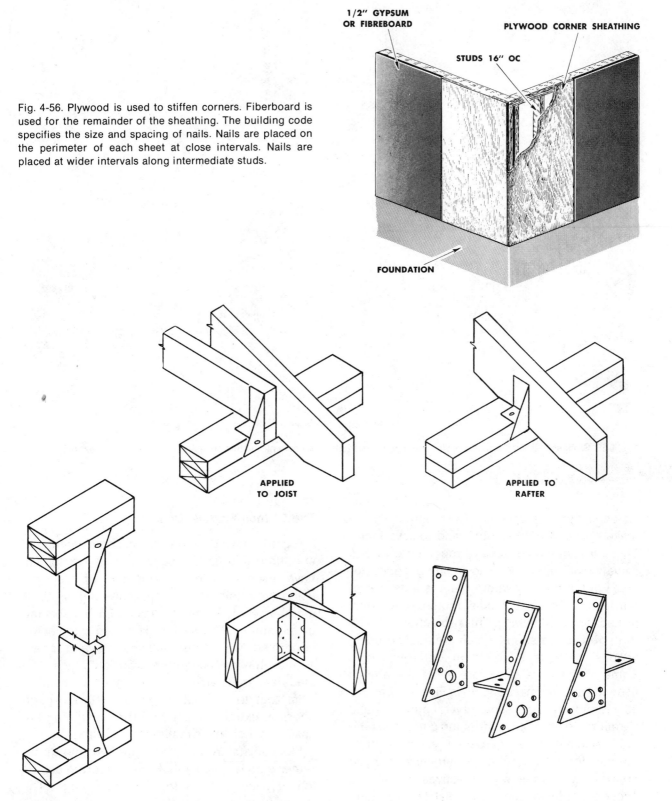

Fig. 4-56. Plywood is used to stiffen corners. Fiberboard is used for the remainder of the sheathing. The building code specifies the size and spacing of nails. Nails are placed on the perimeter of each sheet at close intervals. Nails are placed at wider intervals along intermediate studs.

1/2" GYPSUM OR FIBREBOARD

PLYWOOD CORNER SHEATHING

STUDS 16" OC

FOUNDATION

APPLIED TO JOIST

APPLIED TO RAFTER

Fig. 4-57. Framing anchors provide solid nailing at intersections of members. Anchors are required for special applications. (Timber Engineering Co.)

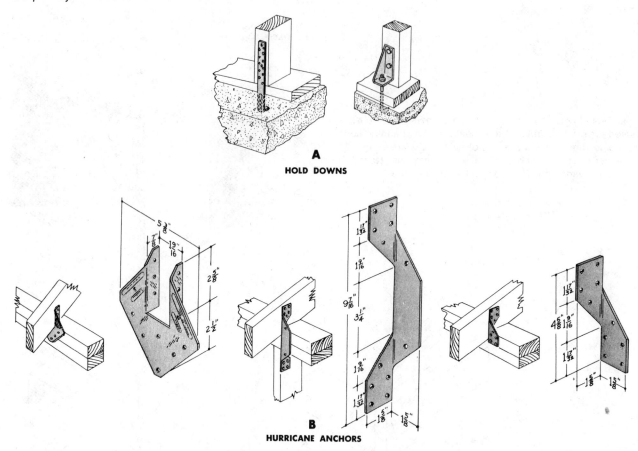

A
HOLD DOWNS

B
HURRICANE ANCHORS

Fig. 4-58. Steel accessories provide protection against hurricane and seismic disturbances (earthquakes). (Simpson Co.)

ed earlier in this chapter. Their purpose is to make sound structural joints and to save labor. There are additional types of rough hardware to serve other functions. The building codes for areas of the country where there is a danger of hurricanes or earthquakes require additional protection for buildings. This protection is in the form of ties and straps which are applied using various methods at joints in the framework. Some of the metal hardware and the methods used to apply them are shown in Figs. 4-57 and 4-58. Framing anchors serve several purposes and permit rapid assembly of framing parts. Toe nailing is eliminated. As a carpenter, you must study the building code for the community where you are working and follow its directives closely. The anchors are fastened with special 1¼ inch long nails which are supplied with the hangers.

The 24 Inch Module System

Experiments, designed to cut costs of material and labor and use plywood to its full potential, have resulted in the development of a system using a module of 24 inches. See Fig. 4-59. (A *module* here is a unit of measure for the spacing of members.) The system is based on placing floor joists, wall studs, ceiling joists, and rafters so that they are aligned in a vertical plane. The loads are transferred from member to member from roof to foundation. Full sheets of rough flooring, exterior finish, and roof sheathing are used uncut or with a minimum of cutting.

Floor joists are placed in line over the girder rather than lapped so that rough flooring can be joined over joists to the best advantage. The rough floor is ¾ inch tongue and groove plywood which is nailed and glued to the joists to

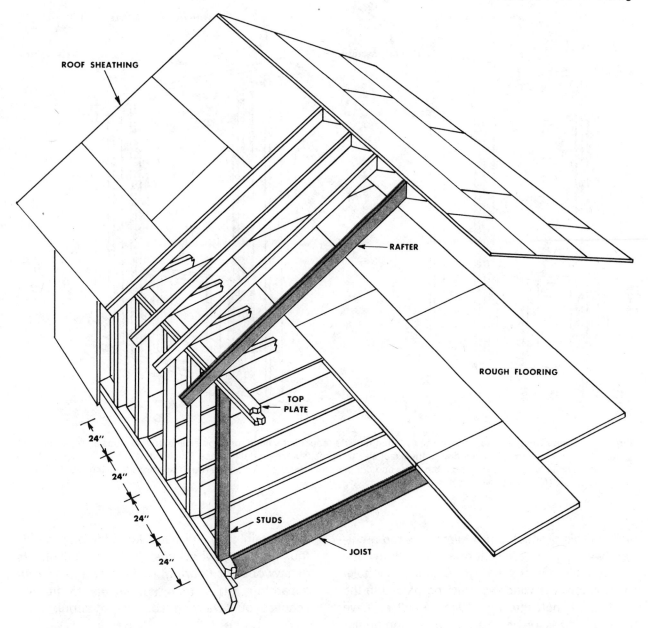

Fig. 4-59. The 24 inch module framing system produces a series of in-line frames that use the framing members and plywood skin to the best advantage. (American Plywood Assoc.)

make a strong firm floor. Full sheets are used, staggering the end joints.

Sheathing is made with ½ inch plywood. This provides strength to the wall and backing for exterior finish material. Plywood can also serve the combined function of siding and sheathing. When a single plywood sheet serves both pur-

poses, place vertical battens (narrow strips) to cover the joints. Battens are not needed when patterned plywood is used because this type of material has a shiplap edge and presents a finished surface. Framing material is most economically used when the openings are placed to fall on the 24 inch module spacing. See Fig.

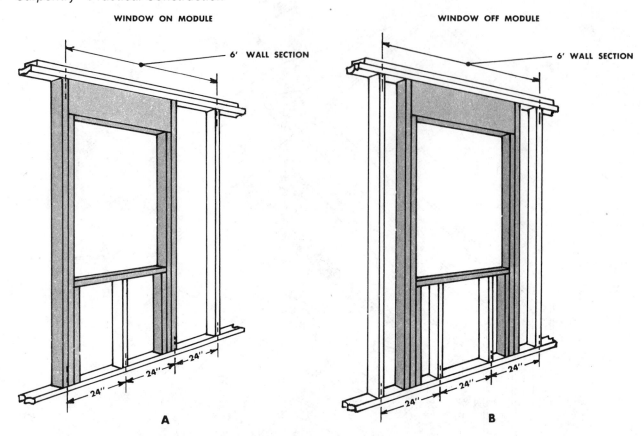

WINDOW ON MODULE   **WINDOW OFF MODULE**

6' WALL SECTION   6' WALL SECTION

24"   24"   24"   24"   24"   24"

A   B

Fig. 4-60. Placing windows and doors on the 24 inch module reduces the need for cripples and double studs. In (A) above, the required rough opening for the window is the exact size to fit the 24 inch modular space. The window is also located so that it falls on stud spaces. In (B), the new double studs and cripples are introduced when the rough openings for windows do not fall on stud spaces. (American Plywood Assoc.)

4-60A. This procedure however limits the architect who must chose windows of a width which fit the openings created by the 24 inch module. In addition, the windows must be placed in the wall at 24 inch intervals. When windows have different sizes or are placed with nonmodular spacing, additional studs are introduced. See Fig. 4-60B.

Ceiling joists are placed so that they line up with the studs to receive ceiling finish material. The ceiling finish material also fits the module. Roof rafters are fastened alongside the joists. Trusses when used are placed directly over the studs.

In order to take full advantage of the 24 inch module system, builders and architects must design the structure with great care. They must place openings and choose windows and doors which minimize the need for additional studs, thus reducing overall cost. If you use full sheets of plywood for flooring, wall material, and roof sheathing rather than cut sheets to fit odd spaces, you save both labor and material.

Framing with Metal Structural Members

The steel and aluminum industries have developed framing systems which in some instances are competitive in price with wood. In highrise and commercial structures, metal partitions, of the type shown in Fig. 4-61, have largely replaced wood for nonbearing walls. Metal framing has various advantages. It does not warp, shrink, or rot, and it is not affected by termites. New tools, such as power screwdrivers, make assembly much more easy. Some parts are welded together quickly, using the electric arc welding

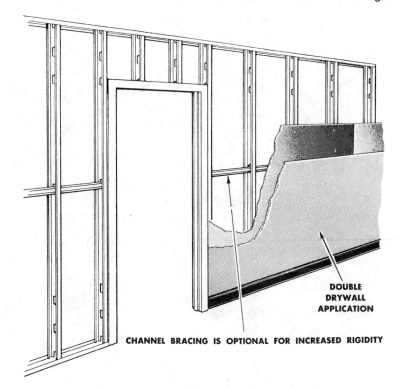

Fig. 4-61. Steel frame partition systems provide a satisfactory base for drywall. Steel studs are fastened to channels in the floor and ceiling. The studs have holes which allow for the passage of conduit or pipe. (National Gypsum Co.)

DOUBLE DRYWALL APPLICATION

CHANNEL BRACING IS OPTIONAL FOR INCREASED RIGIDITY

technique. The studs, joists, and other items are available in stock sizes. They are made with minimum weight and are adaptable to fit the needs of almost any design problem.

Metal tracks are fastened to the floor and ceiling with nails, screws, or powder activated fasterners, depending on the material used for the floor. The studs are snapped into place on 24 inch centers and fastened with screws or crimped (bent) to stay in position. Gypsum board for the drywall is fastened to the studs and channels with screws driven by a power screwdriver. Frames for doors are included in the wall where needed.

Steel and aluminum framing is currently being used successfully in residential construction. See Figs. 4-62 and 4-63. The metal framed building however requires careful engineering and the approval of local building authorities. The metal framework is not as adaptable as wood framework because the members cannot be as easily cut and fastened on the job. Great care must be exercised when you cut the members. When certain parts are cut, they lose some of

their strength. Some members permit passage of pipes and ducts. Others present a problem which can only be solved by careful planning. In event of a major fire, metal members have a tendency to bend, and the entire structure can collapse. However a wood frame structure is flammable and likewise subject to collapse.

Some typical metal framing members are shown in Fig. 4-64. The most common type of floor joist is made in the shape of the letter C.

Nailable metal joists are also manufactured. Nails fastening the plywood subfloor are driven through the material into the slots in the joists and are held firmly. Bridging consists of continuous steel straps which are fastened to the top and bottom surfaces of the joists. Plywood subflooring is applied over the joists and fastened with self-drilling, self-tapping screws. Aluminum joists are also available. Aluminum members have some added advantage. They can be cut with hand power saws which use combination blades. They can also be nailed with hand or power driven nails. Spiral shank nails are suggested for structural members.

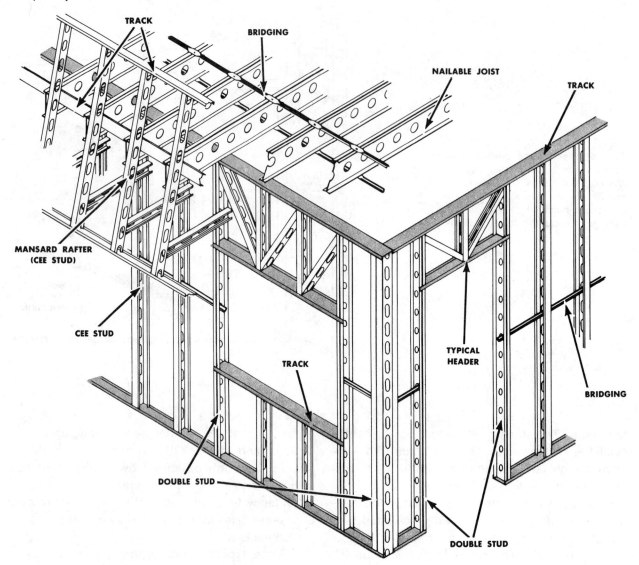

Fig. 4-62. Framing an exterior wall with a steel frame. The illustration shows the arrangement of members for the steel frame. (Wheeling-Pittsburgh Steel Corp.)

Load bearing walls are made using top and bottom tracks (channels) into which the studs are fastened. The steel studs are supplied in the shape of channels or the letter *C.* Aluminum studs are in the shape of a letter *I.* Holes are provided in most of the studs (both steel and aluminum) for the passage of pipe or conduit. The studs are placed on 24 inch centers to accommodate plywood or other wall covering material. The plywood sheets (or diagonal straps or wires) counteract any tendency toward racking

(distortion across the diagonal of the wall). Metal window and door frames are made as integral parts of the framework. Wood liners are provided in the metal framework for wood windows. Drywall finish is applied to the studs with self-tapping screws. Steel joists are used to support floors in residential construction, as shown in Fig. 4-65. The C type joist is supported on the foundation wall and tack welded to supporting steel beams at the center of the building. One advantage of aluminum joists is shown in Fig. 4-

Fig. 4-63. Aluminum studs are used to frame walls and partitions. The members are cut and fastened using conventional power hand tool methods. Studs are spaced 24 inches O.C. (Alcoa)

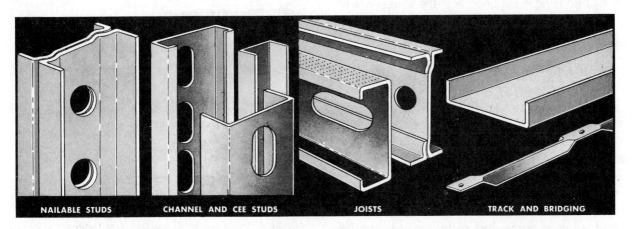

NAILABLE STUDS   CHANNEL AND CEE STUDS   JOISTS   TRACK AND BRIDGING

Fig. 4-64. Steel framing members are used to build light construction framework. (Wheeling-Pittsburgh Steel Corp.)

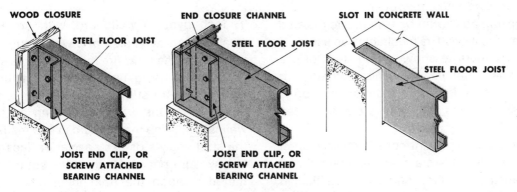

WOOD CLOSURE

STEEL FLOOR JOIST

JOIST END CLIP, OR SCREW ATTACHED BEARING CHANNEL

END CLOSURE CHANNEL

STEEL FLOOR JOIST

JOIST END CLIP, OR SCREW ATTACHED BEARING CHANNEL

SLOT IN CONCRETE WALL

STEEL FLOOR JOIST

Fig. 4-65. Steel channel floor joists are used in buildings with light construction. (Inland Ryerson)

Fig. 4-66. Plywood rough floor is applied over aluminum joists using nails and a pneumatic nailer. Note the opening for a stairway. (Alcoa)

66. Subflooring is installed with the same ease as when wood joists are used.

Stud Walls for Drywall Application

Manufactures of gypsum products have invested great sums of money into research and experiments to perfect drywall systems which are both practical and economical. The walls must provide excellent surfaces for interior finish, meet fire rating standards, and also stay within the limits for sound transmission. As is true in all phases of building, good workmanship and a certain amount of knowhow are essential for those mechanics who do the work. Partitions for drywall application can be made using conventional wood studs, metal studs, or several thicknesses of gypsum board (known as a solid partition or *sandwich construction*).

The gypsum board can be fastened to the stud wall in various ways. The mechanic can use nails or screw fasteners, as shown in Fig. 4-67, or use a glued-nailed technique. (The methods for applying wallboard, finishing joints, and applying trim at openings and corners are discussed in Chapter 7, *Interior Finish.*)

The two most important considerations when selecting a drywall system are fire rating and sound transmission. Fire ratings are established to comply with underwriter's tests. One layer of

1-1/4" USG DRYWALL SCREW - TYPE W - BUGLE HEAD

1-1/4" GWB-54 ANNULAR RING NAIL

1-7/8" 6d GYPSUM WALLBOARD NAIL CEMENT COATED

Fig. 4-67. Three representative types of fasteners are used for applying gypsum drywall. (U.S. Gypsum Co.)

$5/8$ inch gypsum wallboard, with fire resistant qualities, on each side of a wood stud partition provides a one hour fire rating. Fire resistance is doubled by using two layers of wallboard on each side of the studs. Sound transmission is cut down in various ways, depending on how the wall is built. The mechanic can use insulation in the stud spaces or use special devices to hang the gypsum board. Much of the sound is transmitted through the framing members and the wall covering itself.

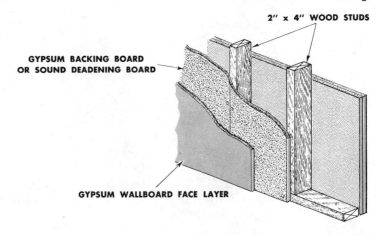

Fig. 4-68. A conventional stud wall is used for a two layer application of drywall. The sound deadening quality of the wall is vastly improved when sound deadening gypsum board is used for the first layer. (U.S. Gypsum Co.)

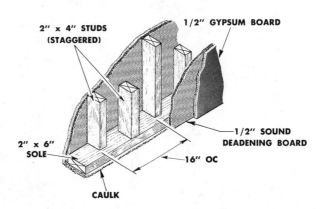

Fig. 4-69. Staggered stud construction cuts sound transmission. (National Gypsum Co.)

Using wood studs with drywall. The common base for drywall in residential construction is formed with wood studs on 16 inch centers, using conventional framing techniques. The mechanic can use various methods to lower sound transmission, if that has a high priority in the building. 1.) The inner layer of a two layer application can be made of sound-deadening gypsum board, as shown in Fig. 4-68. 2.) An insulating wool blanket or batts can be installed in the stud space. 3.) A staggered stud partition, as shown in Fig. 4-69, can be erected. 4.) A double stud partition can be used. See Fig. 4-70. 5.) Resilient channels can be applied horizontally, as shown in Fig. 4-71. The mechanic often uses two or more of these methods in combination to further decrease the noise transmission level.

Using metal studs for drywall. A typical non-load bearing steel frame partition is shown in Fig. 4-72. The studs are fastened to tracks (chan-

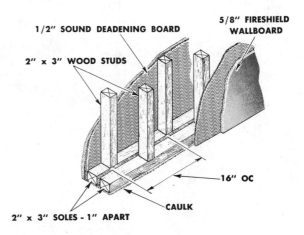

Fig. 4-70. Double stud wall separates the two wall surfaces completely. (National Gypsum Co.)

nels) at the top and bottom. Additional studs are used to frame openings. Holes or knockouts are provided at the top and bottom of the studs to

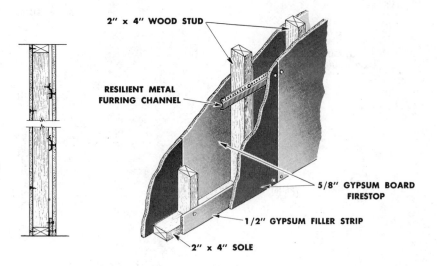

2" x 4" WOOD STUD

**RESILIENT METAL
FURRING CHANNEL**

**5/8" GYPSUM BOARD
FIRESTOP**

1/2" GYPSUM FILLER STRIP

2" x 4" SOLE

Fig. 4-71. A resilient channel is fastened to the studs to permit the gypsum board to float. This construction cuts down on sound transmission by isolating the gypsum board from the studs. (Georgia-Pacific Corp.)

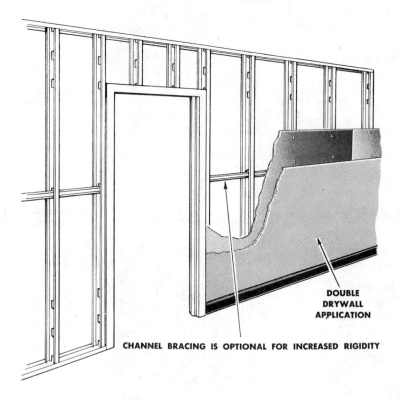

**DOUBLE
DRYWALL
APPLICATION**

CHANNEL BRACING IS OPTIONAL FOR INCREASED RIGIDITY

Fig. 4-72. The first layer of gypsum board (drywall) is fastened to the steel studs with self-tapping screws. The second layer is adhered to the first layer using drywall compound. (National Gypsum Co.)

allow passage of pipes or conduit. Some studs have hole patterns which differ from those shown in Fig. 4-72. Some studs can have open webs. Figure 4-73 shows how the studs and tracks join at the floor or ceiling. It also shows how base and drywall finish are applied.

A chase wall can be used when a large diameter pipe requires a thicker wall. See Fig. 4-74. Two metal stud walls are erected and tied together at intervals. Rectangular pieces of gypsum wallboard are fastened to the studs with screws to join them together. See Fig. 4-74.

A solid gypsum partition without studs is shown in Fig. 4-75. The coreboard is fastened to

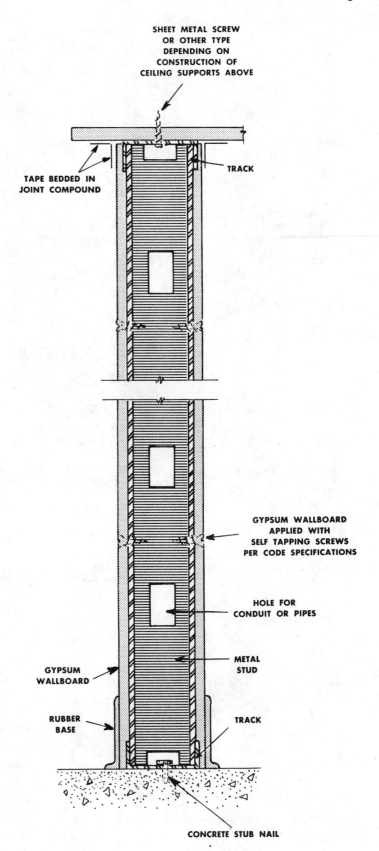

Fig. 4-73. Tracks engage the metal studs at the floor and at the ceiling. The gypsum wallboard is fastened to metal studs with self-tapping screws.

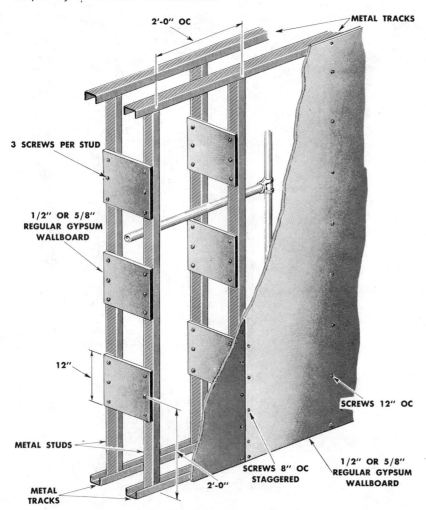

METAL TRACKS

2'-0'' OC

3 SCREWS PER STUD

1/2'' OR 5/8''
REGULAR GYPSUM
WALLBOARD

12''

METAL STUDS

METAL
TRACKS

2'-0''

SCREWS 8'' OC
STAGGERED

SCREWS 12'' OC

1/2'' OR 5/8''
REGULAR GYPSUM
WALLBOARD

Fig. 4-74. A chase wall provides adequate space for piping or ducts. (The Celotex Corp.)

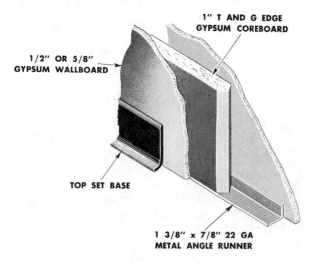

1'' T AND G EDGE
GYPSUM COREBOARD

1/2'' OR 5/8''
GYPSUM WALLBOARD

TOP SET BASE

1 3/8'' x 7/8'' 22 GA
METAL ANGLE RUNNER

Fig. 4-75. Two inch solid gypsum drywall has a high fire resistance rating. (U.S. Gypsum Co.)

the floor and ceiling with metal angle runners. Finish gypsum wallboard is applied to both sides.

Framing Used with Plaster Wall Finish

Although most interior finish is done with drywall or wall paneling, you must have some idea of plastering so that you can do the framing properly. The traditional procedure is to apply two or three coats of plaster over ⅜ inch rocklath to a total thickness of ¾ or ⅞ inch. (Rocklath is made of two surfaces of special paper which enclose a gypsum core. It resembles gypsum board used for drywell except for the sheet size and the type of paper covering.) The rocklath is nailed to the walls and ceiling.

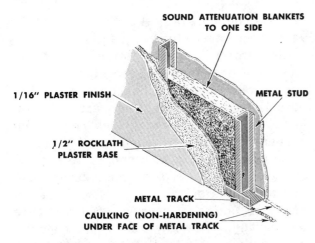

SOUND ATTENUATION BLANKETS TO ONE SIDE

1/16" PLASTER FINISH

METAL STUD

1/2" ROCKLATH PLASTER BASE

METAL TRACK

CAULKING (NON-HARDENING) UNDER FACE OF METAL TRACK

Fig. 4-76. A metal stud wall serves veneer plaster finish. (U.S. Gypsum Co.)

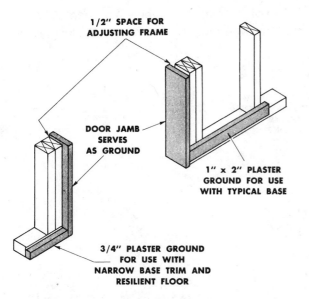

1/2" SPACE FOR ADJUSTING FRAME

DOOR JAMB SERVES AS GROUND

1" x 2" PLASTER GROUND FOR USE WITH TYPICAL BASE

3/4" PLASTER GROUND FOR USE WITH NARROW BASE TRIM AND RESILIENT FLOOR

Fig. 4-77. Plaster grounds serve as screeds for plaster thickness and a nailing base for the trim.

One type of plaster finish called *plaster veneer* is similar to drywall. It has a thin coat (¹/₁₆ to ³/₃₂ inch thick) of plaster over the surface. See Fig. 4-76.

Before the rocklath is applied, strips of wood (called *grounds*) are put in place. The grounds serve three purposes. They provide a vertical surface against which the plasterer stops the plaster. They are a thickness guide because they are made the same thickness as the dimension for the plaster and rocklath together. They also serve as a nailing base for wood trim members. See Fig. 4-77. The best practice is to use a 1 × 2 inch strip of wood for the base ground. It should be placed high enough to serve as a nailing base for the wood base mold.

Since the main purpose of plaster grounds is to provide a guide for the plasterer to follow, it is important that you use special care when placing the grounds. You should use a straightedge and a level so that the position of the grounds is correct and true. Any extra time you use to make certain the grounds are set accurately is offset by the advantages gained later when you are nailing on the trim. It takes much less time and patience to fit the trim to the plaster if the grounds are accurately placed. Window and door frames are generally made to such a width that, once they are placed in the openings, the jamb can serve as a ground. When baseboard trim and trim around openings are applied, the grounds are completely concealed.

SPECIAL FRAMING PROBLEMS

The building of a modern frame house involves the labor of many mechanics other than carpenters. These include plumbers, pipe fitters, sheet metal workers, electricians, tile setters, brick and stone masons, etc. The contractor for each trade is only responsible for the work of that trade. However, all mechanics must cooperate in every respect since they all have the responsibility of giving the owner a top quality house. With proper cooperation among the mechanics, the finished house will conform to the specifications and agreements signed by the owner and the general contractor.

Carpenters are regarded as the key persons in construction since they provide the building framework into which the work of other mechanics must fit. As a carpenter, you must have a working knowledge of the requirements for the other trades involved in building the house. The building must retain its structural strength after the mechanics who install the pipes, heat and cooling ducts, electrical wiring and fixtures, and other special features complete their work.

Fire Prevention

Eliminating potential fire hazards in a structure is of course the job of all mechanics. However, the burden rests most heavily on the carpenters because they build the basic framework. As a carpenter, you must adhere strictly to the applicable building code which is primarily designed to promote structural and fire safety. You should be aware of the fire ratings required for walls, floors, and roofs and follow them closely.

Many fires have started in the basement of a residence and spread to the attic and roof before the occupants were aware of the danger. Such a catastrophe could be caused by the neglect of carpenters who failed to block drafts with fire stops at the proper places in the floor and wall structures. A fire could also result from the failure of the builders to pack noncombustible material around pipes or heat ducts in wall spaces which connect with the basement.

The carpenters must know how to frame around chimneys and fireplaces. See Fig. 4-78. The Uniform Building Code requires that combustible material shall not be placed within 2 inches of smoke chambers or chimneys when built entirely within a structure, or within 1 inch when the chimney is built entirely outside the structure." Local building codes may require that this space be filled with noncombustible insulation. Cracks can develop in chimneys once the building settles. Any combustible materials or open space near such cracks constitute potential fire hazards.

Plumbing Pipes

One of the major problems a carpenter has to contend with is providing space for supply and

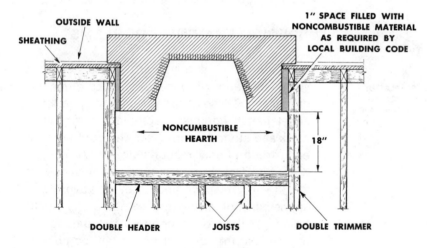

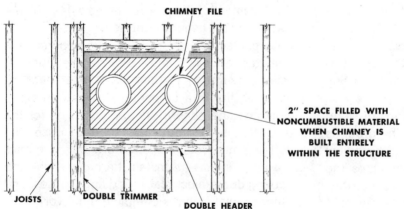

Fig. 4-78. A 2 inch open space is left in the framing of the floor around chimneys. This space is filled with noncombustible material.

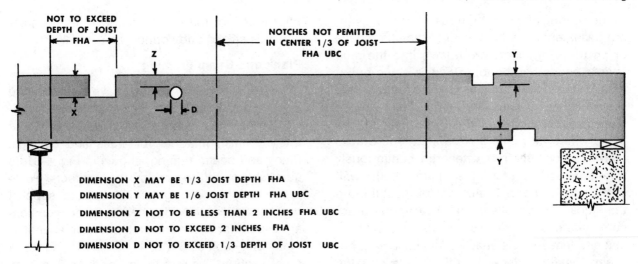

DIMENSION X MAY BE 1/3 JOIST DEPTH FHA

DIMENSION Y MAY BE 1/6 JOIST DEPTH FHA UBC

DIMENSION Z NOT TO BE LESS THAN 2 INCHES FHA UBC

DIMENSION D NOT TO EXCEED 2 INCHES FHA

DIMENSION D NOT TO EXCEED 1/3 DEPTH OF JOIST UBC

Fig. 4-79. Notches and holes in joists must be made to comply with the restrictions of the building code which applies. (Note: *FHA* is the Federal Housing Authority and *UBC* is the Uniform Building Code.)

waste water piping. In some cases, piping in the walls extends from the basement or ground floor, up through the wall, and is vented through the roof. Horizontal piping, particularly under the bathrooms, must be installed without impairing the structure. Building codes restrict cutting joists to allow for the passage of piping so that the structure remains strong. See Fig. 4-79. When holes are cut, they should not exceed 2 inches in diameter. They should be a minimum of 1/6 of the depth of the joist, from the top or bottom edges of the joist. The joist is considerably weakened when the top and bottom portions are cut. Joists should not be notched in the center one third of the span. Notches shall not exceed 1/6th of the depth of the joist, except as specified by the code.

Additional plumbing problems. A partition in back of fixtures in the bathroom must often be widened to 6 inches in order to provide room for a soil stack and piping. Wall hung lavatories require backing or additional framing members so that they are firmly supported. See Fig. 4-80.

Heating ducts. The sheet metal worker calls on carpenters to provide openings in the floor and walls for heating and air conditioning ducts. The carpenters should be provided with a duct layout showing the size and location of diffusers, ducts, and other related equipment before work

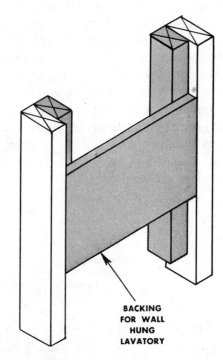

BACKING FOR WALL HUNG LAVATORY

Fig. 4-80. Backing is placed flush with the face of studs where heavy objects such as lavatories or other fixtures are to be hung.

begins on the framework. As the building is erected, the holes made in floors, walls, and partitions must be made in a manner which preserves the strength of the building.

The heating and cooling units for houses without basements can be placed in the attic space or on the roof. Ducts drop down into the stud spaces opened at the top. Ducts in some buildings are buried in the slab. In such situations, the stud spaces in walls and partitions must be placed so that they coincide with the stubs at the place where they emerge from the floor. Stacks in two story buildings often run continuously from the basement to diffusers high on the wall of the second floor. The carpenter must line up the studs of the walls for the two floors to accommodate these stacks. Provisions must be made to leave ample space in the basement between the floor under partition walls and the beams or girders where horizontal runs join vertical stacks. When for some reason the diffusers in the walls cannot be placed to come within stud spaces, the studs must be cut, and additional studs nailed in position. In many cases stacks in walls are brought up to the next floor level, where they are run horizontally to a new position in a second-story partition wall. In such cases, the floor joist must line up with the wall studs. The carpenters solve these and similar problems by doubling joists, introducing new studs, and tying partitions together with metal straps where soles and plates are cut. Carpenters must frequently improvise so that they maintain a structure which is strong and sound.

Plank and Beam Framing

Trends in modern design have been toward large open areas, with few partitions, and extensive use of glass on outside walls and wood as interior finish material for walls and ceilings. Plank and beam framing, shown in Fig. 4-81, is admirably suited to provide these effects. Posts are usually set at regular intervals to support roof beams and the ridge beam. These members are usually large dimension timbers or laminated members, such as 3×8, 4×8, 4×10 inches in cross section. The roof beams support 2 inch or 3 inch (nominal size) decking. The floor can be made from 2 inch planking that is supported by floor beams. This structural system is dependent on the thick planking which serves a structural function. The spacing between beams can be 6 feet, 8 feet, or even greater, depending on the load requirements.

Houses which use this framing method must be carefully designed. There are limitations caused by the regular intervals at which the posts are located. Partitions and openings must fall into this pattern. The carpenter must work especially carefully because the fitting of every

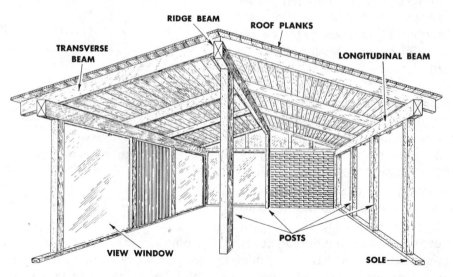

Fig. 4-81. Plank and beam framing requires careful placement of posts and beams. The roof planks must be strong enough to span the distance between beams.

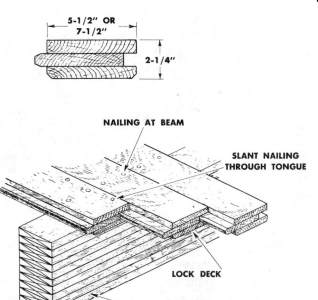

Fig. 4-82. Laminated decking and beams replace solid decking and solid beams for some applications.

Fig. 4-83. Laminated beams, decking, and posts provide great strength. (Potlatch Corp.)

member in the walls and roof is exposed to view from inside the house.

Plank and beam framing is generally con-structed using solid decking, 1½ or 2½ inches thick, and large dimension solid timbers for posts and floor and roof beams. However, adaptations of other materials have been developed for use with plank and beam framing and other framing systems. Both laminated decking and laminated beams are available. Laminated decking consists of several pieces of plank glued together in an electronic gluing process which joins the pieces together and provides great strength. The decking comes in 5½ or 7½ inch widths and 2¼, 2⅝ inch, or greater thicknesses. It is available in lengths from 6 to 16 feet long (at one foot intervals) and is end matched. See Fig. 4-82.

Laminated beams and trusses have been used for many years in heavy construction but are usually made to order for specific jobs. However, laminated beams are now available in a number of sizes for use in light construction. They are made up of a number of planks glued together to make up the desired dimensions. See Fig. 4-82. Laminated members have structural qualities which far surpass those of solid members. They range in size from approximately 3 × 3 inches to approximately 11 × 21 inches. The moisture content of the laminating lumber is carefully controlled so that the members resist shrinkage when put to use. The material is selected so that it has excellent appearance. Posts ranging from approximately 3 × 3 to 5 × 8 inches are made in the same way. A laminated post and laminated decking and beams are shown in Fig. 4-83.

Wood Framing for a Masonry Building

Wood members are used for the internal structure of a masonry building in fundamentally the same way as for a frame building. However, when a brick house or one designed to use other masonry units is built, timing becomes a factor because the carpenter must be ready to set window and door frames as soon as the walls rise to the designated floor level. (The setting of window and door frames is discussed in Chapter 6, *Exterior Finish*.)

Joists are installed so that they rest on the masonry and on a steel beam or wood girder in the same manner as used for a frame building. The ends of the joists are cut on an angle to make a *fire cut,* as shown in Fig. 4-84. In event of a severe fire, the joists will burn through or fail

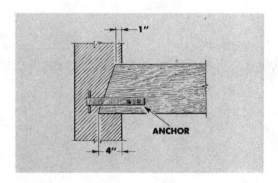

Fig. 4-84. The fire cut on joists permits the joists to fall without destroying the wall.

and be more likely to fall into the building without destroying the wall. If joists were not provided with a fire cut, the upper corner of the joist would act as a lever, forcing the wall to fall outward and endangering the lives of persons in the vicinity. Anchors are provided on joists in accord with the building code which governs construction in the area. Anchors here are metal straps which are fastened a little above the bottom of the joist. In case of fire, they will pull out of the wall relatively easily and thus do as little damage as possible to the wall itself. A short bar passes through the anchor and is embedded in the wall.

A typical sectional view of ordinary brick construction is shown in Fig. 4-85. The first floor joist rests on a row of brick on a 12 inch wall and on an S or W beam. The rough floor is laid as soon as the joists are in position in order to provide a platform to work on. When the masonry reaches a height so that the second floor joists can be installed, the carpenter makes the bearing partitions and erects them, using the same tilt up method used for a frame house. Joists are then put in place so that they rest on the masonry and the double plate of the bearing partition.

The joists for the ceiling at the roof generally have a smaller dimension than the joists for the floors because their main function is to carry the weight of the ceiling and attic insulation. Short struts rise from the bearing wall plate to support the roof joists at their centers. These struts vary in length so that they provide a slope to the roof for drainage.

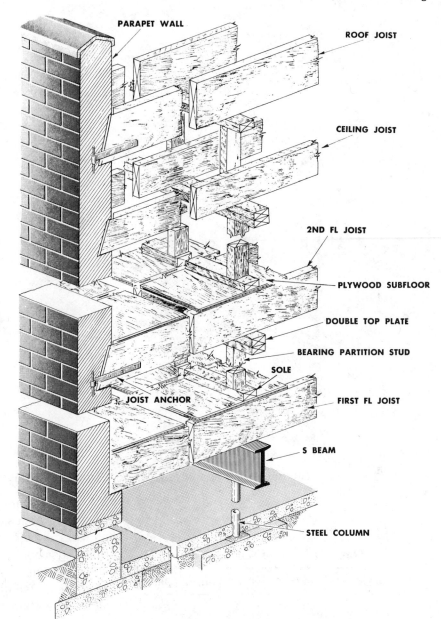

Fig. 4-85. Typical ordinary construction for a masonry building consists of wood joists and wood bearing partitions.

Labels on figure: PARAPET WALL, ROOF JOIST, CEILING JOIST, 2ND FL JOIST, PLYWOOD SUBFLOOR, DOUBLE TOP PLATE, BEARING PARTITION STUD, SOLE, JOIST ANCHOR, FIRST FL JOIST, S BEAM, STEEL COLUMN

Wood Framing for a Masonry Veneer Building

Basically, a masonry veneer house is a frame house with a skin of brick or stone. A few problems which should be given careful consideration involve the shrinkage of lumber, tying the masonry wall to the frame wall, and providing for the control of condensation within the wall.

The foundation is usually made with a top wide enough to support the sill of the wood frame-work, and a shelf to support the brick wall. Platform (western) framing is generally used. See Fig. 4-86. A ¾ or 1 inch space is provided between the frame wall plywood sheathing and the masonry wall. Corrugated galvanized ties are placed in the mortar joints and fastened to the wall to tie the two parts together. They are made so that they allow for a slight amount of vertical adjustment. The Uniform Building Code requires that the ties be placed so that the brick wall is

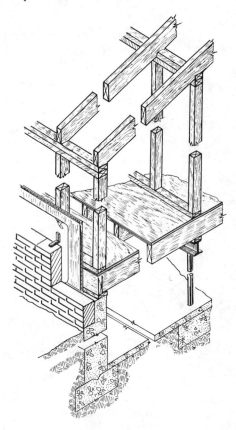

Fig. 4-86. A brick veneer house uses platform framing.

anchored to studs. Suggested tie spacings are given in Table 4-1.

TABLE 4-1. TIE SPACINGS.

Horizontal	Vertical (3″ Brick)	Vertical (Other)
32″ O.C.	12″ O.C.	16″ O.C.
24″ O.C.	16″ O.C.	20″ O.C.
16″ O.C.	24″ O.C.	24″ O.C.

The space between the walls provides a means for ventilating the wall and prevents the accumulation of moisture. Sheathing paper and flashing protect the wood. Weep holes allow the condensed moisture to escape. See Fig. 4-87. Some building codes require that asphalt impregnated fiberboard be used instead of an air space between the masonry and wood walls.

Structural Design

Every carpenter should have some knowledge of structural design for a variety of reasons. Architects' drawings of residences and other small buildings usually don't show the arrangement of structural members in floors, walls, and roofs. Much remodeling work is done without the services of architects. The carpenter in many instances must decide on the size and placement of members and make other decisions which enhance the strength of members and the assembly of members. Almost all temporary construction (such as, scaffolding, barricades, sidewalk sheds, form bracing, shoring, etc.) is built by the carpenter and should follow good principles of design and applicable safety codes.

However, the design of the building, the size and spacing of columns, girders, posts, joists, and other supporting members are all strictly the job of the architects and designers. It is their responsibility to plan a safe and sound structure.

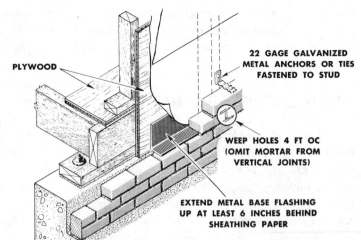

Fig. 4-87. Special precautions are used in a brick veneer wall to provide for condensation.

PLYWOOD

22 GAGE GALVANIZED METAL ANCHORS OR TIES FASTENED TO STUD

WEEP HOLES 4 FT OC (OMIT MORTAR FROM VERTICAL JOINTS)

EXTEND METAL BASE FLASHING UP AT LEAST 6 INCHES BEHIND SHEATHING PAPER

Furthermore, a carpenter or a builder cannot be expected to be qualified to make the calculations needed to determine the correct information. It is sufficient if the carpenters and builders know how wood members react under certain loads and have some understanding of how to read tables of maximum allowable spans for wood floor joists, ceiling joists, and rafters.

Types of Stresses

There are four types of stress which act on or in structural members. These are *tension, compression, shear,* and *bending.* See Fig. 4-88.

Tension is a stress in a structural member which tends to stretch the member or make it longer. An example of structural tension is vertical support for a suspended ceiling.

Compression is a stress in a structural member. It is caused by a load which tends to compress a structural member or make it shorter. An example of compression is a wood post that supports a girder.

Shear results when two forces act on a body in opposite directions in parallel adjacent planes (like a pair of shears), tending to make the fibers of the two portions slide past one another. For example, the end of a joist resting on a brick wall is under stress which tends to cut the end of the joist off. The weight on the joist is one force, and resistance of the wall to change is the other force.

Shearing stress parallel to the grain results when the strain is such that it tends to make the member split lengthwise. For example, when a wood beam is excessively loaded, it bends and tends to split along its length.

Bending results when external and internal forces act on a horizontal member, tending to

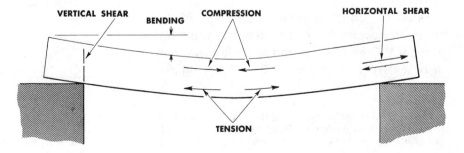

Fig. 4-88. Stresses in a simple beam are shown above.

VERTICAL SHEAR

BENDING

COMPRESSION

HORIZONTAL SHEAR

TENSION

cause it to deflect. For example, girders and joists are subject to bending of their own weight as well as the live loads imposed by people, furniture, etc.

USING ENGINEERING DATA

This part of the chapter points out the importance of following the working drawings and specifications. The architect or an engineer makes complex calculations to arrive at structural information. The carpenters on the job must not change the size of members or species of wood without good authority. The carpenters must have a genuine respect for specific directions.

Carpenters, however, must develop a sense of structural values and limitations. They are responsible for solving problems which cannot be shown in detail on the working drawings. They also must use good judgement when they are responsible for remodeling jobs, building scaffolding, and in other situations which do not involve an engineer or an architect.

Interpreting Stress Tables

Testing machines have been used to determine the load bearing characteristics and the point of failure of wood structural members. Each species and grade of wood generally used for structural purposes has been tested. The results have been compiled into tables which are included in building codes. The building commission in each community requires strict observance of the limitations on the spans of members. Table 4-2 *(Working Stress for Joists and Rafters)* and Table 4-3 *(Allowable Spans for Floor Joists)* are typical pages taken from *Dwelling Construction Under the Uniform Building Code.*[1]

Table 4-2 gives information on some of the common species of wood used for structural purposes. The *allowable unit stress* is expressed in pounds per square inch. The actual figures

---

[1]Dwelling Construction Under the Uniform Building Code, (International Conference of Building Officials, 5360 S. Workman Mill Road, Whittier, Calif. 90601)

were determined by extensive experimentation. The *modulus of elasticity* is the relationship of the unit stress to unit elongation (length). When a member is placed under stress, it elongates at a uniform rate until it reaches its limit of elasticity. After this point has been reached, the member no longer returns to its original length. In Table 4-2, the modulus of elasticity is expressed in millions of pounds per square inch. For example, the modulus of elasticity or Douglas fir—larch is actually 1,500,000 pounds per square inch.

Table 4-3 gives the allowable span for floor joists. Other tables provide information for roof members. The modulus of elasticity is shown as column headings. The size and spacing of the members is shown at the left. Two numbers appear in each box. The number above is the maximum span allowed, expressed in feet and inches. The number below is the allowable unit stress in bending, which is related to a column appearing in Table 4-2.

The problems which appear below show you how to use these tables to determine the maximum size or span for floor joists in given situations.

Problem 1: What size structural grade Douglas fir is needed to span 15'-0", with floor joists 16" O.C.? (Note: *Structural grade* is the grade commonly used for rough framework. It must comply with the American Lumber Standards of the National Bureau of Standards.).

Table 4-2 indicates that structural grade Douglas fir members have a bending stress limit of 1200 and a modulus of elasticity of 1.5. Now if you refer to Table 4-3, under the Modulus of Elasticity column headed *1.5,* you will notice the following information:

Size	Spacing	Allowable Span	Allowable Bending Stress
2 × 8	16" O.C.	12'-7"	1200
2 × 10	16" O.C.	16'-0"	1200

Therefore, the 2 × 10 inch joist, which can span up to 16 feet, meets the requirements here.

TABLE 4-2. WORKING STRESSES FOR JOISTS AND RAFTERS—VISUAL GRADING.

These "F_b" values are for use where repetitive members are spaced not more than 24 inches. For wider spacing, the "F_b" values should be reduced 13 percent.

Values for surfaced dry or surfaced green lumber apply at 19 percent maximum moisture content in use.

Species and Grade	Size	Allowable Unit Stress In Bending "F_b"			Modulus of Elasticity "E" 1×10^6 psi
		Floor Members	Roof Members		
			Snow Loading	No Snow Loading	
BALSAM FIR (Surfaced dry or surfaced green)					
Construction		800	920	1000	.9
Standard	2×4	450	520	560	.9
Utility		200	230	250	.9
Studs		600	690	750	.9
No. 1 & Appearance	2×6	1150	1320	1440	1.2
No. 2	and	950	1090	1190	1.1
No. 3	wider	550	630	690	.9
CALIFORNIA REDWOOD (Surfaced dry or surfaced green)					
Construction		950	1090	1190	.9
Standard	2×4	550	630	690	.9
Utility		250	290	310	.9
Studs		700	800	880	.9
No. 1		1700	1960	2120	1.4
No. 1, Open grain	2×6	1350	1550	1690	1.1
No. 2	and	1400	1610	1750	1.3
No. 2, Open grain	wider	1100	1260	1370	1.0
No. 3		800	920	1000	1.1
No. 3, Open grain		650	750	810	.9
DOUGLAS FIR—LARCH (Surfaced dry or surfaced green)					
Construction		1200	1380	1500	1.5
Standard	2×4	675	780	840	1.5
Utility		325	370	410	1.5
Studs		925	1060	1160	1.5
No. 1 & Appearance	2×6	1750	2010	2190	1.8
Dense No. 2	and	1700	1960	2120	1.7
No. 2	wider	1450	1670	1810	1.7
No. 3		850	980	1060	1.5
DOUGLAS FIR SOUTH (Surfaced dry or surfaced green)					
Construction		1150	1320	1440	1.1
Standard	2×4	650	750	810	1.1
Utility		300	340	380	1.1
Studs		875	1010	1090	1.1
No. 1 & Appearance	2×6	1650	1900	2060	1.4
No. 2	and	1350	1550	1690	1.3
No. 3	wider	800	920	1000	1.1

cont'd

TABLE 4-2. WORKING STRESSES FOR JOISTS AND RAFTERS—VISUAL GRADING. CONT'D.

EASTERN HEMLOCK—TAMARACK (Surfaced dry or surfaced green)

Construction		1050	1210	1310	1.0
Standard	2×4	575	660	720	1.0
Utility		275	320	340	1.0
Studs		800	920	1000	1.0
No. 1 & Appearance	2×6	1500	1720	1880	1.3
No. 2	and	1200	1380	1500	1.1
No. 3	wider	725	830	910	1.0

TABLE 4-3. ALLOWABLE SPANS FOR FLOOR JOISTS 40 LBS. PER SQ. FT. LIVE LOAD.

DESIGN CRITERIA: Deflection—For 40 lbs. per sq. ft. live load. Limited to span in inches divided by 360. Strength—Live load of 40 lbs. per sq. ft. plus dead load of 10 lbs. per sq. ft. determines the required fiber stress value.

Joist size (IN)	spacing (IN)	Modulus of Elasticity, "E", in 1,000,000 psi													
		0.8	0.9	1.0	1.1	1.2	1.3	1.4	1.5	1.6	1.7	1.8	1.9	2.0	2.2
2×6	12.0	8-6 720	8-10 780	9-2 830	9-6 890	9-9 940	10-0 990	10-3 1040	10-6 1090	10-9 1140	10-11 1190	11-2 1230	11-4 1280	11-7 1320	11-11 1410
	16.0	7-9 790	8-0 860	8-4 920	8-7 980	8-10 1040	9-1 1090	9-4 1150	9-6 1200	9-9 1250	9-11 1310	10-2 1360	10-4 1410	10-6 1460	10-10 1550
	24.0	6-9 900	7-0 980	7-3 1050	7-6 1120	7-9 1190	7-11 1250	8-2 1310	8-4 1380	8-6 1440	8-8 1500	8-10 1550	9-0 1610	9-2 1670	9-6 1780
2×8	12.0	11-3 720	11-8 780	12-1 830	12-6 890	12-10 940	13-2 990	13-6 1040	13-10 1090	14-2 1140	14-5 1190	14-8 1230	15-0 1280	15-3 1320	15-9 1410
	16.0	10-2 790	10-7 850	11-0 920	11-4 980	11-8 1040	12-0 1090	12-3 1150	12-7 1200	12-10 1250	13-1 1310	13-4 1360	13-7 1410	13-10 1460	14-3 1550
	24.0	8-11 900	9-3 980	9-7 1050	9-11 1120	10-2 1190	10-6 1250	10-9 1310	11-0 1380	11-3 1440	11-5 1500	11-8 1550	11-11 1610	12-1 1670	12-6 1780
2×10	12.0	14-4 720	14-11 780	15-5 830	15-11 890	16-5 940	16-10 990	17-3 1040	17-8 1090	18-0 1140	18-5 1190	18-9 1230	19-1 1280	19-5 1320	20-1 1410
	16.0	13-0 790	13-6 850	14-0 920	14-6 980	14-11 1040	15-3 1090	15-8 1150	16-0 1200	16-5 1250	16-9 1310	17-0 1360	17-4 1410	17-8 1460	18-3 1550
	24.0	11-4 900	11-10 980	12-3 1050	12-8 1120	13-0 1190	13-4 1250	13-8 1310	14-0 1380	14-4 1440	14-7 1500	14-11 1550	15-2 1610	15-5 1670	15-11 1780
2×12	12.0	17-5 720	18-1 780	18-9 830	19-4 890	19-11 940	20-6 990	21-0 1040	21-6 1090	21-11 1140	22-5 1190	22-10 1230	23-3 1280	23-7 1320	24-5 1410
	16.0	15-10 790	16-5 860	17-0 920	17-7 980	18-1 1040	18-7 1090	19-1 1150	19-6 1200	19-11 1250	20-4 1310	20-9 1360	21-1 1410	21-6 1460	22-2 1550
	24.0	13-10 900	14-4 980	14-11 1050	15-4 1120	15-10 1190	16-3 1250	16-8 1310	17-0 1380	17-5 1440	17-9 1500	18-1 1550	18-5 1610	18-9 1670	19-4 1780

[1] The required extreme fiber stress in bending, F_b, in pounds per square inch is shown below each span.

[2] Use single or repetitive member bending stress values (F_b) and modules of elasticity values (E), from Table No. 25-A-1.

[3] For more comprehensive tables covering a broader range of bending stress values (F_b) and Modulus of Elasticity values (E), other spacing of members and other conditions of loading, see the Uniform Building Code.

[4] The spans in these tables are intended for use in covered structures or where moisture content in use does not exceed 19 percent.

Problem 2: What is the maximum allowable span for eastern hemlock joists with 2 × 10 inch dimensions, spaced 16″ O.C.?

Table 4-2 indicates that eastern hemlock members have a bending stress limit of 1050 and a modulus of elasticity of 1.0. If you refer to Table 4-3 under the Modulus of Elasticity column headed *1.0,* you will note that the allowable span is 14′-0″. The bending stress limit for a 14′-0″ span is 920, which is well below the limit of 1050 set in Table 4-2.

Therefore the maximum allowable span for the eastern hemlock joists is 14′-0″.

FRAMING FOR ENERGY CONSERVATION

The search for ways to conserve energy has brought about considerable experimentation on the part of manufacturers of insulating materials and architects and builders. One solution is to use styrofoam (polystyrene) sheets in the place of sheathing. Another is to make structural changes which permit the use of thicker layers of insulation.

Increased insulation value is added to outside walls by substituting one inch thick rigid sheets of styrofoam for conventional insulation board or plywood sheathing. The sheets of styrofoam have tongue and groove edges which enable them to fit tightly. They are applied to fit up to the top of the top plate, carefully cut out at openings and extended down over the foundation. See Fig. 4-89. Wood or metal horizontal siding is applied directly over the foam board and nailed through it into the studs. The styrofoam sheathing which is exposed below the bottom piece of siding is covered with protective material, such as mineral fiber (cement asbestos) sheets.

The same styrofoam sheets are applied over the ceiling to the underside of ceiling joists or trussed rafters. Drywall finish is applied to the

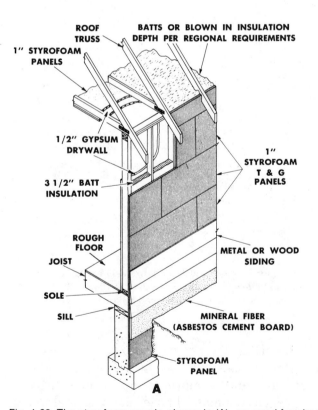

Fig. 4-89. The styrofoam panels, shown in (A), are used for sheathing and over the ceiling to prevent heat loss. The continuous envelope of tongue and groove styrofoam insulation (B) extends from roofline to frostline and provides energy savings. The insulation extends below the grade. (Dow Chemical Co.)

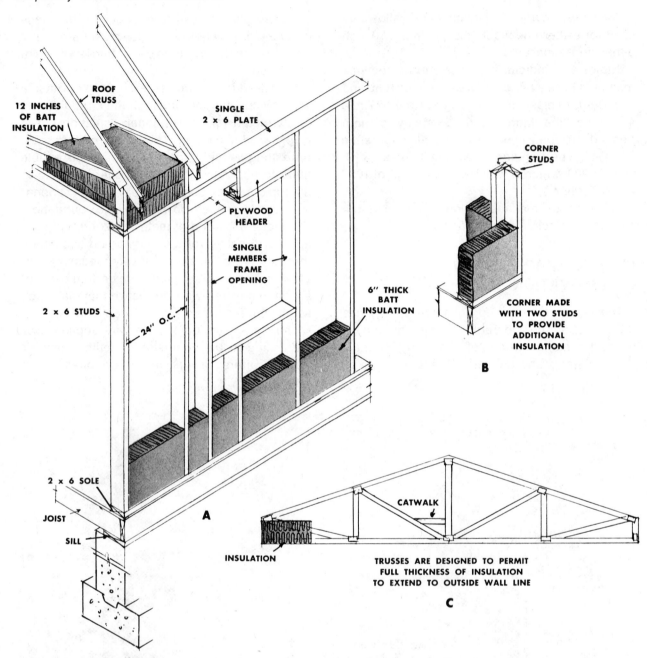

Fig. 4-90. Innovations in framing provide for greater thicknesses of insulation in exterior walls and over ceilings. Batts 6 inches thick are compressed into the spaces between 2 × 6 inch studs. Trusses are designed to make room for a 12 inch thickness of insulation out to the wall line. Applications shown here include the framing and insulation for wall and ceiling (A), at corners (B), and with trusses (C).

styrofoam sheets with nails or screws driven into the wood members above.

One advantage of increasing the insulation value of the home in this manner is that conventional framing discussed in this chapter can be used throughout. The insulation within the wall

is the usual 3½ inch batts of fiberglass or other material. Insulation over the ceiling is made up of batts and/or blown-in type insulation. The thickness or depth of the insulation is prescribed for the particular region.

Another solution for minimizing heat loss

through walls and roofs uses 6 inch insulation batts in the exterior walls and a full 12 inches of insulation over the ceiling. However, this method requires some changes in the rough framing. Studs are increased from 2 × 4 inch to 2 × 6 inch and placed on 24 inch centers. Because of the added structural value of the members, only single members are used at each side of openings. Headers are simplified, and only two studs are used at corners. See Fig. 4-90. Headers are made with plywood sheets filling the header opening. They are installed on the outside of the wall so that insulating batts can be applied behind them. Two studs instead of the usual three are used at corners so that insulation can extend into the corner.

Trusses used to support the roof are designed to provide an opportunity for the full thickness of the insulation to extend to the outside wall. Two thicknesses of 6 inch batts or one thickness of 6 inch batt and 6 inches of blown-in material can be used. The thickness would vary as required for the particular region.

QUESTIONS FOR STUDY AND DISCUSSION

1. What are some of the advantages of platform framing?
2. What are some distinctive features of plank and beam framing?
3. What is a box sill?
4. How can the wood members near the ground be protected from termites?
5. What is a built-up girder?
6. How is framing done around a stairwell?
7. When are joists under partitions doubled?
8. How are plywood sheets arranged when they are laid as rough flooring?
9. What is solid bridging? What is compression type bridging?
10. What are the advantages of floor trusses?
11. How is a channel made for a wall and partition intersection?
12. What principles are involved when you frame for a window opening?
13. What determines the size of members used for headers over openings in walls?
14. What are some advantages of using plywood for exterior wall sheathing?
15. Why is the nailing schedule so important for fastening plywood sheathing?
16. When is diagonal bracing required in outside frame walls?
17. What requirements must be followed when you are installing wall bracing?
18. Explain the 24 inch module system of framing.
19. Name the advantages of using metal studs instead of wood studs?
20. Give the procedures for fastening gypsum wallboards to steel studs.
21. What is the fire rating for gypsum wallboard?
22. What is suggested for cutting down sound transmission in a wood stud wall with drywall application on both sides?
23. What are plaster grounds?
24. How are floors framed around chimneys?
25. Where may holes be drilled in joists for pipes?
26. How is laminated decking used in plank and beam framing?
27. What is a fire cut on a joist?
28. How is a brick veneer wall protected from condensation?
29. What framing techniques can be used to save energy?
30. What are tension, compression, shear and bending stresses?

CHAPTER 4-1 PRACTICE IN WALL AND FLOOR FRAMING

The subject of rough framing covering each of its elements is included in detail in Chapter 4, *Wall and Floor Framing.* In order to apply this background to the building of houses, we have chosen two structures to analyze the carpenter's work in actual situations. The preliminary steps used to prepare for the job, the layout and cutting of the members, and the assembly of members are covered. The first house to be studied is House Plan A, shown in the working drawings at the end of Chapter 1, *Preparing for the Job.* You are already familiar with these drawings because they were used earlier in Chapter 3-1, *Practice in Concrete Formwork,* to provide information for the construction of footings and foundation walls and placing the slab.

You should understand from the outset that there are various ways to build a house from a particular set of working drawings. Conventional good framing practices, discussed in Chapter 4

(Wall and Floor Framing), establish general procedure. The architect does not usually show how the framing is done beyond providing the information shown on the section drawing of the working drawings. (See Sheet 5, Section A-A.) Therefore the framing may also vary with the background and experience of the builder, regional practices, and the limitations set up by the local building code. Some builders use components such as wall and partition sections which are made away from the jobsite. The procedures which follow are not intended to suggest a particular best way to do the job. They simply serve to point out how the job can be efficiently done and to explain why some of the steps are taken. Whatever method is chosen, you must study the prints of the working drawings carefully and try to visualize how the building framework should look at various stages, such as shown in Fig. 4-91 and Fig. 4-92.

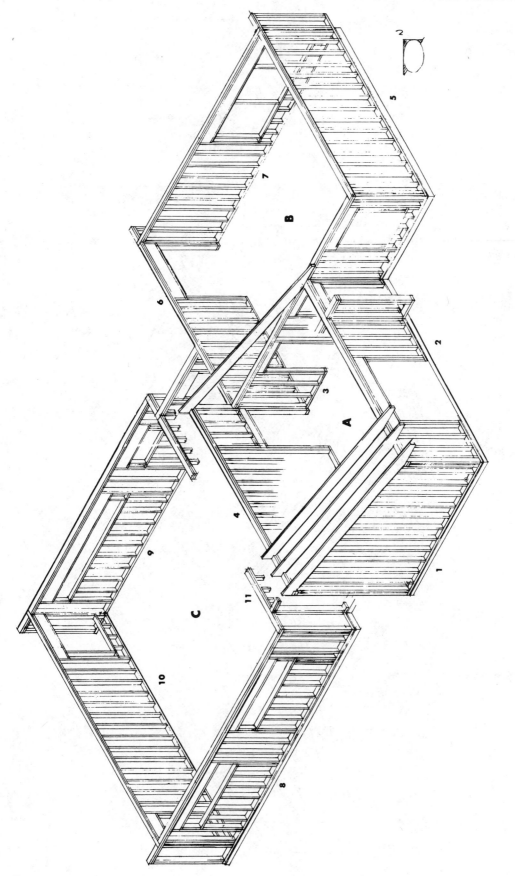

Fig. 4-91. The rough framing requires the accurate placing of studs and the placing of rough openings.

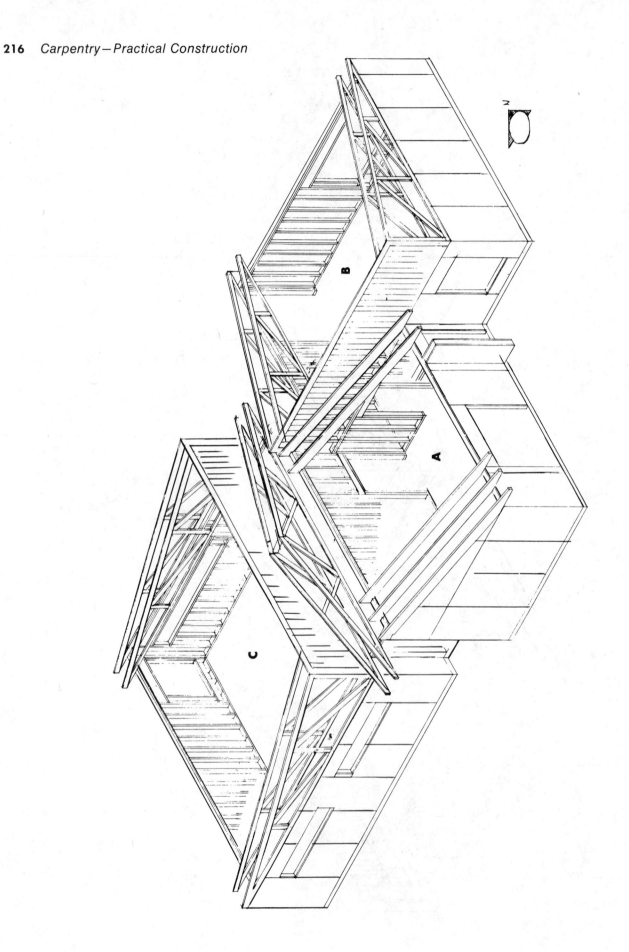

Fig. 4-92. This view of the framing for House Plan A includes the sheathing, trusses, and knee walls.

FRAMING HOUSE PLAN A

Framing House Plan A

Reading the working drawings. Very little information about the framework is given on the working drawings. You must be able to adapt the general procedures outlined in Chapter 4 to specific situations. These include building a structure with dimensions that are consistent with the drawings, placing openings where they are required, making ceilings at the right height, and building all other features as indicated. The time you spend studying every item on the working drawings which in anyway concerns your job pays off. It saves time in the long run and gives you the confidence to make decisions as the work progresses.

Exterior walls. One approach to building the house is to temporarily forget the windows and doors and imagine the house as a box consisting of outside walls and a roof. House Plan A (Fig. 4-91) designates various areas (or volumes) by using letters *A, B,* and *C.* The sheets of the working drawings most useful at this point are Sheet 1 (the Floor Plan) and Sheet 5 (Section A-A). This plan view (Floor Plan) shows the length and location of all exterior walls. (You have already used these dimensions when you set up the building lines in Chapter 3-1.) You will find the location of the section line for Section A-A on Sheet 1, the First Floor Plan. This line is an east-west line, passing through the bedroom area, the Entry, and the Family Room. Sheet 5 (Section A-A) shows the height of walls, measuring from the slab to the top of the top plate, is 8'-1". The trusses over areas *B* and *C* are shown in Fig. 4-92. The ceilings in these areas are hung from the trusses at the 8'-1'' height.

The ceiling in the family room (area *A* on Fig. 4-91) is sloped with rafters which support the ceiling rather than the trusses. The wall marked *4* in Fig. 4-91 is extended up to carry the upper end of the rafters by using long studs the entire distance. Looking at Section A-A, Sheet 5 of the working drawings, you will find the dimension from the slab to the top of the top plate is 12'-1".

The other walls which extend above the 8'-1" ceiling line at the face of trusses over areas *B*

and *C* do not need framework because the vertical truss members, instead of studs, provide the nailing base for the sheathing. Sheet 5, Section A-A of the working drawings, shows this dimension for area *C* is 7'-0" above the top of the top plate.

Windows and doors. Windows and doors must be located in two directions—vertically, above the slab, and horizontally, from building corners. The Window Schedule on Sheet 6 of the working drawings shows that there are two types of windows. They are marked *1* and *2.* The Door Schedule on the same sheet shows that all doors, except the louvered kitchen doors (these are swinging cafe type doors), are 6'-8" high. Sheet 3, Front and Left Side Elevations, shows that the rough openings for windows and doors are 6'-10½'' above the slab. You will be able to calculate the rough opening height once we discuss the layout of wall sections.

Locating windows and doors horizontally requires a study of Sheet 1 (the Floor Plan) and Sheets 3 and 4, which show the various elevation views. All of the exterior doors, including number *4* in the Dining Room and number *5* in the Family Room, are centered on the walls. Windows marked *1* in the bedroom area are shown on the front and rear elevation views. They are shown in a continuous ribbon with a plywood panel equivalent to one sash, filling in a blank space. You will center the whole ribbon of windows in the wall, measuring on the outside of the wall. Windows marked *2* in the bedrooms 1 and 2 and in the North wall of the living room are placed so that the casing fits against the wall projection without space for any siding. See Sheet 1 of the working drawings. The window marked *2* on the South side of the living room is placed 2'-0'' away from the outside corner. This is one instance where it is okay for you to scale a print. (If you do not have an architect's scale, you can use your pocket rule. One quarter inch on your rule is equal to one foot on the actual structure. One sixteenth of an inch on your rule is equal to 3 inches.)

Partitions. Partitions are located with dimensions to the building lines of outside walls and to centers of partitions. In a few cases, dimensions are given to the face of the finished partitions, particularly at closets. These dimensions must be carefully followed when chalk lines are snapped on the floor so that all of the room sizes, doorways, and closets will work out correctly. You must refer to the Door Schedule to find the width of doors so that you can calculate the rough opening. Then you must check how the door is to swing and how the frame is placed in the opening with an allowance for trim.

Ceiling and roof structure. Sheet 5, Section A-A of the working drawings, shows the ceiling over area *C* is hung from the roof trusses. You can assume that the same is true for the ceiling over area *B*. See Fig. 4-92. The ceiling for area *A* is formed by fastening gypsum dry wall sheets to the bottom of the rafters.

The roofs over areas *B* and *C* are supported by pre-engineered trusses made away from the job in a truss fabrication plant. Your only concern is to put them in place and fasten them. The rafters over the family room (area *A*) are simple shed roof rafters. Chapter 5 discusses rafter layout.

Procedure for sills. The sills are important because they establish the faces of exterior walls and the location of partitions. Figure 4-93 shows where the sills are placed for areas *A, B,* and *C*. The numbers for wall parts shown in Fig. 4-93 are arranged in sequence in the order they are to be erected. Another sequence may be preferred. However, you must decide how the job is to be done so that the corners work out when the sections are tipped up into position. All of the wall or partition sections can be erected in one piece provided there are three people available to lift

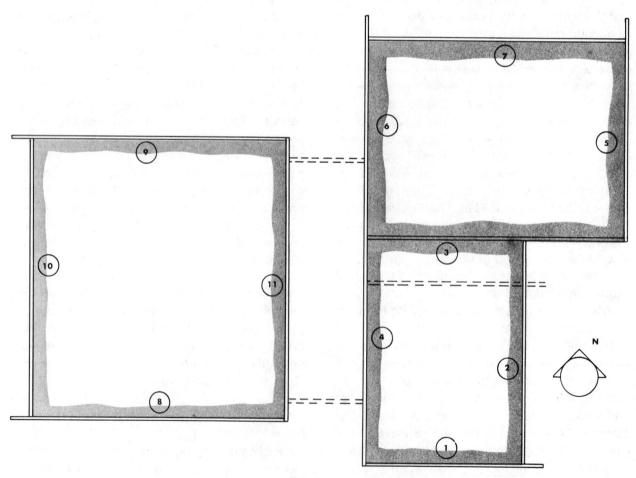

Fig. 4-93. Exterior walls are built in sequence, as shown above, so that the corners work out for erection.

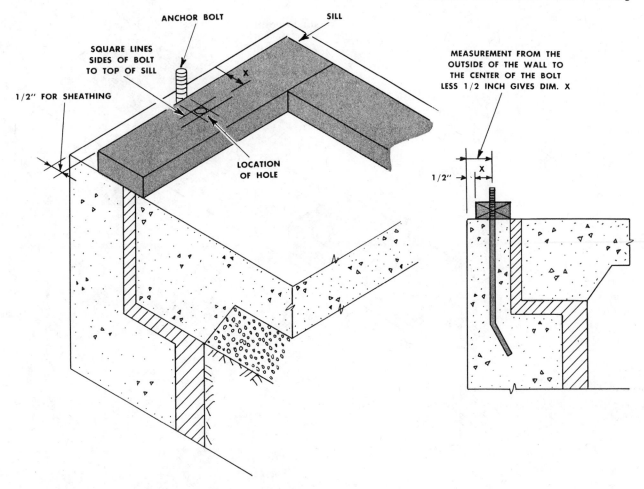

SQUARE LINES SIDES OF BOLT TO TOP OF SILL

ANCHOR BOLT

SILL

1/2" FOR SHEATHING

LOCATION OF HOLE

MEASUREMENT FROM THE OUTSIDE OF THE WALL TO THE CENTER OF THE BOLT LESS 1/2 INCH GIVES DIM. X

X

1/2"

Fig. 4-94. Anchor bolt holes must be located for each anchor bolt. The detail to the right provides a section view of the anchor bolt in place.

them into place. Anchor bolts used to tie down the sills have been placed along the outside walls and along bearing partition footings when the concrete was placed. They are ½ inch diameter and spaced a maximum of 6 feet O.C.

Wall section *1*, shown on Fig. 4-93, is 18 feet long. Choose a straight piece of 2 × 4 inch material and place it against the anchor bolts, as shown in Fig. 4-94. Square a line across the sill on both sides of each bolt. Measure the distance from the center of a bolt to the face of the concrete wall. Then subtract ½ inch to allow for the sheathing. This gives the distance to the center of the bolt hole. Measure each anchor bolt location to compensate for the fact that they are not in perfect alignment. (You can make a simple

device to do the job without measuring for each bolt. See Fig. 4-95. A piece of metal or a stick with two holes can be used. Place the sill tight against the bolts.)

Measure the distance from the center of the wall to the sill piece. Subtract ½ inch from the thickness of the sheathing and ¼ inch for one half the diameter of the bolt. Mark this measurement on the marker. At one point drill a hole 1/16 inch larger than the bolt diameter and drive a nail at the other point. (If you use a metal piece, drill a small hole for a nail instead.) When you use the marker shown in Fig. 4-95, slip it over each bolt in turn and mark the center by tapping the nail. Remove the sill piece from the wall. Place it on a pair of sawhorses and bore the an-

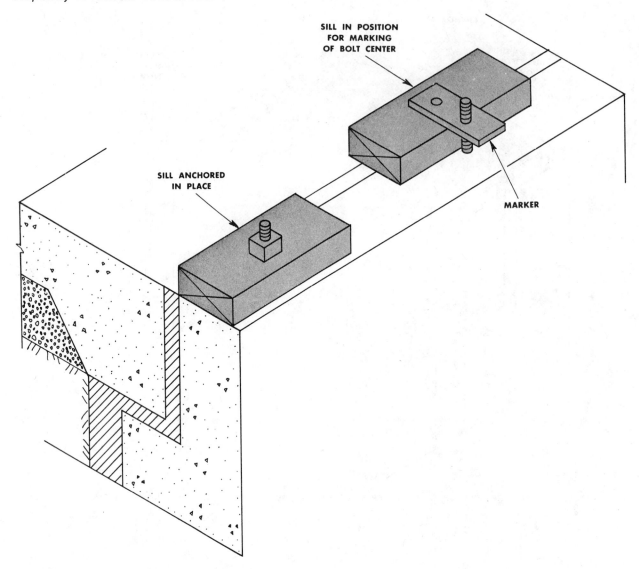

SILL IN POSITION
FOR MARKING
OF BOLT CENTER

SILL ANCHORED
IN PLACE

MARKER

Fig. 4-95. An anchor bolt marker is used to mark sills for anchor bolt location. After the marker is placed, the mark for the center of the bolt is made with a hammer blow on a nail.

chor bolt holes. Choose sill stock for all of the other walls and bearing partitions. Cut them to proper length and bore them to match the anchor bolt locations. Wall sections *4* and *6,* although in line, are separate walls for convenience. Sills for walls *3, 7, 8, 9, 10,* and *11* are too long to be made of one piece. They require two pieces held in line by a scab. (A *scab* is a short piece nailed to the two parts of a butt joint.) Place all sills over anchor bolts to make sure that the corners work out as planned by the erection

sequence. Check the building measurements again after the sills are dropped in place.

Remove sill pieces, spread mortar on the slab at the sill locations, and replace the sills. It may be necessary to drive wedges under the sills at several places to support them until the mortar sets. Check the elevation of the sills using a builders' level at several points to see that they are level all around. Sight along each sill to see that it is straight. Fasten the sills firmly using nuts with washers.

Studs. Some of the things to keep in mind regarding studs are their length, corner arrangement, and spacing. You must also consider the studs in terms of providing for openings, and providing for intersection with secondary partitions.

Stud Length. Study Section A-A, Sheet 5 of the working drawings. You will find that the wall height from the slab to the top of the top plate is 8'-1". To find the length of the studs, subtract 4½ inches for the thickness of the sill and the two top plates.

$$(8'\text{-}1'' - 4\tfrac{1}{2}'' = 7'\text{-}8\tfrac{1}{2}'')$$

All of the studs are 7'-8½" long with the exception of those for wall *4,* (Fig. 4-93), where the studs extend up to support the rafters. The wall height for wall *4* is 12'-1". Subtract 4½ inches for the thickness of the sill and two top plates. (12'-1" − 4½ = 11'-8½") Cut studs for wall *4* exactly 11'-8½" long, using 12 foot material.

Laying out a Master Stud Pattern. Make a master stud pattern (also called a *story pole*) to locate door and window openings. This pattern saves a lot of measure time and insures that accurate dimensions are used. Choose a straight piece of 2 × 4 inch material. Make sure that one end is square. Mark and square a line 8'-1" from the end. Mark off 1½ inches from the one end and two spaces of 1½ inches from the other end and then draw lines. See Fig. 4-96. These

lines represent the sill and plates. Measure 11½ inches from the bottom of the top plate and draw a line to represent the bottom of the header. All doors are 6'-8" high. See Sheet 4-96. The head jamb is 1½ inch stock with a ½ inch rabbet for the door. See Fig. 4-97. Allow ¾ inch for the threshold and ¾ inch for fitting. The jamb here is 1" (1½" − ½" = 1"), and the rough opening is 6'-10½" (6'-8" + ¾" + ¾" + 1" = 6'-10½"). Mark the master stud pattern and draw a line 6'-10½" from the end. See Fig. 4-96.

The types of windows are designated either *1* or *2*. Refer to Sheets 3 and 6 of the working drawings. Sheet 3 shows that for all windows there is a dimension of 6'-10½" for the rough opening above the slab. This coincides with the rough opening for the doors. The windows are packaged windows. (*Packaged windows* are shipped to the job with sash, glass, and hardware installed and with the exterior casing in place.) Vertically, the rough opening only needs to be ¾" more than the frame measurement. (Manufacturers supply this information for their products. Allowing for rough openings has already been discussed in Chapter 4.) Measure the window frames and add ¾" to the vertical dimension for the rough opening. Using the measurement, measure down from the header on the master stud pattern and locate the subsills for the two types of windows. See Fig. 4-96.

Measure 1½ inches and draw the lines representing the subsills. The header over windows

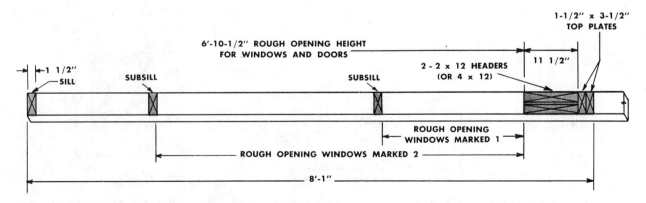

Fig. 4-96. Making a stud pattern (story pole) saves time in stud layout.

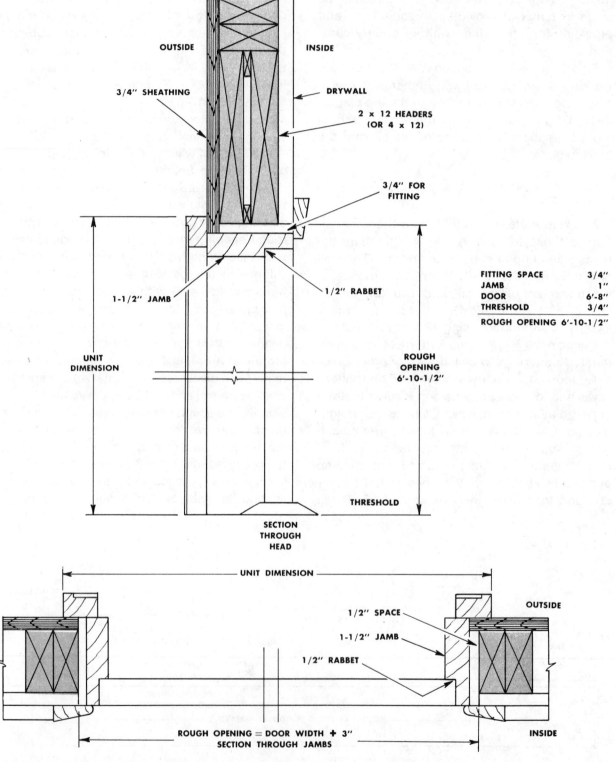

OUTSIDE

INSIDE

3/4" SHEATHING

DRYWALL

2 x 12 HEADERS
(OR 4 x 12)

3/4" FOR
FITTING

FITTING SPACE 3/4"
JAMB 1"
DOOR 6'-8"
THRESHOLD 3/4"

ROUGH OPENING 6'-10-1/2"

1-1/2" JAMB

1/2" RABBET

UNIT
DIMENSION

ROUGH
OPENING
6'-10-1/2"

THRESHOLD

SECTION
THROUGH
HEAD

UNIT DIMENSION

OUTSIDE

1/2" SPACE

1-1/2" JAMB

1/2" RABBET

INSIDE

ROUGH OPENING = DOOR WIDTH + 3"
SECTION THROUGH JAMBS

Fig. 4-97. Here the architect has established the rough opening to accomodate 12 inch headers. In other applications the carpenter establishes it.

and doors works out to be 11½ inches, allowing for the use of two 2 × 12 (or one 4 × 12) inch members for the header.

Arranging the Studs. As a general rule, studs on both outside walls and interior partitions must be placed 16 inches O.C. There may be a few exceptions. Adjustments for placing openings for windows and doors are made by introducing additional studs, rather than changing the spacing. Studs at corners are arranged so that double studs are placed first at the end of the wall to be erected. This permits the adjoining wall to be tilted into place, with the single corner stud in a position to be nailed into the two other studs. The 16 inch spacing of studs is arranged so that a full sheet of 4 foot wide plywood sheathing is able to cover the corner studs and reach to cover one half of the third stud from the corner.

Wall *8* is chosen as an example of stud layout and framing for openings. See Fig. 4-98. (Also see Fig. 4-93 to see the location of wall *8* in relation to the other walls.) Choose pieces of 2 × 4 inch stock and cut them to the same length required to make the sill. Two pieces which total 28 feet are needed to make up the bottom member of the double plate. (It is easier to mark the plate rather than the sill because the sill is already anchored.) Starting from the right-hand end, mark off the double corner members on the edge of the stock with crosses. Measure 4 feet from the corner to locate the center of a stud. Draw ¾″ lines on each side of the center point and mark a cross, indicating that it is a stud. Measure 16 inch dimensions from one of these lines to locate studs along the full length of the sill. Mark crosses for each stud location so that the studs are 16″ O.C. The total measurement for rough openings for the windows plus the blank space is 20′-4″. See the Front Elevation, Sheet 3 of the

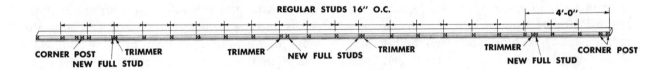

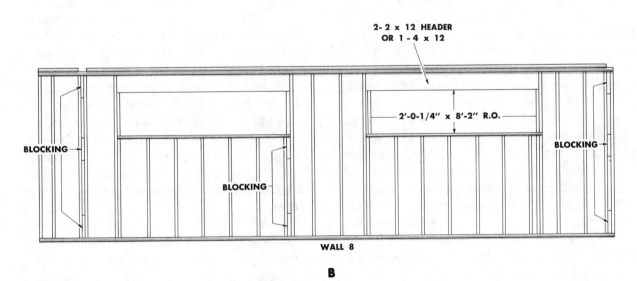

Fig. 4-98. The position of each stud is marked off on the sill and a plate.

working drawings, for a view of the windows. They are centered on the wall. The length of the wall is 28'-0". The length required for the window rough opening is 20'-4".

$$28'\text{-}0'' - 20'\text{-}4'' = 7'8''$$

$$7'\text{-}8'' \div 2 = 3'\text{-}10''$$

Measure 3'-10" from both ends of the wall and locate trimmer studs to frame the rough openings for the windows. Each pair of windows is in a single frame and requires a rough opening of 8'-2". Measure 8'-2" and mark the location of the trimmers alongside the rough openings. Mark positions for additional full studs alongside the trimmers. Mark location of the two studs which form the corner for the west wall. All of the markings are then transferred to the sill.

Lay the plate on the floor opposite the sill and arrange the studs as marked. Cut trimmers and headers as shown in Fig. 4-98. The headers works out to be exactly 11½ inches wide. Cut two pieces of 2 × 12 (or one piece of 4 × 12) inch lumber the designated length. (2 × 12 inch members require ½ inch filler blocks between them so that the assembled header is 3½ inches thick.) Tack boards across the studs near the bottom to keep them in line until the wall is raised into position. Measure the assembly diagonally to see that the wall is square. Tack two diagonal braces across the assembly to keep it square. Tilt the wall up into place. Tack three long 2 × 4 inch members to the sides of studs, one near each end of the wall and one at a middle stud to serve as diagonal bracing to keep it from falling. (Tack them to the studs near the top plate.) Extend them down to stakes driven into the ground about 4 feet out from the wall. Make sure that the wall is plumb before driving the nails into stakes. Line up the position of each stud with the markings on the sill and toenail each one in place.

Frame and erect the opposite wall (wall *9*, Fig. 4-93) and erect it in the same way. Build wall *10*, swing it up into place, and nail the corners together. The top plate is made so the corners lap over each other. See Fig. 4-98. Splices in the plates should fall over studs or headers. The two top plate splices must be staggered at points at least 4 feet apart. Check the Door Schedule for

the door widths and add 3 inches for the rough opening.

Frame and erect the other walls using the same procedure. Wall *4* is built using long studs to support the rafters over the family room. Wood blocking is provided at the 8'-1" height for the application of interior gypsum wallboard. See Sheet 5 of the working drawings.

Interior partitions. A suggested sequence for erecting interior partitions is shown in Fig. 4-99. The main considerations are to provide room for assembling the partitions on the floor, tilting them up conveniently, and tying them together at intersections of walls and other partitions. A channel type intersection, discussed in Chapter 4, is usually possible.

Snap chalk lines on the floor to locate the two sides of all of the partitions marked *12* through *22* on Fig. 4-99. Check room sizes and verify they agree with the dimensions given on the Floor Plan, Sheet 1 of the working drawings. Measure at two places near opposite exterior walls to make sure that the walls are parallel to these walls and the corners are square. Cut all of the sole members and lay them in place. Cut plate members the same length, making provision for lapping corners. Lay the plate members on top of the sill pieces. Mark off stud spaces in the same manner used for the outside walls. The rough opening width allowance (over the actual door width) for doors may be less than 3", depending on the type of jamb used. Begin assembling the longest partition first. Instead of fastening the sill to the floor first, assemble the whole partition including the sill. Swing the partition into place and position it over the chalk lines on the floor. If the partition needs temporary support until the other partitions are placed, use horizontal diagonal members tacked between the top plates of the exterior walls and the partition plate. Fasten the interior partition sills to the slab, using powder activated studs a maximum of 6 feet O.C.

Frame and erect secondary partitions, using channel type intersections. Tie top plates together by lapping. Plumb the partitions before nailing and fastening them to the slab. Short partitions for closet walls are framed, using the same procedures as for other partitions.

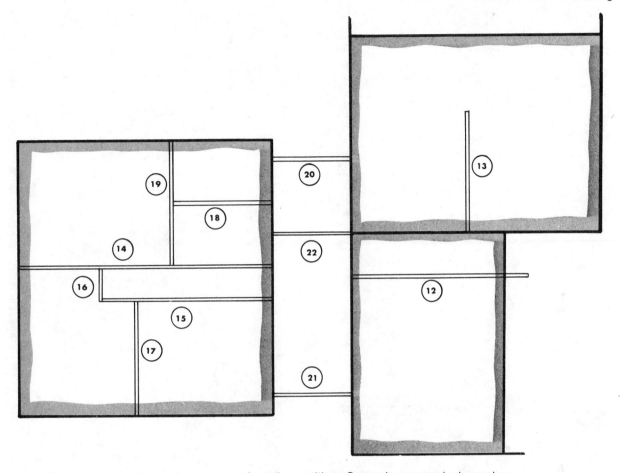

Fig. 4-99. The carpenter decides on the sequence of erecting partitions. One such sequence is shown above.

Trusses and rafters. The erection of trusses and the cutting and installing of rafters is discussed later, in Chapter 5.

Sheathing. Install full size sheets of 4 × 8 foot plywood sheathing to the exterior of the framework. Use the largest sheets you can to fit the odd spaces. The studs have been placed 16 inches O.C. so there is a minimum of cutting. Use 6d nails, spaced 6 inches O.C. on edges and 12 inches O.C. on intermediate studs (or as required by the local building code).

HOUSE PLAN C

Framing House Plan C — Platform Framing

The second house to be studied in detail is House Plan C. It has two stories, a basement, and an attached garage. It is built using platform framing. You should already be familiar with some of the features of this house since you used the working drawings for a study of its foundation formwork in Chapter 3-1, *Practice in Foundation Formwork.* The working drawings are shown at the end of that chapter, pp. 143 to 148.

Reading the working drawings. The first step is to determine the sizes of joist, studs, and other framing members and to discover how the architect intends to have them used in this particular house. The Section View on Sheet 6 of the working drawings shows these details. You are expected to use the conventional procedures discussed in Chapter 4. You should be able to visualize various stages of the construction, as shown in Figs. 4-100, 4-101, and 4-102.

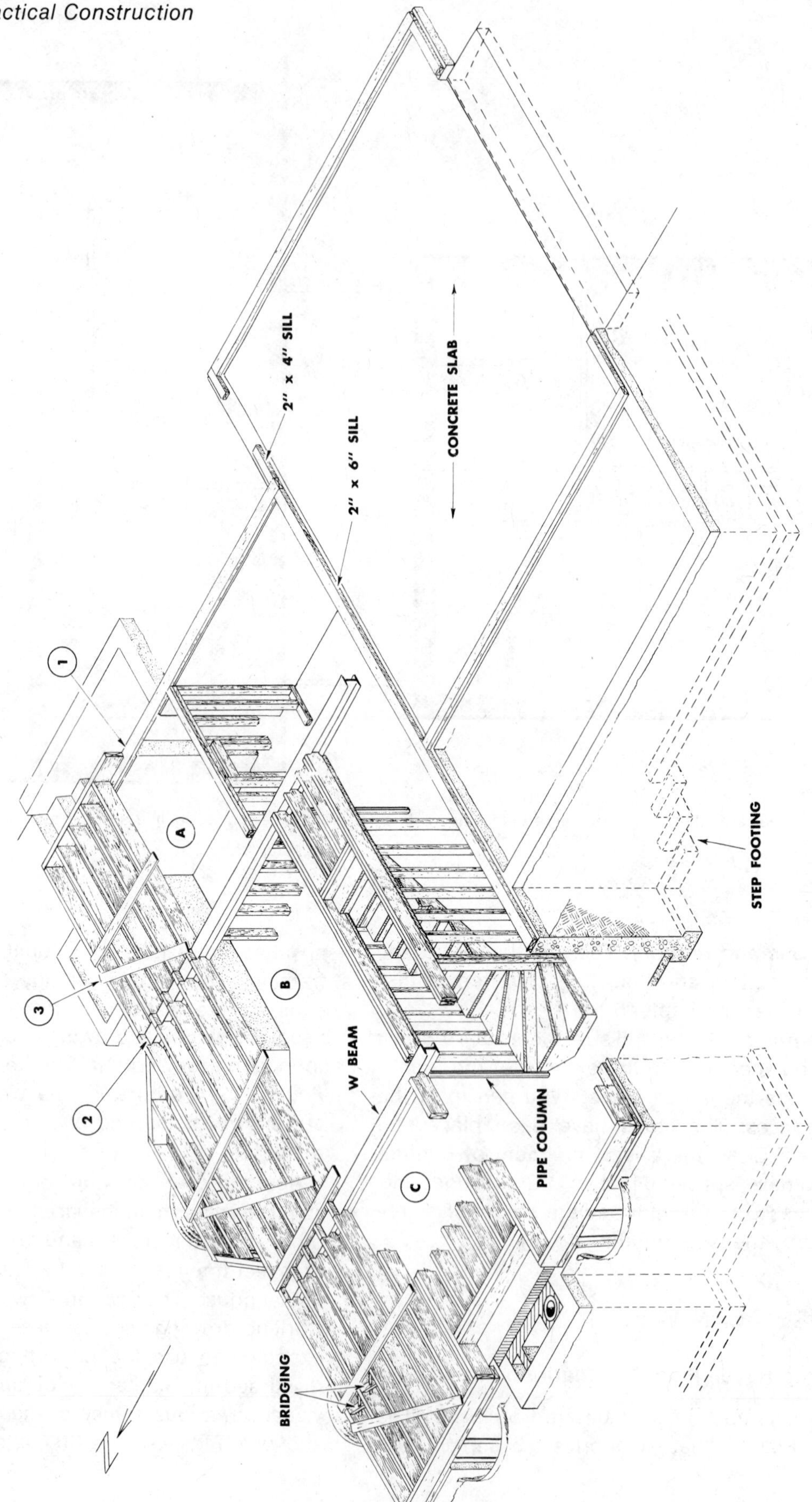

2" x 4" SILL

2" x 6" SILL

CONCRETE SLAB

STEP FOOTING

1

A

3

2

B

C

W BEAM

PIPE COLUMN

BRIDGING

N

Fig. 4-100. House Plan C is framed with platform framing. This isometric view shows the box sill, framing at girders, the openings at the stairs and the chimney, and how the joists are arranged.

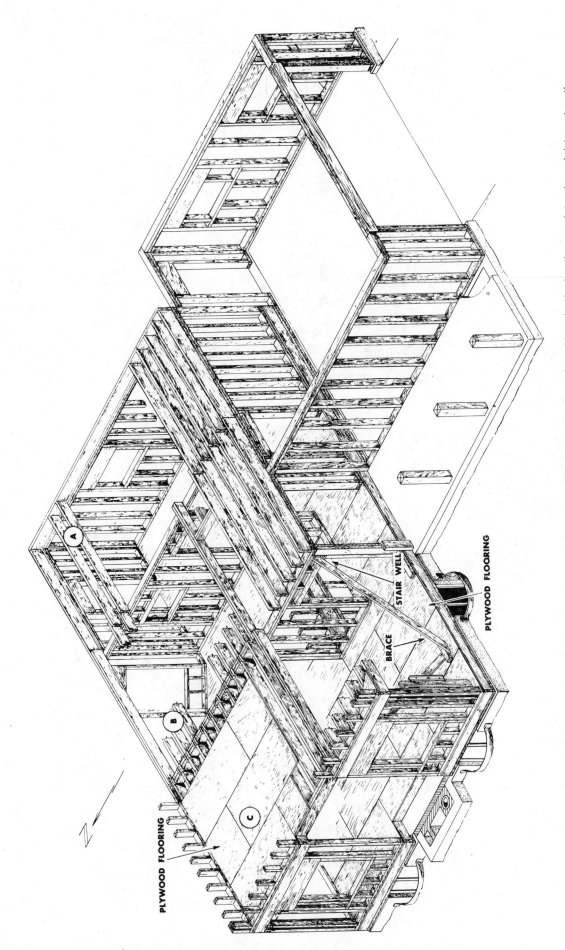

PLYWOOD FLOORING

PLYWOOD FLOORING

BRACE

STAIR WELL

Ⓐ

Ⓑ

Ⓒ

Fig. 4-101. Framing House Plan C with platform framing permits the erection of sections of walls and partitions by tipping them up into place. Joists rest on the top plates. Joists are hung from the girder over living room opening with metal joist hangers.

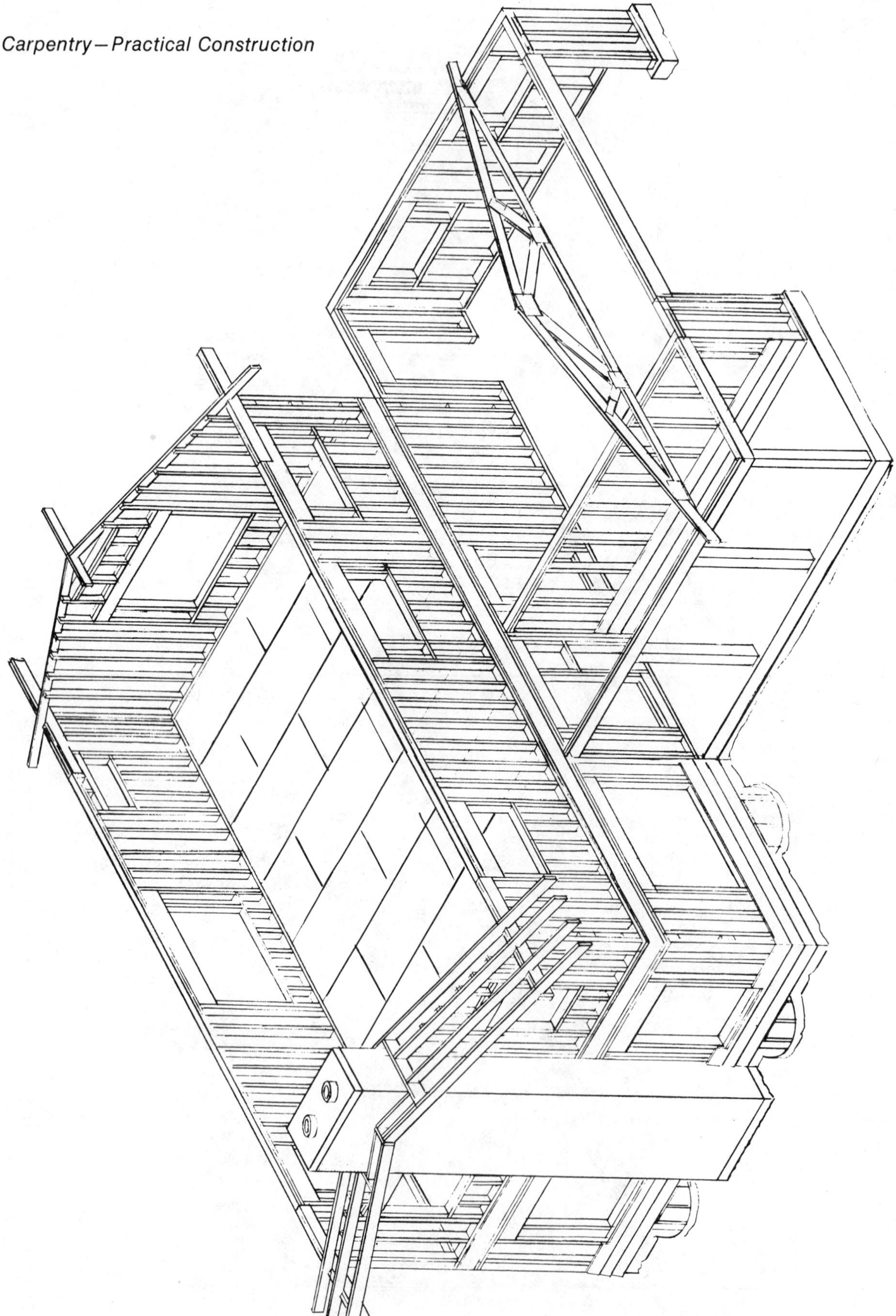

Fig. 4-102. Platform framing provides a platform on which to assemble walls and partitions. Partitions on the second floor are erected after the roof is framed.

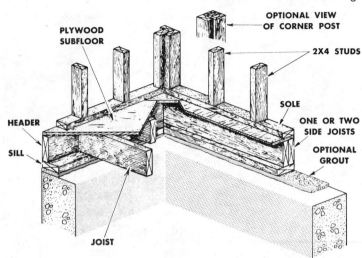

Fig. 4-103. The box sill provides support for the joists.

First Floor Framing

(Note: It is assumed at this point that the concrete foundations is completed and that termite shields have been placed over the top of the foundation walls.)

Sill framing. Begin by setting the sills on the east wall shown at *1,* Fig. 4-100. Then follow around the south and west sides, leaving the placing of the sills on the north side until the last. The type of sill construction used here is shown in Fig. 4-103. This is a typical box sill. A single 2 × 6 inch member is used around the foundation. The sill pieces must be carefully lined up and accurately measured because they establish the faces of the building.

Sheet 1 of the working drawings shows that the west wall is 23′-2″ long. The east wall of the building, excluding the garage, has the same dimension. Select two straight 2 × 6 inch planks 12 feet long. Lay one of them against the anchor bolts with one end of the plank set back ½ inch from the north face of the foundation wall. This provides a space for sheathing when it is applied to the north wall. Butt the other plank against the end of the first plank.

Mark all of the bolt hole centers, using the same method as described for the House Plan A on p. 219. Mark the end of the second piece so that, when cut, it will be ½ inch from the face of the south wall.

Remove the sill pieces and place them on sawhorses. Bore holes for anchor bolts and cut the second piece to the marked length. Continue the same procedure for the sills around the building. Put all of the pieces in place to verify that they fit. Make sure that they line up by sighting them from the corners. Check the dimensions of each side and measure across opposite corners in both directions to verify that the enclosed area is a perfect rectangle.

Remove the sill pieces and place a layer of grout over the foundation wall to serve as a bed for sills. Replace the sill pieces. Level them with the use of a builders' level or a transit level, as discussed in Chapter 2, *Leveling Instruments and Site Work.* Fasten the sills with washers and nuts placed over the bolts. (Some builders use a sill seal in the place of grout. It is made of insulation material enclosed in waterproof kraft paper. See Fig. 4-104.)

Framing the joists. The joists are designed to bear the load of the flooring, the partitions, the plumbing fixtures, etc, and the added live load of furniture and people. The joist framing on the steel wide flange beam presents a special problem because the beam is not to project down below the bottom of the joists. If the recreation room ceiling were to be finished, it would be a flat ceiling. The joists are to be cut out to rest on the lower flange of the beam and are to be fastened together in pairs by means of a metal

Fig. 4-104. A sill seal fills in irregularities between the top of the foundation wall and sill. (Conwed Corp.)

dog running under the beam. See Fig. 4-105.

A header, which is a piece of the same dimensions as the joists, is placed along the east and west walls, lining up with the outer edge of the sill. It is toenailed in place. The header serves to line up and fasten the joists and later serves as a nailing base for the sheathing. The outside joists on the north and south walls are put in place to line up with the outer edge of the sill and are toenailed in place. All of the joist are then set, plumbed, and nailed fast. The last joists to be put in place are those at the bay.

Procedure for Framing Joists. Lay out the joist pattern for area *A,* Fig. 4-100. To find the length of the joists, measure from the inside of the header to the middle of the wide flange beam and deduct one half of the thickness of the web of the beam. Make a template so that the joist fits properly into the beam as shown in Fig. 4-105. Cut as many joists as are required for area *A.*

The placement of the joists depends on the size of 4 × 8 foot sheets of plywood subflooring. Use full sheets rather than cutting them, if this is

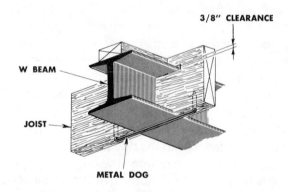

Fig. 4-105. The joists are cut to fit the steel beam so that the ceiling below is flush.

possible. In order to make the full (or half) sheet come to the edge of the outside joist and share one half of the top edge of a joist 8 (or 4) feet away, place the second joist 16 inches away from the outside of the first joist. Place all other joists on 16 inch centers across the building. See Fig. 4-106. Measure joist spaces and mark them on the sill for the east wall. Transfer the markings to

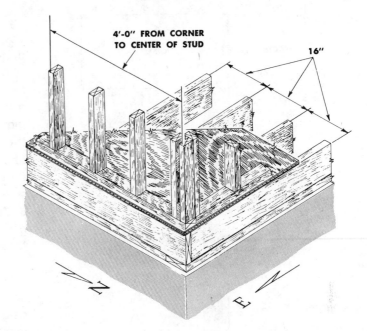

4'-0" FROM CORNER TO CENTER OF STUD

16"

Fig. 4-106. The corner of a building with platform framing is arranged so that the walls can be tilted up into place. Joist spacing in this application depends on the dimensions for the rough flooring.

the two sets of beams and to the sill on the west wall.

Line the joists up with the marks on the sill (marked *1*) and beam (marked *2*), Fig. 4-100, and plumb the ends. When the joists are in a perfectly plumb position, drive nails through the header into the ends of the joists. Also toenail the joists into the sill. Use a diagonal strip and a straight strip as shown at *3* in Fig. 4-100, tacking the strips to the joists to keep them in line until rough flooring is laid.

Lay out a pattern joist for area *B* (Fig. 4-100), with cutouts to fit the beams. Cut the joists and place them in a position lining them up with the marks on the beams. Again, fasten them in alignment with the use of temporary strips. Nail metal ties or dogs, shown in Fig. 4-105, between the joists in area *A* and area *B*.

Study the plans to find the location of the stairwell. Frame double trimmers on each side of the stairwell and a double header at the head of the stair. (Stairwells are discussed later in Chapter 7.) Frame short joists to fill in this space.

Cut the joists for area *C* and install them in the same manner as for area *A*. Provide double joists and headers at the fireplace and hearth. These details have already been covered in Chapter 4.

Use temporary strips to hold joists in line. Plumb and nail joists to the header at the west side of the building. Fasten metal dogs to joists which rest on steel beams.

Subflooring and bridging. Before starting to lay the subflooring, check all joists carefully to make sure that each joist is plumb and straight. Also make sure that every joist is nailed securely in position. This final check helps to insure straight outside walls and full joist strength.

When you are laying the subfloor, begin on the east side and proceed toward the west. (The plywood is 4 × 8 foot sheets of interior grade, ¾ inch thick.) Lay the subfloor so that the face grain is at right angles to the joists. See Fig. 4-101. Remove the diagonal braces which hold the joists in line. Then proceed with the laying of the subfloor.

Install compression type metal bridging between joists at the center of each span. (If solid bridging is used, it can be installed before the subfloor is laid so that it can be nailed with greater ease. The pieces of bridging must be accurately cut so that the bridging fits between the joists and holds them firmly in alignment.)

Exterior wall framing. The first step in framing walls and partitions is to study the working

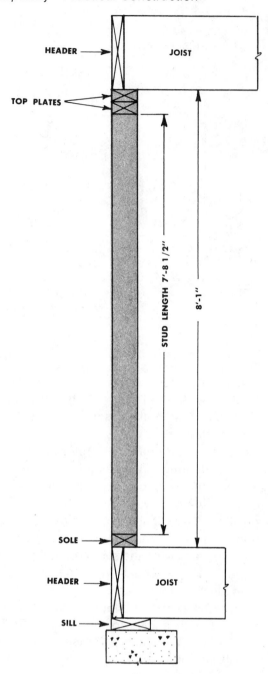

Fig. 4-107. The stud length is determined by subtracting the sole and plate thickness from the joist dimension.

drawings to find the floor-to-ceiling height and the height of various openings above the sub-floor. The length of the studs (See Fig. 4-107.) can be determined by calculation or by making a stud pattern. See Fig. 4-108. When there is a vari-

ety of window types and several different window heights, it is advisable for you to make stud patterns.

Procedure for finding stud lengths. Study the section views on Sheet 6 of the working drawings. Stud length can be determined by using simple math calculations. First, determine the dimension from the top of the first floor joists to the bottom of the second floor joists. In this case it is 8'-1". Next find the dimensions for the sole and top plates. Here, they are 1½" (sole) and 3' (top plate). Add these two dimensions together. (1½" + 3" = 4½".) Subtract this total from the dimension determined above. (8'-1" − 4½" = 7' 8½".) See Fig. 4-107.

Make a stud pattern to help you locate headers and subsills for windows at the designated heights. It also helps if you measure the right lengths for studs at openings. There are several different sizes of windows in the house. The position of the subsills varies, depending on the height of the windows. The top jamb for all windows is at the same height, 6'-8" above the sub-floor. Although only one window is shown on the stud pattern (Fig. 4-108), subsills for other windows can also be shown on the same stud pattern.

To layout the master stud pattern, choose a straight piece of 2 × 4 inch material and place it over the sawhorses. Square one end and cut it on the line. Measure ¾ inch and draw a line to represent the plywood subfloor. Measure 6'-8" to locate the line for the bottom of the top jamb. Measure 1 inch for the jamb and ¾ inch for the fitting space. Draw lines representing the jamb and space. (Both the jamb thickness and fitting space vary, depending on the particular type of windows used.) The last line drawn indicates the location of the bottom of the header. It also provides information on the length of trimmer studs. Determine the jamb dimensions (height of the window frame) from the manufacturer's literature or by measuring the frame. Measure this dimension from the top of the head jamb line to locate the top of the subsill. (The window frame generally rests on the subsill.) Determine the jamb dimensions for the other windows and locate the subsill on the same stud pattern. Use the

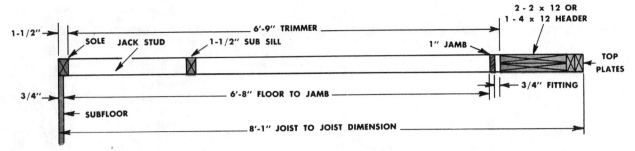

Fig. 4-108. A stud pattern helps locate the header and subsills for window openings. It also gives the lengths of trimmers and jack studs.

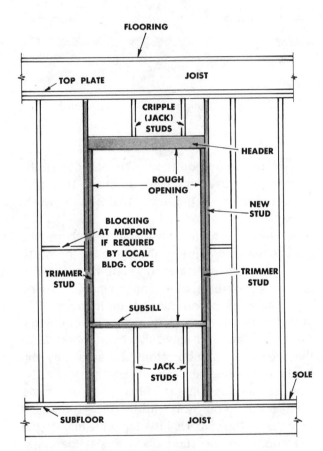

Fig. 4-109. The various studs used for framing a window. A stud pattern is used for measuring the length for these members.

stud pattern by laying it alongside the members that are to be cut to frame the openings. The pattern indicates the length of trimmer studs and the jack studs to be used to support the subsills. See Fig. 4-109.

You can also make a stud pattern for the door. The important items are the threshold, the jamb thickness, the rabbet in the jamb, and the fitting space. This has been discussed in connection with House Plan A on p.221.

Procedure for framing exterior walls. Study the elevation drawing on Sheet 4 and the First Floor Plan on Sheet 2 of the working drawings to gather the information needed on the east wall, which is built first. See Fig. 4-110.

Cut the soles for the wall and mark them for the location of studs and openings. See Fig. 4-110. Cut the lower member of the top plate the same length. Transfer the markings to the top plate. Cut the required number of studs and arrange them on the floor so that they line up with the markings on the sole and the top plate. Cut the jack studs, the headers, and the subsills under windows then assemble the wall. The east wall can be built in one piece. The south wall, however, has to be made in two or three assemblies. The north and west walls are broken up into smaller sections because of the fireplace and bay.

True up the east wall, which you have assembled on the floor, by measuring diagonally across the corners. Tack temporary braces in place to keep the wall from racking. If you sheathed the wall before tipping it into place, the temporary bracing is unnecessary. (Three helpers are needed. Two tip the wall up into position, and the third braces it with diagonals and nails it to blocks on the floor.) Line up the wall sections

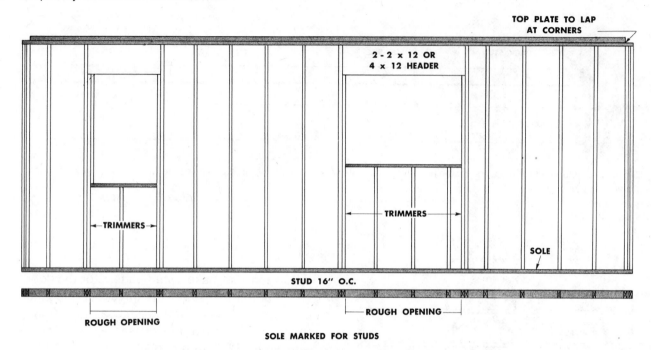

Fig. 4-110. Studs are marked on the sole. The markings are transferred to a top plate before the section is assembled.

with the outside of the header and the edge of the subfloor below.

You can choose the order in which you build and erect the other sections. If you build parts of the north and south walls in sequence next, they would help brace the east wall. Check all of the walls after they are in position to verify they are plumb and in line.

Applying sheathing.   Apply full sheets of plywood sheathing at all exterior corners. This procedure makes it unnecessary to install diagonal bracing (except in those parts of the country where building codes require it). You can install the fiberboard sheathing for the remainder of the areas later.

Assembling and erecting partitions.   Snap lines on the floor, showing the location of all partitions, and mark the openings in the partitions. Tack all partition soles in place. Cut the lower member of the double top plate and temporarily tack it on top of the soles. See Fig. 4-111. Where interior partitions intersect, it is important to decide which one is installed first because the top plates lap when they tie into each other. After the plates have been tacked to the soles, mark

the stud locations on them. The partition at the kitchen has been chosen to show this procedure. See Fig. 4-111 and Sheet 2 of the working drawings.

Beginning with the base line (which in this case is the inside line of the studs for the north wall), mark the sole and top plate member with an *X* for each stud space. (The top member of the top plate is nailed on later.) Mark the letter *T* to indicate the location of trimmers at openings. After the plate and sole have been marked, lift the top plate member from the sole. Lay the studs on the floor in line with the marks on the sole. Lay the top plate member at the other end of the studs in position with the marks identical to those on the sole. Nail the top plate member to the studs with two 16d nails. Transfer the markings for jack studs over openings to the headers by laying them alongside the sole in the proper location. Assemble the trimmers to the plate and install headers. True up the frame by adjusting the studs to fit the marks on the sole, then nail the studs to the sole. After all studs, trimmers, and headers are assembled, nail the second member of the plate on top of the one nailed to the studs,

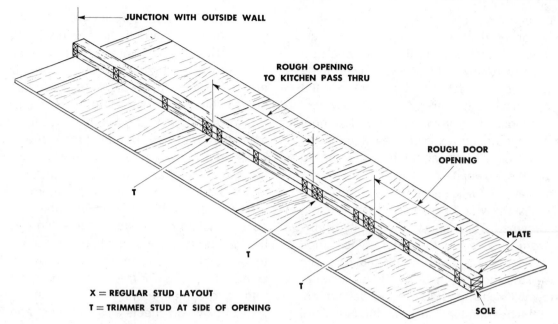

JUNCTION WITH OUTSIDE WALL

ROUGH OPENING
TO KITCHEN PASS THRU

ROUGH DOOR
OPENING

PLATE

T

T

T

SOLE

X = REGULAR STUD LAYOUT
T = TRIMMER STUD AT SIDE OF OPENING

Fig. 4-111. Openings in partitions and location of studs are marked on sole and plate. Studs are marked at openings to locate jack studs over opening and to maintain the stud spacing in the partition.

unless the second member is needed to tie two partitions together. Before the partition is raised into place, nail two 1 × 6 (or 2 × 4) temporary braces to the ends of the section so they extend to the floor at an angle after the partition is in place. The temporary braces prevent the wall from falling and keep it plumb. After the wall is raised, fasten it through the sole with 16d nails. When the partition is plumb, nail the brace to blocks fastened to the floor.

Second floor framing. Figure 4-101 shows how the joists are supported for the second floor. The joists over area *A* are supported on the top plate of the exterior wall and the top plate of the partition. The joists over area *B* are supported on the same partition and are hung from a girder with joist hangers. The First Floor Plan, Sheet 2 of the working drawings, shows information about this girder. The joists over area *C* are hung from the girder and rest on the top plate of the exterior wall. Joists for areas *A* and *B* lap each other over the partitions. Joists for areas *B* and *C* are in line at the girder.

Procedure for Framing Second Floor. Place headers over the east and west walls flush with the outside of the top plates. Toenail them into

the top plate. Measure the length of the joists for area *A* from the inside of the header to the top plate of the partition. Add the width of the partition top plate so that the joists have full bearing on the partition. (*Full bearing* means that the joist rests on the full 3½ inch bearing surface of the top plate.) Cut as many joists for area *A* as required.

Measure the joist length for area *B* providing full bearing (3½ inches over the kitchen partition). Cut as many joists as required.

Measure the joist length for area *C*, measuring from the face of the girder to the inside of the header on the outside wall. Cut as many joists as required.

Before placing the second floor joists in position, check the bearing partition and corner posts again to make sure they are plumb. Also check the walls to make certain they are straight and thoroughly braced. Lay out the joist positions on the top plates of the east and west walls and on the bearing partitions.

Begin by placing the second floor joists and headers in area *A*, Fig. 4-101. Select straight joists for the outside and place them in their respective positions, flush with the face of the walls. To raise the joists to the second floor position, push one end of each joist up over the bearing partition first, then raise the other end and step up on to a sawhorse or a ladder. Lift the joists over the wall plate, leaving them in a flat position until all joists of area *A* are on the wall.

Place the joists in a plumb position. Nail them to the header on the exterior wall and to the bearing partition plate, sighting each one to make sure the crowned side is up.

(Note: Material for joists is cut from the log at the mill. As the wood seasons, some of it becomes crowned. This means it has a slight curvature from one end to the other. When such lumber is used as joists, you should make certain that it is placed so that the curvature is up rather than down. Once the joists sag or deflect because of the bearing load on them, the curvature tends to even out and the tops become level.)

Laying subflooring for second floor. The subfloor for the second floor is to be laid using 4 by 8 foot sheets of plywood. Lay them so that they are flush with the outside of the headers and the joists at the north and south walls. Start alternate rows with half sheets so that joists are staggered. See Fig. 4-101. Cuts sheets whenever required so that the end joints between the sheets fall over joists. See Fig. 4-101.

Second floor wall and partition framing. The procedure for framing the walls and partitions of the second floor follows the same routine which is used for the first floor. It would be advisable to do the north and south walls first. These are straight wall sections, whereas the east wall has a gable end and a louver and the west wall has a gable end and a chimney problem. See Fig. 4-102.

The wide extention of the roof at the gable ends requires support for the rafters. This can be achieved in several ways. The one shown in Fig. 4-102 requires a special top plate for the north and south walls. Two 2 × 6 members are used in place of the last 4 feet of wall plate on each corner. They are extended to support the rafters.

The east and west walls have gable ends. You must use longer studs as you approach the center. (Chapter 5, *Roof Framing*, discusses the details of laying out rafters and cutting gable end

members.) To frame the east wall, mark the position of each stud and the jack studs on the sole. Also mark a 2 × 4 inch piece, which is temporarily tacked in a horizontal position about six feet above the sole to keep the upper ends of the studs in line. Lay a pair of rafters on the studs at the correct slope and mark the studs for cutting so that they fit the rafters. Frame the header, jacks, and the subsill for the window. Frame a horizontal member for support of the sill for the louver. Nail the whole assembly together.

Nailing sheathing. The sheathing should now be nailed on all walls where the openings have been framed. Nail the sheathing in place up to the height which can be reached by a worker standing on a sawhorse. Install approved scaffolding to complete the job. The sheathing consists of plywood sheets at the corners and insulating fiberboard sheathing for the remainder of the wall space. Make the sheathing extend into the fireplace and chimney opening from 2 to 4 inches. This procedure enables you or the bricklayer to cut the sheathing to fit the brickwork. This insures a straight tight joint. Cut the sheathing flush with the studs at all door and window openings. Cut the sheathing flush with the bottom of the rafters on the east and west sides. On the eaves sides (north and south) of the house, carry the sheathing up to the bottom of the top plate. This allows for fitting of the rafters to the plates.

Garage and entry framing. The garage can either be framed at the same time as the rest of the building or after the main part of the building has been covered with sheathing. The framing procedure is the same as for the rest of the building. Frame the east and west walls on the garage slab. Tilt them up, bracing them back to stakes driven in the ground. Frame the south wall, installing the girder over the doorway. Set the posts and install headers for the entry way. Install pre-engineered trusses to support the roof sheathing.

Bay window framing. After the outside walls have been sheathed up to the second floor level, you can frame the bay window of the dining room. See Fig. 4-101 and the working drawings, Sheets 5 and 6. Headers are used in the place of top plates. Spike together two 2 × 6 inch planks with ½ inch spacers to make the headers 3½ inches thick. Miter the header pieces so that they fit the shape of the bay at the floor. Fit the inner corners against the wall studs. Cut six 2 × 4 inch studs to the required length and erect them at the corners of the bay window. Place and nail the header in position. Frame the subsill and jack studs below the window then sheath the walls.

Second floor partitions. It is the usual practice to delay work on second floor partitions until the rafters have been installed and the roof is completed. The open floor space permits ease in handling material for the roof. It is also to your advantage to complete the roof of the house as quickly as possible so that you have a dry place in which to work in case of inclement weather. After the roof is installed, build and install the partitions, following the same procedure used on the first floor.

Framing odds and ends. Up to this time, only the most essential part of the framework of House Plan C has been erected. Our aim has been to erect the skeleton of the building and then cover it with sheathing. Much still remains to be done on the inside of the building before the framing is completed. Minor partitions must be set, plumbed, and tied securely in place. The rough stairs must be erected and the insulations put in place.

All these various odds and ends of framing can be postponed until those times when weather conditions prevent outside work from being done. However, if weather conditions are consistently good, then the carpenter should go ahead and complete the interior framing so that other mechanics, such as electricians, plumbers, and heating and air conditioning workers, can proceed with their work.

QUESTIONS FOR STUDY AND DISCUSSION

House Plan A

1. How is the ceiling supported over areas *B* and *C*?
2. Where on the working drawings do you find information on the horizontal and vertical location of windows?
3. How are the partitions located in relation to the dimensions given on the floor plan of the working drawings?
4. How are outside wall sections anchored to the foundation?
5. What is the purpose of a master stud pattern?
6. How are exterior wall sections braced down to the ground?
7. What difference is there in the assembly of outside walls and partitions regarding the sills?
8. Why is the sequence of erecting partitions so important?

House Plan C

9. How are first floor joists supported?
10. How are first floor joists framed around the stairwell?
11. How should the subfloor be laid?
12. What type of bridging is used? Where is it placed?
13. What types of sheathing are used? Where are they used?
14. Explain the procedure in laying out the framework for a partition.
15. What is the procedure for erecting the second floor joists?

CHAPTER 5
ROOF FRAMING

ROOF SUPPORT STRUCTURES

The main purpose of the roof is to protect the building from such weather conditions as rain, snow, heat, and cold. Another purpose of the roof is to enhance the appearance of the building so that the structure, the details of exterior finish, and the roof are architecturally in harmony.

The roof frame furnishes the base to which the roofing material is attached. The frame must be made strong and rigid in order to withstand the wind and snow and other loads peculiar to the locality, and to support the weight of the roofing and the workers who apply it.

Single support members, called *rafters,* are arranged at equal intervals along the walls and extend up from the top of the wall to the ridge. The carpenter must learn how to lay out, cut, and erect the rafters. This involves determining the exact length required for each member and making complex cuts at each end. Solid bearing at the wall and accurate fitting where rafters meet are essential.

A truss is made from a combination of horizontal, sloped, and diagonal structural members, usually constructed to form a triangular shape. The truss is also designed according to engineering principles so that it can support loads over wide spans. A trussed rafter is a truss designed to support the ceiling as well as the roof. The trusses which the carpenter deals with most and which are used in residential construction are made of wood. They are made of solid plank or timber members which are fastened together by rigid metal connectors, nailing plates, or other type fasteners. Carpenters also install large laminated trusses. These trusses are made in a press. Most of the trusses and trussed rafters used in residential construction are made in factories rather than on the job.

Sloped roofs for large structures are usually supported by trusses made of steel or wood. Large wood trusses are made in two ways. Some are made of heavy planks or timbers which are bolted together. Others are structures made with a process in which many boards are laminated and glued together. They are made into the desired shape in huge presses. (Note: The discussion of large wood trusses comes under the subject of Heavy Timber Construction and is beyond the scope of this book.)

From a structural standpoint, the design of a roof involves various engineering concepts. The size and spacing of the rafters depends on the span of the building, the slope of the roof, the type and grade of lumber, and the load which has to be carried.

Long rafters, if they are to be unsupported at their centers, must be made from large dimension material. They can be spaced at closer intervals to provide more bearing capacity. The species of lumber used in each case has a bearing on the size and spacing of members. Each type of wood has its own structural characteristics.

Regional variations are an important factor in roof design. Provisions must be made in some areas for heavy snow loads. Roofs built in areas which are subject to high winds must be made strong enough to withstand these conditions. Here, special attention is given to anchoring the roof to the building. The wracking effect of seismic disturbances (earthquakes or tremors) must be counteracted in other areas of the country by adding strength and flexibility.

When designing the roof, the architect must closely observe all applicable directives from both local and national building codes which apply to the particular community. The codes are based on engineering data and include a strong margin for safety. Trusses are designed by architects and engineers and manufactured under close supervision. Manufacturers are required by local building ordinances to certify the load bearing capacity of the trusses.

If a large number of houses with similar plans are built at the same time, it is usually less expensive to frame the roof with roof trusses. This is particularly true when the houses have gable roofs. Pre-cut factory built roof parts can be used for hip roofs, intersecting roofs, and dormers. Individual custom built homes are usually framed most economically if conventional roof framing techniques are used.

You should first learn about simple roofs—the names of the various rafters, how to lay them out, and how to cut and erect them. You can then increase your knowledge so that you understand the layout for the common, jack, hip, and valley rafters needed for more complex roofs. You must know how to cut these rafters accurately so that they fit together correctly. Basically, you should develop your ability to visualize the whole roof so that you can understand how the various planes intersect once the roof is finished. The working drawings rarely include a roof plan. You must learn how to frame a roof, using the information supplied on the floor plans (which give the shape of the building) and the elevation views (which show the contour of the roof).

STYLES OF ROOFS

If carefully designed, the roof adds greatly to the beauty of a building. Contemporary houses use a number of roof styles. The shape of the house plan and the requirements of the climate are some of the limiting factors. Some owners may want the building to have the flavor of a particular architectural period, such as Dutch Colonial or Cape Cod. The roof is very important in bringing out these special effects. A few common types of roofs used in the construction of houses are: shed, gable, hip, gambrel, and mansard, all shown in Fig. 5-1.

Shed Roof

The simplest type of roof is the shed or lean-to. See Fig. 5-1. The shed roof consists of a single plane surface, with one side raised to a higher level. It is used over fine homes and commercial buildings, particularly in warm climates. Generally it is used for small sheds, porches, or other structures where appearance is not a matter of primary importance.

Gable Roof

The gable roof has two sloping surfaces, as shown in Fig. 5-1. Two surfaces of the roof come together in the middle of the roof at the ridge, forming triangular gables at each end. It is the most common of all types of residential roofs and is particularly adaptable to buildings with rectangular floor plans. The slope of the gable roof can vary from an almost flat surface to one which is very steep, depending on such factors as climate, snow load, and desired architectural effect.

Hip Roof

The hip roof has four sloping sides. The line where two adjacent sloping sides of the roof meet is called a *hip*. If the plan of the building is rectangular, and the roof has the same slope on all four sides, a ridge occurs. See Fig. 5-1.

Gambrel Roof

A variation of the simple gable roof is the gambrel roof, which has two different slopes on

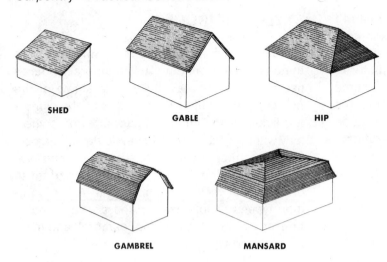

SHED GABLE HIP

GAMBREL MANSARD

Fig. 5-1. Five basic styles of roofs are used in building today.

each side. See Fig. 5-1. This type of roof has been used on barns for many years because the enclosed volume provides the best storage space for hay. It is used for residential work in the Dutch Colonial design because it provides more attic space than either hip or gable roofs.

Mansard Roof

The historic mansard roof is used on some custom built houses influenced by French architecture. It has two slopes extending upward from each side of the roof. The lower part is quite steep, whereas the upper part has a low slope. This type of roof has the advantage of providing additional space in attic rooms. See Fig. 5-1.

The modern adaptation of the mansard roof retains the steep slopes of the lower portion of the roof and has a flat roof above. See Fig. 5-2. It is popularly used as the main structure of roofs for houses and also for porches and bays. It is used in apartment buildings and commercial buildings mainly to provide an architecturally decorative effect at the roof.

Combination Roofs

When the plan view of a building is not a square or rectangle, the possibilities for a variation in roof shape are increased. Generally the roof has one basic style, such as gable or hip. Intersections with hips and valleys develop where the roof parts meet.

A combination of two or more types of roofs is often used to enhance the appearance and character of a building. See Fig. 5-3. Also a particular

Fig. 5-2. A modified mansard roof has a steeply sloped section and a flat section.

Fig. 5-3. On this residence, a mansard roof is used in combination with a hip roof which appears over the garage. (Ponderosa Pine Woodwork)

requirement, such as a need for attic space in one section of the house, may dictate that the architect use a combination of different types of roofs.

ROOF TRUSSES AND TRUSSED RAFTERS

A wood truss is an assembly of structural members used to support a roof over a wide span. The members for trusses used in residential construction are generally 2 × 4 or 2 × 6 inch material, or a combination of the two sizes. They are arranged with a horizontal lower member or lower members (chord), sloped members (top chords) to support the roof, and diagonal members (web) to provide stiffness and transfer the load from one member to another. The load on the roof and the weight of the truss itself is transferred vertically to the wall. The size of the members, the arrangement of web members, and the type and size of connection devices at intersections are determined by engineering calculations.

The loads on a truss are the live loads of winds, snow, seismic disturbance, and the weight of the workers. The dead loads are roofing, sheathing, and the weight of the truss itself.

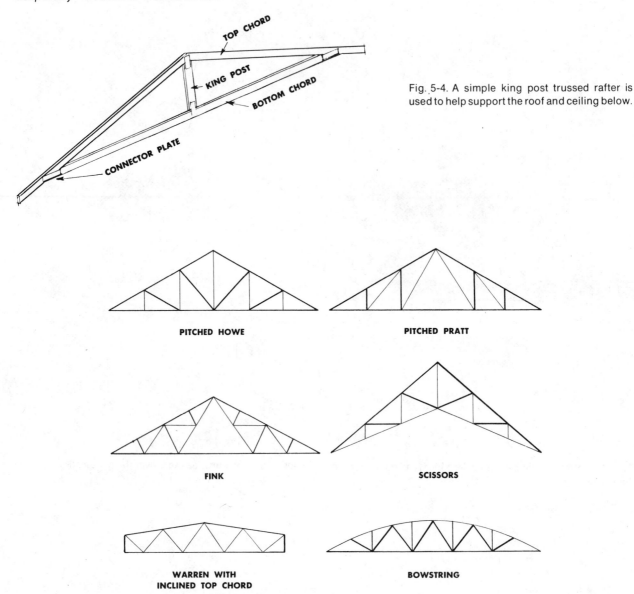

Fig. 5-4. A simple king post trussed rafter is used to help support the roof and ceiling below.

Fig. 5-5. Heavy trusses are made of wood in the various shapes shown.

A trussed rafter is a truss with chord members which also serve as rafters and ceiling joists. See Fig. 5-4. The chord members are subject to bending stress in addition to direct stress. A simple trussed rafter, since it has only a few members, must have members which are strong enough to carry the weight of the sheathing and roofing. The sloped members are considered to be rafters. Because the ceiling is hung from the bottom horizontal members, these members must be designed to withstand bending stress or deflection within the allowable limit.

Wood trusses have been used for hundreds of years to support huge roofs spanning large areas. Wood trusses are still found to be practical in both cost and construction for supporting the roofs of large structures. These trusses are spaced at wide intervals, ranging from 8 to 20 feet apart, and can be made using various shapes. See Fig. 5-5.

The manufacture of trusses as building components is now common throughout the construction industry. A large percentage of homes have roofs supported by trusses. This is particu-

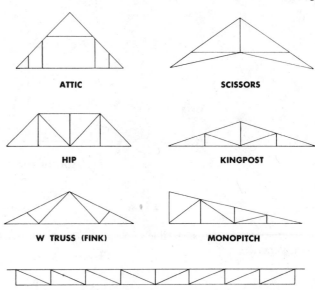

ATTIC

SCISSORS

HIP

KINGPOST

W TRUSS (FINK)

MONOPITCH

FLAT TRUSS

Fig. 5-6. Roof trusses used for houses and small buildings are designed to meet the particular needs of the structure. The W truss is most commonly used in residential work.

larly true for project buildings where the roofs have relatively simple shapes. The arrangement of members in the truss can be made so that the truss can be used to serve a variety of purposes. See Fig. 5-6.

Of primary importance is the question why trusses should be used in place of conventional roof framing. Trusses have several advantages. They can be erected quickly, thus putting the building under roof in a relatively short period of time. They provide an economical use of labor. The load of the roofing and roof structure is transferred to the outside walls. This allows the carpenter to erect partitions after the roof is completed and also allows the designer to place partitions wherever desired, without being concerned about bearing partitions. If properly designed, trusses can be made with a greater clear span distance than that possible with conventional rafters. By meeting engineering standards, the truss manufacturer provides a product which has economy of material because it is designed for its particular application. Truss fabricating plants are located in most parts of the country. The finished trusses are transported to the location by truck without great expense. The advantages of trusses diminish when used for single houses or on buildings with irregularly shaped roofs.

Design of Trusses

As a carpenter, you should know that the design for all trusses is the job for an engineer. The engineer selects the material (both the size and the type) and the truss arrangement and designs the method and the exact specifications for joining the members together. Each span, slope, and load problem requires a different solution. Factory-made trusses are manufactured according to precise engineering data under controlled circumstances. Each truck load of trusses is certified by the manufacturer that the trusses will perform under prescribed loads. Job-made trusses must conform to the specifications shown on the architect's or engineer's drawings, both in regard to the material used and the method for assembling them.

The most common truss design for relatively short spans, such as used in residential work, is the W truss. See Fig. 5-6. Other truss shapes, shown in Figs. 5-5 and 5-6, are designed to meet specific architectural requirements. One type creates roof space which can be used for attic storage or for rooms. The hip type can be adapted for buildings with hip or mansard roofs. The flat truss, shown in Fig. 5-6, is made of 2 × 4 inch material arranged so that the top and bottom chords are flat when the truss is in position on

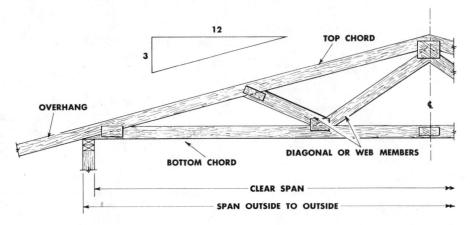

Fig. 5-7. A W truss is assembled in single plane. The connection between the various members is done using metal clips which have barbs.

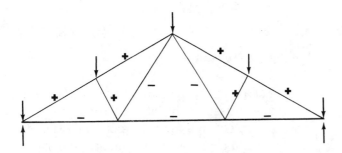

Fig. 5-8. The load on a truss is considered to be concentrated at panel points. Some members are in tension (signified by the − symbol) and others are in compression (signified by the + symbol).

the building. These trusses are discussed in Chapter 4, *Wall and Floor Framing.* See pp. 172–174.

You should know about the various parts of trusses, how they are assembled, and how they are erected. You should understand the reasons for the different designs so you can work with them intelligently. The names of the various parts of a truss are shown in Fig. 5-7.

Figure 5-8 shows how the truss members work together. The loads are considered concentrated at the panel points (indicated by arrows) and transferred through the members to the supports at the walls. The members carry different loads. Some members in large trusses are designed to carry great loads, and they are made larger than other members. All of the members in smaller trusses, such as those used in small home construction, are usually the same size. Trusses which carry large loads or cover greater

spans are increased in size so that all members are 2 × 6 inch instead of 2 × 4 inch. The factor of safety has been taken into consideration so that all members in each truss are sufficient to bear their share of the load.

The members of a truss are joined together in various ways. Members are usually joined together with metal connector plates. One type of connector plate, shown in Fig. 5-9, is a steel plate made with rows of barbs. Another type, used for light trusses, is called a *truss clip.* See Fig. 5-10. The nailing plate, shown in Fig. 5-11, consists of a flat plate with evenly spaced holes for nails. When barbed connector plates or clips are used, the engineer specifies the types and size so that the number of barbs is sufficient for the loads on the intersection. When nailing plates are used, the engineer specifies the number and size of the nails used at each intersection. For example, when a truss is made on the

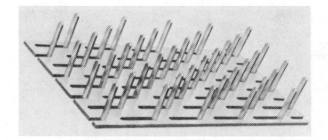

Fig. 5-9. Connector plates are used to hold truss members together. (Struct O Matic, Inc.)

Fig. 5-10. Truss clips have the effective holding power of 20 to 60 nails. (The Panel-Clip Co.)

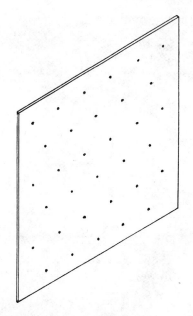

Fig. 5-11. Nailing plates are used to connect truss members. The size and number of nails required are both specified by the engineer. (Struct O Matic, Inc.)

job, using nailing plates at the intersection of members, ten 8d nails are sufficient to hold two members in place. Twenty 12d nails may be required for nailing two other members.

Trusses are usually spaced on 24 inch centers, although this can be changed to provide for different load conditions. The usual slope (cut) is 2, 3, 4, 5, or 6 inch rise per foot for manufactured trusses. A certain amount of deflection will occur once the trusses are put in place and loaded. It is possible to counteract this deflection by arching the bottom cord a certain amount, for example,

½ inch. This enables the bottom cord to become level once it is in position. This arching is called *camber*.

Prefabricated Trusses

The first step in manufacturing a truss is the design. The engineer gathers information on the span and slope of the roof, the spacing between trusses, the projection, and the various loads, such as the sheathing, roofing, as well as snow and other loads. The engineer calculates the size of members and, on the basis of loads at each connection, chooses the type of connector plates used for each intersection. The material, which is specified in both species and grade, is cut to length in the shop and placed in a jig. See Fig. 5-12. (A *jig* is a clamping device used to hold the members in place until they are fastened together.) Connector plates are placed on both sides of the truss. The truss is then passed between huge rollers or under pneumatic presses which force the prongs on the connector plates deep into the wood.

A truss made with connector plates is called a *single plane truss.* All of the members are in

Fig. 5-12. A jig holding device is used to assemble trusses. The operator is using a pneumatic press to fasten connector plates. (Hydro-Air Engineering, Inc.)

Fig. 5-13. Single plane trusses are conveniently stacked for loading.

the same plane so that the trusses can be conveniently stacked for delivery to the job. See Fig. 5-13.

Job-Built Trusses

Trusses are built right on the job in locations where prefabricated trusses are not readily available, or in situations where the architect or engineer designs the trusses for a special application. The type of truss made on the job is often the same as a prefabricated truss. It is a single plane truss made with wood members and connector plates (or clips). The individual members can be held in position in a temporary jig or laid out on the floor. The connector plates (or clips) are fastened in various ways. The barbs can be hammered into the wood, as shown in Fig. 5-10. A pneumatic nailer, shown in Fig. 5-14, can also be used. Pneumatic truss presses are sometimes employed. See Fig. 5-15.

Another type of job-built truss, shown in Figs. 5-16 and 5-17, is designed to use plywood gusset plates. (A *gusset plate* is a flat piece applied to connect members together.) The engineer specifies the thickness and type of plywood and the size, number, and spacing for the nails.

Fig. 5-14. A power nailer is used to fasten truss members with a nailing plate. The long member is the bottom chord of the truss. (Bostitch Div. of Textron, Inc.)

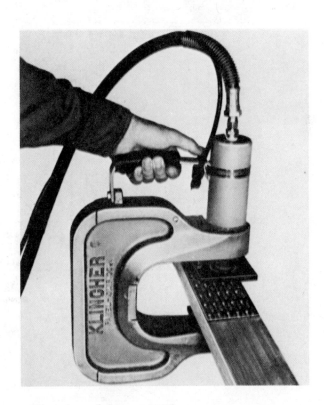

Fig. 5-15. A pneumatic press uses clips to fasten truss members. (The Panel-Clip Co.)

The third type of job-built truss is the split ring type, shown in Figs. 5-18, 5-19, and 5-20. When the split ring connector is used, circular grooves are cut into the two members to be joined. A 2½ inch ring is used for 2 inch lumber. A 4 inch ring is used for heavier trusses made from 3 inch or thicker material. The ring is embedded half of its depth into each member and is then brought up

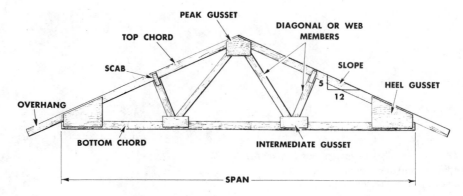

Fig. 5-16. A job-built truss is made using gusset plates. The gusset plates are nailed or are glued then nailed in place.

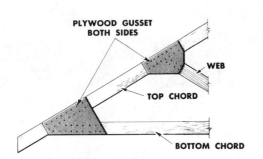

Fig. 5-17. Plywood gussets are glued and nailed in place. They are used to make strong but lightweight trusses. The size, number, and spacing of the nails is an important factor.

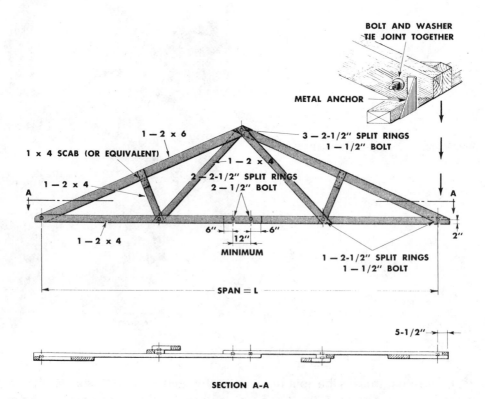

Fig. 5-18. Split ring connectors are used in the assembly of a truss. Section A-A, below, is a plan view which shows how the lower chord and diagonal members meet and are assembled with the split ring connectors. The detail shown at the top right shows the metal anchors with which the truss is fastened to the plate.

Fig. 5-19. A split ring, shown at left, fits into a groove cut in the members, as shown at the right. (Timber Engineering Co.)

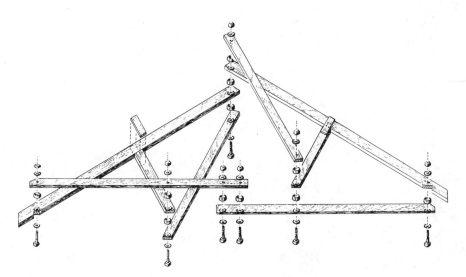

Fig. 5-20. This exploded view shows how a truss is assembled using split rings. (Timber Engineering Co.)

tight with a bolt which passes through the connection. See Fig. 5-19. The split ring is designed to transfer the stress at each connection.

One advantage of the split ring method of fastening trusses is that the trusses can be made on the job with a minimum of equipment and with greater ease than required for other systems. The split ring also allows for greater shrinkage and swelling of the lumber without causing slippage at the connecting points. This type of truss can be transported knocked-down and later quickly assembled on the job. (These trusses are partially assembled or in separate parts ready to be put together.)

Applying Trusses to Hip and Combination Roofs

Trusses are most often used on gable roofs, when the units are alike, and on parts of hip and combination roofs, when there are many units of the same size and shape. The parts of the roof where hip and valleys occur are sometimes framed with conventional rafters or with pre-cut fill-in rafters. However, truss manufacturers make trusses designed for almost any type of problem. Trusses can be use to effect the transition between parts of a combination roof, as shown in Fig. 5-21. The main roof is sheathed, then progressively larger trusses are used to fill in the space and support the roof for the addition. Trusses can be manufactured to solve many varied and complex roof problems. Some types of trusses are generally only used for projects which require a reasonable quantity of the same type so that costs can be held down. An example of the versatility of truss construction is shown in Fig. 5-22. Here the progressively shorter trusses are used in place of jack rafters.

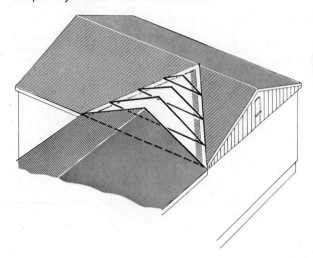

Fig. 5-21. Modified trusses may be used at the intersection of an L shaped building.

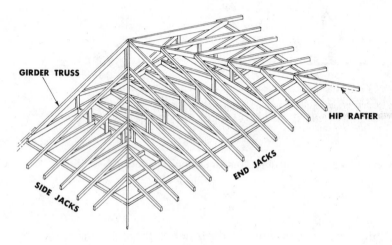

GIRDER TRUSS

HIP RAFTER

END JACKS

SIDE JACKS

Fig. 5-22. Trusses are used to frame the hip end of a roof. (Truswal)

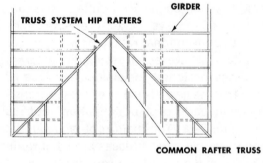

TRUSS SYSTEM HIP RAFTERS

GIRDER

COMMON RAFTER TRUSS

Erecting a Truss Roof

Trussed rafter and lightweight trusses are usually erected by a crew of three workers. Two are perched on the wall or on scaffolds to fasten the trusses. The trusses are placed upside down hanging on the two walls. The third swings them up into place from below. See Fig. 5-23. Larger trusses require a crew of four, with one worker located at the ridge. The gable end truss is installed first. It is braced down to the ground with wood braces. The braces are fastened to stakes

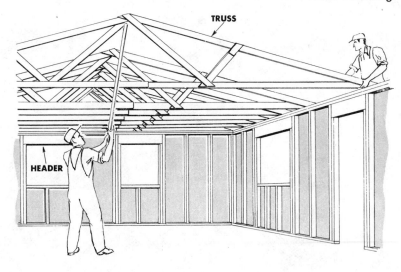

Fig. 5-23. Trusses are swung into place (worker on the ground), then lined up and fastened (worker above). Each truss is secured in place prior to setting the next one. (Braces between the trusses are not shown.)

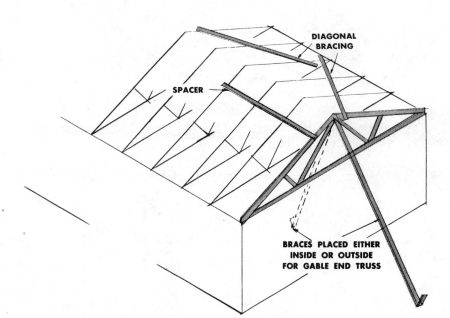

Fig. 5-24. Adequate bracing is needed to ensure safety while erecting the trusses. The gable end truss is held in place with a wood brace which extends down to a stake on the ground. Guy wires are used when the brace required would be too long.

on either the outside or the inside of the building so that the truss remains plumb and does not overturn. See Fig. 5-24. Guy wires are used for larger trusses in place of very long wood braces. The other trusses are installed after the installation of the gable end truss. Spacers and diagonal bracing are nailed in place to keep the trusses in line. See Fig. 5-24. Diagonal cross bracing is often required, extending from the end wall up to the top of an intermediate truss.

When openings for chimneys or dormers must be made in the roof, care must be taken to not change the 24 inch spacing any more than necessary. When a large opening is required, special trusses must be designed by a registered engineer or architect to provide adequate support.

CONVENTIONAL ROOF FRAMING

The terms used to designate the members of a roof are shown in Figs. 5-25 and 5-26. You

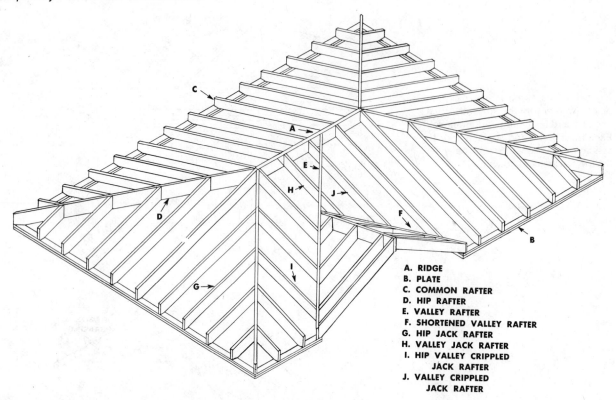

A. RIDGE
B. PLATE
C. COMMON RAFTER
D. HIP RAFTER
E. VALLEY RAFTER
F. SHORTENED VALLEY RAFTER
G. HIP JACK RAFTER
H. VALLEY JACK RAFTER
I. HIP VALLEY CRIPPLED
   JACK RAFTER
J. VALLEY CRIPPLED
   JACK RAFTER

Fig. 5-25. It is important that the carpenter know the names of the various members to understand roof framing.

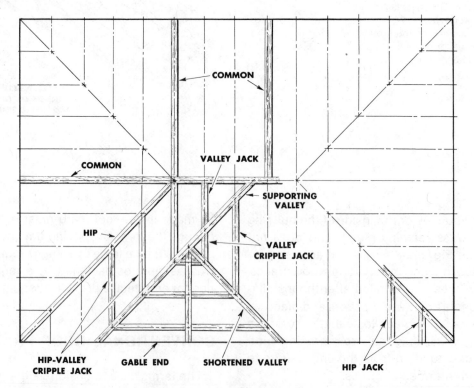

Fig. 5-26. Rafters are named according to their position in the roof and their cuts.

should become familiar with the names of these different members or parts of a roof, so you can readily identify them.

Roof Framing Members

Ridge. The highest horizontal roof member is the ridge. The ridge helps to align the rafters and tie them together at the upper end.

Rafter plate. The framing member at the top of the wall upon which the rafters rest is known as the *rafter plate* or *double top plate.*

Rafters

The sloping structural members of a roof designed to support roof loads are called *rafters.* Rafters are spaced at regular intervals in order to equally distribute the loads from sheathing, the roofing, and such other factors as snow. In a complex roof of the type shown in Fig. 5-25, some of the rafters are cut shorter to fit into a particular place. Some of the load is transferred to the diagonal rafters and then to the plate. There are various kinds of rafters, including common, hip, valley, jack, and cripple jack.

Common rafter. The common rafter extends from the ridge to the plate and is at right angles to both of them. Neither end has a miter cut because the ends fit squarely against the ridge and fascia boards.

Hip rafter. The roof member extending diagonally from the exterior corner of the plate to the ridge is known as a *hip rafter.* The hip rafters form the hips where adjacent slopes of the roof meet.

Valley rafter. The rafter extending diagonally from the plate to the ridge at the line of the interior intersection of two roof surfaces is called a *valley rafter.* It has the name valley rafter because it is located where adjacent roof slopes meet to form a hollow or valley. When two roofs intersect so that the ridges are at the same elevation, the valleys are the same length. When the ridges are at different elevations, such as shown in Figs. 5-25 and 5-26, several arrangements of rafters may be used. One valley rafter, called a *supporting valley rafter,* is cut to fit from plate to ridge. The other valley rafter, called a *shortened valley rafter,* is cut to fit from the plate to the supporting valley rafter.

Jack rafter. Jack rafters are similar to common rafters except that one end of a jack rafter is cut with a miter cut to fit against a hip or a valley rafter. Carpenters use the term *hip jack rafter* to designate those rafters which end against the hip rafter. A hip jack rafter is shown at (G) in Fig. 5-25. A valley jack rafter is shown at (H).

Cripple jack rafter. A rafter which extends from a hip to a valley rafter or between two valley rafters is called a *cripple jack rafter.* This rafter touches neither the ridge of the roof nor the rafter plate of the building. Both ends have miter cuts. A hip valley crippled jack placed between a hip and a valley rafter is shown in Fig. 5-25 (I). A valley crippled rafter is placed between two valley rafters. See (J), Fig. 5-25.

Roof Framing Definitions

Overhang. The *overhang* is the portion of the rafter which extends beyond the outside edge of the plate or wall of the building. It is a part of the rafter itself. The term *projection* is used for the horizontal measurement of the overhang. See Fig. 5-27A. When the part which projects beyond the wall is a separate piece nailed to the rafter, it is called a *lookout.* See Fig. 5-27B.

Bird's-mouth. The cutout near the bottom of the rafter (when there is an overhang) which fits over the rafter plate is called the *bird's-mouth.* A bird's mouth is shown in Fig. 5-27C. Some carpenters might call this a *seat cut,* although a seat cut is actually the horizontal cut of the bird's mouth. The bird's mouth is made with two saw cuts, the seat cut and the vertical cut called the *heel cut.* (A rafter without an overhang has a seat cut and also a heel cut.) See Fig. 5-27C.

Rise, run, and line length. The underlying principle of roof framing is to relate all layout problems to the three sides of a right triangle. The simplest type of common rafter, shown in Fig. 5-28A, illustrates how these three sides are related in a roof. The horizontal side is called the *run,* the vertical side is called the *rise,* and the hypotenuse is called the *line length.* The basis for layout continues to be the right triangle, even when rafter layout becomes more complex, as shown in Fig. 5-28B.

The Run. The base of the right triangle used

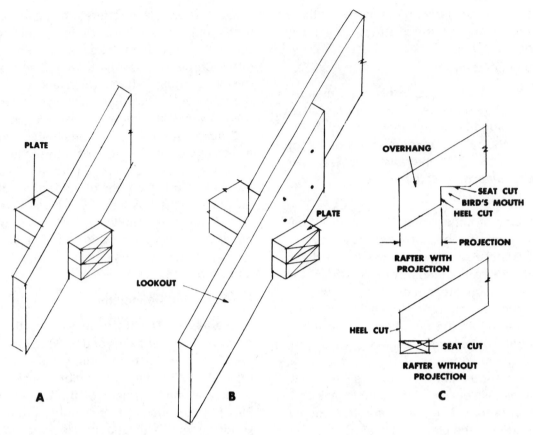

Fig. 5-27. The overhang is the part of the roof which extends beyond the plate projection. The horizontal dimension for the overhang is called a *projection*. A lookout is a part added on to a rafter.

for rafter layout is the dimension running from the outside of the plate to the center line of the roof. The horizontal dimension is plumbed from the center line of the ridges rather than from the side of this member.

Total Rise of Rafter. The altitude of the right triangle (or theoretical total rise) is the vertical (or plumb) distance between the top point of the heel cut on a simple rafter and the top edge of the rafter extended to the center line of the ridge. See Fig. 5-29A. The theoretical total rise can also be described as the vertical (or plumb) distance between the top plate corner and a line drawn parallel to the top edge of the rafter, extended to the center line of the ridge. See Fig. 5-28B.

Theoretical Total Length of Rafter (Line Length). The hypotenuse of the triangle is mea-

sured along the top edge of the rafter and extended to the center line of the ridge. See Fig. 5-29A. The hypotenuse of the triangle can also be considered as a line from the plate corner drawn parallel to the edge of the rafter and extended to the center line of the ridge. See Fig. 5-29B. (Note that the hypotenuse is not the center line of the rafter.)

The span. Distance between rafter plates (from outside to outside) of the building is known as the *span*. See Fig. 5-29A. The span distance is found on from the working drawings by studying the plan view for the floor which is to be located immediately below the roof. When a building is under construction, the span measurement can be found by actually measuring the dimension from the outside of one plate to the outside of the plate on the other side of the building.

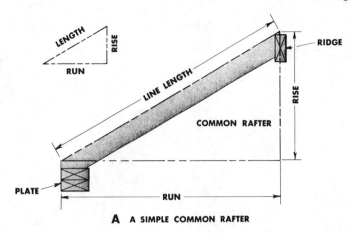

A A SIMPLE COMMON RAFTER

Fig. 5-28. Rafter layout is based on the concept of a right triangle.

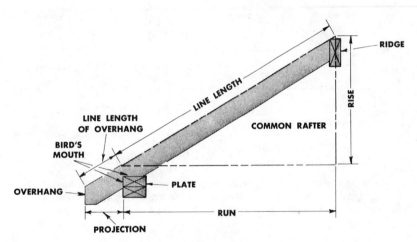

B A COMMON RAFTER WITH OVERHANG

Unit Measurements

Unit measurements play an extremely important role in rafter layout. They are shown on elevation or section views of the working drawings by a small triangle. See Fig. 5-30A. They are called *unit run, unit rise,* and *unit length.* The unit length is derived from the unit run and the unit rise. See Fig. 5-30B. The relationship between these three units, expressed in units of rise per foot of run, is called the *slope.* By the using unit measurements, you can determine the total (or line) length. The cut at the ridge, the cut at the rafter plate, and the overhang of the rafter are always laid out using unit measurements.

Unit run. The unit run for all common and jack rafters is one foot.

Unit rise. The rise in inches that the rafter extends in a vertical (or plumb) direction for every foot of unit run is the *unit rise.* The slope of a roof is expressed in terms of unit rise. The rafter, shown in Fig. 5-30, is described by the triangle as a 6 inch rise per foot (of run).

Unit length. The bridge measure or the hypotenuse of the right triangle, formed by the unit run (12 inches) and the unit rise (6 inches), is the *unit length.* See Fig. 5-30. This is obtained by measuring or bridging between the horizontal (the run) and the vertical (the rise) measurements on two sides of a framing square.

The cut. The cut of a roof is another expression for the unit rise and run. Carpenters refer to a roof as having a "cut of 8 inches per foot".

Pitch. It is obsolete to describe the slope of a

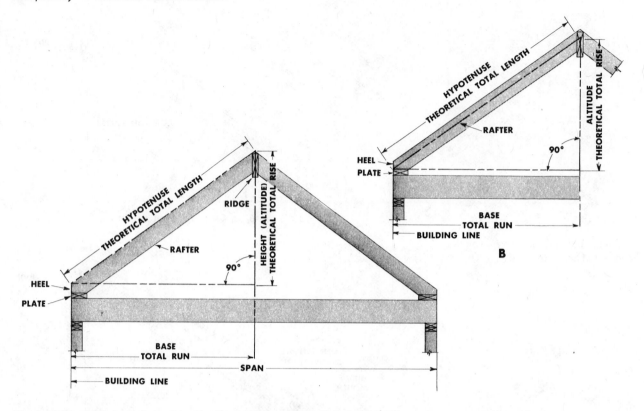

Fig. 5-29. The basic triangle representing the rafter is made up of the base (total run), the altitude (total rise), and the hypotenuse (total length). A triangle based on the point at the top of the heel cut is shown in (A). The triangle based on the point at the top of the plate is shown in (B). The triangles for both (A) and (B) are identical.

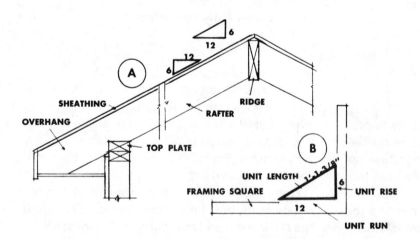

Fig. 5-30. Unit measurements are used in rafter layout. They are unit run, unit rise, and unit length. The small triangle in (A) is used to describe the relationship among them. The unit length is devised from the unit rise and the unit run, as shown in (B).

roof with the term *pitch.* (This term is included here so that if you encounter it in the field, you will understand how to convert it into the figures for unit rise and unit run.) The pitch is a ratio between the total rise and total span of a roof. It is expressed in a fraction, such as ¼ pitch or ½

pitch. A roof with a 7 foot total rise and 28 foot span would have a ¼ pitch roof. The way to convert from pitch to unit rise is to multiply the fraction by 24. For example, if the pitch of a roof is given as ⅓, you multiply ⅓ by 24 to arrive at the unit rise. (⅓ × 24 = 8.) Therefore, the unit rise

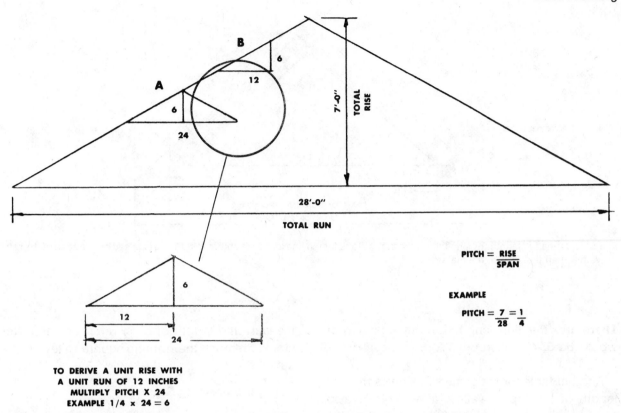

PITCH = $\dfrac{\text{RISE}}{\text{SPAN}}$

EXAMPLE

PITCH = $\dfrac{7}{28} = \dfrac{1}{4}$

**TO DERIVE A UNIT RISE WITH
A UNIT RUN OF 12 INCHES
MULTIPLY PITCH X 24
EXAMPLE 1/4 x 24 = 6**

Fig. 5-31. This illustration shows how to convert from pitch to unit dimension. Pitch is based on an isosceles triangle, as shown in (A). Unit dimensions are based on a right triangle, as shown in (B).

here is 8 inches per foot. Figure 5-31 shows why the number *24* is used. Pitch is based on an isosceles triangle. Unit rise is based on a right triangle with a 12 inch base.

RAFTER LAYOUT METHODS

Several different methods are used to determine the length of rafters and to lay out the cuts. They are discussed in the sections which follow.

Using the Full Length Rafter Table

Tables are available which give the line lengths of all common, hip, and valley rafters for roofs with slopes of ½ inch rise per foot or run, up to 24 inches rise per foot of run, in ½ inch increments.[1] The line length of the rafter is laid out on the rafter stock, and a framing square is

[1] A. Riechers, *Full Length Roof Framer*. Box 405. Palo Alto, California 94302

used to lay out the cuts. With the assistance of the rafter tables, a carpenter is able to layout all of the rafters for a roof very easily and quickly.

The book of tables is arranged so that there are two pages of information and tables for each slope (such as an 8 inch rise per foot). The tables are based on the span of the roof, including the overhang on either side. See Fig. 5-32. A model roof used as a learning tool later in this chapter has a slope of 8 inches rise per foot and a span of 9'-6". Information on the table is specifically for rafters which have a slope (unit measurement) of 8 inches rise per foot. The table indicates the following:

Span in Feet	Length in Feet and Inches	Span in Inches	Length in Inches
8	4'-9¾"	5	3
9	5'-4⅞"	6	3⅝
10	6'-0⅛"	7	4¼

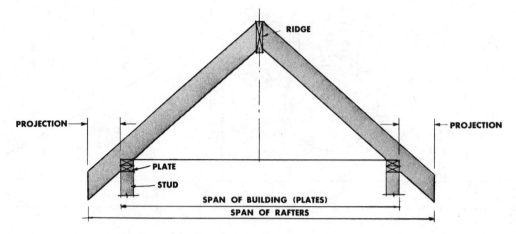

Fig. 5-32. In the full length framing tables, the dimensions for the line length of common rafters are based on the span of the rafters rather than the span of the plates.

Therefore, the line length for the model roof would be 5'-4⁷⁄₈" (9 ft.) + 3⁵⁄₈"(6 inches), or 5'-8¹⁄₂".

Another table for the same slope gives the line length of the hip or valley rafter. This is also based on the span of the roof. Information on the table indicates the following concerning hip rafters:

Span in Feet	Length in Feet and Inches	Span in Inches	Length in Inches
8	6'-3"	5	3⁷⁄₈
9	7'-0³⁄₈"	6	4³⁄₄
10	7'-9³⁄₄"	7	5¹⁄₂

The line length for a roof with a span of 9'-6" would be 7'-0³⁄₈" + 4³⁄₄", or 7'-5¹⁄₈".

Another table for the same slope shows the length of jack rafters. The most common use for this table is to find the common difference in length of a series of jack rafters spaced 16 or 24 inches O.C. The following items are shown for a slope of 8 inch:

Jack rafters spaced 16 inches O.C., common difference 1'-7¹⁄₄"

Jack rafters spaced 24 inches O.C., common differences 2'-4⁷⁄₈".

When you use these full length rafter tables,

measure the length of each rafter on the rafter stock, then use the framing square to layout the cuts.

Using the Framing Square

The framing square is handy for finding line lengths for two framing methods. One method is the step off method, and the other method uses the rafter table stamped on the side of the square.

Many of the layout operations calling for the framing square use the square in the manner shown in Fig. 5-33. Using the blade for the run (unit horizontal measurement) and the tongue for the rise (unit vertical measurement), and by measuring diagonally between these two

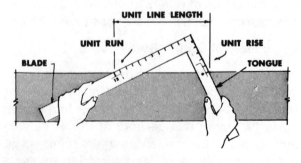

Fig. 5-33. The framing square is held on the unit run and the unit rise to "bridge the square". Plumb cuts are marked along the tongue, and level cuts are marked along the blade.

points, you can derive the unit line length. This is called *bridging the square.* The procedure for laying out the cuts on rafter ends using the framing square is covered in connection with each specific type of rafter.

The step off method. You can find the line length of a rafter by using the framing square set to the cut (rise per foot of run) of the rafter. By marking and stepping off the same number of steps as there are feet of run, you can automatically determine the line length. The roof, shown in Fig. 5-34A, has a run of 5'-0" and a cut of 8 inches. To find the line length, lay a board on a pair of horses and begin with a line which represents the center line of the ridge. Lay out 5 steps (one step for each foot of run). The last step locates the top point of the heel cut and the total length. See Fig. 5-34B. Most carpenters lay out rafters upside down because it is more convenient to hold the framing square in this manner. See Fig. 5-33. Carpenters must work accurately to avoid an accumulation of errors.

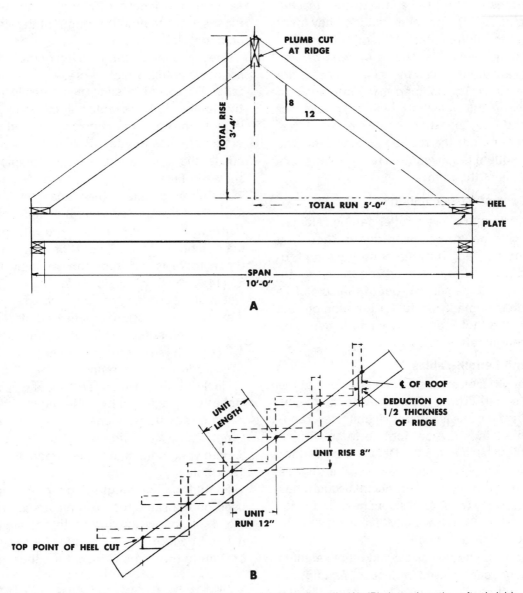

Fig. 5-34. A simple common rafter with a 5'-0" run and a slope of 8-12 is shown in (A). (B) shows how the rafter is laid out by taking five steps with the framing square.

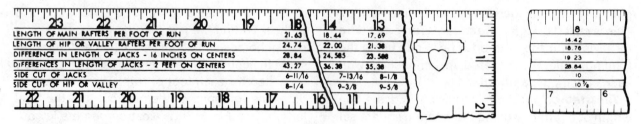

	21.63	18.44	17.69		
LENGTH OF MAIN RAFTERS PER FOOT OF RUN	21.63	18.44	17.69		
LENGTH OF HIP OR VALLEY RAFTERS PER FOOT OF RUN	24.74	22.00	21.38		
DIFFERENCE IN LENGTH OF JACKS - 16 INCHES ON CENTERS	28.84	24.585	23.588		
DIFFERENCES IN LENGTH OF JACKS - 2 FEET ON CENTERS	43.27	36.38	35.38		
SIDE CUT OF JACKS	6-11/16	7-13/16	8-1/8		
SIDE CUT OF HIP OR VALLEY	8-1/4	9-3/8	9-5/8		

Fig. 5-35. Tables stamped on the blade of the framing square supply information for establishing the line length and cuts for rafters.

Using The Taping Method

Carpenters often erect the common rafters and the ridge and then use a tape to measure the line lengths for hip and valley rafters. They measure from the ridge to the plate, or between the intersecting points. The taping method can be the most accurate of all if the measurements are made carefully. This method takes into account variations in the building which may have developed during construction. However, it requires that part of the roof be erected first. The framing square is still required to lay out the cuts at the ends of the rafters.

Using Geometric Calculations

The line length of every rafter can be calculated by using geometry. Using this method, you derive the line length by squaring the length of the total rise and total run, adding them together, then taking the square root of the sum. This approach is primarily useful for pointing out the basis for the tables on the framing square and the full length rafter tables.

Using Unit Length Tables

Unit length tables are shown on the face of the framing square. These tables provide the means to calculate the line length for all unit rises, ranging from a 2 inch rise per foot up to an 18 inch rise per foot. See Fig. 5-35. Data is included for common, jack, hip, and valley rafters. This method involves multiplying decimal quantities and converting them to feet, inches, and fractions of inches. The possibility of making a mathematical error is quite high.

The framing square has several other valuable features that can be used for measuring and laying out rafters. The square has scales divided into $\frac{1}{8}$, $\frac{1}{16}$, $\frac{1}{32}$, $\frac{1}{10}$, $\frac{1}{100}$, and $\frac{1}{12}$ of an inch.

These features will be discussed later. The square also includes data for calculating board feet and the length of diagonal braces, and a means for obtaining the length of the sides of an octagon. See Fig. 5-36.

The inch marks along the top of the table serve as index points which represent the cut of the rafter. For example, all of the figures listed below the number 8 on the table are related to a cut of 8 inches. The cut for common rafters and jack rafters is 8 inches rise for 12 inches of run. To calculate the line length of the common rafter shown in Fig. 5-34, you find the index number 8 in the framing square, then locate the line under it which is designated *Length of Main (common) Rafter per Foot of Run*. Here you will notice that the number is 14.42. This is the unit line length. Since there is a 5 foot run, multiply 14.42 by 5. (14.42 × 5 = 72.10) Referring to Table 1, *Decimal Equivalents of an Inch*, you will note that .10 inch is approximately equivalent to $\frac{1}{8}$ inch. Therefore the line length of the rafter is 6'-0 $\frac{1}{8}$". (72 inches are equal to 6 feet.)

The common difference between the lengths of jack rafters, which are evenly spaced 16 inches O.C., is found by following the line on the framing square designated as *Difference in Length of Jacks 16 inches O.C.* to the index point 8. Here you notice that the common difference is 19$\frac{1}{4}$ inches.

To find the line length of a hip or valley rafter used on the same roof as the common rafters in Fig. 5-34, follow the line on the framing square marked *Length of Hip or Valley Rafters per Foot of Run*[1] to the index point 8. Here the unit length

---

[1]The line designated *Length of Hip or Valley Rafter per Foot of Run* should read *Length of Hip or Valley Rafter per Foot of Run of the Common Rafter Related to the Hip Rafter.*

TABLE 5-1. DECIMAL EQUIVALENTS OF AN INCH.

4ths	8ths	16ths	32nds	64ths	to 2 places	to 3 places	4ths	8ths	16ths	32nds	64ths	to 2 places	to 3 places
				1/64	0.02	0.016					33/64	0.52	0.516
			1/32		0.03	0.031				17/32		0.53	0.531
				3/64	0.05	0.047					35/64	0.55	0.547
		1/16			0.06	0.062			9/16			0.56	0.562
				5/64	0.08	0.078					37/64	0.58	0.578
			3/32		0.09	0.094				19/32		0.59	0.594
				7/64	0.11	0.109					39/64	0.61	0.609
	1/8				0.12	0.125		5/8				0.62	0.625
				9/64	0.14	0.141					41/64	0.64	0.641
			5/32		0.16	0.156				21/32		0.66	0.656
				11/64	0.17	0.172					43/64	0.67	0.672
		3/16			0.19	0.188			11/16			0.69	0.688
				13/64	0.20	0.203					45/64	0.70	0.703
			7/32		0.22	0.219				23/32		0.72	0.719
				15/64	0.23	0.234					47/64	0.73	0.734
1/4					0.25	0.250	3/4					0.75	0.750
				17/64	0.27	0.266					49/64	0.77	0.766
			9/32		0.28	0.281				25/32		0.78	0.781
				19/64	0.30	0.297					51/64	0.80	0.797
		5/16			0.31	0.312			13/16			0.81	0.812
				21/64	0.33	0.328					53/64	0.83	0.828
			11/32		0.34	0.344				27/32		0.84	0.844
				23/64	0.36	0.359					55/64	0.86	0.859
	3/8				0.38	0.375		7/8				0.88	0.875
				25/64	0.39	0.391					57/64	0.89	0.891
			13/32		0.41	0.406				29/32		0.91	0.906
				27/64	0.42	0.422					59/64	0.92	0.922
		7/16			0.44	0.438			15/16			0.94	0.938
				29/64	0.45	0.453					61/64	0.95	0.953
			15/32		0.47	0.469				31/32		0.97	0.969
				31/64	0.48	0.484					63/64	0.98	0.984
1/2					0.50	0.500	1					1.00	1.000

is 18.76. Again you multiply this number by 5. (18.76 × 5 = 93.80). Therefore the line length is 93.80 inches, or 7'-9¹³/₁₆ inch.)

Finding the line length of the rafter by the taping method. After you have erected some of the common rafters and the ridge of the roof and braced them securely, you can tape the line length of the hip and valley rafters. This method is discussed later in this chapter under *Hip Rafters*, p. 279.

PLANNING THE RAFTER LAYOUT FOR A ROOF

Generally a set of working drawings does not include a plan view of the roof unless the roof is very complex. You must learn to visualize the roof from the information on the floor plan view of the house, which shows the shape and dimen-sions of the building, and from the elevation views, which show how the roof looks from the various sides. The section view gives information on the projection (horizontal measurement) at the eaves and the bird's-mouth. With this infor-mation, you must be able to build the roof with all of its supporting and secondary members. See Fig. 5-37. A beginner in carpentry may find it

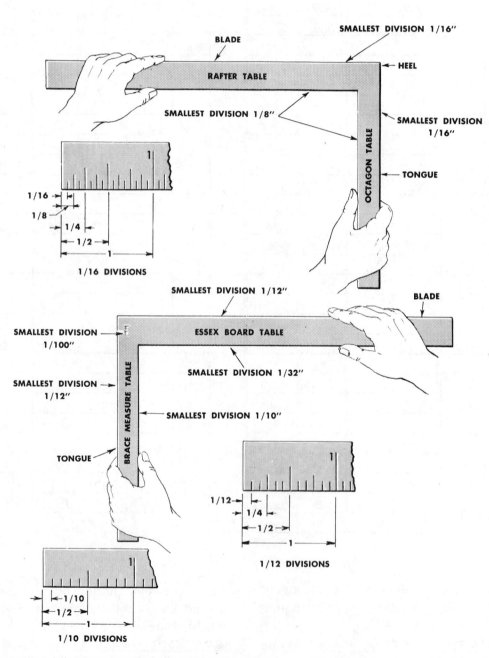

Fig. 5-36. The carpenter's framing square includes many useful features.

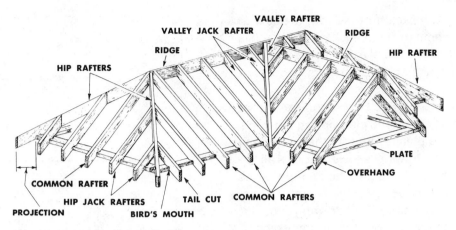

Fig. 5-37. This isometric view of a roof shows the various rafters in position.

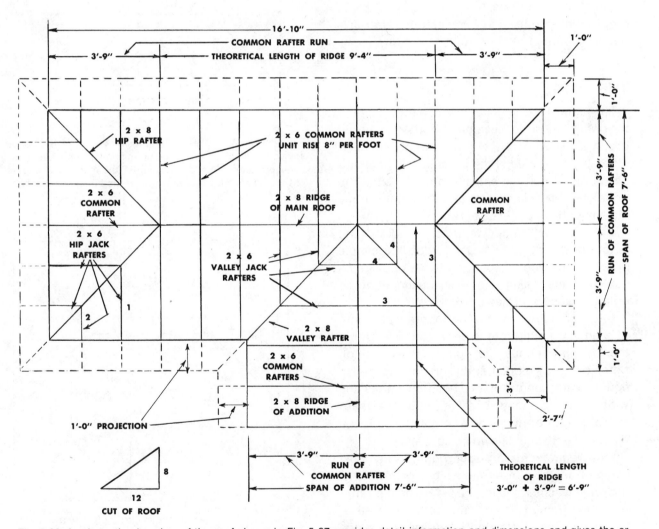

Fig. 5-38. A schematic plan view of the roof shown in Fig. 5-37 provides detail information and dimensions and gives the arrangement of the rafters.

helpful to make a sketch of the roof to scale to see how the members intersect. Such a sketch is shown in Fig. 5-38. The sketch can be done directly over the floor plan of the working drawings.

The roof shown on Fig. 5-37 is the same roof as shown by the schematic plan view in Fig. 5-38. Each line on the plan represents the center line of a rafter. This drawing shows the overall dimensions of the building and the amount of the projection. (The *projection* is the horizontal distance of the overhang.) The drawing also includes the location of the major framing members and the problem that must be solved at the intersection of the hips and the ridge and at the intersection of the valley and the ridge. (Note: The roof shown in Figs. 5-37 and 5-38 is a model roof which is be used to demonstrate rafter layout procedures.)

THE RIDGE

The highest framing member of a roof is called the *ridge*. The ridge piece has a double function. It serves to line up the rafters opposite each other on either side of the ridge so that they are spaced to match the spacing on the plate. The ridge piece likewise serves as a means for firmly fastening the upper ends of rafters.

Gable roof ridge lengths. Finding the length of the ridge on a gable roof is a simple process. The ridge piece for a two-slope simple gable roof is the same length as the length of the building plus the projection at the gables.

Hip-roof ridge length. The theoretical length of a ridge on a hip roof is equal to the length of the building minus the run of the common rafter on each hipped end. This length is also equal to the length of the building minus the width of the building. See Fig. 5-39. The theoretical length of ridges of a complex roof is derived in the manner shown in Fig. 5-40.

True lengths for hip-roof ridge. The most common method of framing hip-roof rafters at the ridge is illustrated in Fig. 5-41. The hip rafter is shown framed against the common rafters. When you use this method, you must add one half the thickness of the common rafter to each

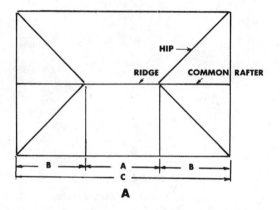

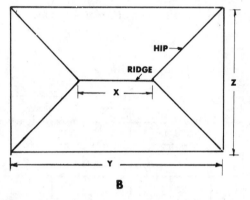

Fig. 5-39. In (A) above, the theoretical length of the ridge equals the length of the building minus the run of the end common rafters. (A = C − 2B) In (B), the theoretical length of the ridge is equal to the length of the building minus the width. (X = y − Z.)

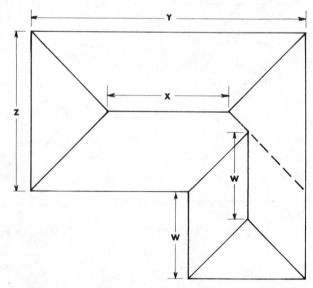

Fig. 5-40. The ridge length of complex roofs is related to the plate dimensions. (X = Y − Z) The plate and ridge length (both marked *W*) are equal.

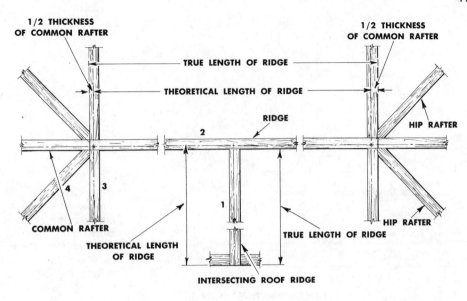

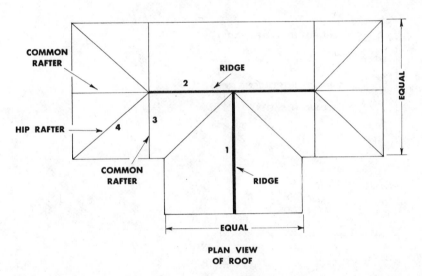

Fig. 5-41. The true length of ridge members must allow for the thickness of other members at intersections, as shown in the top view. The bottom plan view show the position of the rafter on the roof.

end of the theoretical length of the ridge. The length of the ridge is the theoretical length plus ¾ inch at each end (one half the thickness of the common rafter).

Intersecting roof ridge. When the span of an intersecting roof is the same as that of the main roof, as shown on the plan view in Fig. 5-41, the intersection of ridges is a simple process. The ridge of the intersecting roof is shortened ¾ inch (one half the thickness of the ridge stock).

When the span of the addition or intersecting roof is smaller than that of the main roof, the ridge of the addition is lower and does not meet the ridge of the main roof. See plan view Fig. 5-42. One method frequently used is to frame the whole main roof as though there were no secondary roof. After this roof is sheathed, the end of the ridge for the secondary roof is cut to fit the slope of the roof sheathing. The secondary roof is then built to rest on the sheathed main roof. This method of framing is discussed later in this chapter under *Blind Valley,* p. 292.

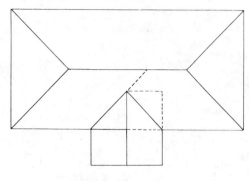

PLAN VIEW OF ROOF

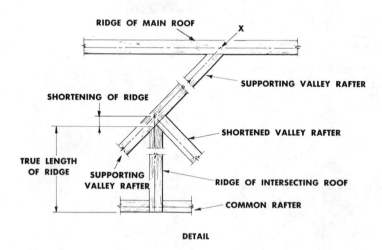

RIDGE OF MAIN ROOF

X

SHORTENING OF RIDGE

SUPPORTING VALLEY RAFTER

TRUE LENGTH
OF RIDGE

SHORTENED VALLEY RAFTER

SUPPORTING
VALLEY RAFTER

RIDGE OF INTERSECTING ROOF

COMMON RAFTER

DETAIL

Fig. 5-42. The ridge of the main roof intersects a supporting valley rafter at point *X.* The shortened valley rafter butts against the supporting valley. The plan view at the top shows the location of the valley rafters in relation to the whole roof.

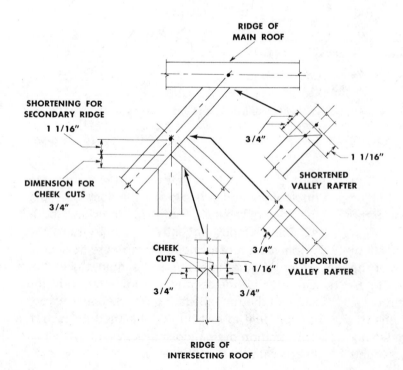

RIDGE OF
MAIN ROOF

SHORTENING FOR
SECONDARY RIDGE
1 1/16"

3/4"

1 1/16"

SHORTENED
VALLEY RAFTER

DIMENSION FOR
CHEEK CUTS
3/4"

CHEEK
CUTS

3/4"

1 1/16"

3/4"

3/4"

SUPPORTING
VALLEY RAFTER

RIDGE OF
INTERSECTING ROOF

Fig. 5-43. The shortening and cheek cuts allow rafters to fit properly.

One method used, especially when maximum attic space must be provided, is shown in Fig. 5-42. One valley rafter, called a *supporting valley rafter,* is carried up to the ridge of the main roof. This is shown by the dotted lines on the plan view and point X on the detail view. The supporting valley rafter is shortened 1 1/16 inches (one half the 45° thickness of the ridge stock) with miter cuts of 3/4 inch on each side of the shortening point. See detail of cuts in Fig. 5-43. The shortened valley is shortened 3/4 inch to butt against the supporting valley. The ridge of the intersecting roof is shortened 1 1/16 inch and provided with cheeks marked 3/4 inch back on each side. (Cheek cuts are diagonal sidecuts.)

COMMON RAFTERS

When drawn on a plan view or viewed from above, the common rafter is that member of the roof which extends at a right angle from the rafter plate to the ridge. The common rafter takes its name from the fact that it is the simplest rafter and is the basis for the layout of other rafters. A common rafter does not have mitered cuts on either end.

Layout Methods for Common Rafters

Three methods can be used to lay out common rafters. These are the full length rafter table method, the unit length method, and the step off method.

Full length rafter table method. If a set of full length rafter tables are available, the carpenter can use this method to determine the line length. This method is both accurate and quick. A framing square set at the cut of the rafter is used to lay out the cuts to mark the lines which indicate the cuts that are to be made.

Unit length method. Another method for laying out common rafters is the unit length method which uses the unit length rafter table on the side of the framing square. The line of information on the table designated *Length of Main (Common) Rafter per Foot of Run* gives the line length using the even number inch increments in rise per foot between 2 and 18 inches. By multiplying the number found in the table by the

number of units of run for the whole rafter, you can derive the total length. After the total length is marked on the rafter stock, use the framing square to lay out the cuts for the bird's mouth, the tail cut, and the plumb cut at the ridge.

Step off method. The step off method is often used because it does not require calculations or obtaining a book of full length rafter tables. The framing square is set to the unit rise and unit run and placed on the rafter stock. By repeating the marking of each step as many times as there are units of run in the total run, you can derive the total length. If you lay out the steps carefully, the full length dimension which results will be accurate. This method requires more time than some of the other methods, but it is best for learning rafter framing because no short cuts are used. Once you have mastered the subject of rafter framing, you should use the method which produces the most accurate results in the shortest period of time.

Finding Rafter Information

Gathering information about the unit rise, the span, and the total run for the rafter is basic to any layout. The unit rise per foot is usually indicated by a triangle adjacent to the roof slope that appears on one of the elevation views, or on the rafter itself on a section view of the working drawings. See Fig. 5-44. If the slope of the roof is described by a pitch instead of a unit rise, multiply the pitch by 24 to obtain the unit rise. For example, if the pitch is given as 1/3, multiply 1/3 by 24. (1/3 × 24 = 8). Therefore the unit rise is 8 inches per foot.

The total run of a simple gable roof would be one half of the span. To find the total run of common rafters on an irregularly shaped roof, you must note the spans of the various parts of the building as indicated on the floor plans, then study the shape of the roof as shown on the elevation views. (Note: The span of a roof is generally considered as the dimension from the outside of the plate on one wall to the outside of the plate on the opposite wall. The span as used in the full length rafter tables includes the overhang and is equal to the outside dimension of the two plates plus the projection on either side

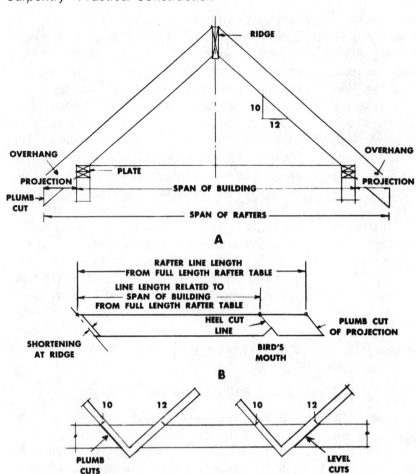

Fig. 5-44. This illustration provides the details for finding the line length and establishing points on a common rafter using the full length rafter table.

of the roof. This is shown as the span of the rafter on Fig. 5-44.) The carpenter should mark some of the information on the rafter stock before starting so that it is available for ready reference.

Layout Using the Full Length Rafter Table Method

(Note: The roof shown in Figs. 5-37 and 5-38 serves as a model to demonstrate procedure. It is fully discussed later in Chapter 5-1, *Practice in Roof Framing*.) The main advantage of using the full length rafter table method is that the line length of the whole rafter is easily found by referring to a table. You need the information about the cut of the roof (rise in inches per foot of run) and the span of the rafters from the plumb cut on the end of the overhang of one rafter to the

plumb cut on the overhang on the end of the opposite rafter. See Fig. 5-44A.

By adding the quantities which apply for the cut of the particular roof, as indicated on the full length rafter table, you can determine the line length of the rafter. An example of readings taken from a table are shown on p. 257. When laying out the rafter, mark the line length on the top edge of the rafter stock. See Fig. 5-44B. (Plumb and level cuts are made by holding the framing square against the stock, using the cut of the rafter as two points of reference. See Fig. 5-44C.) Plumb cuts are marked along the tongue side of the square. One line is marked to represent the centerline of the ridge and another is marked to represent the tail cut on the rafter. See Fig. 5-44B. Level cuts and measurements are marked by holding the framing square in the same position. However measure on the blade side. The shortening at the ridge is measured and marked, then the bird's mouth is located. See Fig. 5-44B. (The layout of cuts is covered under *End and Bird's Mouth Cuts for Common Rafters* in the next column.)

Layout Using the Unit Length Method

The unit length table is found on the side of the framing square, as discussed on p. 260. The unit length measurement (corresponding to the unit rise for the roof) is multiplied by the number of feet of run to obtain the full length dimension.

The first step in the layout process is to mark off the line length along the top edge of the rafter. See Fig. 5-45. Using the cut of the rafter on the framing square (the unit rise and unit run), draw lines which represent the center line of the ridge and the heel cut line (or building line). Laying out the heel cut or bird's mouth and shortening the rafter at the ridge is done with the framing square. (The layout of cuts is fully covered later in this chapter.)

Layout Using the Step Off Method

A piece of stock is chosen to be used for the rafter pattern. Lay this piece of stock across a pair of sawhorses so that the top edge is away from you. Lay the square on the stock with the heel away from you, as shown in Fig. 5-46. This is done because it is more convenient to work with it in that position. (The carpenter must be aware that the rafter is laid out upside down from the position it takes when eventually installed in the roof.)

The framing square is set against the top edge

of the rafter in the position shown in Fig. 5-47A. It is set with the unit rise on the tongue and the unit run (12 inches) on the blade. The first line to be marked along the tongue is the center line of the ridge. As many steps are marked off as there are feet of run for the rafter. The last point to be marked (point *X* in Fig. 5-47A) is for the outside of the plate (the building line). It is important that all of the lines marked when stepping off the rafter be marked accurately. A sharpened pencil or knife can be used. Framing square clips, shown in Fig. 5-46, are time savers.

When the total run of the rafter is designated by an even number of feet, all of the steps are equal, having a unit run of 12 inches. When the total run is not in even feet, for example 14'-6" or 11'-2", the rafter has an odd unit. In order to avoid leaving out this odd unit, lay it out first. Place the framing square on the stock using the cut of the rafter. See position *1* in Fig. 5-48. Draw a line along the tongue side to represent the center line of the ridge. Reading the number of inches in the odd unit along the blade side, make a mark, shown as point *A* in Fig. 5-48. Now shift the square so that the tongue side is directly on this line, then draw a line through the point. This point is the starting point you use for laying out all the full units of run for the rafter.

End and Bird's Mouth Cuts for Common Rafters

Shortening rafter at ridge. The ridge is usually made of 2 inch thick (1½") material. The rafter layout is based on a theoretical triangle which makes no allowance for the ridge member. Therefore, the ridge cut on the common raft-

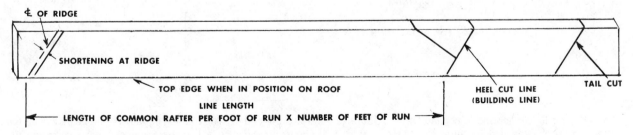

Fig. 5-45. The line length is derived from the unit run table on the framing square. The cuts are laid out using the framing square. (Note: The rafter is in the layout position.)

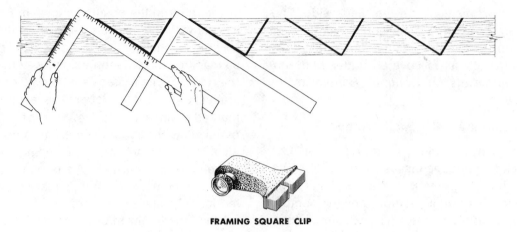

FRAMING SQUARE CLIP

Fig. 5-46. Carpenters usually lay out rafters upside down from the position the rafter will take on the roof. It is easier to hold and read the framing square in this position. (This is called the *layout position*.) Framing square clips (one is shown below) add accuracy and speed to the job.

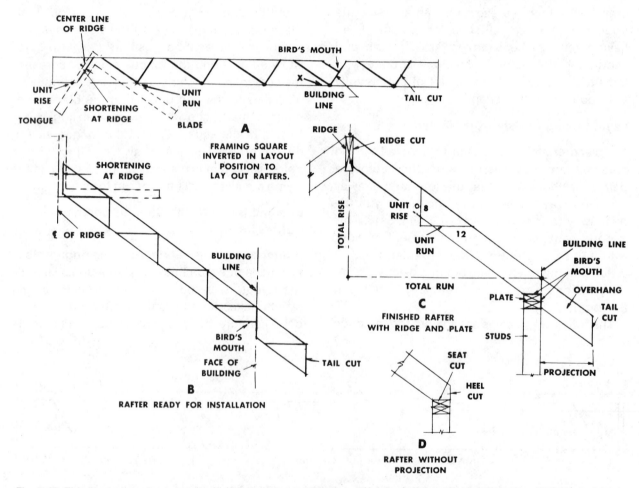

Fig. 5-47. This illustration shows common rafter layout using the step off method. The framing square should be placed on the rafter stock so that the markings are made on the far side of the square, as shown in (A). The carpenter should visualize how the rafter will look once it is in position for installing on the roof. See (B). The rafter will fit at the ridge and plate (C), if it is laid out correctly. A rafter without a projection is finished with a heel and seat cut, as shown in (D).

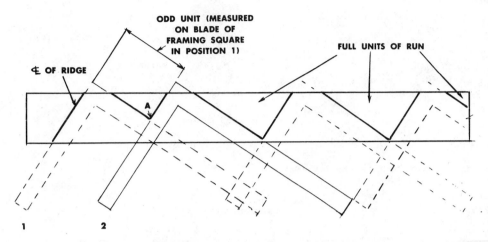

Fig. 5-48. The measurement for the odd unit is taken from the center line of the ridge.

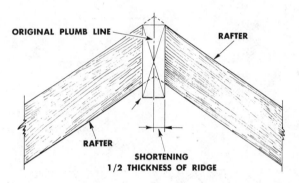

Fig. 5-49. The shortening at the ridge is made equal to one half the thickness of the ridge, as indicated by moving the framing square from position 1 to position 2. The detail at the top shows the rafters in position against the ridge.

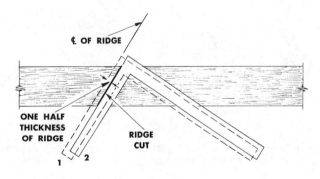

er must be shortened one half the actual thickness of the ridge stock where the common rafter rests against it.

Place the framing square in position *1* along the line representing the center line of the ridge. Use the cut of the rafter. See Fig. 5-49. The square is then shifted to position *2* (shown in Fig. 5-49). which is one half the thickness of the ridge away from position *1* measured on a level line parallel to the blade. A line drawn along the tongue becomes the ridge cut line.

Laying out the heel and the bird's mouth. The rafters are finished at their lower end in accord with the architectural style of the house. This is

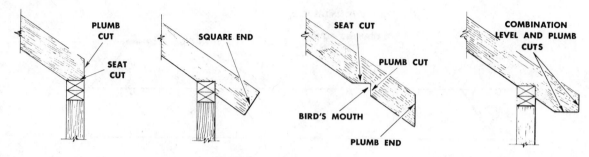

Fig. 5-50. The rafter end can be finished in several ways.

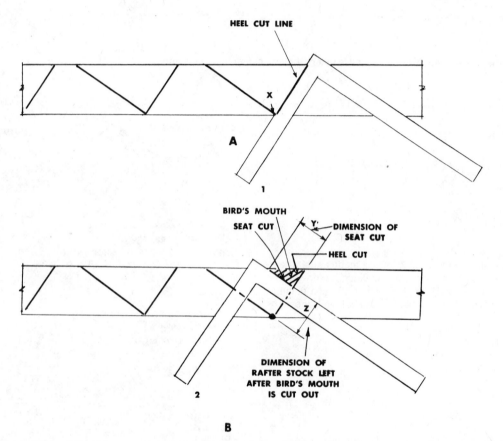

Fig. 5-51. The bird's mouth for the common rafter is laid out in two steps. Step 1 is shown in (A) above, with the framing square in position one. Step 2 is shown in (B) below, with the framing square moved to position two.

shown on a section view of the working drawings. Some of the various ways of finishing a rafter are shown in Fig. 5-50. Some buildings have no projections. When this is the case, the lower end of the rafter is terminated with a plumb heel cut (in line with the face of the wall) and a seat cut where the rafter rests on the plate. Generally however, most rafters are designed to have an overhang.

The heel (or plumb) cut of the bird's mouth is drawn with the framing square held in position *1*, as shown in 5-51A. This location coincides with the line at the outside of the plate.

Make the seat cut by laying the framing square

on the rafter and using the two dimensions of the cut (the rise per foot of run) as the points of reference. The square is moved sideways until the designated seat cut dimension is obtained. This is shown as dimension *Y* in Fig. 5-51B. This dimension should be sufficient to provide solid bearing for the rafter on the plate. It can be as large as 3½ inches for a 2 × 4 inch plate. It is important at the same time to leave enough stock to support the overhang, shown as dimension *Z* in Fig. 5-51B.

Laying out the overhang. Most rafters are designed to have an overhang so that the roof provides additional protection from rain and the sun. The amount of stock remaining at the heel after the bird's mouth is cut is important. This part of the rafter must be strong enough to support the overhang, the roofing, in some areas the weight of snow, and a worker who may have to repair the roof.

The relationship of overhang to the whole rafter differs, depending on the layout method used. When the full length method is used, the overall length of the rafter includes the overhang. When the other methods are used, the rafter is laid out to the plate line. The overhang although part of the rafter is considered separately.

With the full length method, the total length of the rafter is determined first. See point *X* in Fig. 52A. The tail cut line is drawn through this point, using the framing square set at the cut of the raft-

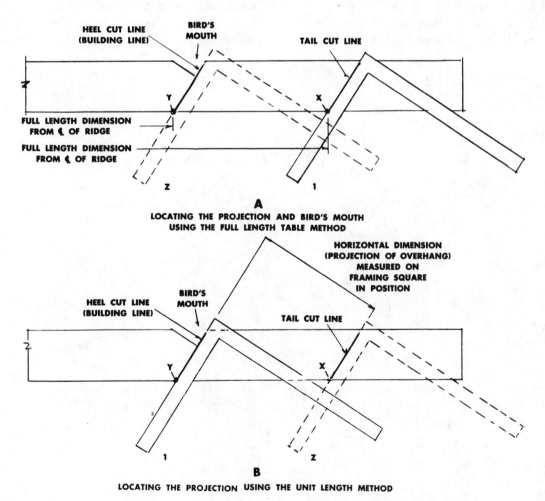

A
LOCATING THE PROJECTION AND BIRD'S MOUTH USING THE FULL LENGTH TABLE METHOD

B
LOCATING THE PROJECTION USING THE UNIT LENGTH METHOD

Fig. 5-52. The overhang can be located by using either of the alternative methods, shown in (A) and (B).

er. The heel cut of the bird's mouth is determined by obtaining the full length dimension from the center line of the ridge to the heel cut line. This dimension is obtained from the full length table. Use the span of the building and not the span of the roof.

When the unit length or step off method is used, the overhang is considered as though it were added on to the length of the rafter beyond the heel line of the bird's mouth. See Fig. 52B. If the projection is more than a foot, the framing square is moved as many times as required. In laying out the overhang part of the rafter, place the framing square in position at the building line, position *1* in Fig. 5-52B, then slide it to a position where the horizontal dimension of the projection is found on the blade of the square. The line drawn through this point becomes the tail cut line.

The tail cut can be finished in several ways. Three methods are shown in Fig. 5-50. When the finish for the overhang of a roof rafter is not shown on the working drawings, the carpenter or builder must decide on the type of design to use. The choice is governed by such considerations as the type of cornice, the width of the fascia, and the method used for supporting the overhang members. Some carpenters install the rafters to permit the overhang to project beyond the required dimensions. They snap lines to mark the projection and cut the rafters to the line.

Ladder Construction at Gable End

When there is a projection of the roof over a gable end, the construction is modified to provide support for the last pair of rafters and the roof sheathing. There are several ways to do this. When the projection at the gable ends is very wide, the method shown in Fig. 5-53 is often used. This is done by constructing an assembly resembling a ladder, which rests on the top

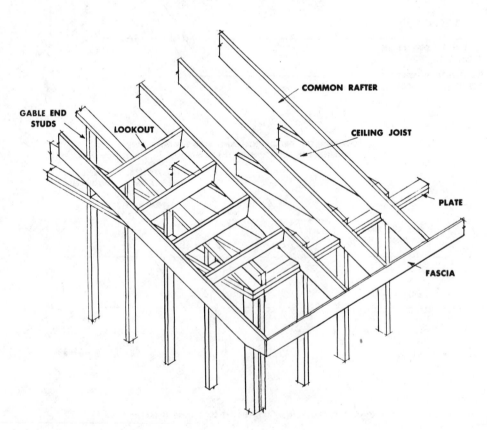

Fig. 5-53. Ladder construction is used for the projection at gable ends.

member of the gable end. The assembly is fastened to the pair of rafters set in one rafter space from the end of the building. The cantilever action of the horizontal pieces provides the support.

HIP RAFTERS

When two roof surfaces slope upward from the right angle external corner of two walls of a building, these two surfaces come together in a sloping line known as a *hip.* If both roof surfaces incline upward with the same slope, the two sides of the roof are said to be *equal pitch.* If the surfaces incline upward with different slopes, they are said to be of *unequal pitch.* An equal pitch hip roof is illustrated in Fig. 5-37. The rafters which extend diagonally from the corners of the building to the ridge are called *hip rafters.*

When you look directly down on a plan view of a roof, such as shown in Fig. 5-38, a hip rafter is the diagonal of a square. The square is formed by the theoretical total run of the common rafters and the plate lines. The diagonal of this square is the total run of the hip rafter. Each line on the plan view represents the total run of the respective rafters.

Since the unit of the common rafter is 12

inches, the unit run of the hip rafter is the diagonal of a 12 inch square, or 16.97 inches. See Fig. 5-54. The number 16.97 is so close to 17, that for all practical purposes it is quite satisfactory to use the number 17. Therefore for every 12 inch unit of common rafter run on an equal pitch roof, the hip rafter has a unit run of 17 inches. Figure 5-55 shows the relationship between the length of common and hip and valley rafters for slopes ranging from 1 to 24 inches rise.

To find a plumb cut of the hip rafter, take the unit run of 17 inches on the blade of the framing square and the unit rise of the common rafter (for that particular roof) on the tongue of the square. (A *plumb cut* is any vertical cut on the finished rafter when it is in position on the roof.) The square is laid on the rafter stock in the position shown in Fig. 5-56, with the tongue near the left end of the piece. A line drawn along the outside edge of the tongue gives the plumb cut of the hip rafter. A line drawn along the outside edge of the blade of the square gives a level cut of the hip rafter.

The basic right triangle used to describe the relationship between run, rise, and line length of a common rafter can be adapted to a hip rafter also. See Fig. 5-57A. The run however is the diagonal of a square base and the line length of a diagonal of the prism across opposite corners.

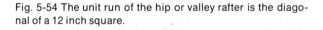

Fig. 5-54 The unit run of the hip or valley rafter is the diagonal of a 12 inch square.

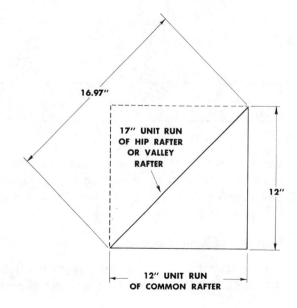

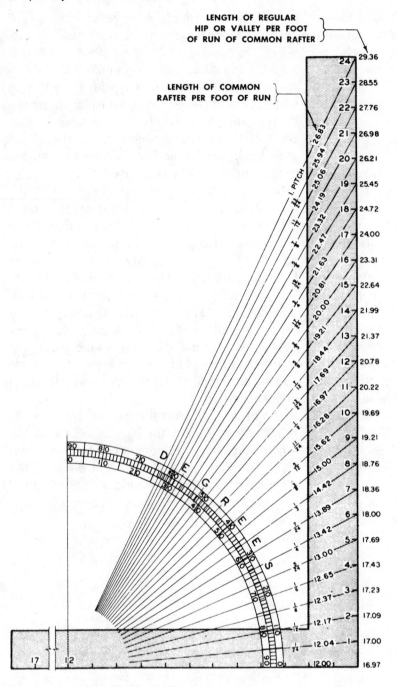

LENGTH OF REGULAR
HIP OR VALLEY PER FOOT
OF RUN OF COMMON RAFTER

LENGTH OF COMMON
RAFTER PER FOOT OF RUN

Fig. 5-55. This graph shows the unit rise per foot for related common and hip rafters.

See Fig. 5-57B. The rise is the same as that of the common rafter to which the hip is related. See Fig. 5-57C. Figure 5-58 shows how actual rafters relate to the right triangles.

Laying Out a Hip Rafter

The carpenter can use one of several methods for laying out the hip rafter. These are the full length rafter table method, unit length method, step off method and taping method.

Finding rafter information. The information needed to lay out related common rafters is the same basic information used for the hip rafter as well. The cut of the hip has the same unit rise. Instead of one foot, the unit run is 17 inches. Run dimensions for an odd unit or projection are

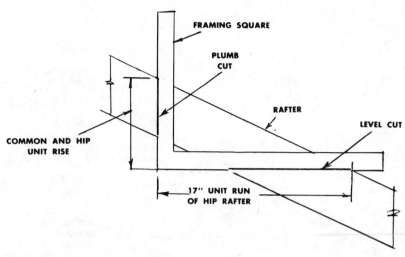

Fig. 5-56. Plumb and level cuts on the hip rafter are made using the cut of the hip. The unit rise is equal to the unit rise of the common rafter. The unit run is 17 inches. (Note: The rafter and framing square are not in layout position but in the position the rafter will actually take on the roof.)

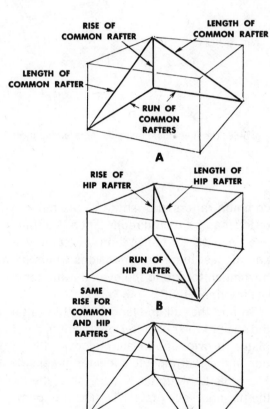

Fig. 5-57. The length of a common rafter is the diagonal of a rectangle, as shown in (A). The length of a hip rafter is the diagonal of a square prism, as shown in (B). The common and hip rafters have the same rise. See (C).

based on the diagonal of a square, with the dimension for the sides of the square derived from the common rafter.

Using the full length rafter table method. The main advantage of using the full length raft-

er table method is again its speed, once the carpenter knows how to read the tables correctly. A table is provided for each roof slope, beginning with 1 inch rise per foot and continuing up to 24 inches rise per foot. Tables covering lengths of

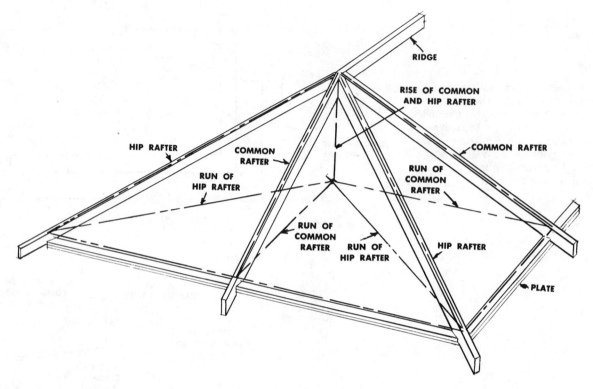

Fig. 5-58. The theoretical triangles representing the run, rise, and the line length of rafters are the basis for layout.

common rafters and length of the hip or valley rafters can be read directly when the span is an even number of feet. When the span is in feet and inches, there are provisions which indicate how much to add to the overall rafter length for the fractional unit.

To find the full line length of a hip rafter, add the span of the building from the outside of both plates to *twice* the measurement of the projection of the overhang, to allow for either side. See Fig. 5-59. Look up the line length of the rafter in the hip and valley table. To find the location of the heel cut (bird's mouth plumb cut), consult the table for the line length of the particular span of that building, not including overhang. Information about cuts is discussed later in this chapter under *Making Cuts on Hip Rafters,* p. 279.)

Using the unit length method. The unit length is found on a line on the side of the framing square designated *Length of Hip or Valley Rafter Per Foot of Run.* Slopes between 2 and 18 inches are shown in one inch increments. Each

number at the top along the rule itself serves as an index for a unit of rise. The reference *Per foot of Run* refers to the run of the common rafter related to the hip, rather than the run of the hip itself. By multiplying the appropriate number by the number of feet of run on the common rafter,

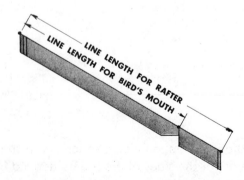

Fig. 5-59. The line length for the hip rafter is taken from the full length rafter tables. The line length for the bird's mouth is taken from the same table.

you can obtain the theoretical line length of the hip rafter. When there is an odd unit of run, convert the run of the common rafter into feet and decimals of a foot units before you multiply by the framing table factor. See Table 5-2 for the *Table of Decimal Equivalents of a Foot.*

Once the line length is known, the dimension is marked along the top edge of the rafter. The framing square is then used to lay out the bird's mouth, the overhang, and the shortening at the ridge and the side cuts. All of these cuts are shown on Fig. 5-60.

Using the step off method. The framing square is laid on the stock in position *1,* shown in Fig. 5-61, using the cut of the hip (unit rise of the common rafter and unit run of 17 inches). Draw lines along each side. Line *A* along the tongue side represents the plumb line at the center of the ridge. Using the cut of the hip, move the square as many times as there are units of run in the total run of the common rafter. Line *B,* position *3* in Fig 5-61, represents the last step, which is the plumb line above the corner of the building.

When the run of the common rafter is not in full foot units, an odd unit results for both the common and the hip rafter. You obtain the odd unit for the hip rafter by bridging the framing square, using the odd unit for the common rafter on the tongue and the blade. See Fig. 5-62. When laying out the rafter, the odd unit is usually laid out first. See Fig. 5-63, position *1.*Full steps of the framing square follow. See Fig. 5-63, position 2.

Using the taping method. The taping method provides an almost foolproof method for determining the line length of hip rafters. This method can also be used on roofs that have slopes which do not work out to an even number of inches or half inches so that the full length and unit tables can be used. Roofs with unusual features, such as ridges that are not level or are curved, can be framed using this method.

The roof framework must be partially completed before you begin working on the hip rafters. Once sufficient common rafters and the ridge are in place and the structure is rigid, you can measure the dimension from the corner of the ridge to the plate corner. See Fig. 5-64.

Each corner is measured in turn. This is an excellent procedure since it provides for variations in the building which could cause this dimension to vary. You must note, however, that the measurement is not the theoretical line length but a diagonal within the member, as shown in Fig. 5-65.

When using this dimension in the layout of the hip rafter, draw a line (*A* in Fig. 5-65) at the cut of the hip to represent the building line. The heel dimension is taken from the common rafter pattern. This point represents the plate corner. Make the taped measurement from this point so that it intersects the top edge of the member, thus locating the ridge corner point. The framing square is used to lay out the bird's mouth, the overhang, and the shortening at the ridge and the side cuts.

Making Cuts on Hip Rafters

Laying out the overhang. When the rafter has an overhang, place the framing square in position *1* at the building line, Fig. 5-66, and mark the point for the run of the projection on the blade side. If the common rafter has a projection which is in full one foot steps, the hip projection also is in full steps, with a unit run of 17 inches for each step. If the projection of the common rafter is an odd dimension, dimension *A* in Fig. 5-66 is the diagonal of a square, based on the odd unit for the common rafter. To save time, carpenters can allow the rafter to extend, then run a line from the common and jack rafters. The tail end is then cut on the line.

Dropping the hip rafter. The layout for the heel cut (building line), seat cut lines, and layout of the bird's mouth is like that used for common rafters. The same basic heel dimension (Fig. 5-67) is used for laying out the common and hip rafters for each roof. If the hip rafter were cut using the same heel dimension, the roof sheathing would not lie flat at the hip.

The center line of the hip rafter is the theoretical line where two roof slopes meet. This is one of the reasons why measurements for hip rafters are taken along the center lines of the top edge of the stock, rather than the edge of the rafter. To prevent the hip rafter from projecting above the

TABLE 5-2. DECIMAL EQUIVALENTS OF A FOOT.

0″	.0000	1″	.0833	2″	.66667	3″	.2500
1/16	.0052	1 1/16	.0885	2 1/16	.171875	3 1/16	.2552
1/8	.0104	1 1/8	.09375	2 1/8	.1771	3 1/8	.2604
3/16	.015625	1 3/16	.0990	2 3/16	.1823	3 3/16	.265625
1/4	.0208	1 1/4	.1042	2 1/4	.1875	3 1/4	.2708
5/16	.0260	1 5/16	.109375	2 5/16	.1927	3 5/16	.2760
3/8	.03125	1 3/8	.1146	2 3/8	.1979	3 3/8	.28125
7/16	.0365	1 7/16	.1198	2 7/16	.203125	3 7/16	.2865
1/2	.0417	1 1/2	.1250	2 1/2	.2083	3 1/2	.2917
9/16	.046875	1 9/16	.1302	2 9/16	.2135	3 9/16	.296875
5/8	.0521	1 5/8	.1354	2 5/8	.21875	3 5/8	.3021
11/16	.0573	1 11/16	.140625	2 11/16	.2240	3 11/16	.3073
3/4	.0625	1 3/4	.1458	2 3/4	.2292	3 3/4	.3125
13/16	.0677	1 13/16	.1510	2 13/16	.234375	3 13/16	.3177
7/8	.0729	1 7/8	.15625	2 7/8	.2396	3 7/8	.3229
15/16	.078125	1 15/16	.1615	2 15/16	.2448	3 15/16	.328125
4″	.3333	5″	.416667	6″	.5000	7″	.5833
4 1/16	.3385	5 1/16	.421875	6 1/16	.5052	7 1/16	.5885
4 1/8	.34375	5 1/8	.4271	6 1/8	.5104	7 1/8	.59375
4 3/16	.3490	5 3/16	.4323	6 3/16	.515625	7 3/16	.5990
4 1/4	.3542	5 1/4	.4375	6 1/4	.5208	7 1/4	.6042
4 5/16	.359375	5 5/16	.4427	6 5/16	.5260	7 5/16	.6093
4 3/8	.3646	5 3/8	.4479	6 3/8	.53125	7 3/8	.6146
4 7/16	.3698	5 7/16	.453125	6 7/16	.5365	7 7/16	.6198
4 1/2	.3750	5 1/2	.4583	6 1/2	.5417	7 1/2	.6250
4 9/16	.3802	5 9/16	.4635	6 9/16	.546875	7 9/16	.6302
4 5/8	.3854	5 5/8	.46875	6 5/8	.5521	7 5/8	.6354
4 11/16	.390625	5 11/16	.4740	6 11/16	.5573	7 11/16	.640625
4 3/4	.3958	5 3/4	.4792	6 3/4	.5625	7 3/4	.6458
4 13/16	.4010	5 13/16	.484375	6 13/16	.5677	7 13/16	.6510
4 7/8	.40625	5 7/8	.4896	6 7/8	.5729	7 7/8	.65625
4 15/16	.4115	5 15/16	.4948	6 15/16	.578125	7 15/16	.6615
8″	.666667	9″	.7500	10″	.8333	11″	.916667
8 1/16	.671875	9 1/16	.7552	10 1/16	.8385	11 1/16	.921875
8 1/8	.6771	9 1/8	.7604	10 1/8	.84375	11 1/8	.9271
8 3/16	.6823	9 3/16	.765625	10 3/16	.8490	11 3/16	.9323
8 1/4	.6875	9 1/4	.7708	10 1/4	.8542	11 1/4	.9375
8 5/16	.6927	9 5/16	.7760	10 5/16	.859375	11 5/16	.9427
8 3/8	.6979	9 3/8	.78125	10 3/8	.8646	11 3/8	.9479
8 7/16	.703125	9 7/16	.7865	10 7/16	.8698	11 7/16	.953125
8 1/2	.7083	9 1/2	.7917	10 1/2	.8750	11 1/2	.9583
8 9/16	.7135	9 9/16	.796875	10 9/16	.8802	11 9/16	.9635
8 5/8	.71875	9 5/8	.8021	10 5/8	.8854	11 5/8	.96875
8 11/16	.7240	9 11/16	.8073	10 11/16	.890625	11 11/16	.9740
8 3/4	.7292	9 3/4	.8125	10 3/4	.8958	11 3/4	.9792
8 13/16	.734375	9 13/16	.8177	10 13/16	.9010	11 13/16	.984375
8 7/8	.7396	9 7/8	.8229	10 7/8	.90625	11 7/8	.9896
8 15/16	.7448	9 15/16	.828125	10 15/16	.9115	11 15/16	.9948

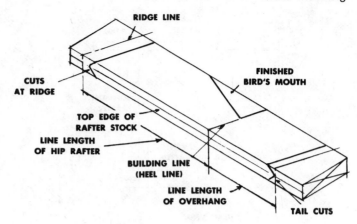

Fig. 5-60. When the unit length method is used, the line length is marked on the top edge of the rafters. Cuts are made using the framing square. (Note: The rafter is in the layout position.)

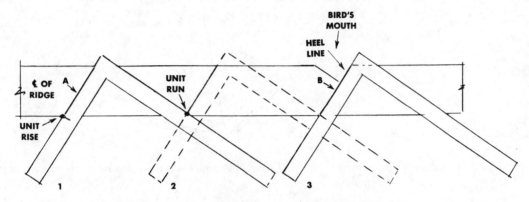

Fig. 5-61. The hip rafter is laid out with the step off method as shown. The unit run is 17 inches.

Fig. 5-62. The length of the odd unit may be determined by bridging the square between two points set at the odd unit run of the common rafter. (An example of 9 inches is provided here.)

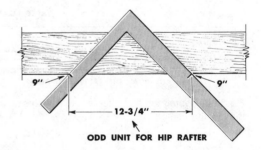

Fig. 5-63. An odd unit results when the run of the common rafter does not come out to full one foot steps. Such an odd unit is laid out first on the hip rafter as shown. The rafter is in the layout position.

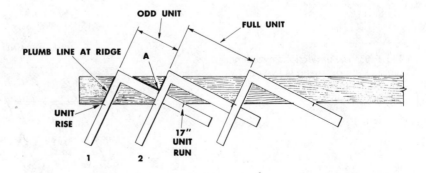

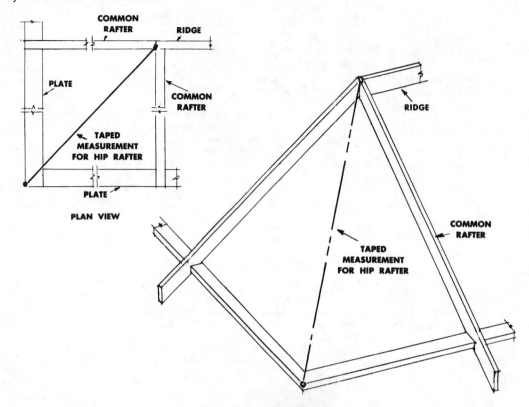

Fig. 5-64. The taping method for obtaining the line length of a hip rafter requires a measurement from the ridge corner to the plate corner.

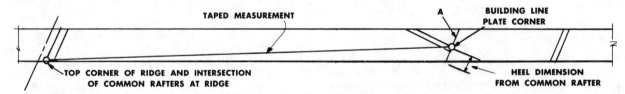

Fig. 5-65. Taping the hip rafter length requires an accurate measurement of the diagonal on the side of the rafter. The rafter is in the layout position.

Fig. 5-66. The projection (level dimension) of the overhang is laid out from the building line. The rafter is in the layout position.

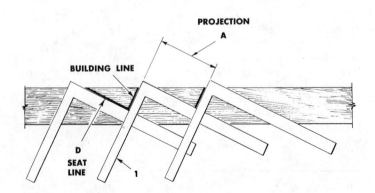

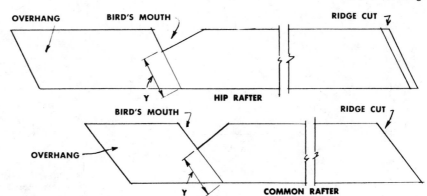

Fig. 5-67. The basic heel dimension (Y) is the same for common and hip rafters on the same roof.

jack rafters, as shown in Fig. 5-68, the top edge of the rafter must be dropped. The amount of drop is important so that the edge of the hip rafters and the top of the common rafters end up on the same plane. See Fig. 5-69.

The amount of drop can be determined by placing the square on the rafter (using the cut of the hip), then drawing a level line and marking one half of the thickness of the hip rafter on the line. See Fig. 5-70. The plumb distance to the edge of the rafter is the amount of drop *(D)* to be used at the seat cut *(E)*.

Some carpenters avoid the above procedure for determining the drop dimension by taking an approximate dimension. This practice is discouraged because the amount of drop varies greatly with the amount of slope. A cut of 3 inches rise in 17 inches of run requires a drop of only ⅛ inch. A cut of 18 inches rise in 17 inches requires a drop of ¾ inch.

Side cuts at heel. If the rafter does not have an overhang, side cuts are usually made to allow the fascia (extending from the row of common rafters) to lie flat and to provide a nailing surface. See Fig. 5-71A. As shown in Fig. 5-71B, a center line is drawn on the top edge of the rafter stock, and the building line *(1)* is squared across the top edge. One half the thickness of the hip rafters is laid off on a level line on the side of the building line toward the ridge. Plumb line *2* is drawn and also squared across the top edge of the piece of stock. Side cuts from line *2* to the center line (at point *X*) are drawn. Cuts are made as shown in Fig. 5-71C.

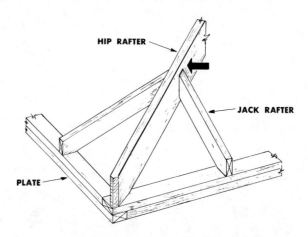

Fig. 5-68. If the heel dimension for both common and jack rafters is the same, the sheathing will not lie flat at the hip. Here you see the jack rafters lie below the top of the hip.

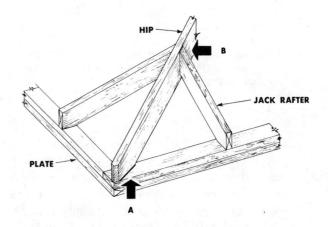

Fig. 5-69. When the hip is dropped, enough stock is cut away at the seat (A) so that the edge of the hip is level with the top of the jacks (B).

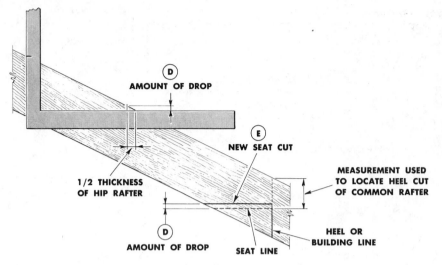

Fig. 5-70. The amount of drop is determined by laying out a level line anywhere on the hip rafter, measuring ½ the thickness of the stock (¾ of an inch) and then measuring the plumb distance at that point. The seat line is raised this amount so that the rafter drops accordingly.

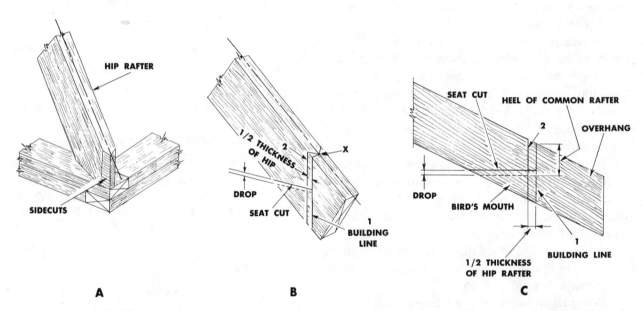

Fig. 5-71. The side cuts are made to provide a nailing base for fascia members when the hip rafter has no projection, as shown in (A). In (B), the cuts at the heel are laid out. When there is an overhang, a straight cut is made at the building line corner instead of the side cuts, as shown in (C).

Laying out the bird's mouth. When there is an overhang on the rafter, sidecuts are seldom made at the bird's mouth. The exception is when the hip is exposed to view from below, and the work must be finished carefully. Usually a plumb cut is made on the building line, line *1* in Figs. 5-71B and 5-71C, so that the bird's mouth does not interfere with the plate corner. The level cut (seat cut) of the bird's mouth is cut on the line representing the drop.

Side cuts at tail. The side cuts at the lower end of the rafter are necessary to accommodate the fascia of the cornice. See Fig. 5-72. Side cuts are determined by measuring one half of the thickness of the hip rafter on a level line. Carpenters can eliminate this step by having the hip

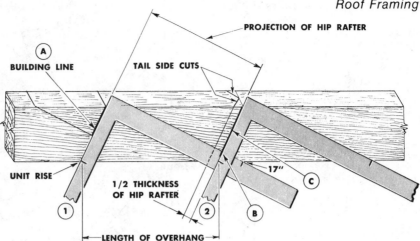

Fig. 5-72. The overhang for the hip is determined by using the diagonal of a square which has sides equal to the unit run of the common rafter projection. Side cuts are necessary. (Note: The rafter is in layout position.)

extend beyond the line of the common and jack rafter ends. A line is extended and snapped to mark the end of the hip. The rafter side cuts are made on these lines.

There are several quick methods for obtaining the side cuts at the heel and tail ends of the hip rafter. One method is provided by the full length rafter framing tables. These tables give two measurements to be used on the framing square for each roof slope. For example, the two numbers for a slope of 8 inch rise per foot are $9^3/_8$ and $8^1/_2$ inches. The bevel is marked as shown in Fig. 5-73A.

The other method uses the line in the table on the framing square designated *Side Cut of Hip or Valley*. A dimension is given for each slope. The dimension is used, with a constant of 12 inches on the framing square, to layout the side cut line. For example, for an 8 inch rise per foot, the two numbers used are $10^7/_8$ and 12 inches. This procedure is illustrated in Fig. 5-73B. The bevel is marked on the 12 inch side.

Shortening hip rafter at ridge. If the rafters and the ridge were merely lines or planes without any thickness, as they appear on the plan view (Fig. 5-38), these framing members would all meet at a point *X*. See Fig. 5-74. Since these framing members do have thickness, the rafters must be shortened at the ridge accordingly.

When the hip rafter is framed against the common rafters, as shown in Fig. 5-74, the shortening of the hip rafter is always one half of the 45 degree thickness of the common rafter, whether the material used for the ridge is 1 inch or 2 inches (nominal size) thick.

To find one half of the 45 degree thickness of the common rafter, lay the framing square across the edge of the rafter stock, using the same figures on each side of the square as shown in Fig. 5-75. A line *(A)* is drawn, and the distance is measured from the edge to the center of the stock. This provides one half the 45 degree thickness of the common rafter. (The dimension will be $1^1/_{16}$ inch for a $1^1/_2$ inch thick rafter.) This dimension is laid out on a level line at right angles to the plumb cut, shown at *A* in Fig. 5-76. A second plumb line *(B* in Fig. 5-76*)* is drawn and is squared across the top edge of the hip-rafter stock.

Making side cuts. Side cuts at the ridge can be made using the methods shown in Fig. 5-73 (either A or B), or with a layout such as shown in Fig. 5-74. After marking the shortening (line *B* in Fig. 5-76), measure one half the hip rafter thickness on a level line. This locates a third plumb line *(C)*. Lines *B* and *C* are squared across the top edge of the hip rafter stock. The side cut is drawn from the point where the line *(C)* intersects the edge of the stock to the center *(X)*. The side cut line for the other side is drawn to intersect at point *X*.

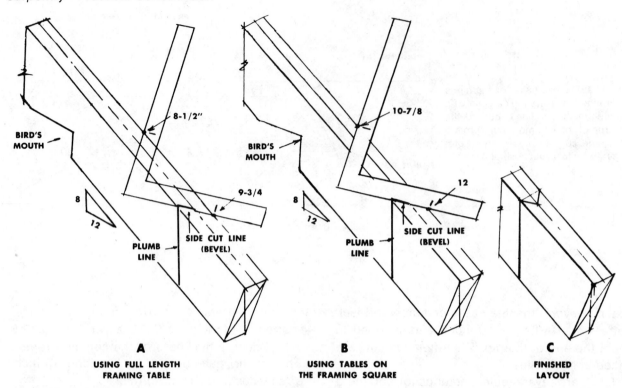

BIRD'S MOUTH

8-1/2″

9-3/4

8

12

PLUMB LINE

SIDE CUT LINE (BEVEL)

BIRD'S MOUTH

10-7/8

8

12

PLUMB LINE

SIDE CUT LINE (BEVEL)

12

A

USING FULL LENGTH FRAMING TABLE

B

USING TABLES ON THE FRAMING SQUARE

C

FINISHED LAYOUT

Fig. 5-73. The framing square is used to draw the side cut lines for the tail cut. In (A), the side cut dimensions are derived from the full length framing table. The dimensions in (B) are obtained from the tables on the framing square. The resulting side cut is the same using either method.

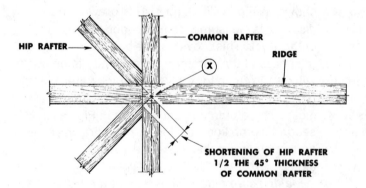

HIP RAFTER

COMMON RAFTER

RIDGE

X

SHORTENING OF HIP RAFTER 1/2 THE 45° THICKNESS OF COMMON RAFTER

Fig. 5-74. The shortening of the hip is one half of the 45° thickness of the common rafter.

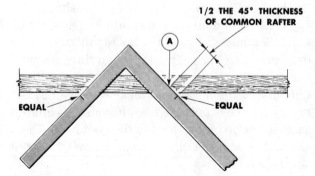

1/2 THE 45° THICKNESS OF COMMON RAFTER

A

EQUAL

EQUAL

Fig. 5-75. Take equal measurements of any amount on each leg of the framing square to mark a 45° diagonal on the common rafter. Measure one half of the diagonal, as shown from (A) to the point indicated here on the blade of the framing square.

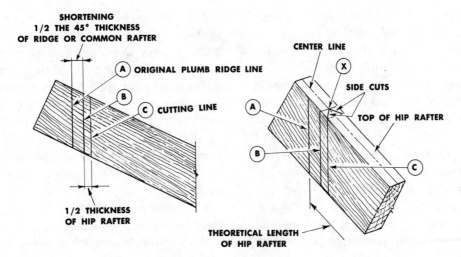

Fig. 5-76. The side cuts are first marked on the side of the hip rafter on a level line, as shown in the left view. In the right view, the points are connected on the top of the rafter to the center (X).

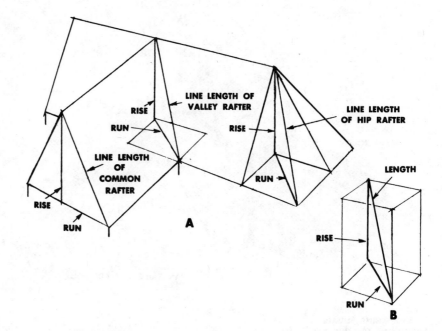

Fig. 5-77. The unit rise and full length rise is the same for related rafters. The run of hip and valley rafters is the diagonal of a square, as shown in (A). The line length of hip and valley rafters is the diagonal of a prism, as shown in (B).

VALLEY RAFTERS

Framing Intersecting Roofs

A valley rafter, as shown earlier in Fig. 5-37, p. 263, is the member which rises diagonally from an inside corner where two plates meet at right angles. The two slopes of the roof meet at this member. The basic right triangle used in conjunction with the hip rafter also applies to the valley rafter. The run of the valley rafter is the hypotenuse of a right triangle whose sides are equal to the run of the common rafter. See Fig. 5-77. The length of the valley rafter is a diagonal of a prism in the same manner as the hip rafter. The rise of the valley rafter is the same as the rise of the related common rafter.

The intersection of two roof slopes can cause several rafter framing problems. The main factor causing differences is the method used in framing the ridge of the intersecting roof.

When the span of the addition is the same width as the span of the main roof, the ridges for each roof meet on the same level, as shown at A in Fig. 5-78. In this case, the common rafter of both the main roof and the addition have the same run. The run of the valley rafter is the hypotenuse of a right triangle whose sides are equal to the run of the common rafters. The valley rafters have double side cuts to fit against both ridges, as shown in the detail at A in Fig. 5-78.

One method of framing a valley rafter is shown in Fig. 5-79. This method is used when the span of the addition is less than the span of the main roof. One of the two valley rafters (called a *supporting valley*) is framed against the ridge of the main roof with a single side cut, as shown in the detail at A in Fig. 5-79. The shortened valley rafter is then framed against the supporting valley rafter with a square cut. The run of the shortened valley rafter can be found by taking the hypotenuse of a right triangle which has sides that are equal to the run of the common rafter of the secondary roof.

For small roofs, such as dormers, you frame the ridge against a header between the common rafters of the main roof. See Fig. 5-80. The run of these valley rafters is found by using the run of the common rafters of the dormers with a 17 inch unit run.

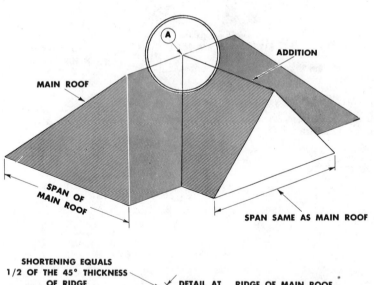

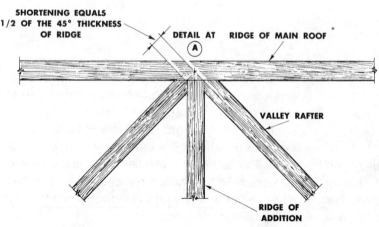

Fig. 5-78. When the span of the main roof and the addition are the same, the ridges and the valley rafters meet. This occurs at (A) at the top. Detail is provided below. The ridge of the addition butts against the ridge of the main roof.

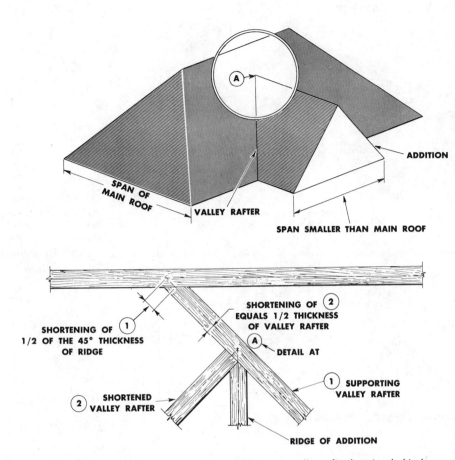

Fig. 5-79. When the main roof span is greater than that of the addition, one valley rafter is extended to become a supporting valley rafter. Detail of (A) at the top is provided below. The supporting valley rafter (1) and the shortened valley rafter (2) are shortened as shown.

Laying Out the Valley Rafter

The general procedure for laying out the valley rafter closely follows the method used for laying out the hip rafter. There are a few exceptions. The cut of each valley rafter has the same unit rise as the common rafter to which it is related. The run is 17 inches for each foot of run on the common rafter. When the run of the common rafter works out to be an even number of feet, the run of the valley rafter is made so it has the same number of 17 inch units. When the common rafter has an odd unit of inches, the valley rafter has an odd unit made up of the diagonal of a square. The square has sides equal to the length of the odd unit on the common rafter.

One difference between the layout method used for hip and valley rafters is that there is no need to drop the valley rafter at the seat cut to prevent the sheathing from buckling. When the cheek cuts at the ridge are properly made, and the member is correctly in place, the center line on the top of the valley rafter is automatically depressed to provide for the sheathing. See Fig. 5-81.

Finding the line length. The same four methods used earlier to find the line length of the hip rafter are used to find the line length of the valley rafter. The full length rafter table, used for both the hip and valley rafters, gives the line length directly. The span of the building over the plates is used to find the length to the heel point. The span of the common rafters, including the projections, is used to find the length to the tail cut point.

The unit length table on the framing square

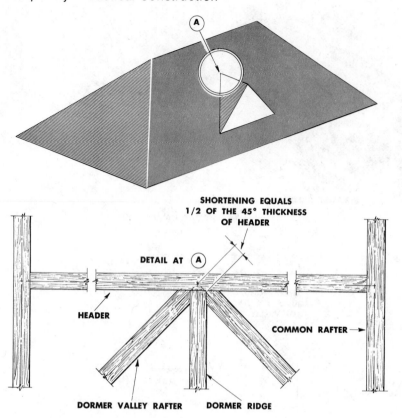

SHORTENING EQUALS
1/2 OF THE 45° THICKNESS
OF HEADER

DETAIL AT (A)

HEADER

COMMON RAFTER →

DORMER VALLEY RAFTER   DORMER RIDGE

Fig. 5-80. A method of framing the valley rafters for a dormer roof is shown here. The detail of (A) at the top is shown below, indicating that a header supports the end of the dormer ridge.

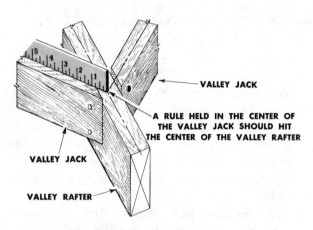

← VALLEY JACK

A RULE HELD IN THE CENTER OF
THE VALLEY JACK SHOULD HIT
THE CENTER OF THE VALLEY RAFTER

VALLEY JACK

VALLEY RAFTER →

Fig. 5-81. Valley rafters do not have to be dropped. A rule is used to make sure the plane of the top of the valley jack rafter terminates on the center line of the valley rafter.

can be used to provide the line length of the valley rafter. The line of unit dimensions designated *Length Hip or Valley Rafter per Foot of Run* is used to give the unit run. This quantity is multi-

plied by the number of units (feet) of run on the common rafter.

The step off method, when applied to the valley rafter, is identical to the procedure used for the hip rafter. The cut is the unit rise for the related common rafter with a unit run of 17 inches.

Taping the line length is a common practice because the part of the roof with the valley rafters is usually the last part of the roof to be assembled. Once the ridges are firmly in place, it is relatively easy to tape the dimension from the point where the ridges intersect to the plate corner.

Laying out the bird's mouth. The bird's mouth is laid out with the same plumb heel dimension used for the common rafter. See Fig. 5-82A and B. This dimension locates the level line for the seat cut. If the building has no projection, the heel line or building line point can be used as the plumb line of the valley rafter. See Fig. 5-82 A and B. However, the plumb cut for the heel does not provide as good as a nailing

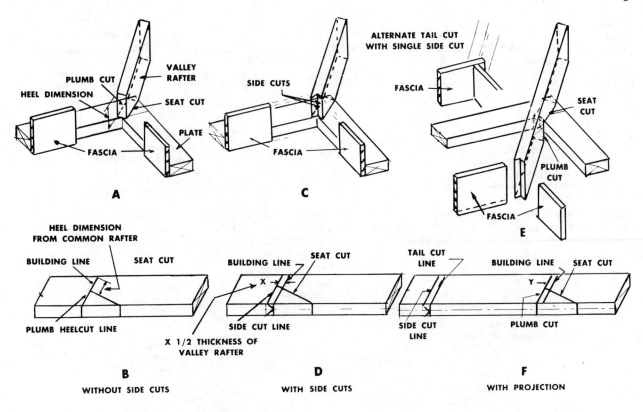

Fig. 5-82. Details of the valley rafter showing the bird's mouth and the tail cut.

base for the fascia boards as the side cuts shown in Fig. 5-82C. To make these side cuts, draw a line (marked *X* in Fig. 5-82D). This measurement is one half the thickness of the stock toward the lower end of the piece. Lines to define the side cuts are drawn on the top or bottom of the rafter. When the rafter has a projection, the line *Y* in both Figs. 5-82F and 5-83 is used as the plumb line for the bird's mouth. The line is cut square on this line and on the seat cut line to form the bird's mouth. See shaded portion in Fig. 5-82E.

Sides cuts at the rafter tail. Side cuts at the end of the overhang are laid out using one half the thickness of the valley rafter on a level line. These side cuts are taken beyond the tail line, instead of toward the bird's mouth, in order to provide nailing for the fascia. See Fig. 5-83. This layout procedure can be eliminated if the rafter is to be marked and cut when in position.

Shortening at the ridge. When the roof is framed in the manner shown earlier in Fig. 5-78, the layout for shortening the valley rafter is identical to that used for the hip rafter. See Fig. 5-83. When the roof is framed as shown in Fig. 5-79, the supporting valley is given a single side cut. Figure 5-84 shows the procedure used for this. The measurement *Y*, which is one half the 45 degree thickness of the ridge, is made from the plumb ridge line to locate the shortening line, as shown at *B* in Fig. 5-84. The cutting line, shown at *C* in Fig. 5-84, is drawn one half the thickness of the valley rafter, from line *C* through the center point *(X)*, to locate the cutting line on the opposite side. The upper end of a shortened valley rafter is cut off with a square end in the same manner as is used for a common rafter. See Fig. 5-79. The shortening is one half the thickness of the supporting valley.

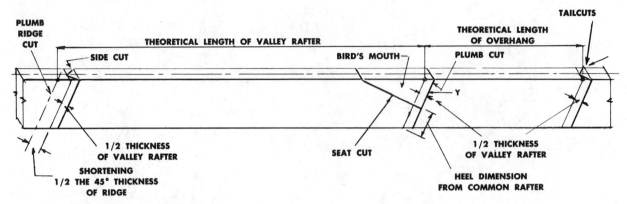

Fig. 5-83. The line length for the valley rafter is laid out in the same way as a hip rafter. The bird's mouth and tailcut require a different layout. The shortening at the ridge and cuts at the ridge are the same as used for hip rafters.

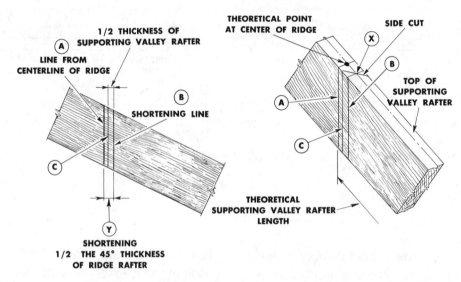

Fig. 5-84. The shortening for the supporting valley is marked first, followed by the side cut line, (C). The cutting line is drawn through (X), as shown in the right view.

THE BLIND VALLEY

A blind valley develops where two roofs, such as for an *L* or *T* shaped building, meet with the minor roof framed to rest on top of the sheathing. See Fig. 5-85. Thus the intersection is not made an integral part of the two roofs. This type of construction can also be used for some types of dormer roofs and for roof saddles which are placed behind chimneys to divert water to each side.

Valley Strip and Ridge

Pieces of 1″ × 4″ or 1″ × 6″ (called *valley strips*) are nailed to the roof sheathing in the place of valley rafters. They provide nailing and help distribute the weight of the roof of the addition. The ridge is cut to fit against the valley strips, using the cut of the common rafters of the main roof. The rafters are similar to valley jack rafters and have an angle seat cut at their lower end.

The length of the valley strip can easily be taped because the main roof at this point has

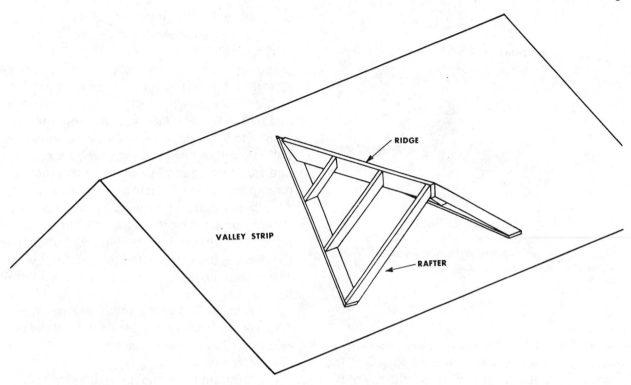

Fig. 5-85. A blind valley roof is built after the main roof has been sheathed.

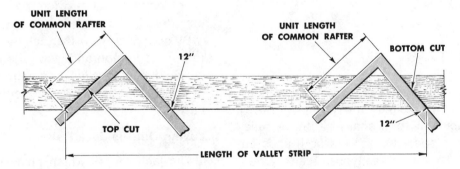

Fig. 5-86. The top and bottom cuts for the valley strip are shown in detail. The rafter is in layout position.

already been sheathed. The top cut of the valley strip is found using the unit length of the common rafter on the tongue of the square and the unit run of 12 inches on the blade. A line is drawn along the tongue side of the square. See Fig. 5-86. The cut at the bottom is made using the same units on the square. Draw the line along the blade side of the square. See Fig. 5-86.

The Rafter

The theoretical length of the longest rafter is obtained by the full length rafter table, the step off, or the line length methods in the same manner as used or other common rafters. Deduct one half the thickness of the ridge stock. The cut at the lower end is first laid out with a plumb line

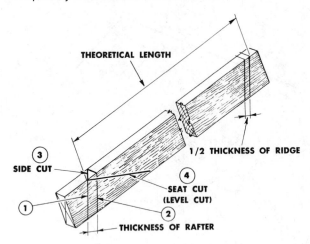

Fig. 5-87. The layout of a rafter for a roof with blind valleys shows the shortening at the ridge and the seat cut at the valley strip.

(2 in Fig. 5-87) drawn with a level dimension from line *1* and equal to the thickness of the member. Lines are squared across the top of the member, and a diagonal is drawn between the points. This line is the side cut line. Draw level line (4 in Fig. 5-87) to complete the layout. Cut the rafter on lines *3* and *4*.

JACK RAFTERS

Generally jack rafters can be considered common rafters, with either the lower or upper end cut diagonally to fit against a hip or valley rafter. Only in special cases are both ends cut diagonally. There are three types of jack rafters, depending on their location in the roof. The *hip jack* extends from the rafter plate to the hip rafter, as shown at *1* in Fig. 5-88. The *valley jack* extends from the valley rafter to the ridge of the roof, as shown at *3* and *8* in Fig. 5-88. The third type of jack rafter is the *cripple jack*. The cripple jack can be classified into two further types, *hip valley cripple* and *valley cripple*. Both of these types are shown in Fig. 5-88. Neither of these cripple jack rafters touches the ridge or the plate. The hip valley cripple extends between the valley and the hip rafters. When the ridges of the two roofs are on different levels, the valley crip-

ple jack is framed from the supporting valley rafter to the shortened valley rafter, as shown at 7 in Fig. 5-88.

The unit run, unit rise, and unit length of all jack rafters are the same as the unit run, unit rise, and unit length of the common rafters on that particular roof. As is the case with the common rafter, the run is the basis for jack rafter layout. The run of any of the various jack rafters is one side of a square. Because of this fact, it is often possible to determine the run by relating it to other rafters, for which dimensions are known, or to dimensions along the plate or ridge. The total run of the hip jack rafter (*1* in Fig. 5-88) is the same as the distance in *2*, which is the distance the jack is set from the corner of the building.

The total run *(3)* of the valley jack is equal to distance *4*, which is the same as the distance the valley jack is set from the point of the valley and ridge intersection.

The total run *(5)* of the hip valley jack rafter is equal to the rafter plate length (*6* in Fig. 5-88). This is the distance the hip valley jack is set from the corner of the building. The total run *(7)* of the valley cripple jack is twice the run of the valley jack (*8* in Fig. 5-88).

By observing how the jack rafters are related to other common and jack rafters, and by measuring dimensions locating them along the plate or ridge, you can determine the total run in each case.

Laying Out Hip Jack Rafters

The jack rafters which meet a hip rafter, as shown in Fig. 5-89, are a series of rafters spaced the same distance apart as the common rafters. Jacks are spaced uniformly to permit proper support of roofing, the nailing of sheathing, and, where applicable, the snow load. Since this spacing is uniform, the length of each successive jack rafter is uniformly shortened. The difference in line length from jack to jack is known as the *common difference of jack rafters*, as shown in Figs. 5-89 and 5-90.

If the roof is carefully planned, there is a minimum number of different jack rafters for each roof. If the jack rafters are equally spaced, start-

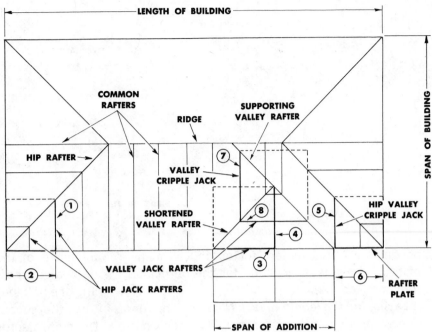

Fig. 5-88. Four types of jack rafters are shown. These include hip jacks, valley jacks, valley cripple jacks, and hip valley cripple jacks.

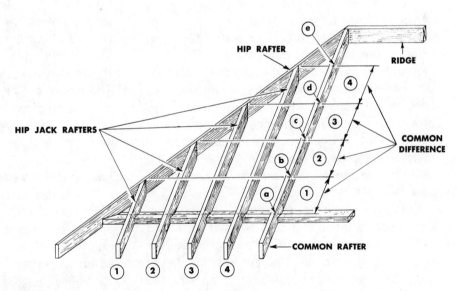

Fig. 5-89. Hip jacks which are evenly spaced along the plate, as shown (1, 2, 3, 4), have a common difference in length.

ing at the plate corner point, they are arranged in right- and left-hand pairs and are the same at corners of the house. These four pairs of rafters can be cut at the same time.

It is good practice to lay out a jack rafter pattern for the longest jack rafter on a common rafter, as shown in Fig. 5-90. The other jacks are marked on this pattern rafter, showing their cutting lines. The bird's mouth and overhang are the same for all these rafters. This pattern rafter

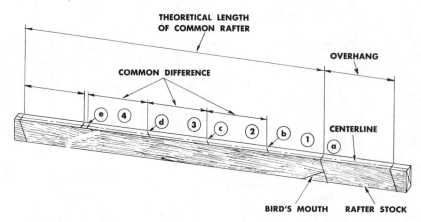

Fig. 5-90. The longest jack rafter is laid out on a common rafter. The other jacks are laid out from the long point using the common difference measurement.

becomes a layout pattern for cutting as many pairs of jacks as are needed for the particular roof under construction.

Laying out the length of jack rafters. The run of each jack rafter is related to its distance from the corner of the plate. The jack rafters are generally spaced 16 or 24 inches O.C.

The line length for the common differences can be obtained from one of the full length rafter tables. A different table is provided for each slope of roof, such as 8 inches per foot. The common differences is given on each table, based on jack rafter spacing from 1 inch to 48 inches. Two lines of numbers on the framing square also provide the common difference dimension information. One line is designated *Difference in Length of Jacks—16 Inches on Centers* and the other is designated *Difference in Length of Jacks—2 Feet on Centers.*

The layout of the jack rafter pattern is made on a common rafter so that cuts can be transferred to the stock for jack rafters. See Fig. 5-90. This layout should include the cut for the bird's mouth, the length of the overhang, and the tail cut. There is no need for laying out the cut for hip jack rafters at the ridge.

Since the jack rafter has a miter cut at its upper end, a center line is required along the top edge of the stock for layout purposes. See Fig. 5-91A and B. Lay out the first common difference dimension from the end of the stock to locate point *X* in Fig. 5-91A and B. Draw a line across the member and a plumb line down the side of the stock. Measure one half the diagonal

thickness of the hip rafter on a level line and draw a second plumb line. Draw a third plumb line one half the thickness of the jack rafter stock from the second line, as shown in Fig. 5-91B. Draw lines across the top of the stock. Draw the cut line through point *Y* in Fig. 5-91B.

Measure the common difference dimension from the long point of the first jack to locate the long point of the second jack. See Fig 5-91A. Draw a plumb line. Use a bevel square to draw the cut line on the top edge of the rafter. Repeat this process for as many jack rafters as needed. The actual cut can be determined by using the dimension given in the framing square rafter table or the full length tables. See Fig. 5-91C for an example.

Laying Out a Valley Jack Rafter

The valley jack rafters, shown in Fig. 5-92, are a series of rafters spaced the same distance apart as the common rafters. This spacing is uniform and is usually measured from the common rafter. The common difference for the valley jacks is obtained in the same way as the common difference for the hip jack rafters.

It is good practice to lay out the jacks on a common rafter so that all of the cuts are marked and ready for transfer to individual jacks. The common difference in length between the valley jacks is determined in the same way used for hip jacks. It is used in the same way once the shortest jack is laid out. The details for laying out the shortest jack are shown in Fig. 5-93. To obtain the shortening plumb line at the

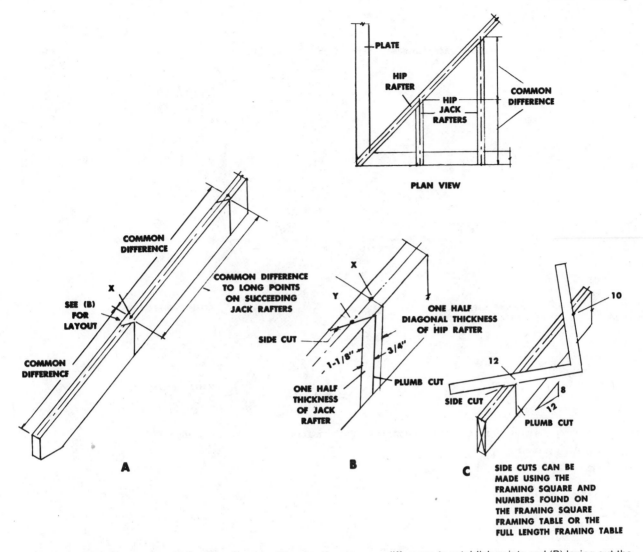

Fig. 5-91. Laying out cuts on hip jack rafters involves (A) using the common difference to establish points and (B) laying out the side and plumb cuts. An alternative method of laying out side cuts is shown in (C).

ridge, measure one half the thickness of the jack rafter stock on a level line. See Fig. 5-93A. The layout of cuts is the same as for hip jacks. See Figs. 5-91B and 5-91C.

Laying Out a Valley Cripple Jack Rafter

The rafter framed between the shortened valley rafter and the supporting (or main) valley rafter is known as the *valley cripple jack*. See Fig. 5-94. The angle of the cut at the top end, where the valley cripple jack fits against the supporting valley rafter, is the reverse of the angle of the cut used at the lower end, where it fits against the shortened valley rafter. The run of the valley cripple jack is one side of a square. The run of the cripple jacks is often related to that of other jacks, as discussed on p. 294.

The details for laying out the shortening and the side cuts of a valley cripple jack are shown in Fig. 5-95A. When the two valley rafters have the same thickness, which is usually the case, you can take off twice the 45 degree thickness of the rafter from one end. This saves one operation. However, you should avoid shortcuts unless you are very familiar with what you are doing.

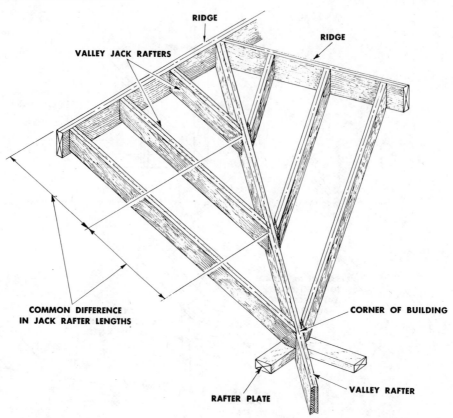

Fig. 5-92. Valley jacks which are evenly spaced along the ridge have a common difference in length.

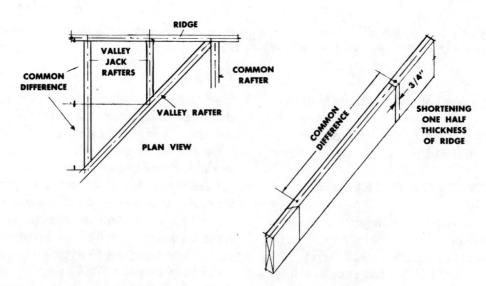

Fig. 5-93. In laying out cuts on valley jack rafters, the common difference is used to locate side cuts. Details for making side cuts are the same here as for hip jacks. See Fig. 5-91.

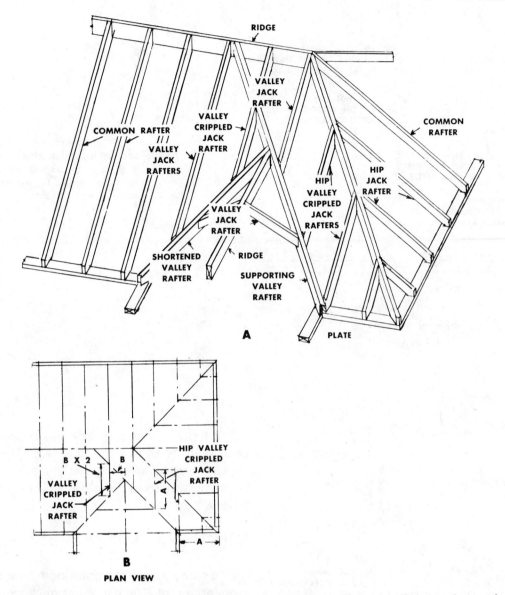

RIDGE

VALLEY
JACK
RAFTER

COMMON
RAFTER

COMMON RAFTER

VALLEY
CRIPPLED
JACK
RAFTER

VALLEY
JACK
RAFTERS

HIP
VALLEY
CRIPPLED
JACK
RAFTERS

HIP
JACK
RAFTER

VALLEY
JACK
RAFTER

SHORTENED
VALLEY
RAFTER

RIDGE

SUPPORTING
VALLEY
RAFTER

PLATE

A

HIP VALLEY
CRIPPLED
JACK
RAFTER

B X 2   B

VALLEY
CRIPPLED
JACK
RAFTER

A

A

B

PLAN VIEW

Fig. 5-94. In (A) above, the crippled jack rafters are identified by their position on the roof. In (B) below, the run of crippled jack rafters is related to other measurements which are available.

The side cuts (one at each end) are laid out in the same way as the side cuts for other jack rafters. However, you must remember that the angles for these side cuts extend in opposite directions, as shown in Fig. 5-95B.

Laying Out a Hip Valley Cripple Jack Rafter

The hip valley jack run is determined by taping or by relating the rafter to other rafters, as discussed on p. 294. When laying out the hip valley cripple jack, use the cut of the common rafter. The theoretical length of the hip-valley cripple jack can be obtained by taping the dimensions from the hip to the valley rafter. The hip valley cripple jack must be shortened one half the 45 degree thickness of the hip and valley rafter stock at both ends. See Fig. 5-96A. If the hip and valley rafters have the same thickness, take off twice this thickness from one end of the rafter, thus saving an operation.

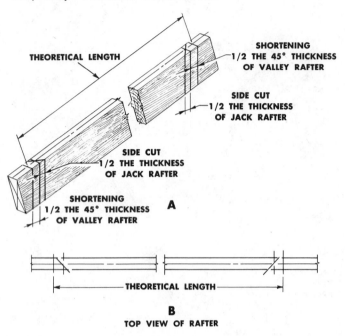

THEORETICAL LENGTH

SHORTENING
1/2 THE 45° THICKNESS
OF VALLEY RAFTER

SIDE CUT
1/2 THE THICKNESS
OF JACK RAFTER

SIDE CUT
1/2 THE THICKNESS
OF JACK RAFTER

A

SHORTENING
1/2 THE 45° THICKNESS
OF VALLEY RAFTER

THEORETICAL LENGTH

B
TOP VIEW OF RAFTER

Fig. 5-95. When laying out the valley crippled jack rafter, cuts are marked in relation to theoretical length points, as shown in (A). In (B), the cuts are made in opposite directions.

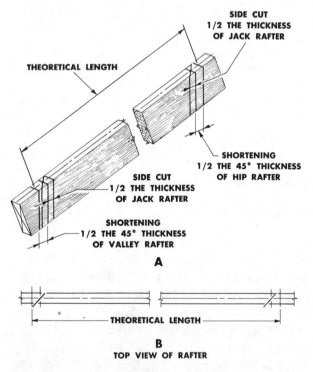

SIDE CUT
1/2 THE THICKNESS
OF JACK RAFTER

THEORETICAL LENGTH

SHORTENING
1/2 THE 45° THICKNESS
OF HIP RAFTER

SIDE CUT
1/2 THE THICKNESS
OF JACK RAFTER

SHORTENING
1/2 THE 45° THICKNESS
OF VALLEY RAFTER

A

THEORETICAL LENGTH

B
TOP VIEW OF RAFTER

Fig. 5-96. When the carpenter lays out a hip valley crippled rafter, the cuts, as shown in (A), are marked in relation to the theoretical length points. The cuts are parallel when viewed from the top, as shown in (B).

The hip valley cripple jack has a side cut on each end, as shown in Fig. 5-96B. These side cuts are parallel to each other and are laid out in the same way used for other side cuts on jack rafters.

ERECTING ROOFS

One of the most hazardous jobs in carpentry is erecting a roof. The carpenter must work at the highest point of the building, lift heavy members, then nail them in place, often in an awkward position. Temporary flooring or planks should be installed and fastened where necessary to provide safe footing. Scaffolds with guard rails should be built whenever the ridge is too high to reach conveniently.

Erecting A Gable Roof

The first thing to do is to lay out the rafter spacing on one of the rafter plates by squaring lines across the plate and marking where the rafter is to be nailed, as shown in Fig. 5-97.

Select a straight piece of ridge stock. The

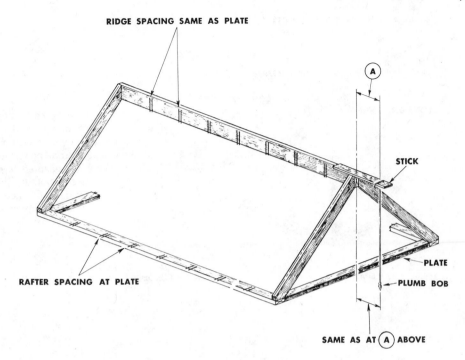

RIDGE SPACING SAME AS PLATE

A

STICK

PLATE

PLUMB BOB

RAFTER SPACING AT PLATE

SAME AS AT (A) ABOVE

Fig. 5-97. Erecting a gable roof begins with marking rafter spaces on the plate and the ridge. (The rafter spacing is not to scale.)

ridge piece is set on edge on the rafter plate, flush with one end of the building. Transfer the rafter spacing to the ridge piece. The length of the ridge piece for a gable roof is equal to the length of the building (measuring from outside to outside of the rafter plates) plus the end projections on both ends. If the ridge is so long that you need more than one piece of stock for the ridge length, make the joint on the center of a rafter.

Select a pair of straight rafters for each gable end. Note: It is a distinct advantage to have three carpenters working together when the four end rafters are erected. One carpenter holds the ridge piece in position, while the other two nail the rafters at the bird's mouths.) Brace the frame with diagonal members between the ridge and a wall or a ceiling joist. Put the intermediate rafters in place by nailing them opposite to each other. Sight the ridge for trueness as the nailing progresses.

Once the rafters are in place, plumb the ridge at the gable end with a straightedge and level, or a plumb bob on a stick fastened to the top of the ridge. The plumb bob should be held at the same distance from the plate (at plate level) as the line is from the ridge at the top. See A, Fig. 5-97. After the gable has been plumbed and some of the sheathing applied, the frame is permanently braced.

Erecting a Hip Roof

On an equal pitch hip roof, the run of the hip rafter is the diagonal of a square formed by the run of the common rafters. The run of the common rafters is equal to one half the span of the building, as shown on the Roof Plan, Fig. 5-38, p. 263.

Measure a distance from the corner equal to one half of the span to locate a point on the plate. This is the center point on the common rafter which supports the end of the ridge. Measure one half the thickness of the common rafter stock on each side of the mark and mark these points. These points mark the plate location for the common rafter. See Fig. 5-98. From these markings lay out the common and jack rafters. Jack rafters are usually equally spaced, starting

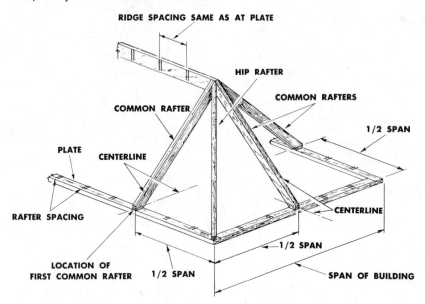

RIDGE SPACING SAME AS AT PLATE

HIP RAFTER

COMMON RAFTERS

COMMON RAFTER

1/2 SPAN

PLATE

CENTERLINE

CENTERLINE

RAFTER SPACING

1/2 SPAN

LOCATION OF FIRST COMMON RAFTER

1/2 SPAN

SPAN OF BUILDING

Fig. 5-98. The first operation in erecting a hip roof is to locate the common rafters which support the ends of the ridge. Spacing for other common rafters is measured from these common rafters on both the plate and the ridge. Jacks can be spaced from the corner or from the common rafters.

from the common rafter, so that if there is an odd space it appears at the corner. Under certain conditions the jacks are laid out so that the spacing begins at the corner.

Select ridge stock and lay the length out to include the portion added at each end. Transfer the spacings for common rafters from the plate to the ridge. Erect a pair of common rafters at each end of the ridge and place one common rafter at the center of each end. See Fig. 5-98.

Add the other common rafters and nail them opposite each other. Check the ridge as the nailing progresses to verify it is straight. Nail the hip and jack rafters in place following the installation of the common rafter. The ridge of a hip roof does not need to be plumbed since it is placed and automatically held in position by the end common rafters and the hip rafters.

Erecting Intersecting Roofs

When a complex roof of the type shown in Fig. 5-99 is framed, the intersecting ridge must be accurately located. If the building does not require attic space throughout the entire roofed area, a simplified procedure involving blind valley rafters is used. See Fig. 5-85.

If the roof to be framed is of the type shown in A or B of Fig. 5-99, the ridge and common rafters for the main roof are installed first, along with as many common rafters as are needed to support the structure. The set of end common rafters and the ridge of the addition are then erected, followed by the valley rafters. The points of intersection of valley rafters must be determined accurately if the roof is to be framed properly. Figure 5-99 shows how these points are located.

If the roof is framed as shown in Fig. 5-99C, the main roof is erected first, omitting the part which intersects with the roof of the addition. The next step is to erect the supporting valley, then the end common rafters, and then the ridge of the addition. The shortened valley is nailed in place at the intersection of the ridge and supporting valley. When all of the hip, valley, and ridge members are in place, the remaining common rafters and jack rafters are installed.

You must locate the intersecting points of the ridge and valley rafters carefully. Also check their location on the plates. The rafter cuts will fit snugly and the rafter lengths will be accurate if you carry out these procedures carefully.

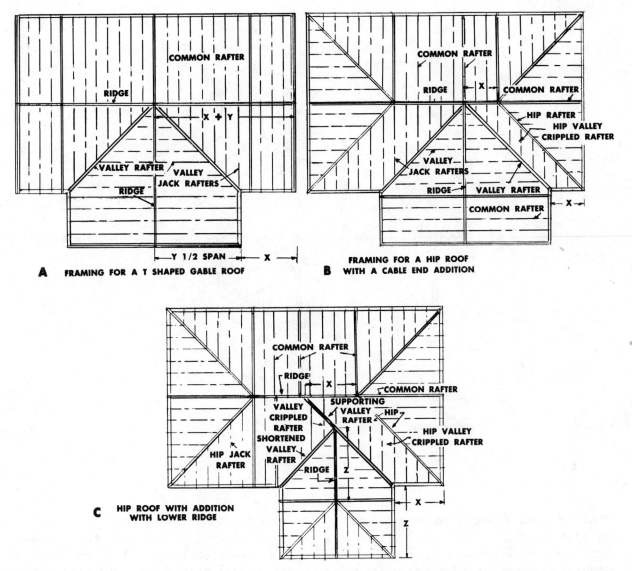

Fig. 5-99. When an intersecting roof is erected, part of the main roof is framed first. The point of support for valley rafters is marked on the ridge.

FRAMING SPECIAL ROOFS AND ROOF PARTS

Roof Framing for a Flat Roof

There are several ways to design the framing members for a flat roof which has a projection on all sides. One method is shown in Fig. 5-100. A double member (called a *trimmer*) serves as an anchor for the lookout rafters which rest on the plate. (A *lookout rafter* helps support the over-hang of a roof.) Metal hangers are used to fasten the inner end of the lookout rafters.

Gambrel Roof

Gambrel roofs are generally used for barns or houses which are designed in the Dutch Colonial architectural style. This type of roof provides additional living space on the upper floor of a two story dwelling.

The method used for framing a gambrel roof is

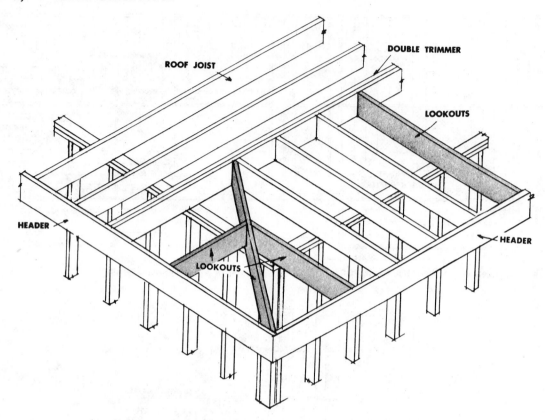

Fig. 5-100. A flat roof is built using cantilever lookouts.

much the same as that used for a gable roof. However, the slope of the roof is broken at a point somewhere between the plate and the ridge, as shown in Fig. 5-101. The slopes of the two parts of the roof and the location of the upper plate vary from roof to roof. The part of the roof below this break makes an angle greater than 45 degrees, generally between a 20 and 24 inch unit rise. See Fig. 5-101.

The gambrel roof can be considered as two separate roofs, with the unit rise of the upper slope considerably less than the unit rise of the lower slope. The main difference in the method used for framing this type of roof occurs at the point where the rafters of the two slopes join. This point can be framed with a plate supported by partitions. The rafters of both slopes of a gambrel roof are cut to fit around the plate. The cuts of each rafter (shown as *1* and *2* in Fig.

5-101 A) are plumb and level cuts, similar to the ridge and seat cuts used for common rafters.

Gambrel roofs have been used for barns for many years because they provided excellent storage space for hay. There are several ways to construct a gambrel roof. The roof must of course be made strong to withstand wind pressure. The purlins, shown in Fig. 5-101B, are made of heavy timbers turned on edge to serve as beams. They are supported by strong posts. See Fig. 5-101B.

Gable End Framing

The studs on the gable end of a roof have a common difference in length. The common difference for studs on 16 inch centers can be found by sliding the square as shown at A in Fig. 5-102. The framing square is placed on a piece of stock to the cut of the roof, with the unit rise

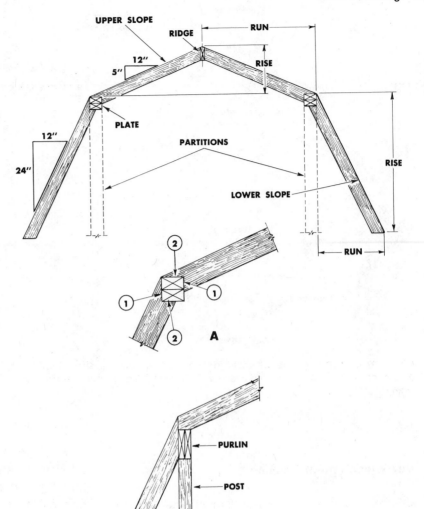

Fig. 5-101. The gambrel roof resembles a gable roof except that it has two slopes. A typical plate, such as used in residential construction, is shown in (A). (B) shows heavy members called *purlins* which are used when the roof must support greater loads.

on the tongue and 12 inches (the unit run) on the blade. A line is drawn along the blade. The square is then moved along this line until it reads 16 inches. The square is held on the line, and the figure on the tongue side of the square is read to find the common difference of the stud lengths.

The angle cut on any stud or board in a vertical position on the gable ends is the same as the ridge (or plumb) cut of the rafter. This cut fits against the rafter and has the same slope as the roof. To find the top cut of a gable stud so that it

fits against the rafter, hold the framing square to the number *12* on the blade and the unit rise on the tongue, as shown at B in Fig. 5-102. A line is drawn along the tongue (or rise side) of the square to locate the correct angle for the cut.

Cut the end angle for a horizontal board (or any framing member in a level position which is fitted against the rafter) in the same way used for the seat cut of the rafter. This angle is also the cut used for horizontal sheathing boards. See C in Fig. 5-102.

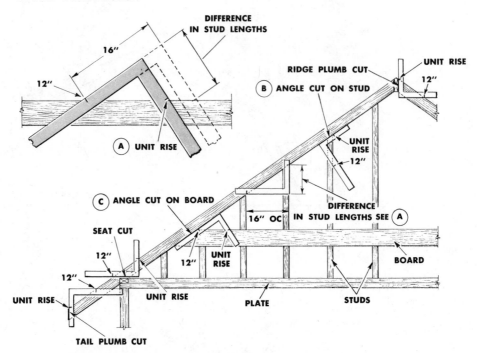

Fig. 5-102. The framing square is useful in laying out the studs and sheathing for gable ends. In (A), it is used to find the common difference in stud length. It is also used for (B) finding the angle cut at the top of the studs and (C) finding the angle cut on the siding.

Bay Window Roof Framing

Some roofs have the features of an octagon or polygon because of the unusual shape of the floor plan of the building. The roof over a bay window brings out this problem. It is advisable to make a sketch layout, either to scale or full size, so that you can decide on the best arrangement of members and find the information you need about the cuts. The run of each rafter can be determined by scaling the sketch.

Bay Window Roof Plan

The roof plan (Fig. 5-103) shows the location and length of the run for all the rafters needed for the bay roof. Although there are several ways of placing the members, this is a typical arrangement. There is always a potential problem with nailing the members. When the members are placed too close to each other, it is difficult to fasten them. The roof layout should either be made at a scale, such as 3 inches equal 1 foot, or

full size. When the bay is a part of an octagon, the distances *o-a* and *o-b* are equal. The run of the hip rafter on a polygon roof should bisect the angle at the plate.

The unit run and unit rise of common rafters are established on the architect's drawing. The unit rise and total rise of the hip are the same as those used for the common rafter. However, since the run of the hip is not the diagonal of a square, the unit run differs from the usual 17 inch unit. The unit run of the hip for the octagon roof is 13 inches for every 12 inches of common rafter run. See Fig. 5-104. The miter cut at the plate of an octagon can be found by taking 5 inches on the tongue of the square and 12 inches on the blade, as shown in Fig. 5-104. The line for the miter cut is drawn along the tongue side of the square.

You can determine the necessary dimensions and cuts of the rafters by making a plan view drawing to scale, as shown in Fig. 5-105. You should make this scale drawing as large as is

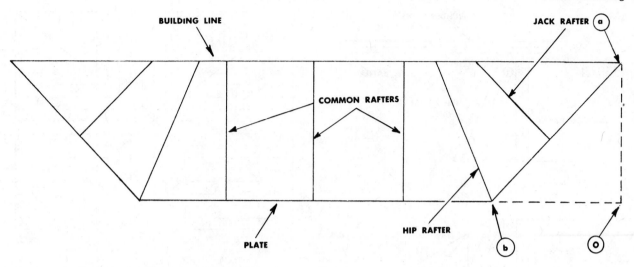

Fig. 5-103. A bay roof plan laid out to scale shows the run and the center line of the rafters.

Fig. 5-104. An octagon hip rafter has a run of 13 inches for each 12 inches on a common rafter.

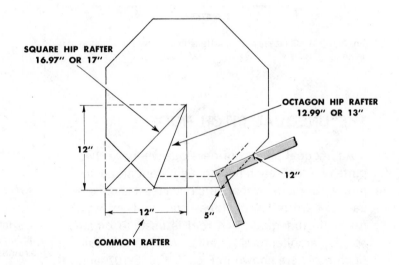

convenient. Draw the outline of the corner of the bay window roof first, followed by the center line for each rafter. The wall rafter is fastened to the building and supports the ends of the sheathing. Draw each rafter the full width of the rafter stock so that the actual dimension of the level cuts can be obtained. See *1, 2, 3,* and *4* in Fig. 5-105. The total run of each rafter is scaled from the sketch shown in Fig. 5-103. The unit rise of the bay roof is found on the working drawings of the build-

ing. Once the information about the unit run, the total run, and the side cut is available, you can lay out the common, jack, and wall rafters. If the bay is part of an octagon, the cut is the unit rise of the common rafter and a unit run of 13 inches. See Fig. 5-104. The amount of drop needed for the hip is very slight since the plate angle at the corner is not 90°. A drop of ¼ inch is sufficient unless the cut of the rafters is more than 6 inch rise per foot.

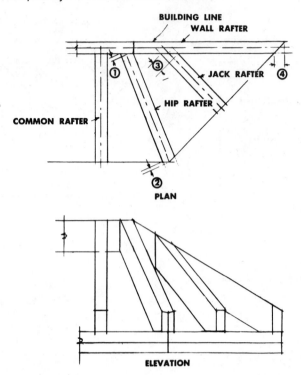

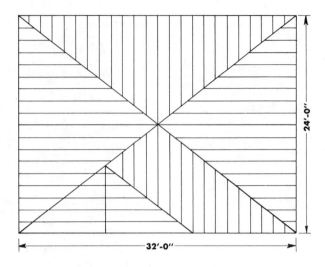

Fig. 5-106. An unequal pitch roof is required whenever a rectangular roof comes to a point.

Fig. 5-105. A full size layout of rafters gives exact information on cuts.

THE UNEQUAL PITCH ROOF

An unequal pitch roof develops whenever two parts of a roof, each having a different slope, intersect. Sometimes it occurs because it is necessary to have different slopes on a roof. At other times the unequal pitch roof is used to create special architectural effects. Various unequal pitch roofs are shown in Figs. 5-106, 5-107, and 5-108. A common unequal pitch roof is required when a rectangular house is designed with a roof which comes to a point. See Fig. 5-106. Other unequal pitch roofs a carpenter may encounter involve two roofs which intersect, such as shown in Figs. 5-107 and 5-108.

The roof in Fig. 5-107 is a T shaped roof, with different spans on the two parts of the roof. Both ridges are at the same elevation. The roof in Fig. 5-108 is an L shaped roof, with different spans on the two roof parts and intersecting ridges. Dormers often require an unequal pitch roof because the dormer roof has a low slope com-

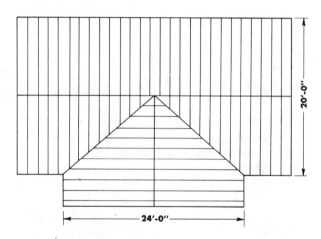

Fig. 5-107. An unequal pitch roof is necessary when there is a slight difference in spans between two parts of a roof, and the ridges and plates are kept at the same elevation.

pared to the steeper slope of the main roof. See Fig. 5-109.

The usual procedure for framing unequal pitch roof is to frame the two parts of the roof as outlined earlier in this chapter except for the unequal pitch hip, valley, and jack rafters. Because this problem can become very complex, most builders fall back on the taping method to determine the length of hip and valley rafters. They make a layout to scale (Fig. 5-110) and superim-

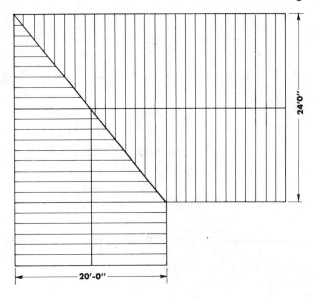
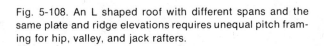

Fig. 5-108. An L shaped roof with different spans and the same plate and ridge elevations requires unequal pitch framing for hip, valley, and jack rafters.

pose the hip (or valley) rafter and two jack rafters at full size (Fig. 5-111) in order to determine the exact shape of the cuts. One major problem in laying out unequal pitch hip rafters is that the unit run of 17 inches cannot be used because the unit run is the diagonal of a rectangle and not the diagonal of a square. See Fig. 5–110. The layout not only gives the carpenter the shape of the cuts as they appear in a plan view, but gives the run of the various members. See Figs. 5-110 and 5-111.

New problems regarding dropping arise when you lay out an unequal pitch hip rafter. Since the roof slopes at a different angle on each side of the hip, the amount of dropping is not the same for each side of the rafter.

The jacks, if evenly spaced along the plate or ridge, do not line up in a herringbone pattern. See Fig. 5-110. (The *herringbone pattern* is the usual arrangement of common rafters so that they come out exactly opposite each other at the ridge. Jacks are in a herringbone pattern when they meet hip or valley rafters, directly opposite each other.) The jacks, however, can be laid out using the same cut as the common rafter on the same side of the roof. (Each side will be different). The line length for jack rafters can be taped.

One other problem you must solve is peculiar

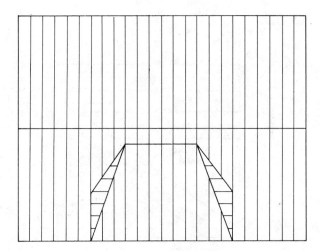

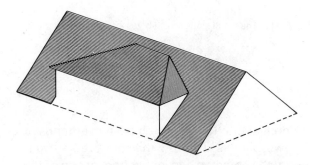

Fig. 5-109. A dormer requires unequal pitch rafters because the slope of the dormer roof and the slope of the main roof are different. The roof is shown in plan view at the top and in an isometric view at the bottom.

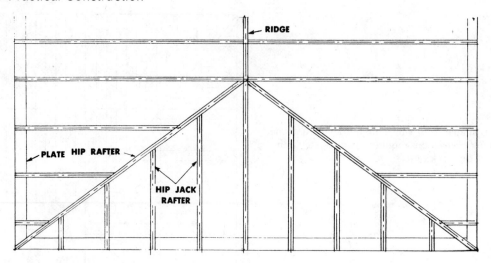

Fig. 5-110. In a plan view, the hips on an unequal pitch roof do not extend from corners to make 45° angles. Likewise, the jack rafters do not meet opposite each other at the hip rafters.

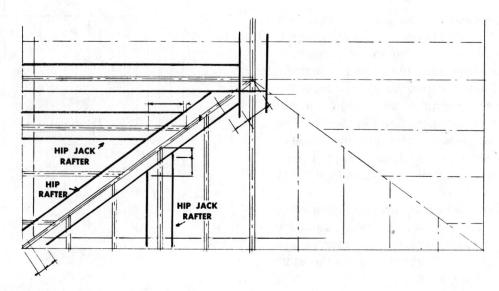

Fig. 5-111. The small scale layout gives information on the run of each rafter. The large scale layout shows the cuts.

to unequal pitch roofs which have overhangs. If the two pitches used on the roof are very different, the common and jack rafters of one of the roofs in many instances do not rest with sufficient bearing on the plate. You must therefore build up the plate by adding another member. This brings the plate up enough to make a satisfactory bird's mouth.

Space does not permit a detailed discussion on how to frame unequal pitch roofs. The subject, however, is discussed in Ch. 5–1, including suggestions for solving layout problems.

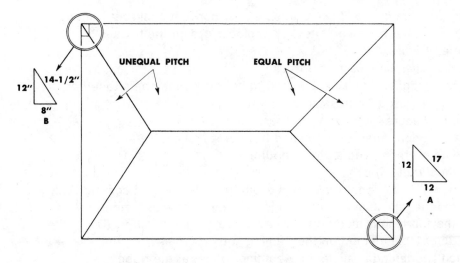

Fig. 5-112. The unit run of an equal pitch roof hip rafter is 17 inches per foot of run of the common rafter, as shown at (A). However the unit run of an unequal pitch hip rafter is not 17 inches per foot of run. In the example, shown at (B), it is 14½ inches.

SAFETY

It is very important to follow safe practices when you are building a roof. You are working at the highest point of the building, assembling rafters and roof parts which are not firmly fastened until they are all in place. This is a special potential hazard.

In many parts of the country the roof slope is so low that carpenters walk on the ceiling joists when they put the roof together. If this is done, temporary planks should be fastened in place. When roofs have a greater slope, so that the ridge is out of the reach of the carpenter, scaffolds with guard rails are required. Adequate temporary bracing is essential to hold the ridge and other members in place until the whole roof is framed. Hard hats are also required. Safety belts are essential for carpenters when they are working on high open structures.

Erecting trusses is a special hazard because they have an awkward size. Previously we explained the value of using braces and guy wires to keep the trusses from overturning. Scaffolding may be necessary for the workers who fasten the trusses at the plate. Large trusses are often erected with a crane. The carpenter must follow good safety practice and fasten the trusses with ropes or slings before they are lifted into place. The carpenter must also be alert to the problem of swinging them into position.

QUESTIONS FOR STUDY AND DISCUSSION

1. What types of loads are roofs subject to?
2. Give the names for five types of roofs used for houses.
3. Describe the modern mansard roof.
4. What is a truss?
5. What is a trussed rafter?

6. What are the advantages of using trusses over conventional roof framing?

7. Why is it necesary that an architect or engineer design the trusses?

8. Describe a W Truss.

9. Why are web members used in a truss?

10. What type of connector plates are used to hold truss members together?

11. Describe the jig used for trusses.

12. State the procedure for erecting trusses.

13. How is a common rafter distinguished from other rafters?

14. Describe an overhang and a bird's mouth.

15. What is the *cut* of a roof?

16. Explain how the unit length rafter tables are used to layout a common rafter.

17. What is the theoretical length of a ridge for a hip roof with dimensions of 24 × 32 feet?

18. Explain the full length method of laying out a common rafter. (Use either the full length table method or the rafter table method on the framing square.)

19. How much shortening is made for a common rafter at the ridge?

20. Explain the procedure for laying out a bird's mouth on a common rafter.

21. Explain how the 17 inch unit run of the hip rafter is derived.

22. What is the odd unit for run for a hip rafter?

23. How is the taping method applied to laying out a hip rafter?

24. Describe the procedures for dropping a hip rafter.

25. What determines how much a hip rafter should be shortened at the ridge?

26. Describe the purpose of supporting and shortening valley rafters?

27. Describe blind valley construction.

28. What is the *common difference* as applied to jack rafters?

29. What is a crippled jack rafter?

30. State the purpose of a large scale layout for an unequal pitch roof.

CHAPTER 5-1
PRACTICE IN ROOF FRAMING

EQUAL PITCH ROOF

The opportunity to work on cutting the members and erecting a roof may occur infrequently, particularly for the beginning carpenter. However, when it does arise, it is helpful if the carpenter has some experience beyond the theoretical level. Some procedures, such as arranging rafters, laying out individual rafters, and cutting and erecting the members can best be taught by actual practice on the job. However, practice can be provided using the roof shown in Figs. 5-113 and 5-114 and the material in this chapter. In some training programs with shops, it is possible to construct this roof or one similar to it as a shop project. The same operations can also be performed by any one individual with a relatively few supplies and tools. To do the layout work using this chapter as a guide, all you need is a framing square, a 6 or 10 foot pocket tape, a pencil, and a 2 × 6 inch plank which is 10 feet long. All of the layout steps for each rafter can be done on this single plank without even cutting it.

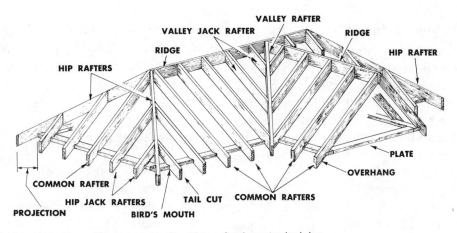

Fig. 5-113. A model hip roof with an offset is used to illustrate rafter layout principles.

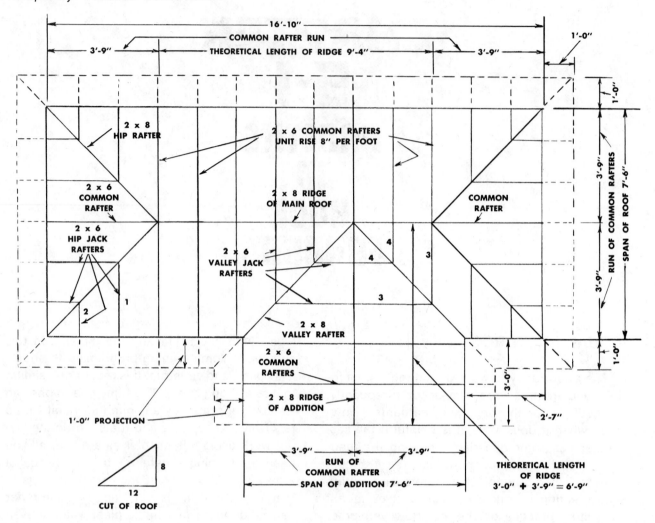

Fig. 5-114. A schematic plan view of the roof shown in Fig. 5-113 provides detail information and dimensions and gives the arrangement of the rafters.

Sequence of Cutting Members

The ridges and a few common rafters are the first members needed to erect the roof. This particular roof requires a main ridge and a secondary intersecting ridge, for the addition to the main roof. They are cut first. The ridge and key common rafters which support the ridge are the first to be erected. Hip and valley rafters are cut next, followed by the jack rafters.

The Ridge

The main ridge. The theoretical length of the ridge is equal to the length of the roof *minus* the span of the roof. See Fig. 5-115. However, add to the theoretical length one half the thickness of the common rafter on each end.

Length of Building	16'-	10"
Span of Building	− 7'-	6"
Theoretical Length of Ridge	9'-	4"
One Half the Thickness of the Common Rafter		³⁄₄"
One Half the Thickness of the Common Rafter	+	³⁄₄"
Actual Length of the Ridge	9'-	5¹⁄₂"

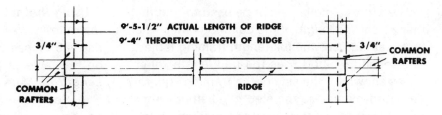

Fig. 5-115. The actual length of the ridge equals the theoretical length plus half of the thickness of the common rafter at each end.

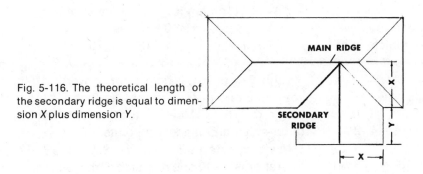

Fig. 5-116. The theoretical length of the secondary ridge is equal to dimension *X* plus dimension *Y*.

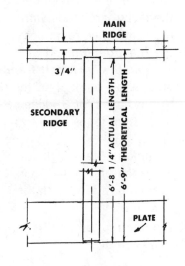

Measure a length of 9 feet, 4 inches at a point away from the end of the piece of stock. Draw plumb lines to represent the theoretical length. Measure an additional ¾ inch on each end and draw two more lines to represent the actual length.

The Secondary Ridge

The theoretical length of the secondary ridge, as illustrated in Fig. 5-116, is equal to one half of the span of the addition plus the amount the addition extends beyond the main roof. The actual length is obtained by subtracting one half of the thickness of the ridge.

One Half the Span of the Addition	3'- 9"
Amount Addition Extends Beyond the Main Roof	3'- 0"
Theoretical Length of the Secondary Ridge	6'- 9"
Subtract one half the thickness of the ridge	— ¾"

Actual Length of the Secondary Ridge   6'-8¼"

Measure 6 feet, 9 inches on the stock then subtract ¾ inch.

The Common Rafter

The common rafter is shown on both Figs. 5-113 and 5-114. The first thing to do is to find out information about the number of rafters, then the slope and the run dimensions. This information is provided here on Fig. 5-114. (Ordinarily this information, except for the number of rafters, is available on working drawings.)

Rafter information. (Note: The information below is found in Fig. 5-114.)

Total Number of Common Rafters: 19
Number of Common Rafters with Overhang: 16
Span of Roof: 7'-6"
Run of Rafter: 3'-9"
Unit Rise: 8 inches per Foot of Run
Horizontal Dimension of Projection: 1'-0"

One of three methods can be used to find the theoretical line length of the common rafter. The first method uses the full length framing tables. The second method uses the unit length rafter tables shown on the side of the framing square. The third method is the step off method. After the line length is determined and marked on the stock, the cuts at the ridge, the bird's mouth, and the tailcut are made with the framing square in the same way, regardless of how the length of the line is determined.

Using the full length framing table. If you have access to a book of full length rafter tables, refer to the pages for *8 Inches Rise per Foot of Run* to find the following information:

Common rafter length for
  a span of   7'-0" = 4'-2½"
Common rafter length for
  a span of   6" =   3⅝"
Common rafter length for
  a span of   7'-6" = 4'-6⅛"

Measure a dimension of 4'-6⅛" along the top edge of the rafter and mark two points. See Fig. 5-117.

Using the unit length table. The total run of the rafter is 3'-9", or 3.75 one foot units. The unit rise is 8 inches per foot of run. Refer to the line on the table of the framing square designated *Length of Main (Common) Rafter per Foot of Run.* See Fig. 5-118. The number 14.42 appears on the line below the number *8.* This is the unit run or hypotenuse of a triangle with an 8 inch altitude and a 12 inch base.

14.42" (unit length) × 3.75 (units of run)
= 54.07 inches
Total Line Length: 14.42 × 3.75 = 54.07

Table 5-1, p. 261, is helpful for converting decimal fractions to fractions of an inch. The line length is 4'-6¹/₁₆". Measure the line length on the rafter stock and mark two points.

Using the step off method. Since the run of the rafter is 3'-9", one odd step with a 9 inch run and three full steps with 12 inch runs are required. Place the framing square in position *1,* (Fig. 5-119A), using the cut of the rafter (8 inches on the tongue and 12 inches on the blade). Draw a line *(a)* along the tongue side. This is the center line of the ridge. While the framing square is in

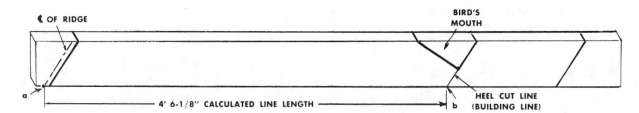

Fig. 5-117. The line length method for laying out the common rafter makes use of either the full length framing tables or the unit length tables on the framing square. Points *a* and *b* are then marked on the rafter stock. The rafter is in layout position.

	21.63	18.44	17.69
LENGTH OF MAIN RAFTERS PER FOOT OF RUN	21.63	18.44	17.69
LENGTH OF HIP OR VALLEY RAFTERS PER FOOT OF RUN	24.74	22.00	21.38
DIFFERENCE IN LENGTH OF JACKS - 16 INCHES ON CENTERS	28.84	24.585	23.588
DIFFERENCES IN LENGTH OF JACKS - 2 FEET ON CENTERS	43.27	36.38	35.38
SIDE CUT OF JACKS	6-11/16	7-13/16	8-1/8
SIDE CUT OF HIP OR VALLEY	8-1/4	9-3/8	9-5/8

8
14.42
18.76
19 23
28.84
10
10 ⅞

Fig. 5-118. Tables stamped on the blade of the framing square supply information for establishing the line length and cuts of rafters.

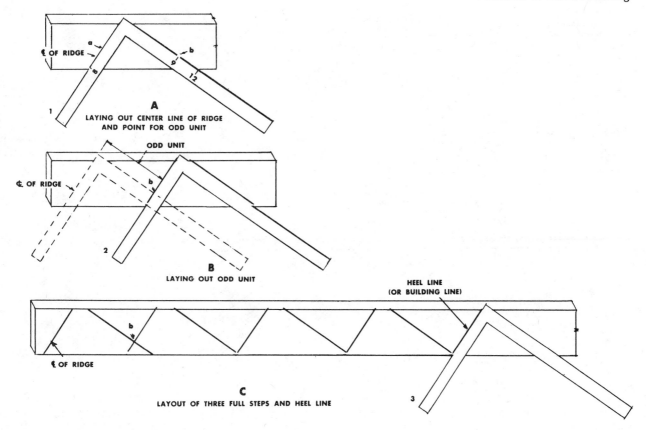

Fig. 5-119. The step off method of laying out the common rafter includes (A) determining the position of the odd unit, (B) laying out the odd unit, and (C) laying out the rafter length.

this same position, mark point *b* at the 9 inch point on the blade. This measurement is the odd unit of run. Shift the framing square to position *2*, with the tongue on point *b*. See Fig. 5-119B. Draw lines along both sides of the framing square to indicate the first full step. Shift the framing square two more times so that three full steps are marked. See Fig. 5-119C. Shift the framing square to position *3*. Mark heel line.

Laying out the bird's mouth and the over-hang. The stock remaining after the bird's mouth is cut away must be strong enough to support the overhang. Measure 2½ inches from the edge of the stock on the heel cut line so that you can locate the seat cut line. See Fig. 5-120A. (Note: On an actual roof, this dimension would depend on the size of the rafter stock and the

load on the projection. Here 2½ inches is considered satisfactory.) Place the framing square in position *4* and draw the seat cut. Shift the framing square to positon *5* (Fig. 5-120B) on the heel cut line, then measure 1' along the blade to locate the tail cut point. This is the horizontal dimension for the projection. Shift the framing square to position *6* to draw the tail cut line.

Shortening rafter at ridge. Place the tongue of the framing square on the center line of the ridge, using the cut of the rafter. See Fig. 5-120C. Shift the framing square to position *7*. Measure a ¾ inch dimension on a level line (parallel to the blade). Draw a line along the tongue. This is the shortening at the ridge, and the line on which the rafter is to be cut. (Note: Fig. 5-121 is a view of the completed rafter layout.)

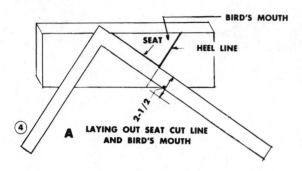

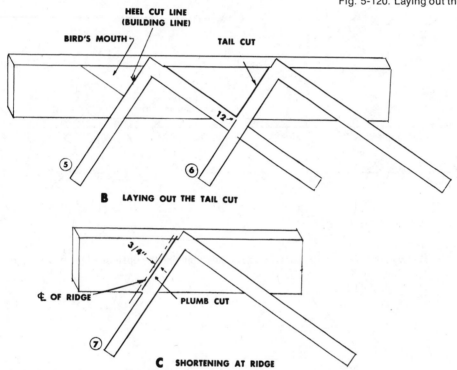

Fig. 5-120. Laying out the cuts on the common rafter.

A   LAYING OUT SEAT CUT LINE AND BIRD'S MOUTH

B   LAYING OUT THE TAIL CUT

C   SHORTENING AT RIDGE

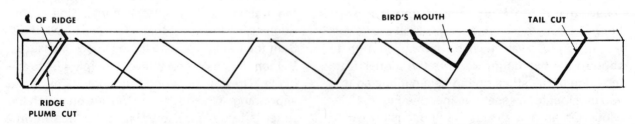

Fig. 5-121. The common rafter is ready to cut out when all of the cuts are marked. The rafter is in the layout position.

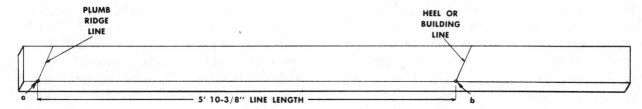

Fig. 5-122. The line length method for finding the length of the hip rafter requires using the full length framing tables or the unit length table. The rafter is shown in the layout position.

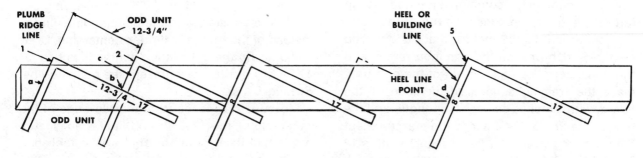

Fig. 5-123. In stepping off the rafter length for the hip rafter, position 1 of the framing square is used to establish the odd unit point, position 2 is the first of three full steps to establish the heel line point, and position 3 is used to mark the heel line.

The Hip Rafter

Rafter information. (Note: The information required to layout the hip rafter depends on the run and unit rise of the common rafter to which it is related.)

Total number of Hip Rafters: <u>4</u>
Span of Roof: <u>7'-6"</u>
Run of Common Rafter: <u>3'-9"</u>
Unit Rise for Common Rafter: <u>8</u> inches rise per foot.
Horizontal Dimension of Projection for Common Rafter: <u>1'-0"</u>

Four methods are used to determine the line length of a hip rafter. Three are similar to those used for common rafters. The fourth is the taping method.

Using the full length framing tables. If the book of full length framing tables is available, you can find the rafter length by referring to the tables designated for a slope of 8 inches rise per foot. See table marked *Hip or Valley Rafter:*

Hip rafter length for a span of   7'-0" = 5'- 5⅝"
Hip rafter length for a span of   6" =   - 4¾"
Total Hip rafter length for a span of 7'-6" = 5'-10⅜"

Measure a dimension of 5 feet 10⅜ inches away from the ends and make two marks (*a* and *b* in Fig. 5-122) on the top edge of the stock.

Using the unit length table. The unit length method requires the use of the table on the side of the framing square and a simple calculation to find the theoretical line length of the hip rafter.

Run of Common Rafter: <u>3'-9"</u>
Unit Rise of Common Rafter: <u>8 inch rise per foot</u>

To determine the unit length, refer to the line designated *Length of Hip or Valley Rafter Per Foot of Run* under number *8* in the rafter table. Here you will find the number *18.76.* (See the framing square table shown in Fig. 5-118.) The total run of the common rafter is 3 feet, 9 inches, or 3.75 feet. Multiply 18.76 by 3.75. This equals 70.35, or 70.35 inches. (70.35 inches equal 5 feet, 10⅜ inches.) Measure this dimension along the top edge of the rafter stock. See Fig. 5-122.

Using the step off method. Place the framing square on the stock at the cut of the hip (8 inches on the tongue and 17 inches on the blade). See *1* in Fig. 5-123. Draw a line (line *a*). This line represents the plumb ridge line, which is the theoretical intersection of the common rafter and the hip rafters and the ridge. Since the run of the common rafter is 3 feet, 9 inches, there is one odd step (a 9 inch unit on the common rafter) and three full steps. To find the run of the odd unit, place the framing square on the face of the board, with 9 inches on both the tongue and the blade. See Fig. 5-124. Record the measurement between the two points. The measurement is 12 ¾ inches. Referring to Fig. 5-123 first, mark point *b* along the blade, then shift the square to position *2* and mark line *c*. Mark three full steps

(17 inch run) to locate the building line (point *d*). Move the framing square to position *5* to locate the heel line. Draw this line. See Fig. 5-123.

Using the taping method. The roof must be partially completed, with some of the common rafters in place supporting the ridge before the taping method can be used. Additional bracing may be required to keep the ridge firm and in line. Using a steel tape, measure from the plate corner to the point of intersection of the faces of the two common rafters. This taped dimension shown on Fig. 5-125 is 5'-11". When laying out the hip rafter, start with the building line (line *a*) instead of the theoretical ridge center line. Draw line *a*, using the cut of the hip (8 inches rise for 17 inches of run) to represent the building line. See Fig. 5-125.

Measure 2½ inches from the top edge of the rafter along the heel line *(a)* to establish the bird's mouth at point *b.* This was established earlier for the common rafter. See p. 317. Measure 5 feet, 11 inches from point *b* on a diagonal, back toward the top edge of the rafter, to locate

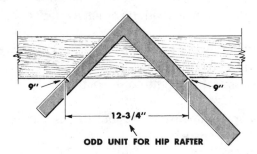

Fig. 5-124. The length of the odd unit may be determined by bridging the square between two points set at the odd unit run of the common rafter. The example shown here is 9 inches.

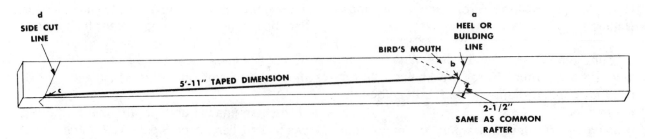

Fig. 5-125. The taped line length dimension is transferred to the hip rafter. The first step is to establish the bird's mouth at point *b.* Then the line length is measured diagonally to locate point *c.* The rafter is in the layout position.

point *c.* Draw line *d* with the framing square through point *c,* using the cut of the hip rafter (8 inch rise for 17 inches of run). This is the side cut line on the hip rafter at the ridge. See Fig. 5–125.

Laying out the bird's mouth and the overhang. (Note: The procedure for laying out the bird's mouth and the overhang below is followed regardless which method you use for determining the length of the rafter.) Measure 2½ inches on the heel line (*a* in Fig. 5-126) to locate the seat cut point *(b),* then move the square to draw the

seat cut line *(c).* The 2½ inch dimension is the same measurement used for the common rafters. See Fig. 5-120A, p. 318.) The hip rafter must be dropped so that the sheathing lies flat. Place the framing square on the face of a piece of material at the cut of the hip, draw a level line along the blade, and measure and mark ¾ inch (one half the thickness of the hip rafter). See Fig. 5-127. Measure the plumb distance to determine the amount of drop. Here this dimension is ⅜ inch. Measure this amount at the seat cut and draw the new seat cut line, shown at *d* in Fig. 5-126.

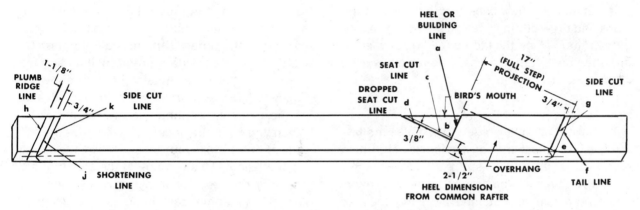

Fig. 5-126. Layout of the bird's mouth, overhang, and shortening at the ridge for a hip rafter. The steps involve include making the bird's mouth by locating line *c* (the seat cut line), making the cuts at the end of the overhang *(f and g),* shortening the hip rafter at the ridge, and marking side cuts *(j and k).*

Fig. 5-127. The amount of drop is determined by laying out a level line anywhere on the hip rafter. Measure ½ the thickness of the stock (¾ of an inch) and then measure the plumb distance at that point. The seat line is raised this amount so that the rafter drops accordingly.

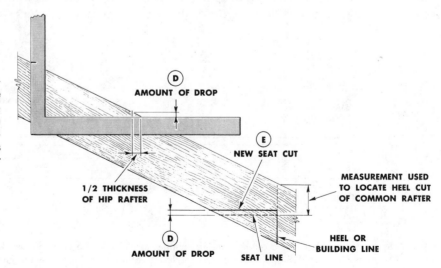

The projection (horizontal) dimension for the overhang of the common rafter is one foot. This requires a full step with a 17 inch run on the hip rafter. Shift the framing square to the heel line *(a)* and mark the full step to locate the tail cut point *(e)*. Shift the framing square again and mark the tail line at *f* on Fig. 5-126. Side cuts are required for the finished tail cut. Measure ¾ inch back from the tail line *(f)* to locate the side cut line *(g)*.

Shortening at the ridge.   Return to the plumb ridge line (line *h* in Fig. 5-126) and measure 1⅛ inches on a level line to locate the shortening line (line *j*). (The 1⅛ inch dimension is one half the 45° thickness of the ridge.) Side cuts are required at the ridge. Measure ¾ inch to locate the side cut line *(k)*. Draw lines across the top of the stock and draw cutting lines on the opposite side of the rafter. When the taping method is used to find the line length, the base line for measuring is line *j*, the shortening line.

Valley Rafters

The layout for valley rafters resembles the layout for hip rafters in many respects. The main difference is in the cuts and the fact that valley rafters don't have to be dropped. The theoretical line length can be obtained by using the line length method, the step off method, or by taping.

Laying out the valley rafter.   Obtain the theoretical line length of the rafter by using the line length, the step off, or the taping method. This is done in the same manner as used for hip rafters. The line length obtained for hips is applicable for related valley rafters. The line length for the hip rafter is 5 feet, 10⅜ inches. See p. 319. Measure and mark this dimension on the edge of the stock (points *a* and *b* in Fig. 5-128). Draw plumb lines through these points, using the cut of the valley (8 inch rise for 17 inches run). Measure 2½ inches, which locates the heel point *(c)*. Draw a level line *(d)* to represent the seat cut. Measure ¾ inch (one half the thickness of the valley rafter) from the building line and locate the plumb cut for the bird's mouth. Notice that this dimension is measured toward the overhang to avoid interference with the plate corner. See line *y* in Fig. 5-129.

The overhang is measured from the building line. Measure a full step with the framing square to locate point *f*. A plumb line *(g)* drawn along the tongue of the square represents the theoretical end of the rafter. If the fascia is to meet the valley rafter to provide a proper nailing surface, the end of the rafter must be cut as shown in Fig. 5-128. Draw a plumb line ¾ inch in advance of the line representing the end of the rafter. When you continue lines *g* and *h* across the bottom of the rafter, you have marked the points for the tailcut. This is made in the shape of a *V*. There is no need to drop the valley rafter because it is automatically in position to allow the sheathing to lie flat.

Jack Rafters

Hip jack rafters are similar to common rafters. The upper ends, however, have a miter cut to fit against a hip rafter. Valley rafters are similar to common rafters, with their lower ends having a

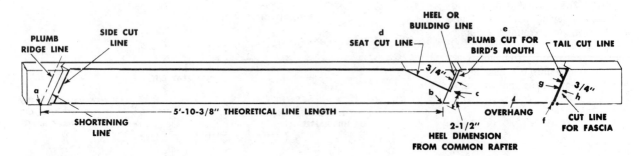

Fig. 5-128. The valley rafter is laid out in a manner similar to the hip rafter. The cuts at the bird's mouth and rafter tail are different. The steps involved include measuring the line length of the rafter *(a, b)*, laying out the bird's mouth *(e, d)*, laying out the tail cut *(h, g)*, and laying out the cuts at the ridge, which are identical for those for a hip rafter.

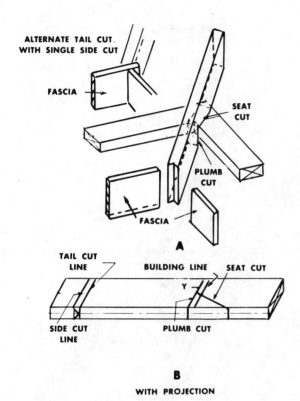

ALTERNATE TAIL CUT
WITH SINGLE SIDE CUT

FASCIA

SEAT
CUT

PLUMB
CUT

FASCIA

A

TAIL CUT
LINE

BUILDING LINE

SEAT CUT

Y

SIDE CUT
LINE

PLUMB CUT

B

WITH PROJECTION

Fig. 5-129. The bird's mouth of the valley rafter is cut to fit over the plate corner, as shown in (A). In (B), line Y is placed ³/₄ inch from the building line toward the lower end of the rafter.

miter cut to fit against a valley rafter. The roof shown in Figs. 5-113 and 5-114 has only hip and valley jack rafters. Some jobs require crippled rafters, which are rafters with miter cuts.

Laying out hip jack rafters. Select a straight common rafter pattern. Draw a center line along the top edge. The bird's mouth, the overhang, and the tail cut are the same for both the common rafter and the jack rafter. Refer to the column designated *Difference in Length of Jacks on 16 inch Centers* on the framing square. See Fig. 5-118, which appeared on p. 316. The number given for an 8 inch unit rise is 19¼ inches. When the jacks are evenly spaced along the plate 16 inches O.C. and the cut is an 8 inch rise per foot of run, the difference in length of jacks measured along the center line of the pattern rafter is 19¼ inches. The jacks for the roof shown in Figs. 5-113 and 5-114 are spaced 16 inches O.C., beginning with the common rafter.

The space from the shortest jack to the corner measured along the plate is only 13 inches.

On the common rafter pattern, measure a distance of 19¼ inches from the theoretical center of the ridge *(a)* to point *b*. See Fig. 5-130A. Draw a line across the top and a plumb line down one side to locate line *c*. Measure 1⅛ inches (or half the diagonal thickness of the hip rafter) and draw plumb line *d*. Measure ³/₄ inch (one half the thickness of the jack rafter) and draw plumb line *e*. Extend line *d* across the top of the rafter and draw a line through points *X* and *Y* which locates the side cut on the opposite side of the rafter. This is the pattern for the longest jack rafter.

Measure 19¼ inches from point *X* to locate point *Z*. With a bevel square set to the angle of *XY*, draw line *ZW*. This determines the cut line for jack rafter *#2*. The pattern rafter is now marked for both jack rafters *#1* and *#2*.

Transfer the markings to other stock, making four sets of right-and left-hand pairs of rafters.

Laying out valley jack rafters. Valley jack rafters are laid out on a common rafter pattern, starting with the longest jack. The layout uses the same common difference and cutting line procedure as used for hip jack rafters. See Fig. 5-131. The difference between the two rafters is that the valley jacks are cut on a diagonal at their lower ends.

Choose a straight common pattern rafter which has the shortening at the ridge marked on the top and side. See Fig. 5-131. Draw a center line on the top edge of the rafter. From the point which represents the center of the ridge *(c)*, measure 38½ inches to locate line *d*. See Fig. 5-131. (38½ inches is twice the common difference, 19¼ inches.) Square the line to the edge and draw a plumb line on the side. Draw another plumb line 1⅛ inches from line *d* to establish line *e*. (The 1⅛ inch dimension is one half the diagonal thickness of the valley rafter.) Draw another plumb line ³/₄ inch from line *e* to establish line *f*. (The ³/₄ inch dimension is one half the thickness of the jack rafter.) Square lines across the top at the rafter and draw line *XY*, extending it to the opposite edge. A plumb line through this point gives the cutting line on the opposite side.

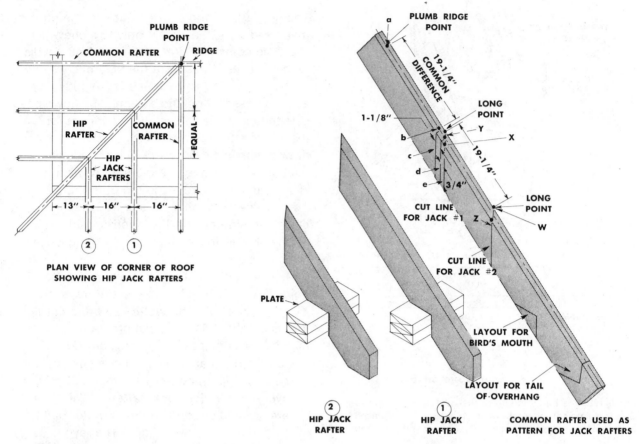

Fig. 5-130. Hip jack rafters are laid out on a common rafter pattern. The longest jack rafter is layed out first. The subsequent rafters are laid out from long point to long point.

To layout valley jack *#2,* measure 19¼ inches from point *X* to locate point *Z.* Draw a plumb line through point *Z* to provide the cutting line on one side of the valley jack rafter. Locate the cutting line on the opposite side by placing a bevel square on line *XY* and transferring it to point *Z.*

The pattern rafter is now complete for valley jack rafters *#1* and *#2.* Transfer the markings to other stock and cut two right- and left-hand pairs of rafters.

UNEQUAL PITCH ROOF

The following section provides you with a general background so that you know how to approach roofs which have unequal pitch or have sections with an unequal pitch. The plan view of the roof used for this study is shown in

Fig. 5-132A. (Note: The difference in pitch between the two parts of the roof shown in Fig. 5-132 is greater than that which is usually found in actual practice on the job. These pitches, however, were chosen to stress the layout problem and show the cuts in greater contrast.)

Laying Out and Building the Equal Pitch Portion of the Roof

When you are faced with an unequal pitch problem, first frame as much as you can of the roof which does not involve unequal pitch hip, valley and jack rafters. For an example, see Fig. 5-132.

The roof can be considered as two separate roofs. One has a gable end and a slope of 18 inches rise per foot, and the other has a hip end with a slope of 12 inches rise per foot. Using the

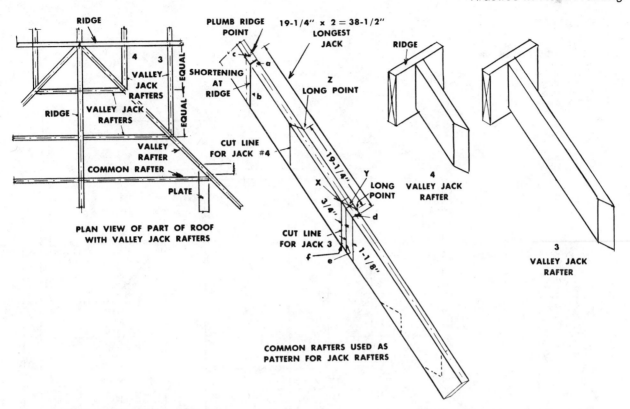

Fig. 5-131. Valley jack rafters are laid out on a common rafter pattern. The longest jack rafter is laid out first. Subsequent jacks are laid out from long point to long point.

instructions given earlier in Chapter 5, lay out all the common rafters for the part using the 18 inch unit rise, the 12 inch unit run, and a total run of 2 feet, 8 inches. Make the deduction for the ridge. Use a 3 inch heel cut. Make all of the common rafters for the low slope portion of the building, using a unit rise of 12 inches, and a total run of 4 feet. Erect the common rafters and the ridges at this time. Brace the ridges where they meet, so that they keep in the proper place. Lay out, then cut and install the hips and jacks for the low slope end.

Making a plan layout for hip and hip jack rafters. A layout drawn to scale of the unequal pitch parts of the roof is essential to determine the relationship between members, the runs for the members, and to help arrive at the shape of side cuts. In analyzing the model roof used in this chapter, we will consider the hip jack portion first. See Fig. 5-133.

First draw the plates, starting at the corner and extending beyond the common rafters which support the ends of the ridges. See Fig. 5-133. Next lay out the center lines for the common rafters, the hip, and the jacks. Space the jacks 16 inches O.C. along each plate. Notice that even though the center lines are measured the same distance from the corner, they do not meet at the hip. After the center lines have been laid out, proceed to lay out the common rafters, the ridge, the hip, and the jacks, in that order. Use stock of the exact thickness (1½ inches). Figure 5-134 shows how the full size sketch of the members is drawn directly on top of the layout shown in 5-133. The sketch is important because it gives the exact dimensions of the cheeks for the hip or valley rafter and the cuts for the jacks. The full size portion is drawn again to show the measurements for the side cuts more clearly. See Fig. 5-135.

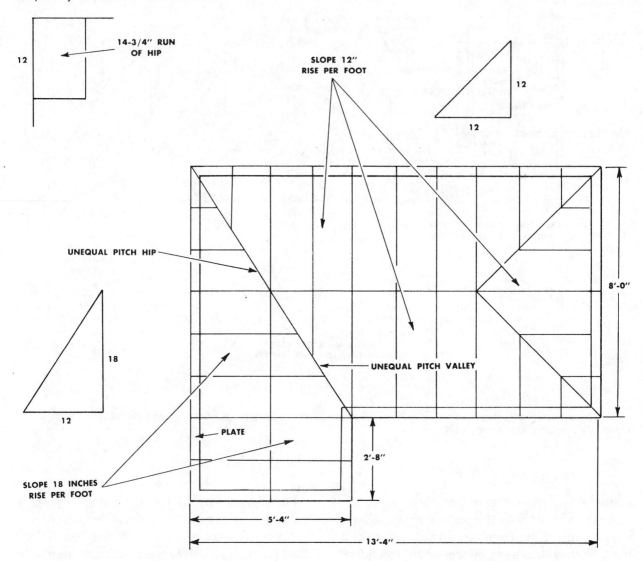

Fig. 5-132. The plan of a roof with an unequal portion should be laid out to scale to find the location and run of the rafters. (Note: The two slopes are made radically different strictly for the purpose of demonstration.)

Layout of the Hip Rafter

Tape the dimension from the intersection of the common rafters and the ridge to the plate corner, as shown in Fig. 5-136A. Lay out the rafter length as discussed under *Using the Taping Method*, p. 279.

Laying out the plumb cut of the hips (heel). Before you consider making any side cuts or dropping, you should make the plumb cut (heel). The top edge of a fascia which goes around the roof must be level and line up with the top of the

plumb cut on all of the rafters. Therefore, you must use the plumb cut at the plate (heel), which has been established on the common rafter. See Fig. 5-136A. (In this case the plumb cut at the plate of the common rafter is established at 3 inches.)

Making cuts on the hip rafters. The procedure for laying out the seat cut and side cuts at the top plate and the ridge are similar to that used for the equal pitch roof discussed earlier in Chapter 5. One difference is in the width of the size cuts. These dimensions are taken from the

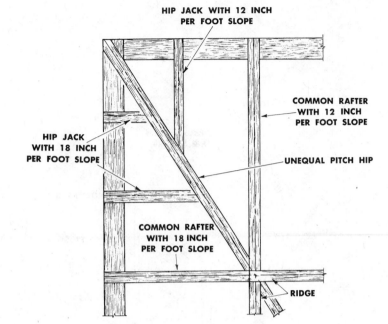

Fig. 5-133. The hip area of the unequal pitch roof is laid out to scale. The jacks do not fall into place opposite each other at the ridge.

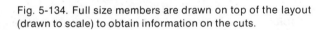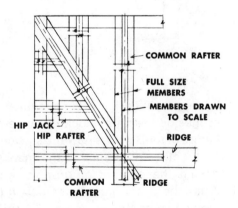

Fig. 5-134. Full size members are drawn on top of the layout (drawn to scale) to obtain information on the cuts.

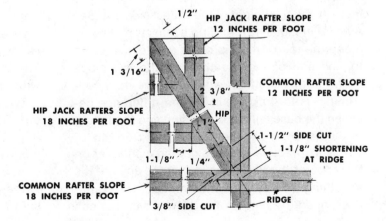

Fig. 5-135. The full size detail of the rafters gives information on the location of the plumb cutting lines.

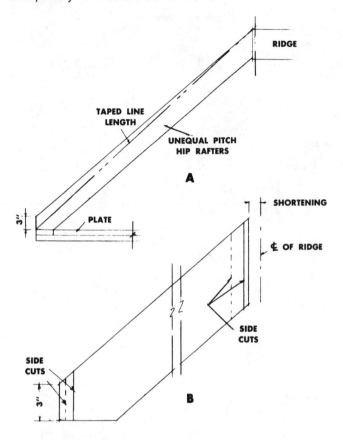

RIDGE

TAPED LINE
LENGTH

UNEQUAL PITCH
HIP RAFTERS

A

3″

PLATE

SHORTENING

℄ OF RIDGE

SIDE
CUTS

SIDE
CUTS

3″

B

Fig. 5-136. Laying out a hip for an unequal pitch roof includes (A) taping the line length and (B) transferring side cuts and shortening from the sketch. See Fig. 5-135.

full size layout (Fig. 5-135) and measured on level lines on the rafter stock to locate plumb lines. Figure 5-136B shows how the side cuts will look.

The other difference is the manner of determining the amount of drop for the seat cut. When there is a great difference in the slope of the two roofs, this becomes a significant problem. The usual procedure is to drop the rafter the difference in the base tail cut measurement (3 inches) and the height of the widest side cut. See A and B in Fig. 5-137. Adding a thin strip of wood along the top of the hip rafter provides support for the end of the sheathing of the low slope roof. Another method, which is more time-consuming, is to drop the rafter the difference between the base tail cut dimension and the height of the narrower seat cut. See C and D in Fig. 5-137. This practice requires cutting a chamfer on the edge of the hip rafter so that the ends of the sheathing lie flat from both directions. The rafter is dropped the height of the lower side cut and

backed the difference between the height of the two side cuts.

Laying Out the Hip Jack Rafters

The peculiar problems which develop during the layout of hip jack rafters are shown in Fig. 5-133. If the rafters are spaced at the same interval along the plate as are the common rafters, they do not meet in a herringbone pattern at the ridge. If, however, they are made so that they meet at the ridge, the spacing along the plate varies with the spacing for the common rafters. The cheek cut does not make an angle of 45° with the hip rafter.

Two types of hip jacks are required for the layout. One set has a cut of 18 inches rise per foot (Fig. 5-138), and the other set has a cut of 12 inches rise per foot. Choose a common rafter pattern for each side. Determine the run of the longest jack rafter by scaling the plan layout, as shown in Fig. 5-133. Use the step off method to

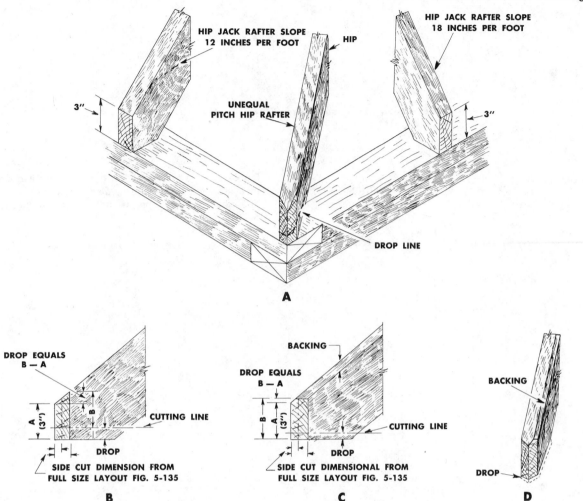

Fig. 5-137. The seat cut on an unequal pitch hip rafter must be dropped to permit sheathing to lie flat. Typical procedures are shown in (A) and (B). Dropping and backing, shown in (C) and (D), are used when the difference in slope on the two sides of the roof is great.

locate point A in Fig. 5-138. (Follow the procedure used earlier in Fig. 5-130, p. 324, except in the present case the measurements for the cut lines are taken from Fig. 5-135.) The shorter jack is developed by establishing the equivalent long or short point, as shown in Fig. 5-130. The jacks for the low slope roof are laid out in a similar manner.

Laying Out the Unequal Pitch Valley Rafter

The problem of laying out the unequal pitch valley rafter is solved by following the same steps used for laying out the unequal pitch hip rafter. It is advisable to make a layout of each

part of the roof where unequal hips, valleys, and jacks develop. In this particular problem, the valley portion and the hip portion of the roof are nearly the same.

The valley rafter and hip rafter are similar in layout. Deductions and side cuts are obtained in the same way. The detail layout in Fig. 5-139 shows the side cuts at the plate. The valley rafter need not be dropped or backed because it does not extend above the other rafters.

Laying out the unequal pitch valley jack rafters. The common rafter pattern is useful for laying out the valley jack rafters in the same manner as the hip jack rafters. A full size sketch

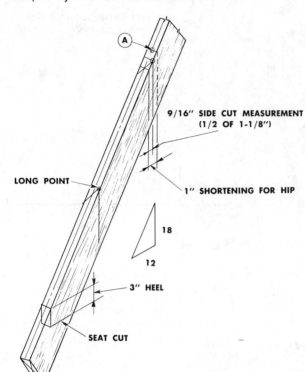

(A)

9/16" SIDE CUT MEASUREMENT
(1/2 OF 1-1/8")

LONG POINT

1" SHORTENING FOR HIP

18

12

3" HEEL

SEAT CUT

Fig. 5-138. The unequal pitch hip jacks are laid out using measurements taken from the plan view shown in Fig. 5-133. The cuts are taken from the full size layout shown in Fig. 5-135.

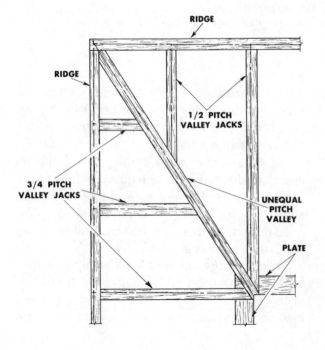

RIDGE

RIDGE

1/2 PITCH
VALLEY JACKS

3/4 PITCH
VALLEY JACKS

UNEQUAL
PITCH
VALLEY

PLATE

Fig. 5-139. The valley area of the unequal pitch roof is laid out to scale to find the run, position on the roof, and the cuts of each rafter.

for the valley portion of the roof should be made to provide information for the deduction at the valley and side cuts.

UNEQUAL PITCH ROOF WITH OVERHANG

Several new problems arise when a roof with an unequal pitch also has an overhang. Instead of using the plate as the base for measurement, use the line representing the top of the fascia which nails to the ends of the rafter. This line must be kept level around the house. Adjustments become necessary in the line lengths, the bird's mouths, and the side cuts. If the change in pitch between one part of the roof and another is great, or if the overhang is wide, the rafters on the steep slope may have a poor bearing on the plate. You must raise the plate for this part of the roof so that the bird's mouth can be properly supported. One other interesting feature of this type of roof is the fact that the hip and valley rafters do not pass over the plate corners. (Note: The roof used to illustrate the layout problems is shown in Fig. 5-140. Again we have used a some-

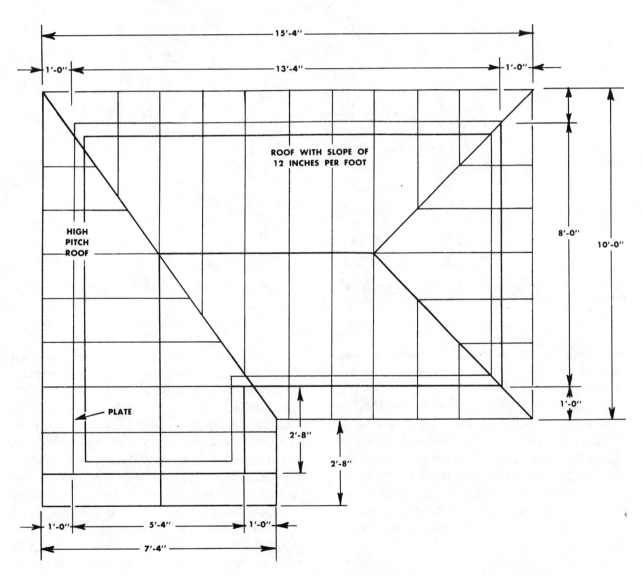

Fig. 5-140. The same building used for the problem shown in Fig. 5-129 is given a 1 foot run overhang on all sides.

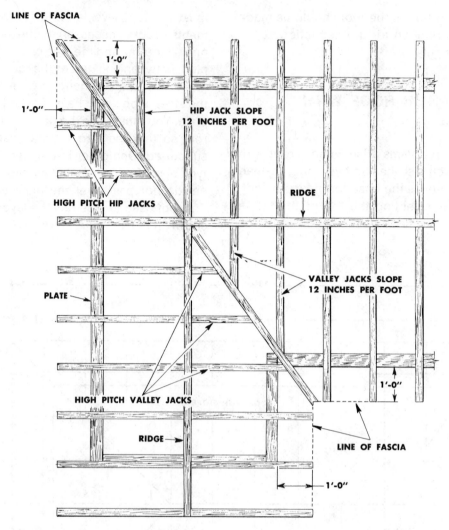

Fig. 5-141. A detailed large scale layout of the rafters gives information on their run and cuts. Hip and valley rafters do not cross plate corners on an unequal pitch roof with a projection.

what extreme change in pitch between the two parts of the roof so that we could illustrate the solutions more clearly.)

The roof shown in Fig. 5-140 is designed to cover the same building as the roof shown in Fig. 5-132, except for the 1 foot overhang. When an equal pitch roof has hips and valleys, the hip and valley rafters are centered directly over the top plate corners. This is not the case, however, for unequal pitch roofs with projections and with fascias maintained in level positions. See Fig. 5-141.

It is advisable to make a scale layout superimposed with full size members, as shown in Fig. 5-

134. This determines the relationship between the hip or valley rafter and the top plate corners, the run of the members, including the jack rafters, and the exact size of the side cuts.

Since the hip and valley rafters do not pass over the top plate corners, sufficient bearing may not have been provided to support the members. (*Bearing* is the area of the seat cut in contact with the top plate.) It may be necessary to support the rafter by adding a block of the required thickness to support the bird's mouth. Figure 5-142 shows the layout of the hip and how a block is added to provide such additional bearing.

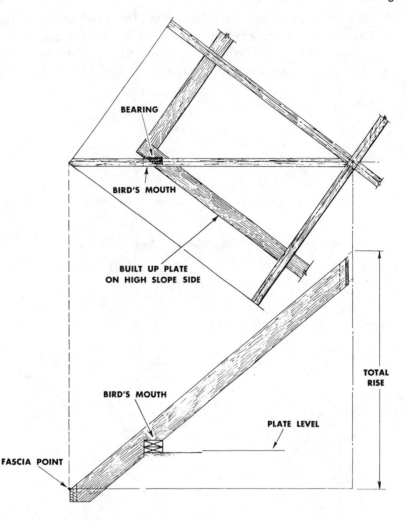

Fig. 5-142. The hip rafter for an unequal pitch roof with a projection can be laid out in detail from the plan view. Blocks are required to raise the plate level to provide adequate bearing.

BEARING

BIRD'S MOUTH

BUILT UP PLATE
ON HIGH SLOPE SIDE

BIRD'S MOUTH

PLATE LEVEL

FASCIA POINT

TOTAL
RISE

QUESTIONS FOR STUDY AND DISCUSSION

1. How do you calculate the length of the main ridge for a hip roof?
2. Explain the procedure for stepping off a common rafter which has an odd unit of run.
3. How much should a common rafter be shortened for the ridge?
4. Explain how to use the framing square table to find the line length of a common rafter.
5. What is the unit of run for a hip rafter?
6. Explain the procedure for dropping a hip rafter.
7. How much do you shorten the hip rafter at the ridge?
8. What is the procedure for taping the hip rafter?
9. How does the bird's mouth for a valley rafter differ from the bird's mouth for a hip rafter?
10. What is the common difference between jack rafters?
11. How is the run of the hip rafter determined?

12. How do you find the line length of the hip rafter using the taping method?
13. Explain the procedure for determining the side cuts and shortening for the hip rafter.
14. How do jack rafters on an unequal pitch roof differ from those on a roof with equal pitch?
15. What is unusual about an unequal pitch roof with an overhang regarding plate corners?

CHAPTER 6
EXTERIOR FINISH

The material in this chapter deals with the roof and outside wall covering, and with the details of architectural woodwork and metalwork which are a part of the exterior of a house. Installing windows and the exterior trim around them is likewise considered a part of exterior finish.

The primary purpose for covering the house is to protect the structure itself and the interior from the elements. Water, dust, and wind must be kept out. The transfer of water vapor and heat through the walls must be controlled and kept at a minimum. Some materials can serve these purposes better than others. Some woods are more durable than others when exposed to weather. Each type of roofing and siding has some particular advantage. Aluminum and steel are today used extensively for siding and exterior trim. The manner in which these materials are installed has a significant bearing on whether they serve their purpose well or not. Installing gutters and flashing, although not always a part of the carpenter's job are included in this chapter so that the builder will be familiar with the problem of protecting the building from water damage.

As a carpenter you should know how to install window and door frames quickly and efficiently. You should likewise be familiar with the pro-

cedures for installing siding and architectural trim, such as cornices and special types of trim at doorways and windows. Trimming details are important because they can add to or detract from the architectural beauty of the house.

You must follow the architect's drawings as closely as possible. Exterior finish items must be constructed so that they are weather tight, structurally sound, and fitted with precision. They are important because they are so visible to anyone viewing the house.

The exterior finish of a building is intended to serve three purposes: (1) to protect the vital parts of the structure (both the framework and the interior); (2) to seal all cracks and crevices to prevent the infiltration and escape of air; and (3) to enhance the appearance of the house. The architect draws the plans with these objectives in mind, but the carpenter must work carefully so that they are carried out.

The principal parts of the outside finish include the cornice trim (or overhang), gutters, roof covering, door and window frames, wall covering and trim, corner treatment, and water table. These items are applied after the walls and roof have been sheathed. Traditional houses, or those with regional characteristics may have porches, canopies, front entrance doorways, special treatment at windows and bays, railings, fences, or decks. Some of these items are manufactured in a mill and installed on the job in units which require little work on the part of the carpenter. However, some of them are built by the carpenter and require careful study of the details shown on the working drawings so that they are properly constructed.

The order in which the work is done depends to some extent upon the type and design of the building, the trim used, and the relation of the different members to each other. The cornice (overhang) is among the first items completed so that the roofing can be applied. Window and door frames are set before siding is applied.

MATERIALS

The materials used for exterior finish vary with the design of the house, the locality, and other conditions. There are many different materials used for exterior finish, such as wood, brick, stone, metal, stucco, and the various manufactured products. However, wood is still the most common material used in most areas.

Wood for Exterior Finish

In choosing the wood for the exterior finish, the architect or carpenter has to consider several factors, such as decay resistance, paint holding quality, and overall appearance (lumber grade). The selection of wood depends to some extent on the kind of wood available in the area.

Woods commonly used for outside finish and generally carried in stock by local lumber yards include pine, cypress, cedar, redwood, and Douglas fir.

Any consideration of the paint-holding quality of the various types of woods not only has to include the species but likewise the density. For example, the paint-holding property of light spring wood is better than that of the heavier summer growths. Edge or vertical grain boards are also found to hold paint better than the flat or plain sawed boards.

Any defect which mars the appearance of wood makes it less desirable for use as exterior trim. A defect in the wood also impairs its dura-

bility and paint-holding property. It is advisable to select only the high grade woods for finishing and trimming the exterior of any permanent building.

Sheathing Paper (Building paper)

The first consideration after the exterior sheathing has been placed is to prevent infiltration of air and dust and to help make the building watertight. The use of insulating fiberboard or treated gypsum board sheathing usually makes further treatment unnecessary. Wood board sheathing and in some areas plywood sheathing must be covered with sheathing paper. Some local codes do not require sheathing. In such localities it is imperative that sheathing paper be applied before the siding is nailed in place. Asphalt saturated felt of low vapor resistance is generally used. The paper must not be waterproof in the sense that it excludes all vapor penetration. If purely waterproof paper were used, it would prevent the escape of water from the interior of the building in winter. The vapor would then condense in the wall and damage the drywall interior finish and cause the wood framing members to rot.

When required, building paper is applied over the whole wall area as soon as possible after the sheathing has been nailed in place. There should be at least a 4 inch toplap at each horizontal joint and a 4 inch sidelap at end joints. Around corners, it should be lapped at least 6 inches from both sides. A sufficient amount of staples should be used to hold it in place until the siding is applied.

Sheathing paper is carefully applied around window and door openings before the frames are placed. This material may be asphalt saturated felt, sisal paper, or waterproof polyethylene film.

Flashing

Sheet metal flashing is installed wherever rain can penetrate behind the siding. (The rain can run down the wall or be driven behind the siding by the wind.) Flashing is applied before the siding is nailed into place. Many building codes require that flashing be applied at the water table (the trim at the bottom of the siding) and at the drip caps over window and door frames. When a building is constructed with brick veneer, the carpenter installs flashing at the base of the sheathing to divert the water to the outside. Otherwise the water would collect inside the wall. (This application was discussed earlier in Chapter 4, under *Wood Framing for a Masonry Veneer Building,* p. 205. Flashing for roofs and at chimneys is discussed later in this chapter.)

EXTERIOR WALL COVERING

Manufacturers of exterior building materials aim for four qualities: beauty, easy maintenance, durability, and low cost. A number of newly developed products for exterior wall covering are available along with the traditional wood boards and wood shingles. Some materials have great resistance to rotting, warping, splitting, and the attack of termites. New products have to a large extent eliminated the need for repainting and other types of upkeep. However, some of these materials are relatively expensive and have their own shortcomings.

Some of the most common types of exterior wall coverings are listed below.

Wood siding is still popular because of its workability, economy, and beauty.

Wood shingles are used to create certain architectural effects. They provide a beautiful wall. Wood shingles are quite inexpensive. The major disadvantage is the cost of installing them. Shakes are hand split wood shingles which present a rugged appearance.

Plywood is manufactured in panels for use as exterior wall coverings.

Wood fiberboard (or hardboard) has been developed for use in several ways as wall covering. This synthetic product is made of wood fiber which has been treated, pressed, then cut into sheets or planks.

Mineral fiber panels and shingles are generally made of asbestos and cement. They are affected very little by weather or age but tend to be brittle.

Aluminum or steel metal siding has achieved considerable acceptance. The baked-on paint finish virtually eliminates painting for many years.

Plastics are also used for exterior wall covering. Wood siding and hardboard covered with a laminated vinyl sheet or a surface coating of vinyl are available. Solid vinyl siding is also manufactured. The solid material is extremely durable. The color of the plastic is not a surface treatment but the color of the material itself. Solid vinyl siding has eliminated much of the maintenance problem traditionally associated with exterior wall covering.

Wood Siding

A great variety of exterior wall coverings have been introduced by the building industry. As noted above, these are manufactured from a variety of materials. Many of these coverings are supplied in large sheets so that they can be quickly applied. However, wood board siding is still commonly used. This type of siding is manufactured in a number of shapes to provide a variety of effects. Woods noted for their resistance to decay, such as Western red cedar and redwood, are especially popular.

Bevel siding came into use early in our Colonial history, and it is still used by builders today. Bevel siding in a residental application is shown in Fig. 6-1. A section view through plain bevel siding is provided in Fig. 6-2. The full width of the butt of each piece of bevel siding casts a deep shadow on the wall, emphasizing the horizontal effect. Bevel siding is generally manufactured in a $\frac{1}{2}$ inch thickness with a 4 inch width, or in a $\frac{3}{4}$ inch thickness with a 5 or 6 inch width. The tip of the siding is $\frac{3}{16}$ inch thick. Wider beveled siding called *Colonial* or *bungalow siding* is 8, 10, or 12 inches wide.

Rabbeted bevel siding (also called *Dolly Varden siding*) is a popular siding pattern which fits tightly against the wall sheathing. It is shown in section view in Fig. 6-2. It is thicker than plain beveled siding. The rabbeted edge makes it possible to be applied with greater speed because it is self spacing. A thicker type of rabbeted bevel siding is called *bungalow siding.*

Drop siding is available with several profiles and with shiplap edges. See Fig. 6-3. Some styles have tongue and groove edges for greater weather protection. Most shapes are made from

Fig. 6-1. Wood bevel siding provides attractive exterior finish. (Western Wood Products Assoc.)

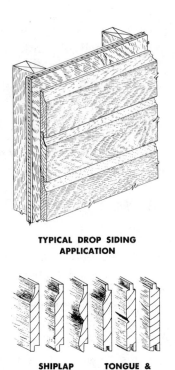

TYPICAL DROP SIDING APPLICATION

SHIPLAP   **TONGUE & GROOVE**

Fig. 6-3. Drop siding fits flush against the sheathing providing good nailing and a comparatively tight wall. It comes in several profiles with either shiplap (bottom left) or tongue and groove (bottom right) edges.

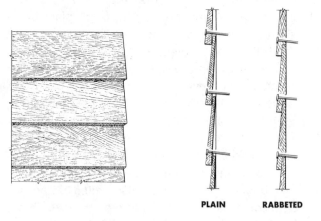

PLAIN   **RABBETED**

Fig. 6-2. Bevel siding is shown here in a traditional siding pattern, both front and section views. (Note: Some manufacturers call rabbeted bevel siding *Dolly Varden siding*.)

Fig. 6-4. Battens emphasize vertical lines. (Western Red Cedar Lumber Assoc.)

¾ inch boards. The width for some styles varies from 6 to 12 inches. Drop siding is particularly useful for applications when sheathing is omitted. The flat back surface fits directly against the studs for nailing.

Board siding can also applied vertically. Using boards and battens is one arrangement. See Fig. 6-4. Boards ¾ inch thick are applied to the wall. There is a narrow space between each board as shown in Fig. 6-5. Strips called *battens* are ap-

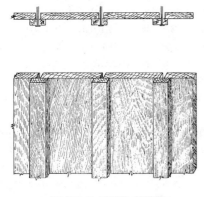

BOARD & BATTEN SIDING

Fig. 6-5. Board and batten siding is applied in a vertical position to produce a bold effect. The section view at the top shows the nailing.

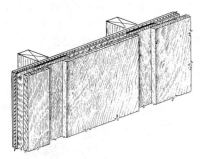

BATTEN & BOARD SIDING

Fig. 6-6. Another way to apply boards and battens is to apply the battens (the narrow boards) to the wall first. An air space is provided in the wall.

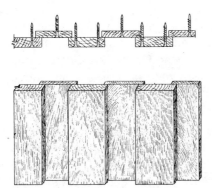

Fig. 6-7. Santa Rosa siding uses two thicknesses of boards, with the thinner board applied to the wall first. Nailing is shown at the top.

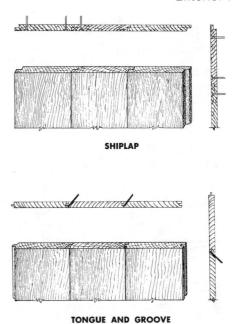

SHIPLAP

TONGUE AND GROOVE

Fig. 6-8. Flush siding can be applied vertically or horizontally. Nailing is also shown.

plied over these openings. Nails are placed to fall in the space between the boards. See Fig. 6-5. An alternative method is to drive the nails in a vertical row so that they pass through the batten and the board on one side of the space. This arrangement allows for some movement in case of shrinkage. A less common variation of the board and batten arrangement is called *batten and board,* shown in Fig. 6-6. The narrow battens are applied to the wall itself, and the wide boards are applied toward the outside. A further variation of this type of board siding is shown in Fig. 6-7.

Some siding patterns are made for vertical or horizontal application. They usually have shiplap or tongue and groove edges which enables them to be relatively weathertight. They are likewise self adjusting when applied. Various patterns are shown in figures 6-8, 6-9, and 6-10.

Applying Wood Siding. Siding cannot be applied until the door and window frames are set. Siding must be lapped so it sheds water and makes the wall covering windtight and dustproof. The minimum amount of lap required for lapped bevel siding is 1 inch. The lap can vary

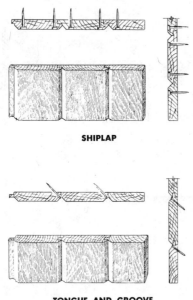

SHIPLAP

TONGUE AND GROOVE

Fig. 6-9. Vee joint siding provides a strong shadow effect. It can be applied vertically or horizontally. Nailing is shown.

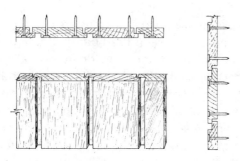

Fig. 6-10. Channel rustic siding is a shiplap type designed to give the effect of board on batten siding. It can be applied horizontally or vertically. Nailing is shown.

slightly if necessary to make the siding line up with the bottom of the window sills and the tops of drip caps at window and door heads. In order that the exposure of the siding will be as uniform as possible, a story pole is laid out for the entire height of the wall, showing the exact height of each piece of siding. See Fig. 6-11.

Procedure. Choose a straight 1 × 2 or 1 × 3 inch strip which is the height of the wall. Measure the siding width. Start at one end of the strip, mark off these units. Some types of siding, such as plain bevel siding, require an overlap.

Make allowance for the overlap in the unit of layout. The lowest board should not extend over the foundation wall more than 1 inch.

Place the strip (the story pole) alongside a window, with the lower end overlapping the foundation one inch. You will note how the units line up with the bottom of the window sill, the top of the window trim, and the top of the wall. In cases when these points nearly line up with the marks on the story pole, you can make small adjustments in the spacing so that the siding joints coincide with these points. You can do this by changing the amount which overlaps the foundation, or by slightly increasing (or decreasing, as required) the overlap between courses of siding. Make new marks on the story pole and scratch out the old ones.

Snap a line around the foundation so that the foundation wall is covered by the siding a minimum of 1 inch at any point along the wall. Check to see that the line is level all around. Hold the story pole in a vertical position, with the lower end on the line, and mark off the units. See Fig. 6-11. Repeat the markings at corners along window and door openings and at intermediate points to accommodate the length of pieces of siding. Snap a line on the sheathing on long walls. The marks provide a guide for locating each piece of sheathing as it is nailed in place. Tack the first pieces in place all around the building and check with a hand level and straight edge to see that they are level. Apply the siding to the wall so that the joints between pieces on long walls do not fall above the pieces nailed previously.

Corners are treated in three ways. Metal corners are used for most corner treatments of beveled siding. See Fig. 6-12A. Some siding is also applied with mitered corners, as shown in Fig. 6-12B. Other siding is applied to butt against corner boards. See Fig. 6-12C.

If the siding does not fit flush against the sheathing, use a miter box to do the cutting. (A power miter box or radial arm saw serves this purpose by speeding up the cutting.) Place a strip in the miter box which gives the siding the correct angle. See Fig. 6-13. It is difficult to make mitered joints that do not open when exposed to changing weather conditions. Plain mitered cor-

Fig. 6-11. A strip of wood called a *story pole* is used to mark off the location of pieces of siding. Marks are made at corners, at intermediate points, and at windows. (Council of Forest Industries, British Columbia)

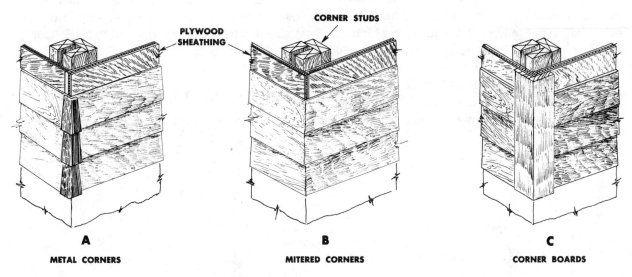

Fig. 6-12. Beveled siding is finished at corners in three ways.

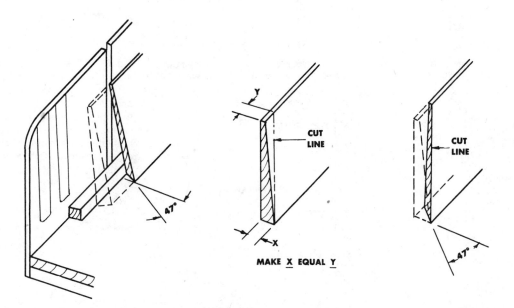

Fig. 6-13. A strip of wood is placed in the miter box to give siding pieces the correct pitch. The alternative method to the right permits the use of a handsaw.

ners must fit closely and stay in place. It is very important to keep properly seasoned lumber dry before it is used on the job. Painting the ends of the joints minimizes their absorbing moisture. It is good practice to cut and fasten the boards by starting at the corners. Fill in the odd length boards in the space between the corners.

Nailing Siding. The nailing of siding is very important. A regular siding nail should be used. 6d nails are used for ½ inch bevel siding, and 8d nail for ¾ inch siding. Either galvanized, aluminum, or stainless steel nails can be used. The latter two types do not rust. Galvanized nails, however, may eventually rust. Using nails which do not rust not only adds to the life of the wall covering, but also prevents stains of rust streaks on the painted surface of the wood. Figure 6-14 shows how to nail some of the typical siding patterns. The clearance shown permits expansion.

Nails are not generally driven through two pieces of siding which overlap.

Wood siding may be obtained pre-primed with paint. It is applied in the conventional manner. It is also provided with a bonded sheath of vinyl plastic. If pre-finished siding of either type is used, it must be applied in accord with manufacturer's directions.

Water Table. The lowest piece of siding generally serves as a drip to divert water which runs down the wall. A water table, which consists of a drip cap and a fascia (a wide horizontal board), is used when the wall surface is stucco or when there is a transition from wood to brick wall facing. See Fig. 6-15.

Sidewall Shingling

Shingles are used for sidewall covering to

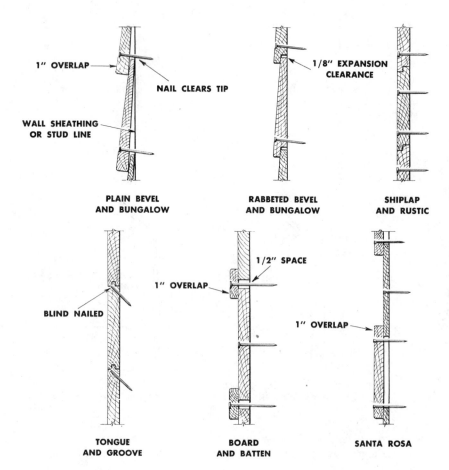

PLAIN BEVEL
AND BUNGALOW

RABBETED BEVEL
AND BUNGALOW

SHIPLAP
AND RUSTIC

TONGUE
AND GROOVE

BOARD
AND BATTEN

SANTA ROSA

Fig. 6-14. The suggested nailing methods for wood siding show the correct overlap and expansion clearance. Note that the nails are placed so that they penetrate only one piece of siding. Bungalow siding is the same as rabbetted bevel siding but it has thicker dimensions. (California Redwood Assoc.)

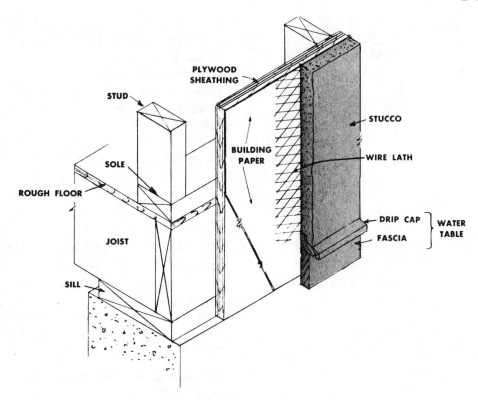

Fig. 6-15. A water table, consisting of a wide board and a drip cap, finishes the bottom of a wall which has a stucco finish and conceals the junction of the sheathing and the foundation.

achieve special effects. They have a long life and require little upkeep.

The customary method used in laying shingles on the side wall is in straight courses. A variation of this can be achieved by staggering the shingles.

Plywood or other wood sheathing is ordinarily used so that the nails will hold. If insulation board sheathing is used, or if the sheathing is omitted entirely, nailing strips fastened to studs are required. See Fig. 6-16.

Laying the shingles in a double course is shown in Fig. 6-17. The deep shadow line obtained in this way adds much to the appearance of the house. Handsplit shakes add character and a soft pleasing appearance to a dwelling. A method of applying handsplit shakes on a side wall is shown in Fig. 6-18.

Applying the sidewall shingles. Shingles must be nailed to a solid base, such as plywood or furring strips. Building paper should first be applied to the wall. Flashing is applied over the drip cap of the water table. See Fig. 6-18. The first course of shingles is doubled or tripled. Following the laying of the first course, carefully study the wall in relation to its total height, the size and height of windows, and the other openings. Next lay out a story pole (shown earlier in Fig. 6-11) with the different courses indicated on it. All the courses should have the same exposure. If possible, the courses should be arranged to line up with the top of the window and door openings. It is also desirable, if possible, to line up the bottom of openings. To make such an alignment may require that you make a slight adjustment in the exposure of the courses. The spacing indicated on the story pole should be marked off for each shingle course at both ends of the wall and at window and door openings. Use a straightedge or snap a chalk line to serve

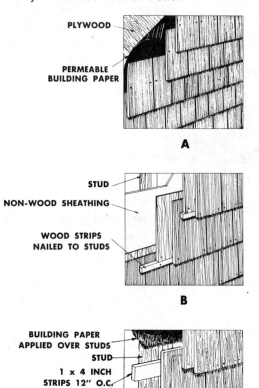

PLYWOOD

PERMEABLE
BUILDING PAPER

A

STUD

NON-WOOD SHEATHING

WOOD STRIPS
NAILED TO STUDS

B

BUILDING PAPER
APPLIED OVER STUDS

STUD

1 x 4 INCH
STRIPS 12" O.C.
(SPACE VARIES
WITH SHINGLE
WIDTH)

1/2" OVERLAP

C

Fig. 6-16. The single course shingles, shown in (A), are laid over the building paper. The nails are concealed as shown. The wood strips, shown in (B), provide a nailing base when non-wood sheathing is used. The shingles, shown in (C), are fastened to 1 × 4 inch strips when sheathing is not used. (Council of Forest Industries, British Columbia)

as a guide line. If you tack a strip of wood to the wall, it is easier to line up the butts before they are nailed down.

Corners are made in several ways. The most attractive but most expensive procedure is to miter the corners. See Fig. 6-19A. A conventional method uses wood strips for outside corners, and a square piece for inside corners. See Fig. 6-19B. A method in which the shingles are cut square to alternately overlap each other is shown in Fig. 6-19C. Metal corner pieces finished to match the stain of the shingles are also available.

Sidewall shake panels are manufactured in widths of 4 and 8 feet, with shakes glued to a backer board. The panels are prefinished in several popular colors of stain. The are likewise available in natural wood. Threaded nails of matching colors are used. The panels are applied rapidly and are easy to handle.

Plywood Exterior Wall Covering

Plywood is versatile as exterior wall covering because of the development of effective waterproof glues. Plywood can be applied in full sheets $\frac{3}{8}$, $\frac{1}{2}$, or $\frac{5}{8}$ inch thick. It has sufficient strength so that conventional sheathing can be eliminated provided the nailing required by the standards of the local building codes are followed. The application shown in Fig. 6-20 requires that nails be spaced 6 inches apart on

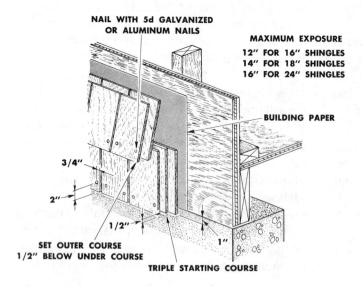

NAIL WITH 5d GALVANIZED
OR ALUMINUM NAILS

MAXIMUM EXPOSURE
12" FOR 16" SHINGLES
14" FOR 18" SHINGLES
16" FOR 24" SHINGLES

BUILDING PAPER

3/4"

2"

1/2"

1"

SET OUTER COURSE
1/2" BELOW UNDER COURSE

TRIPLE STARTING COURSE

Fig. 6-17. Double course shingling adds deep shadow lines. The maximum exposure for the shingles is indicated on the upper right.

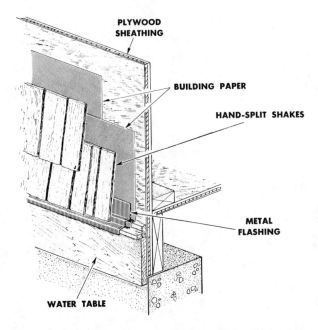

Fig. 6-18. Hand split shakes come in random widths and have irregular surfaces. They give the wall a rugged appearance. A water table and metal flashing can be used.

panel edges and 12 inches apart elsewhere on studs in the field of the panel. 6d or 8d galvanized, aluminum, or stainless steel casing nails are recommended. Box nails are preferred over common nails.

Several types of plywood are supplied by manufacturers for exterior wall finish. The plywood comes in several thicknesses, ranging from 5/16 to 3/4 inch thick. The most commonly used size of sheets is 4 × 8 foot, 5/8 inch thick. This material is supplied with square edges; however, some types have shiplap edges to permit more flexibility in fitting it in place.

A number of surface treatments are available. Several variations in the types of grooving add interest and make the plywood appear like board siding. See Fig. 6-21. The surface can be treated to resemble the texture of smooth sanded or rough sawed lumber. Other sheets are manufactured with a resin coated surface as a prepara-

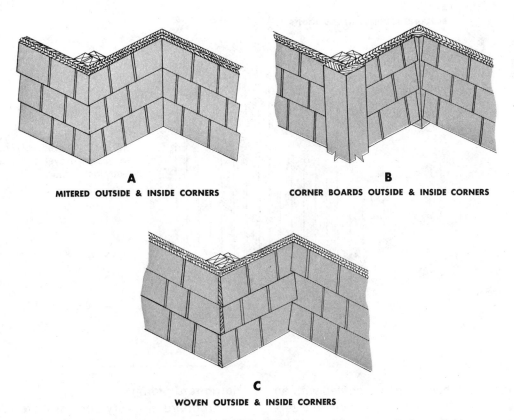

Fig. 6-19. The corner treatment for shingles requires careful fitting. Matching metal corners are also available.

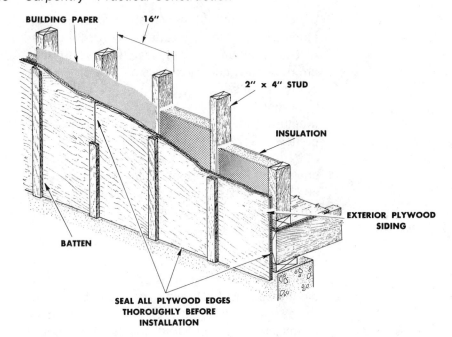

BUILDING PAPER

16″

2″ x 4″ STUD

INSULATION

BATTEN

EXTERIOR PLYWOOD
SIDING

SEAL ALL PLYWOOD EDGES
THOROUGHLY BEFORE
INSTALLATION

Fig. 6-20. Plywood panels are used for exterior finish. The edges are covered with battens. Additional battens are added at stud locations.

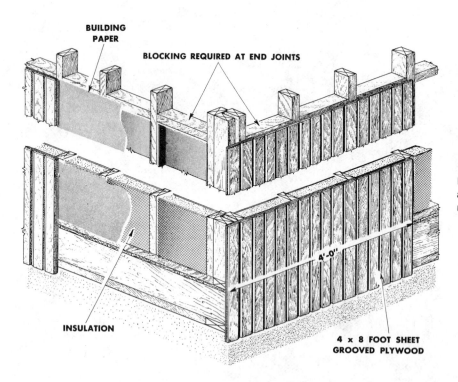

BUILDING
PAPER

BLOCKING REQUIRED AT END JOINTS

INSULATION

4′-0″

4 x 8 FOOT SHEET
GROOVED PLYWOOD

Fig. 6-21. Grooved plywood sheets are fastened over building paper directly to the studs.

tion for painting. Sheets are also coated with an acrylic overlay in color so that no additional surface treatment is required.

Hardboard Siding

Hardboard is a very popular material for wall covering because it does not have knots or other board imperfections. It is dense, which enables it to resist abrasion and denting, and is easily sawed and nailed. Hardboard is made of small wood chips mixed with a binder and pressed into thin sheets using high pressure steam. Hardboard sheets have a variety of applications in the construction of a house. They can be used in large sheets as exterior wall covering. They can also be cut into long strips and applied to the wall as lap siding. See Figs. 6-22 and 6-25.

Hardboard panel siding is available in sheets nominally 7/16 inch thick and 4 feet wide by 8, 9, or 10 feet long. (West of the Rockies, it is available in lengths up to 16 feet.) Panel hardboard is available with V grooves and other grooved patterns, or in plain sheets to which battens are applied. See Fig. 6-22A. Nailing instructions are suggested in Fig. 6-22C. A 1/16 inch space is required whenever two sheets join vertically. Battens are placed over ungrooved panel joints.

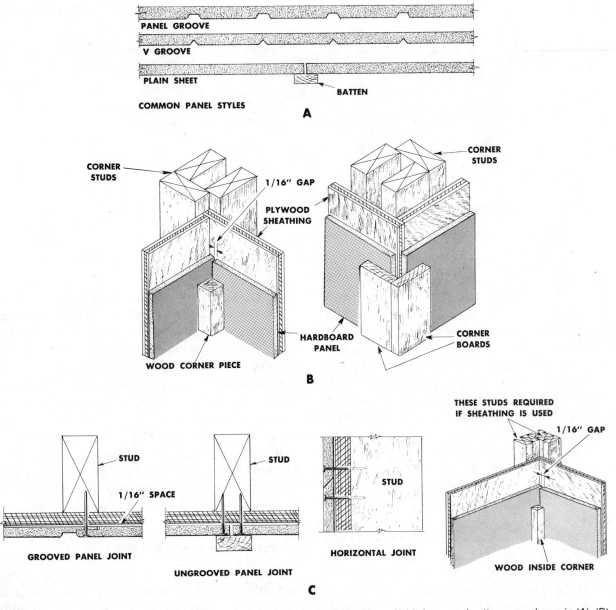

Fig. 6-22. Hardboard panels are applied over plywood sheathing. Hardboard is available in several patterns, as shown in (A). (B) shows the ends of panels are protected at corners with wood pieces. In (C), the edges of panels are firmly nailed with an allowance for swelling.

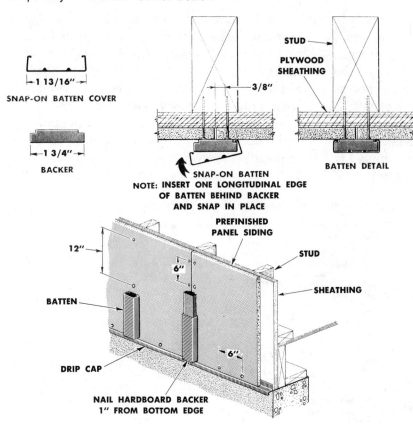

↤ 1 13/16″ ↦
SNAP-ON BATTEN COVER

↤ 1 3/4″ ↦
BACKER

STUD

PLYWOOD SHEATHING

3/8″

BATTEN DETAIL

SNAP-ON BATTEN
NOTE: INSERT ONE LONGITUDINAL EDGE OF BATTEN BEHIND BACKER AND SNAP IN PLACE

PREFINISHED PANEL SIDING

12″

6″

STUD

SHEATHING

BATTEN

6″

DRIP CAP

NAIL HARDBOARD BACKER 1″ FROM BOTTOM EDGE

Fig. 6-23. Prefinished hardboard is applied with snap-on battens which add a finishing touch and conceal nails at panel joints. Backer strips nailed over the joints are concealed with snap-on battens. (Masonite Corp.)

All joints must fall on framing members. Sheets which are prefinished with a vinyl plastic surface can be applied in the conventional manner except that nails and joints are concealed in a special way. See Fig. 6-23. Backer strips of hardboard are nailed to cover the joints, and plastic or metal snap-on batten covers fit over the batten backer strips. The batten covers match the finish on the panels. When the nails are to remain exposed, touch-up paint of the same color, or special precoated nails of the same color are used. Caulking of the same color is also used to cover end joints at windows and other exposed places.

Hardboard lap siding is available in widths of 9 and 12 inches and lengths of 12 and 16 feet. See Fig. 6-24. This siding is generally ⅜ or ⁷⁄₁₆ inch nominal thickness. The commonly used type has a smooth grainless surface, but it is also available with simulated textured patterns which resemble rough sawed planks, shingles, and other

effects. Figure 6-25 gives details for applying hardboard lap siding. Prefinished hardboard lap siding is fastened in several ways. One method is to place nails along the top edge which is concealed by the next piece. The adhesive is applied at intervals along the top edge to hold the bottom of the next piece in place and permit water vapor to escape. See Fig. 6-26. Another method is to use a mounting strip which is an integral part of the siding. The mounting strip presses against the angled top of the lower piece of siding in a wedge action. See Fig. 6-27. Figure 6-28 shows an application of vinyl covered panels and siding.

Mineral Fiber Wall Covering

Mineral fiber material used for wall and roof covering is made of asbestos fiber and Portland cement. It is fireproof and termite proof, does not warp or shrink, and is very durable. The more common type comes in wide shingles which are

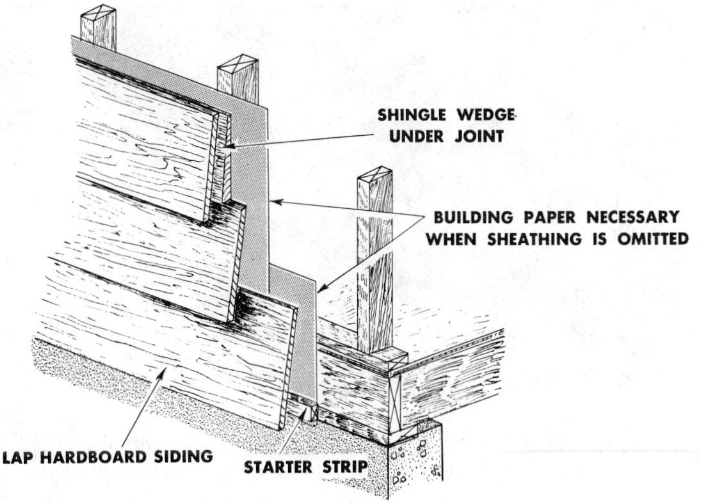

Fig. 6-24. Hardboard is used as lap siding. Joints between pieces are supported by shingle wedges. Metal starter strips are also available.

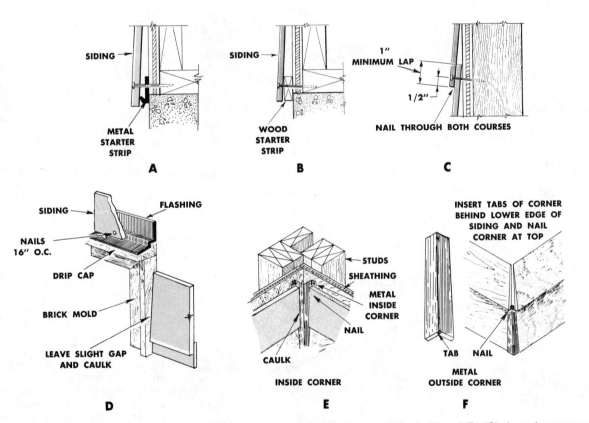

SHINGLE WEDGE UNDER JOINT

BUILDING PAPER NECESSARY WHEN SHEATHING IS OMITTED

LAP HARDBOARD SIDING

STARTER STRIP

SIDING

METAL STARTER STRIP

A

SIDING

WOOD STARTER STRIP

B

1" MINIMUM LAP

1/2"

NAIL THROUGH BOTH COURSES

C

SIDING

FLASHING

NAILS 16" O.C.

DRIP CAP

BRICK MOLD

LEAVE SLIGHT GAP AND CAULK

D

STUDS

SHEATHING

METAL INSIDE CORNER

NAIL

CAULK

INSIDE CORNER

E

INSERT TABS OF CORNER BEHIND LOWER EDGE OF SIDING AND NAIL CORNER AT TOP

TAB **NAIL**

METAL OUTSIDE CORNER

F

Fig. 6-25. Hardboard lap siding in application. Nailing at the bottom of the wall is shown in (A) and (B). (C) shows how courses overlap. Applying the siding at openings is shown in (D). Applying siding at inside and outside corners is shown in (E) and (F).

12″ × 24″ × 3/16″. Another style is 9″ × 32″ × 3/16″. Mineral fiber is also available in panels 4 feet wide, 8, 9, 10, or 12 feet long, and 1/8 inch thick.

Asbetos cement siding is provided with pre-drilled holes. It cannot be nailed without using these holes or making additional holes in advance because the material is dense and breaks if nails are driven through it. Each bundle of sid-

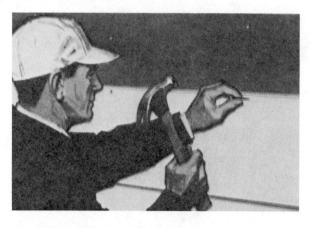

Fig. 6-26. Prefinished hardboard lap siding is applied with nails (left) and adhesive (right). Nails are driven along the top edge. Adhesive is placed at intervals to hold the bottom of the next piece.

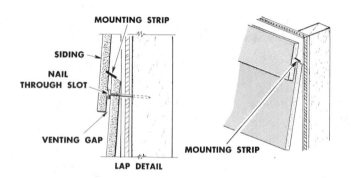

Fig. 6-27. Vinyl covered siding is applied using mounting strips which are an integral part of the siding. The mounting strips space the siding courses automatically. The nails are completely concealed. Note: Use 8d galvanized box head nails. Drive the nails against the surface of the siding, as shown in the left sectional view, to leave a gap for venting. (Masonite Corp.)

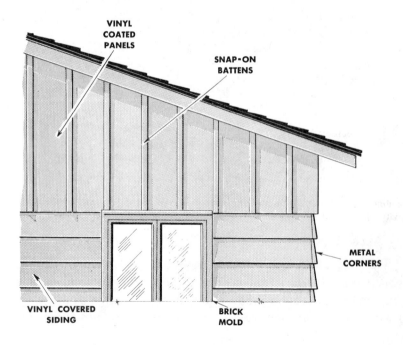

Fig. 6-28. Vinyl covered panels with snap-on battens are shown in the gable end, with vinyl covered siding on the lower wall. Battens and metal corners are finished in color which matches the siding and the panels.

ing is provided with half pieces so that every other course can be staggered from the starting corner. An asbestos board shear tool is used to cut odd lengths.

Mineral fiber (asbestos cement) siding is generally made with a grooved or striated surface so that it resembles shingles. It comes in a variety of colors. A pigmented veneer of Portland cement is applied to the siding prior to the final application of acrylic coating, then it is oven baked. Asphalt backer strips (which help to waterproof the wall) and aluminum nails are furnished with the siding. Nails, inside and outside corners, and cap strips are available in the same color as the siding.

Applying mineral fiber wall covering. A typical installation procedure is shown in Fig. 6-29. A cant strip, which can be a wood lath or other piece approximately $1/2'' \times 1''$, is nailed to the wall in a level position at the bottom of the siding. A level line is snapped $11\frac{1}{2}$ inches above the bottom of the cant strip to serve as a guide to line up the top of the first row of siding. (Other lines should be snapped $10\frac{1}{2}$ inches apart vertically.) The 12 inch material allows for a $1/2$ inch overlap at the bottom of the lowest piece to serve as a drip.

The siding is nailed through the prepared holes at the lower edge with $1\frac{1}{8}$ inch aluminum nails. Backer strips are placed behind each vertical joint. A wood furring strip is then nailed in place at the top edge of the shingle. The nails for the furring strips can be common galvanized shingle nails. They should be driven so that they do not touch the siding. The next row of shingles is then nailed in place. A half piece is used at the corner to start the row. Figure 6-30 shows the use of a shingle backer in an application over wood sheathing. The shingle backer provides additional insulation plus a shadow line.

Prefinished sheets and shingles of mineral fiber are applied using extruded aluminum joint strips, inside and outside corners, and cap strips. These accessories are supplied in the same color as the sheets and shingles.

Metal Siding

Both steel and aluminum metal siding is available. It is used as bevel siding or to simulate vertical boards or boards and battens. The surface can either be smooth or made to look like wood grain. The protective plastic and baked-on finishes are durable and last for many years. The metal siding is available with an insulation backing board which helps to stiffen it and increase the insulation value. The material is easily in-

Fig. 6-29. A typical mineral fiber (asbestos cement) siding shingle application shows cant strip, furring strip, and backer strip. Nails are driven through holes prepared in the siding. (National Gypsum Co.)

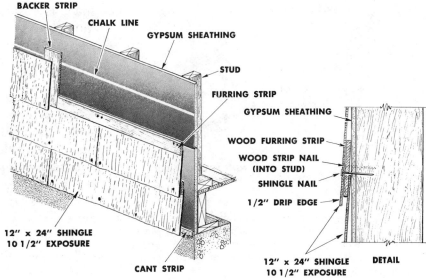

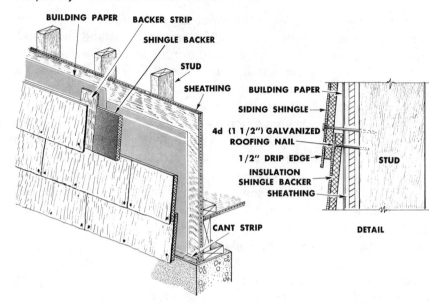

BUILDING PAPER
BACKER STRIP
SHINGLE BACKER
STUD
SHEATHING

BUILDING PAPER
SIDING SHINGLE
4d (1 1/2") GALVANIZED ROOFING NAIL
1/2" DRIP EDGE
INSULATION
SHINGLE BACKER
SHEATHING
STUD

CANT STRIP

DETAIL

Fig. 6-30. Mineral fiber (asbestos cement) siding shingles can be applied over asphalt impregnated shingle backers. Using the shingle backer provides a deep shadow line and saves installation time. (National Gypsum Co.)

Fig. 6-31. Metal siding can be obtained with an insulating backing. (Alcoa Building Products. Inc.)

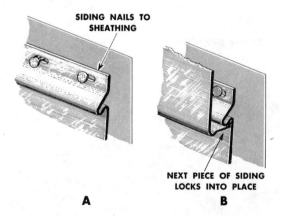

SIDING NAILS TO SHEATHING

NEXT PIECE OF SIDING LOCKS INTO PLACE

A   B

Fig. 6-32. Metal siding is fastened with nails through elongated holes which permit thermal expansion, as shown in (A). In (B), the next piece of siding is pressed into the groove in the previous piece before it is nailed at the top. (Alcoa Building Products, Inc.)

stalled with carpenter's tools plus a hacksaw and tin snips. The low maintenance cost and excellent appearance make it attractive. It is suggested that the lowest course of siding be satisfactorily grounded to eliminate static electricity and reduce the hazard of lightning.

Figure 6-31 shows siding that is provided with an insulating backing. Figure 6-32 illustrates how the siding is nailed and how the pieces interlock. A metal piece called a *butt support* or *back tab* is placed under each horizontal joint. See Fig. 6-33. Special shapes are provided for the starter strip at the bottom of the wall. Outside and inside corners are shown in Fig. 6-34. One shape is designed for fastening at the sides of windows and doors to receive the ends of the pieces of siding. See Fig. 6-34.

Vinyl Siding

Plastic is often used as a protective coating for siding material. It is also now available as solid siding designed to resemble other types of bevel siding and vertical board exterior wall coverings. It is stamped or extruded to a $1/20$ inch thickness and made into boards 12 feet, 6 inches long. Slots for nailing and interlocking lips are integral parts. See Fig. 6-35. All of the necessary accessories such as corners, starter strips, flashing, and window and door trim members are available made from the same material. The siding is

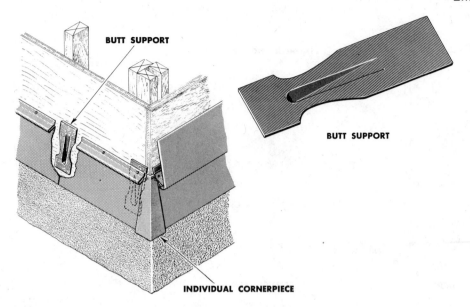

BUTT SUPPORT

BUTT SUPPORT

INDIVIDUAL CORNERPIECE

Fig. 6-33. A butt support is placed under each horizontal joint. A butt support, also called a *back tab,* is shown in application to the left and in the enlargement to the right. (Aluminum Assoc.)

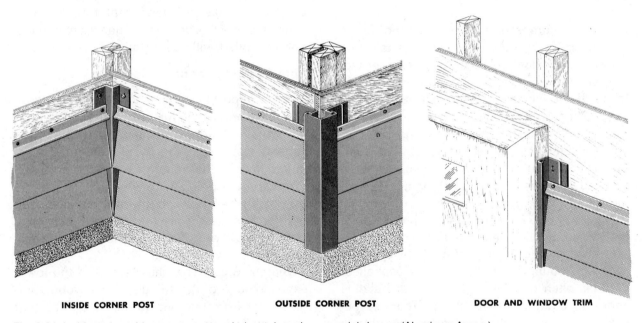

INSIDE CORNER POST

OUTSIDE CORNER POST

DOOR AND WINDOW TRIM

Fig. 6-34. Inside and outside corner posts and trim at doors have special shapes. (Aluminum Assoc.)

supplied with or without insulation backing. The relatively high cost is offset by the fact that the material is virtually indestructible. It does not dent or corrode. It has a color which goes through the material and so it never needs painting. It is worked with tools of the carpenter trade.

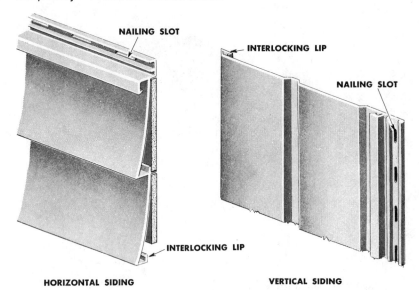

NAILING SLOT

INTERLOCKING LIP

NAILING SLOT

INTERLOCKING LIP

HORIZONTAL SIDING

VERTICAL SIDING

Fig. 6-35. Solid vinyl siding is provided as horizontal clapboards or vertical units. (Bird & Son, Inc.)

ROOF COVERING

The main purpose of any type of roof covering is to keep out the elements, such as rain and snow, and provide protection against other factors, such as wind and dust. If chosen carefully, the roof covering can add materially to the architectural beauty of the building.

Many different kinds of roof covering materials are available, including tile, slate, built-up roofing, asphalt shingles, mineral fiber (asbestos cement) shingles, and wood shingles. Carpenters chiefly work with the last three.

The slope of the roof is a factor in the type of roof which can be used. A flat roof or near flat roof must be a built-up roof made by applying layers of roofing felt bedded in hot asphalt or coal tar pitch over each layer. Local building codes specify the minimum slope required before the use of shingles is allowed. The Uniform Building Code allows composition shingles to be used on roofs with 4 inches rise per foot or greater. The mineral fiber shingles are able to be used on roofs with a minimum slope of 3 inches rise per foot. Wood shingles can be used on roofs with 3 inches rise per foot, but only if specified dimensions are exposed to the weather.

The length of life for any type of roof depends

primarily on the material itself. However, it is to a great extent likewise dependent on the way in which the roof has been laid. The best material, if poorly applied, will give unsatisfactory service.

Roof Covering Terms

A few important terms should be understood before we proceed to further discussion.

Square. Roofing is estimated and sold by the *square*. A square is the amount of shingles required to cover 100 square feet of roof surface.

Coverage. Shingles overlap and, depending on the manner in which they are laid, one, two, or three thicknesses are over the roof at any one place. Thus the roofing is termed *single coverage, double coverage*, etc.

Shingle Butt. The shingle butt is the lower exposed edge of the shingle. Some wood shingles taper from one end to the other. The butt end is the thicker end.

Exposure. Exposure is the distance from the butt of one shingle to the butt of the one above it. This is the portion of the shingle that is exposed to the weather.

Underlayment. An application of saturated felt is placed over the roof surface to protect the roof sheathing until shingles are applied. It is also placed in the valley intersection of two roofs and around the base of a chimney to provide a means to carry off water.

Toplap. The width of the shingle minus the exposure.

Asphalt Shingles

Asphalt shingles are made of asphalt saturated felt coated with mineral granules which give them their color. Light colors are preferred in warm climates because they tend to reflect rather than absorb heat rays. The shingles come in several sizes and shapes. Some of the common types are shown in Table 6-1. The most popular size is the three tab square butt shingle, 12 inches wide by 36 inches long. This size shingle can be applied quickly. It is attractive, relatively inexpensive, and approved for use by most local fire codes.

Figure 6-36 shows some of the popular shingles in greater detail and some of those used for special purposes. The most common type is shown in Fig. 6-36A. It is shown with the self sealing feature of an asphaltum adhesive placed intermittently immediately above the exposure line. After the roofing is laid, the sun warms the adhesive to the point where it melts and adheres to the shingle above it. Anti-stick strips of aluminum foil are placed on the back of the shingles so that they do not stick together before they are placed on the roof. Figure 6-36B shows an asphalt strip shingle with a rustic effect. One edge is made irregular to simulate individual wood shingles. The strips can be laid with either edge toward the weather. A typical three tab hexagonal shingle is shown in Fig. 6-36C. An individual octagonal shingle is shown in Fig. 6-36D. These large 16 × 16 inch shingles give a bold effect to the roof. Figure 6-36E shows a special shingle used where high winds may cause damage. Tabs are provided which slip into slots to form a tight seal against the lifting action of high winds. The shingles are approximately 19 × 20 inches, with a 6½ inch headlap.

The weight of the shingles is important because it is related to the durability of the material. The weight is also a consideration in calculating the dead load on the structure and has a bearing on the size of the rafters. Heavier roofing material is generally considered to be superior.

Asphalt saturated felt used for underlayment and for built-up roofs comes in 36 inch wide rolls. The rolls are either 15 or 30 pounds per square. The 15 pound felt comes in 144 foot rolls. The 30 pound felt comes in 72 foot rolls. Felt used for wood shingles and shakes is 18 inches wide.

The underlayment for asphalt shingles is commonly 15 pound felt. It is laid with the proper sidelap and toplap. See Fig. 6-37. A drip edge

TABLE 6-1. ASPHALT ROOF SHINGLES.

	Shingle type	Shipping weight per square	Packages per square	Length	Width	Units per square	Headlap	Exposure
Strip shingles	2 and 3 Tab square butt	235 lb	3	36"	12"	80	2"	5"
	2 and 3 Tab hexagonal	195 lb	3	36"	11⅓"	86	2"	5"

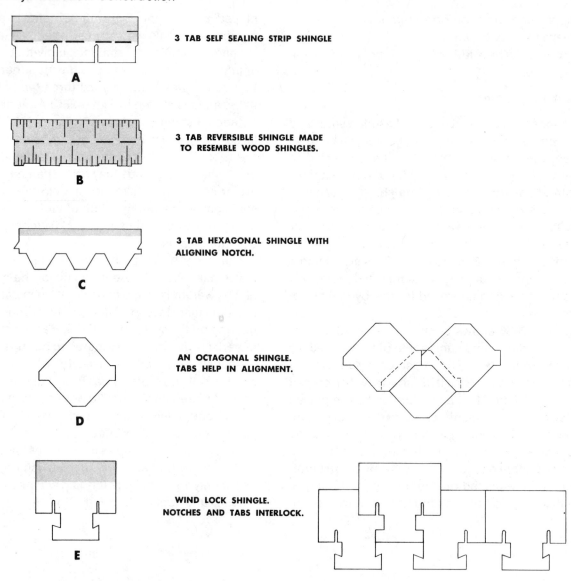

A — 3 TAB SELF SEALING STRIP SHINGLE

B — 3 TAB REVERSIBLE SHINGLE MADE TO RESEMBLE WOOD SHINGLES.

C — 3 TAB HEXAGONAL SHINGLE WITH ALIGNING NOTCH.

D — AN OCTAGONAL SHINGLE. TABS HELP IN ALIGNMENT.

E — WIND LOCK SHINGLE. NOTCHES AND TABS INTERLOCK.

Fig. 6-36. Asphalt shingles are available in several styles. They are made to serve particular needs.

made of galvanized steel is recommended for application to the edge of the roof sheathing both at the eaves and on the rake end (the sloped roof end).The underlayment is placed over the drip edge at the eaves.

Ice often forms along the eaves in cold climates, where the temperature can drop to 0°F (−17.8°C) or lower. As the temperature varies over a period of time, the water (moisture) thaws and freezes intermittently and backs up under the shingles. Instead of running off the roof, the water freezes to form an ice dam. Eaves flashing is placed over the roof boards and the drip edge to prevent this condition from causing damage. Eaves flashing consists of a starter strip of 15 pound felt. It is extended back with additional strips, if necessary, to a line at least 12 inches inside the interior wall line (24 inches for low sloped roofs). See Fig. 6-38. The regular underlayment strips are then laid and cemented to the starter strip with plastic asphalt cement.

Valleys can be flashed by using an open or a

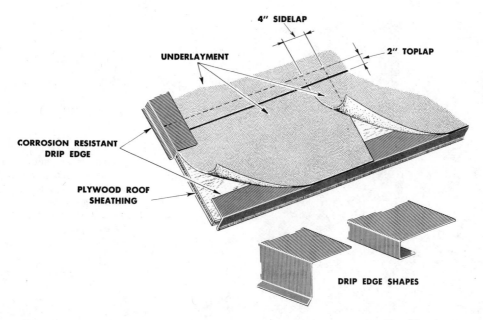

Fig. 6-37. Underlayment for asphalt shingle application must have proper sidelap and toplap. The drip edge appears on the eave and rake sides of the roof.

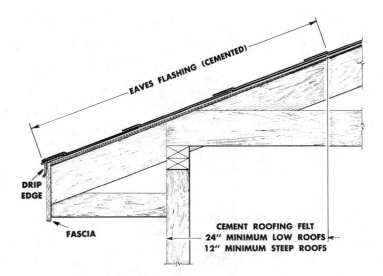

Fig. 6-38. Eaves flashing is used to prevent damage caused by ice dams.

closed method. The open valley method, shown in Fig. 6-39, requires a strip of mineral-surfaced roofing fastened face down to the roof. A second piece which is the same color as the roof shingles is then cemented in place face up. The shingles are cut parallel to the valley line and cemented to the valley flashing strip. Closed valley flashing uses a piece of mineral surface roofing nailed in place in the valley. The strip shingles are then crossed alternately as shown in Fig. 6-40.

Flashing at the chimney requires the use of non-corrosive metal pieces which are bent and soldered to fit. See Fig. 6-41. An apron is placed over the underlayment. The shingles are later laid to extend under the lower edge and over the sides and the top edge of the apron. Base flashing is bent to fit the juncture of the roof and the

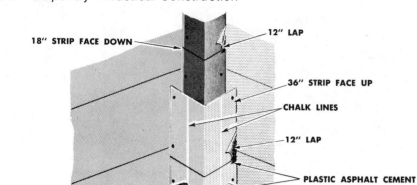

Fig. 6-39. Flashing applied with the open valley method uses strips of mineral surface roll roofing. The first piece is nailed face down. The second piece is cemented to the first piece with the face side up. The shingle ends are cut parallel to the valley line.

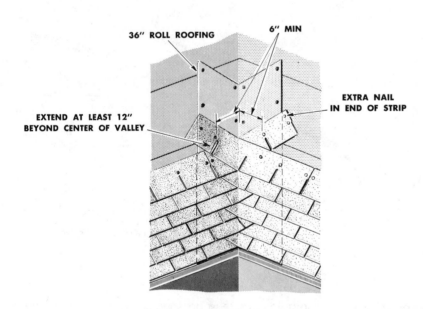

Fig. 6-40. Strip shingles are able to be crossed to provide a closed method of finishing roofing in a valley. A strip of mineral surface roofing is applied first. Shingle strips are extended to cross the valley.

chimney on three sides. The fourth side is protected by a *saddle* (also called a *cricket*) which is made of sheet metal. The saddle is soldered to make a water tight shield which will divert water away from behind the chimney. Cap flashing

(also called *counter flashing*) is placed over the base flashing. The top edge of each piece is bent and built into the chimney joints.

The nails or staples used for applying asphalt shingles should be corrosion resistant, such as

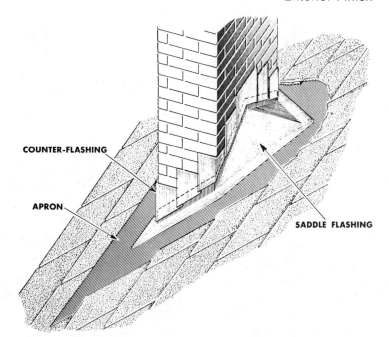

Fig. 6-41. Metal flashing is the most effective way to terminate the roofing at a chimney. The metal parts (shown in the areas shaded in color) should be noncorrosive metal, such as copper or aluminum. If galvanized steel is used, it should be coated with asphaltum paint.

COUNTER-FLASHING

APRON

SADDLE FLASHING

galvanized steel or aluminum. The roofing nails have large flat heads. Shanks of nails can be smooth or threaded. Threaded nails have greater holding power. The galvanized roofing nails should have annular grooves. Aluminum nails should have screw threads.

Applying asphalt shingles. Following the application of the metal drip edge, the underlayment, eaves flashing (when necessary), and the preliminary steps in valley flashing, the roofing operation itself is begun with the application of a starter course. The starter course consists of a 9 inch (or wider) strip of mineral surface roll roofing the same color as the shingles, or strip shingles turned around so that the tabs point up the roof. This arrangement provides a roofing edge without slots. The first regular course is then nailed in place $\frac{1}{4}$ or $\frac{3}{8}$ inch over the drip edge on the eaves and on the rake (at the gable end). The first course of shingles should begin with a whole strip at one end. The second course can be cut so that the lap occurs at the half tab point, if so desired. See Fig. 6-42. A variation on this procedure is to start the second course so that it provides a space smaller than a half tab. This will result in a different effect.

Four nails are used in each shingle. Staples or 1 $\frac{1}{4}$" long nails are used for new roofs. Nails 1 $\frac{3}{4}$ inches long are used when re-roofing is done. A nail is placed 1 inch from each edge, and two other nails are placed near the tab slots. Each nail passes through two thicknesses of shingles and is concealed by the next course when that is applied. Lines are snapped on the underlayment to indicate the location of each course. Some types of felt are supplied with guide lines printed on them. Special precautions must be used to see that the felt is started correctly and adjoining sheets maintain the spacing. The ridges and hips are finished in what is known as *Boston style.* Boston style is a method of overlapping the shingles so that water is shed from one shingle to the next along the ridge or hip. Short shingles are made for this purpose. They are 9 × 12 inches and should be laid with an exposure of 5 inches. See Fig. 6-43.

Built-Up Roofing

(Note: Built-up roofing is generally the job of the roofer rather than the carpenter.)

Built-up roofs are made with 3, 4, or 5 plies of felt. The first layer is laid dry and nailed in place

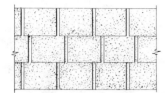

ALTERNATIVE ARRANGEMENT

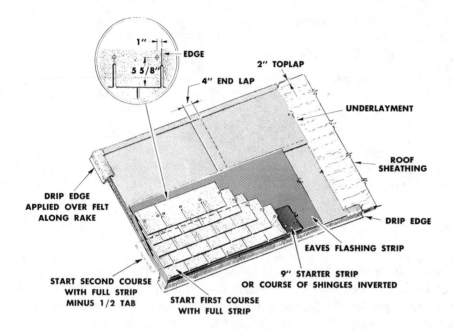

1"
EDGE
5 5/8"
2" TOPLAP
4" END LAP
UNDERLAYMENT
ROOF SHEATHING
DRIP EDGE APPLIED OVER FELT ALONG RAKE
DRIP EDGE
EAVES FLASHING STRIP
9" STARTER STRIP OR COURSE OF SHINGLES INVERTED
START SECOND COURSE WITH FULL STRIP MINUS 1/2 TAB
START FIRST COURSE WITH FULL STRIP

Fig. 6-42. An asphalt shingle roof begins with the application of a drip edge followed by underlayment, eaves flashing strip, or starter strips. Then the first full strip is nailed in place. An alternative arrangement of the strips is shown at the top.

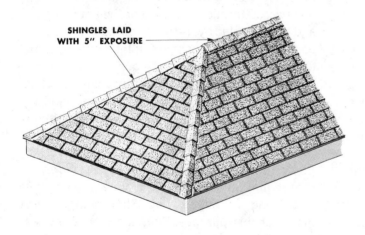

SHINGLES LAID WITH 5" EXPOSURE

Fig. 6-43. A Boston hip and ridge are made by using specially formed overlapping shingles.

with galvanized roofing nails. This layer is mopped with a hot asphalt or hot coal-tar pitch which is spread to meet a certain specification (for example, 25 pounds per 100 square feet). Each succeeding layer of felt is mopped with the same coating. To finish the roof, another layer of felt (called a *cap sheet*) is applied. The top layer of some roofs is covered with a thin layer of aggregate, such as marble chips, pea gravel, or fine slag. (400 pounds of gravel or crushed rock or

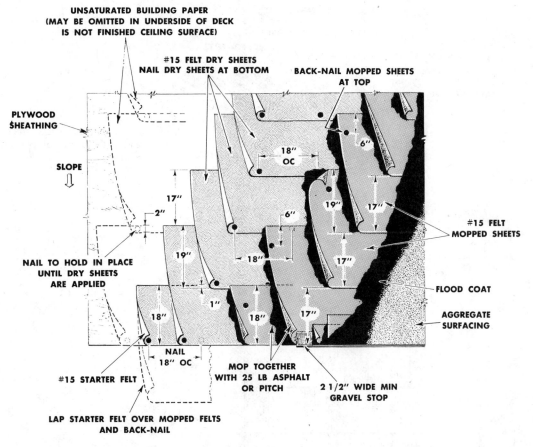

UNSATURATED BUILDING PAPER
(MAY BE OMITTED IN UNDERSIDE OF DECK
IS NOT FINISHED CEILING SURFACE)

#15 FELT DRY SHEETS
NAIL DRY SHEETS AT BOTTOM

BACK-NAIL MOPPED SHEETS
AT TOP

PLYWOOD
SHEATHING

SLOPE

18"
OC

6"

17"

2"

19"

17"

#15 FELT
MOPPED SHEETS

NAIL TO HOLD IN PLACE
UNTIL DRY SHEETS
ARE APPLIED

19"

6"

18"

17"

FLOOD COAT

AGGREGATE
SURFACING

18"

1"

18"

17"

NAIL
18" OC

#15 STARTER FELT

MOP TOGETHER
WITH 25 LB ASPHALT
OR PITCH

2 1/2" WIDE MIN
GRAVEL STOP

LAP STARTER FELT OVER MOPPED FELTS
AND BACK-NAIL

Fig. 6-44. A three ply built-up roof is made of three thicknesses of overlapping asphalt-saturated felt, mopped with asphalt or pitch. The aggregate on the top layer can be omitted.

300 pounds of slag per 100 square feet are typical amounts required.) See Fig. 6-44. The aggregate put on the surface serves to protect the felt and to reflect heat. When the slope of the roof is greater than 2 inches per foot, it is difficult to retain the gravel on the roof. The edge of the roof all around is protected by a metal strip (galvanized steel) called *gravel stop*. It is nailed in place prior to the roofing operation. See Fig. 6-45.

Mineral Fiber Shingles

Mineral fiber shingles are made of asbestos fiber and Portland cement. Because of their mineral content, they are considered highly fire resistant. Workers must be careful when handling and installing them. Workers must also be very careful if they have to walk on them because of their brittle nature. The most popular styles of these shingles are *Dutch lap* and the so-called *hexagonal*. The size for both types is $16'' \times 16'' \times 5/32''$. They are available in several colors. The Dutch lap shingles have a wood grain or striated surface. See Fig. 6-46. Local codes determine the minimum slope allowed for applying mineral fiber shingles. Some codes permit a slope of 3 inches per foot. The material is cut by using a special lever type cutting device. If this device is not available, the shingle is scored with a sharp tool and broken off.

Holes are pre-punched for nails. Galvanized or needle point aluminum nails $1\frac{1}{4}$ inch long are recommended. They should have threaded shanks, especially if the roofing is to be applied over plywood sheathing.

Dutch lap shingles are nailed so that there is

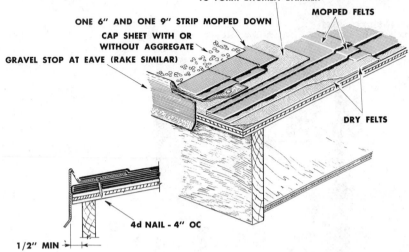

ONE 6" AND ONE 9" STRIP MOPPED DOWN

TURN BACK STARTER FELT
APPROX. 12" OVER MOPPED FELTS
TO FORM BITUMEN BARRIER

MOPPED FELTS

CAP SHEET WITH OR
WITHOUT AGGREGATE

GRAVEL STOP AT EAVE (RAKE SIMILAR)

DRY FELTS

4d NAIL - 4" OC

1/2" MIN

Fig. 6-45. Detail shows the gravel stop and the manner of placing the felt strips.

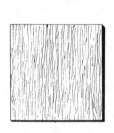

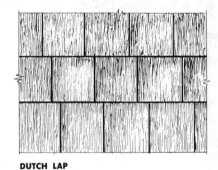

DUTCH LAP

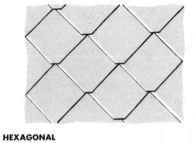

HEXAGONAL

Fig. 6-46. Mineral fiber (asbestos cement) shingles are available in various styles. They are fireproof and have long life.

concealed nailing for two corners. The third corner is held by a storm anchor. See Figs. 6-47 and 6-48. A storm anchor is inserted through each shingle from below before the shingle is placed.

The next shingle is applied. The storm anchor is clinched to hold the two shingles together.

Wood strips must be applied to hips and ridges to provide level nailing surfaces. The ap-

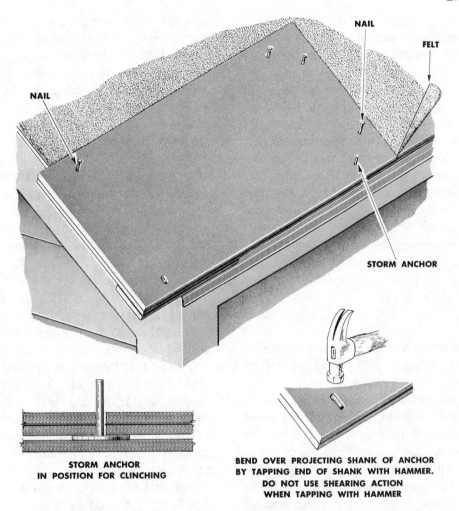

NAIL

NAIL

FELT

STORM ANCHOR

STORM ANCHOR
IN POSITION FOR CLINCHING

BEND OVER PROJECTING SHANK OF ANCHOR
BY TAPPING END OF SHANK WITH HAMMER.
DO NOT USE SHEARING ACTION
WHEN TAPPING WITH HAMMER

Fig. 6-47. Storm anchors hold down the lower corners of mineral fiber (asbestos cement) shingles when laid with Dutch lap. The top view shows the shingle with anchors in application on the roof. The view, left bottom, shows the storm anchor and shingles in a section view. Bottom right shows tapping the anchor down with a hammer.

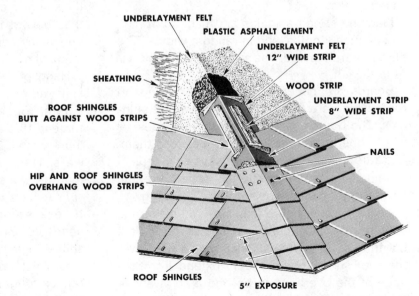

UNDERLAYMENT FELT

PLASTIC ASPHALT CEMENT

UNDERLAYMENT FELT
12" WIDE STRIP

SHEATHING

WOOD STRIP

UNDERLAYMENT STRIP
8" WIDE STRIP

ROOF SHINGLES
BUTT AGAINST WOOD STRIPS

NAILS

Fig. 6-48. Mineral fiber (asbestos cement) shingles are overlapped to form a Boston hip.

HIP AND ROOF SHINGLES
OVERHANG WOOD STRIPS

ROOF SHINGLES

5" EXPOSURE

plication of a Boston hip which uses overlapping pieces at the corner is shown in Fig. 6-48. When the pieces have ben nailed in place at the ridge or hip, the joints are pointed up with plastic asphalt roofing cement. Valleys are flashed with metal. Either the open or closed method is used. With the closed method, individual sheets of flashing are bent to fit the valley and made to rest on top of each pair of shingles meeting at the valley and under the next pair to be applied.

Hexagonal shingles are actually square with corners cut off. See Fig. 6-46. When laid on the roof, they give an effect which is called *hexagonal*. Three edges are covered and nailed with two nails. The front exposed corner is fastened with a copper storm anchor. The finished look is shown in Fig. 6-46.

Wood Shingles

It should be noted at the outset that fire codes in some communities do not permit the use of wood shingles.

Wood shingles have been used as roofing for centuries. Even though they have been replaced to a large extent by material which can be applied more easily, they are still used to create a rustic architectural effect. Wood shingles are generally made of red cedar or cypress because these woods have excellent decay resistance. Two types of wood shingles are available: machine sawed shingles and handsplit shingles called *shakes*.

Machine sawed shingles are made in 16, 18, and 24 inch lengths, and in random widths. The shingles are tapered. The thickness at the butt is $3/8$ to $3/4$ inch.

Boards are generally used as sheathing on roofs which are to be covered with wood shingles. The boards can be laid without spaces, or with a one or two inch space in between them. The attic space area should be well ventilated.

Shingles and shakes must be able to breathe so that they can dry out and avoid rotting. 15 pound saturated felt is required. When there is danger of an ice dam forming, an additional covering of 30 pound felt should be placed over the roof sheathing. It should extend from the edge of the overhang to beyond a point above the place where the overhang and the wall inter-

sect. The nails should be galvanized or aluminum. The amount of weather exposure advisable varies with the type of wood grade and the length of the shingles or shakes. See Table 6-2.

Installing wood shingles. Starting at the eaves, you should double or triple the first course in order to cover the spaces between shingles. See Fig. 6-49. Extend the butts $1\frac{1}{2}$ inches beyond the first sheathing board at the eaves and provide them with a 1 inch projection at the rake. (The *rake* is the edge of the roof at the gable end.) Provide a $1/4$ inch space at each shingle joint to allow for expansion. The joints should be arranged so that there is an offset of at least $1\frac{1}{2}$ inches between the joints on one course and the joints in the course above. Only use two nails in each shingle. Nails should be placed far enough away from the edge to prevent splitting of the shingle ($3/4$ inch minimum) and 1 or 2 inches above the butt line of the next course so that they are concealed.

Shingles for hips and ridges should be arranged using the Boston method. Shingles should be cut and fitted to alternately overlap. Nailing is concealed by the pair of shingles above it. See Fig. 6-50. Valleys are flashed with galvanized or aluminum metal extending up 12 inches on each side of the valley. See Fig. 6-51. The flashing should be crimped (raised) in the center. This prevents water which is running down one slope from crossing and going under the shingles on the other slope. Shingles used at the sides of the valley should be full shingles. The valley corner should be cut off rather than made of two pieces. See Fig. 6-51.

Handsplit shingles (also called *shakes*) are split rather than cut from a log and show the natural rough wood grain. They are manufactured in 18, 24, and 32 inch lengths and in 5 to 18 inch widths. They are split from the log in two ways. They can be *straight split* (which means that the two sides are roughly parallel) or *taper split* (which means they are split so that they have a slight taper). See Fig. 6-52. The term *handsplit and resawn* indicates that a straight split shingle is sawed to make two tapered shingles. Thickness at the butt of handsplit shingles ranges roughly from $5/8$ inch to $1\frac{1}{4}$ inches. Figure 6-52 provides you with an idea how the shin-

TABLE 6-2. MAXIMUM WEATHER EXPOSURE.

Wood shingles		
	Slope of roof	
Grade length	3″ to less than 4″ in 12″	4″ in 12″ and steeper
No. 1 16-inch	3¾″	5″
No. 2 16-inch	3½″	4″
No. 3 16-inch	3″	3½″
No. 1 18-inch	4¼″	5½″
No. 2 18-inch	4″	4½″
No. 3 18-inch	3½″	4″
No. 1 24-inch	5¾″	7½″
No. 2 24-inch	5½″	6½″
No. 3 24-inch	5″	5½″
Tapered wood shakes		
18-inch	7½″	
24-inch	10″	
Straight-split wood shakes		
18-inch	5½″	
24-inch	7½″	

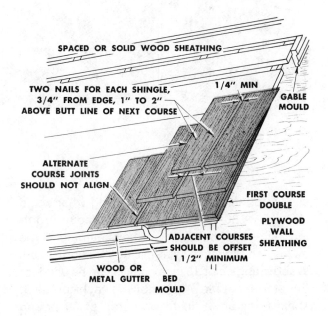

Fig. 6-49. Wood machine sawed shingles are applied in random widths so that joints between shingles are well covered. (Red Cedar Shingle and Handsplit Shake Bureau)

gles are manufactured and enables you to distinguish the differences among the three types. The method used for making handsplit shingles is similar to that used many years ago. The labor and skill involved make them very expensive.

Procedure for applying handsplit shingles is

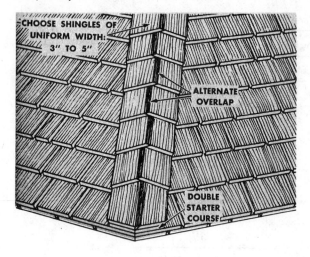

Fig. 6-50. Wood shingles are carefully fitted to make a Boston hip. The *Boston hip* is a method of covering the shingles ending at the hip with short shingles. (Red Cedar Shingle and Handsplit Shake Bureau)

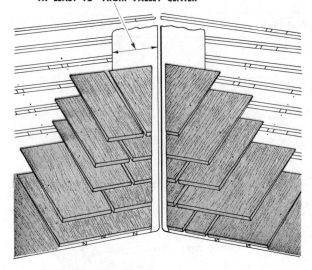

Fig. 6-51. Crimped metal flashing is used to provide a waterproof valley for a wood shingle roof. (Red Cedar Shingle and Handsplit Shake Bureau)

similar to the procedure for applying other wood shingles. A 36 inch wide strip of 30 pound roofing felt should be laid over the sheathing at the eaves line. The first course is doubled or tripled. Wood shingles or 15 inch shakes can be used for this starter course. After applying each course of shakes, lay an 18 inch wide strip of 30 pound roofing felt over the top portion of the shake and extend it out over the sheathing. See Fig. 6-53.

Hips and valleys are covered following the same procedures used for machine made wood shingles. See Figs. 6-50 and 6-53.

THE CORNICE

The exterior finish on a building at the line where the sloping roof meets the vertical wall is known as the *cornice* (or overhang). In addition to its practival value, the cornice can add to the architectural beauty of a building. There are various ways of constructing the cornice, each of which has its own particular advantage. The wide cornice gives greater protection to the building, not only from rain and snow but also from the hot summer sun. The cornice provides

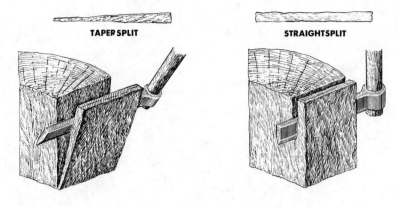

TAPER SPLIT **STRAIGHTSPLIT**

HANDSPLIT AND RESAWN

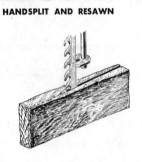

Fig. 6-52. Handsplit shingles called *shakes* are split from a log. Taper split are cut from alternate ends of the log to give a taper to the shingle. Straight split are cut straight down. (Red Cedar Shingle and Handsplit Shake Bureau)

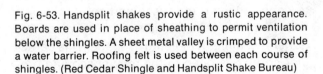

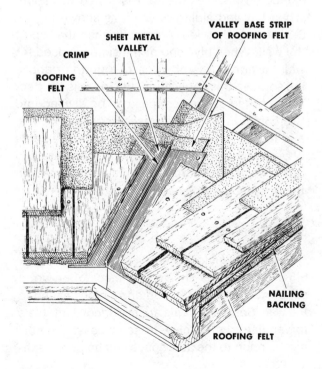

Fig. 6-53. Handsplit shakes provide a rustic appearance. Boards are used in place of sheathing to permit ventilation below the shingles. A sheet metal valley is crimped to provide a water barrier. Roofing felt is used between each course of shingles. (Red Cedar Shingle and Handsplit Shake Bureau)

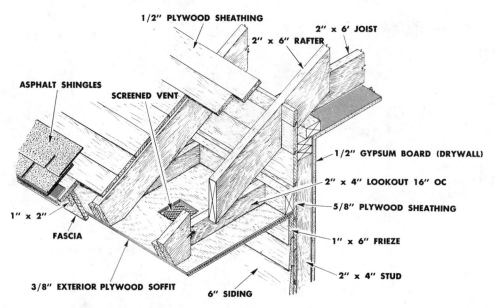

1/2" PLYWOOD SHEATHING

2" x 6" RAFTER

2" x 6' JOIST

ASPHALT SHINGLES

SCREENED VENT

1/2" GYPSUM BOARD (DRYWALL)

2" x 4" LOOKOUT 16" OC

5/8" PLYWOOD SHEATHING

1" x 2"

FASCIA

1" x 6" FRIEZE

2" x 4" STUD

3/8" EXTERIOR PLYWOOD SOFFIT

6" SIDING

Fig. 6-54. The closed cornice used on a modern ranch house is simple in design.

a means for fastening gutters that carry away water as it runs off the roof.

A typical simplified cornice used for a contemporary residence is shown in Fig. 6-54. The board at the outer face directly below the roofing is called the *fascia*. The wide horizontal board extending back to the wall is called the *soffit*. When beveled siding is used, as shown in Fig. 6-54, a board called a *frieze* finishes the top of the wall. When plywood or flat siding is used for wall covering, it is often carried up to or above the soffit. In this application, the frieze is omitted.

In a broad sense cornices can be divided into two common types: the *closed* or *boxed cornice*, and the *open cornice*. The cornice can be made of very few members or it can be elaborate and ornamental. The cornice should harmonize with the architectural design of the building.

Closed Cornice

A simplified closed cornice is shown in Fig. 6-54. The soffit is shown as a piece of plywood supported by the rafter end and a block nailed to the wall. Screened vents are provided at intervals to ventilate the cornice. In some cases the cornice is open to the attic space and helps to venti-

late the whole area under the roof. The frieze is a 1 × 6 inch board rabbeted to receive the top piece of beveled siding. The fascia is another 1 × 6 inch board fastened to the rafter ends and canted for architectural effect. It extends 1/2 inch below the soffit so that the lower corner serves as a drip. A 1 × 2 inch strip is used to close the juncture of the sheathing and the fascia.

There are many variations of the closed cornice used to suit particular requirements. The details of a cornice for a masonry veneer wall are shown in Fig. 6-55. Blocking is necessary to provide nailing for the frieze. The fascia is plowed (recessed) to provide a weather seal for the soffit. A piece of quarter round mold is used to conceal where the frieze and soffit join. The formed metal gutter finishes the cornice. The gutter resembles a molding in shape. This particular cornice is open to the attic space and provides part of the ventilation.

A low pitched roof can have a closed cornice which has the soffit nailed to the rafters. See Fig. 6-56. A gravel stop is necessary because the low slope requires a built-up roof. In this case, ventilation is achieved by a continuous opening in the soffit. The air passes freely under the roof in the attic space.

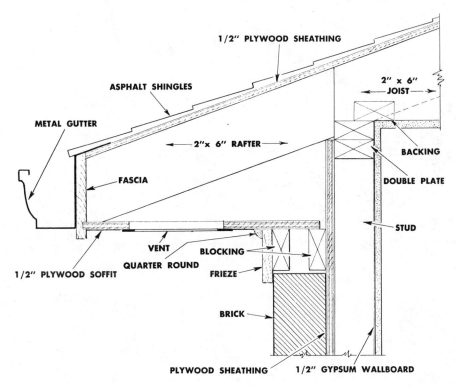

Fig. 6-55. A closed cornice for a brick veneer building is made with basic members. The formed metal gutter adds the finishing touch.

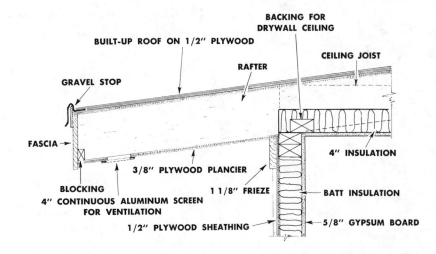

Fig. 6-56. A cornice for a low pitch roof can be designed in several ways. Ventilation can be provided for the space over the ceiling through the soffit vent.

Open Cornice

The rafter overhang is exposed on the open cornice illustrated in Fig. 6-57. Since the under side of the plywood roof sheathing and the ends of the rafters are exposed, it is necessary to use the best side of the plywood and rafters which are clear of blemishes. The frieze and bed molding (when used) must be cut to fit between rafters. See Fig. 6-57A. A cornice used in conjunction with panel siding is shown in Fig. 6-57B. A fascia can be applied over the rafter ends. All types of open cornice require careful cutting and fitting of the wall panels, siding, or frieze be-

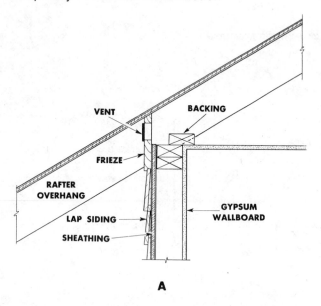

A

Fig. 6-57. The rafters are exposed on an open cornice. The frieze or wall covering, shown in (A), must be carefully fitted between the joists. The fascia, shown in (B), can be omitted.

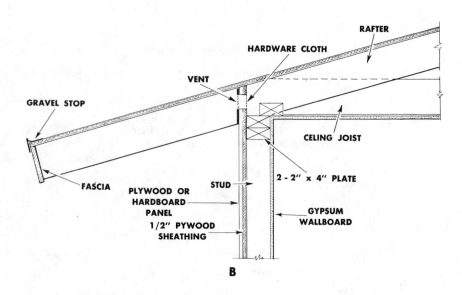

B

tween the rafters so that it is tight against air infiltration or insects and has a good appearance. The open cornice is relatively difficult to paint. There are several ways in which it can be modified to produce unusual effects. See Fig. 6-58.

Almost all of the fancy exterior trim used in past years has been eliminated except when the building is designed in the vein of a particular architectural period or style, such as colonial or Greek classical. The cornice shown in Fig. 6-59

is a typical traditional cornice. It shows all of the members required for an authentic classical design. These members are the *frieze, soffit* (also known as *plancier*) the *fascia,* a *bed mold,* and *crown mold.*

Gable Trim

The edge of the roof at the corners and at the gable end must be treated to form a transition from the cornice. A board equivalent to the fas-

Fig. 6-58. An open cornice can be modified at the fascia and the frieze.

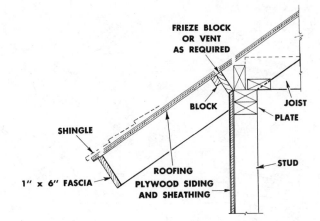

FRIEZE BLOCK
OR VENT
AS REQUIRED

BLOCK

JOIST

PLATE

SHINGLE

STUD

1" × 6" FASCIA

ROOFING
PLYWOOD SIDING
AND SHEATHING

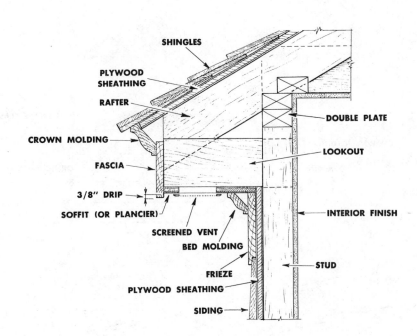

SHINGLES

PLYWOOD
SHEATHING

RAFTER

DOUBLE PLATE

CROWN MOLDING

LOOKOUT

FASCIA

3/8" DRIP

SOFFIT (OR PLANCIER)

INTERIOR FINISH

SCREENED VENT

BED MOLDING

FRIEZE

STUD

PLYWOOD SHEATHING

SIDING

Fig. 6-59. This type of closed cornice is used mainly on colonial or traditional buildings.

cia, called a *verge board,* is often placed along the rake of the roof rising toward the ridge. The last shingles are supported over the drip edge as shown earlier in Fig. 6-42 or on a piece of 1 × 2 inch strip nailed to the last rafter. A soffit is built to enclose the space between the rafter and the wall. See Fig. 6-60A.

The end treatment for the classical cornice shown in Fig. 6-59 is done as shown in Fig. 6-61. All the members of the box cornice are continued around the corner and returned back to the wall. Flashing is required to protect the top. The crown mold and fascia are continued up the rake of the roof.

Metal and Fiberboard Soffits and Cornice Trim

Manufacturers of aluminum, steel, and fiberboard have developed soffits and other cornice trim members which can be assembled quickly and are adaptable to different roof slopes and widths of roof overhangs. Baked enamel or other factory finishes resist chipping, blistering, and peeling and provide long life. The periodic maintenance cost is almost eliminated. Perfora-

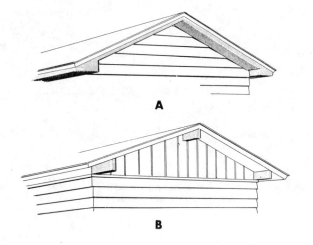

A

B

Fig. 6-60. The junction of the cornice and the gable end trim is made to provide a transition to the gable end trim. A boxed cornice end is shown in (A). Extended timbers at plate and ridge are shown in (B).

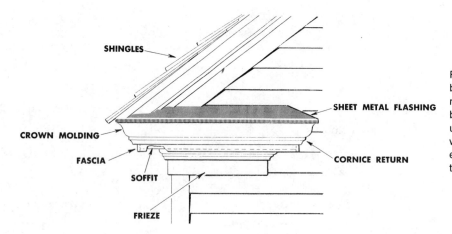

SHINGLES

SHEET METAL FLASHING

CROWN MOLDING

FASCIA

CORNICE RETURN

SOFFIT

FRIEZE

Fig. 6-61. The end treatment of the box cornice includes returning the members around the corner of the building. Some of the members are used on the rake (the upward slope) as well. This type of construction is limited to buildings designed with traditional architecture.

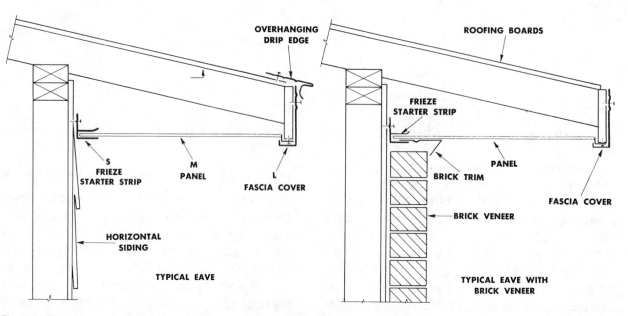

OVERHANGING DRIP EDGE

ROOFING BOARDS

FRIEZE STARTER STRIP

S
FRIEZE STARTER STRIP

M
PANEL

L
FASCIA COVER

PANEL

BRICK TRIM

FASCIA COVER

HORIZONTAL SIDING

BRICK VENEER

TYPICAL EAVE

TYPICAL EAVE WITH BRICK VENEER

Fig. 6-62. The parts of an aluminum soffit system include a frieze starter strip, a fascia cover, and soffit panels. (Rollex Corp.)

Fig. 6-63. Metal soffit pieces slide into place from the end and interlock. This method of assembly eliminates the box at the end of the cornice. (Rollex Corp.)

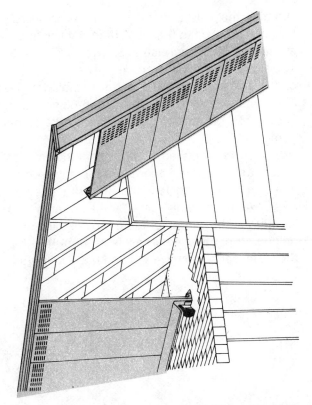

Fig. 6-64. A steel soffit system replaces the wood cornice and the rake trim. A conventional box end treatment at the cornice is shown. (U.S. Steel Corp.)

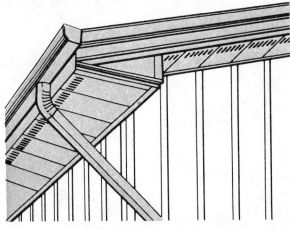

tions in the material provide ventilation for the attic and cornice.

The details of how the parts of a metal soffit system are assembled are shown in Fig. 6-62. A channel track (called a *frieze starter strip* and marked *S*) is fastened to the wall so that it receives the inner ends of the soffit panel pieces (marked *M*). The fascia cover (marked *L*) is bent underneath the fascia board to receive the outer end of soffit panel pieces. Each soffit panel piece is shoved into place from the end of the cornice and interlocks with the adjoining pieces. Figure 6-63 shows how the pieces fit together. Another cornice completely enclosed in metal is shown in Fig. 6-64. The treatment at the end of the cornice includes using a box in a manner which resembles the way a wood cornice is generally terminated. See Fig 6-60 A. A different treatment at the end of the cornice is shown in Fig. 6-63. The siding of the gable end is carried out to the end of the cornice fascia and forms a triangle. This eliminates the box. Both Fig. 6-63 and 6-64 show metal pieces and panels on the rake of the gable end.

Metal trim is applied which covers the ends of rafters and the verge boards on the gable ends. The trim is from 4 to 10 inches in width. See Fig. 6-65. When applying metal soffit systems and metal trim, the carpenter must provide blocking, nailing strips, and a subfascia of wood so that the metal parts can have a sound base. Manufac-

turers of fiberboard panels have also developed soffit systems. The material is prefinished with a durable factory finish. It is adaptable for roofs of various slopes and soffit widths. Ventilation is provided through snap-in plastic vents. The fiberboard products can be cut and installed with ordinary carpenter's tools. See Fig. 6-66.

Gutters and downspouts. Gutters and downspouts are required in most areas where there is sufficient rainfall and slow ground absorption. Some building codes may permit or even require the downspouts to be connected to storm sewers. Generally however downspouts may not be connected to sanitary sewers. Some communities require that the downspouts be connected to

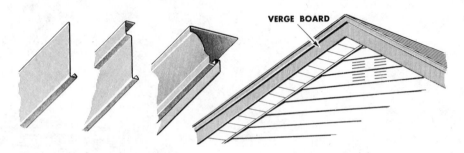

Fig. 6-65. Metal trim is provided to cover the verge board over gable ends. (Alcoa Building Products, Inc.)

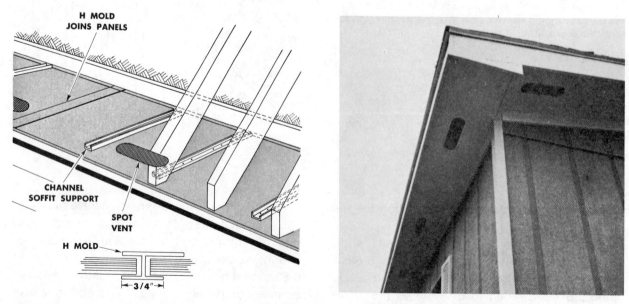

Fig. 6-66. A fiberboard soffit system is constructed with long pieces of fiberboard, joined with H molds, and stiffened by soffit supports. The fiberboard is primed with paint or covered with acrylic film. The left view shows the detail of construction. The right view shows the system in application. (The Upson Co.)

drywells or extended so as to divert water away from the building or on to the ground. Gutters are made of galvanized sheet metal, stainless steel, aluminum, wood, or vinyl. Metal gutters are available with baked enamel finish. The size of the gutters and downspouts is determined by the roof area. A 4 inch gutter is used for a roof with 750 square feet or less of area. A 5 inch gutter is used for roofs with greater areas. A 3 inch downspout is used for roofs up to 1000 sq. ft. in area. A 4 inch size is used for roofs with a greater area. Two types of hangers used are shown in Fig. 6-67. Figure 6-68 shows the various parts used in the assembly of a metal gutter system. The parts slip together and are fastened with pop rivets or sheet metal screws. A mastic can be applied to seal the joints. Usually a slight slope of 1 inch for every 12 to 16 feet of length is provided for drainage purposes.

Wood gutters, shown in Figs. 6-69, are used in some parts of the country. They are made of decay resistant wood and are quite durable. Splices are made by using brass preformed joints or brass strips fastened with brass screws and copper tacks. The downspouts are connected by drilling a round hole and cementing a round metal downspout in place. End caps are wood pieces set in mastic to form a waterproof seal. The gutters are fastened to the fascia with 3″ brass screws. A small airspace is provided by using blocking between the fascia and gutter for ventilation.

Vinyl can provide a permanent gutter and downspout system. The material is solid so that there is no finish problem. It is impact resistant and is cut and assembled with carpenter's tools.

Ventilators.   Even if adequate insulation over the first floor ceiling is provided, a certain amount of water vapor escapes into the attic space. In cold weather, the vapor condenses on the cold surfaces of rafters and roof sheathing and can cause decay. In hot weather and in warm climates, the attic space collects heat through the roof which raises the temperature throughout the building unless the space is adequately ventilated. Several types of ventilators are available which provide solutions for the problem. The most efficient of these (other than using power fans) uses soffit ventilation to provide an input of air at the eaves and louvers in the gable ends, or roof ventilators placed near or at the ridge to provide an output of air. Soffit ventilation is provided by continuous venting, as shown in Fig. 6-70, or the use of spot ventilators.

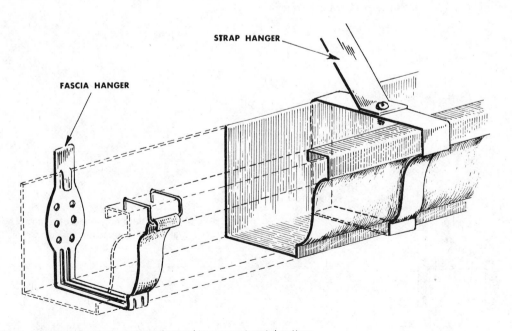

Fig. 6-67. Two types of hangers are commonly used to support metal gutters.

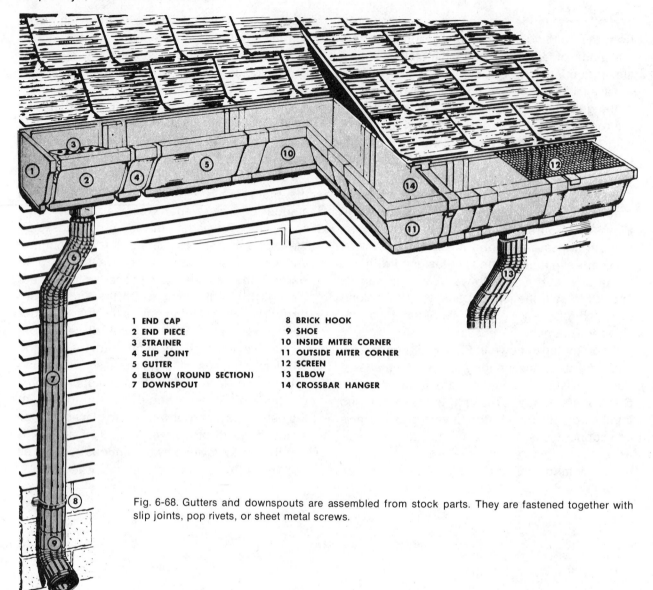

1 END CAP	8 BRICK HOOK
2 END PIECE	9 SHOE
3 STRAINER	10 INSIDE MITER CORNER
4 SLIP JOINT	11 OUTSIDE MITER CORNER
5 GUTTER	12 SCREEN
6 ELBOW (ROUND SECTION)	13 ELBOW
7 DOWNSPOUT	14 CROSSBAR HANGER

Fig. 6-68. Gutters and downspouts are assembled from stock parts. They are fastened together with slip joints, pop rivets, or sheet metal screws.

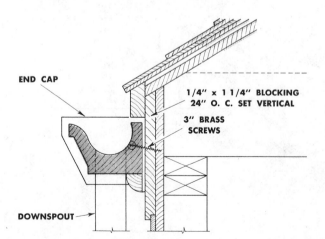

END CAP

1/4" x 1 1/4" BLOCKING
24" O. C. SET VERTICAL

3" BRASS
SCREWS

DOWNSPOUT

Fig. 6-69. Wood gutters are made of decay resistant wood. They are installed using brass joints and screws.

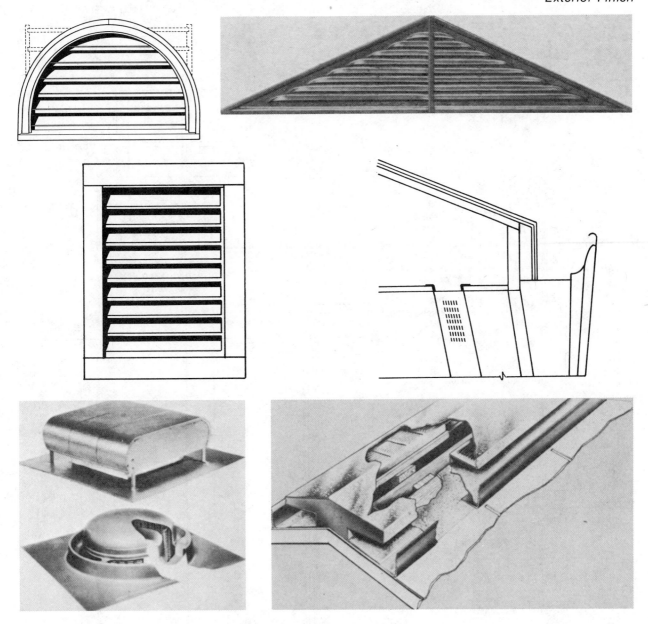

Fig. 6-70. Various types of ventilators are used to provide air circulation in attic spaces.

See Fig. 6-66. Louver ventilators are made with sloped vanes and either triangular or rectangular frames. See Fig. 6-70.

The carpenter provides the rough opening for these ventilators in the same way as he does for a window. The whole ventilator is inserted and fastened from the outside along with the exterior finish. Roof ventilators are fastened to the roof and flashed with shingles. Continuous ridge ven-

tilators provide an escape for the air at the point where the heat has its highest concentration. The size of the ventilators is determined by a ratio of the open area to the area of the building at the eaves. A building with a gable roof requires two louvers high in the gables at opposite ends, with an open area equal to $1/300$ of the building area. A hip roof requires eaves ventilation of from $1/300$ to $1/600$ of the building area represented

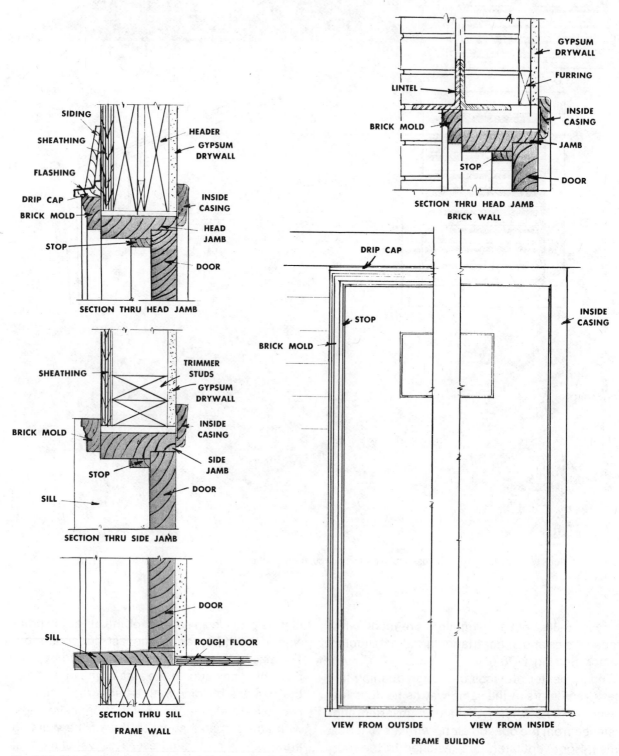

SIDING

SHEATHING

FLASHING

DRIP CAP

BRICK MOLD

STOP

HEADER

GYPSUM DRYWALL

INSIDE CASING

HEAD JAMB

DOOR

SECTION THRU HEAD JAMB

GYPSUM DRYWALL

FURRING

LINTEL

BRICK MOLD

INSIDE CASING

JAMB

STOP

DOOR

**SECTION THRU HEAD JAMB
BRICK WALL**

SHEATHING

BRICK MOLD

STOP

SILL

TRIMMER STUDS

GYPSUM DRYWALL

INSIDE CASING

SIDE JAMB

DOOR

SECTION THRU SIDE JAMB

SILL

DOOR

ROUGH FLOOR

**SECTION THRU SILL
FRAME WALL**

DRIP CAP

STOP

BRICK MOLD

INSIDE CASING

VIEW FROM OUTSIDE

VIEW FROM INSIDE

FRAME BUILDING

Fig. 6-71. Details of the door frame show the frame and trim parts. The same door frame and trim can be used for either frame or masonry buildings.

and another $1/600$ at the ridges. These requirements vary in the warmer parts of the country.

DOOR AND WINDOW FRAMES

Door and window frames are manufactured at a factory or mill. They are then brought to the job ready for placing in the rough openings provided for them. Doors, windows, and frames are often purchased as packaged units. Packaged windows are placed in their frames and are complete with glass, hardware, and exterior casing. The inside casing pieces are provided loose to be installed after the interior finish is in place. Wood door frames are also packaged (prehung). The door is hung in the frame with the butts (hinges) installed. Other doors and frames are sold separately. The frame is supplied complete with exterior casings and sill in place. Doors are fitted and hung on the job.

Metal doors and frames are purchased as complete packaged units, however the frames are not assembled. The parts of the door frames are assembled on the job. They snap together and interlock by bending the tabs over. The side jambs are drilled to receive butts and locks. The doors which are included with the frames have matching butts and lock. For special situations, doors are available which have the frames already assembled and welded at the corner tabs.

The rough openings for door and window frames should be correctly sized and ready to receive them. Framing the rough openings has been discussed earlier in Chapter 4. See pp. 177 to 182.

Wood Exterior Door Frames

The details of door frames may vary, but the construction is generally the same. A typical door frame which can be used in either wood or masonry construction with little modifications is shown in Fig. 6-71. The jambs, head and side, are rabbeted ½ inch to receive the door. Since the outside doors of residences swing inward, the rabbet must be on the inside. When screen door or combination screen-storm doors are used, they fit into a rabbet on the outside of the frame so that they swing outward. The outside doors of most public buildings must swing outward as a safety measure.

The jambs and casings of doors are usually made of durable soft woods, such as white pine. The sills for high quality frames are made from white oak which enables them to withstand wear. White oak is one of the most durable woods. Door sills for masonry buildings are usually made of cut stone or concrete. A metal threshold with weatherstrip features is installed over the sill.

Procedure for setting wood door frames in exterior frame walls. The general practice of using prehung doors has simplified the procedure for installing and hanging doors. Doors and frames arrive on the job with trim fastened in place on one side of the frame and the door installed on its hinges. Square the door in the frame by driving a duplex nail through the jamb into the edge of the door, or by driving a few thin wedges along the lock side between the door and the jamb. This holds the frame and door in alignment. Pull the nail before the door is set in the opening.

It is good practice to tack a strip of heavy building paper 10 or 12 inches wide against the sheathing around the rough wall opening. This prevents infiltration of air. Tack a second piece to go around the corner into the opening. See Fig. 6-72.

Using blocking, adjust the sill and rough framing members beneath the sill so that the sill is at the correct height for the finished floor. A metal threshold must be put in place later. The door must be able to swing with sufficient clearance so that there is allowance for carpeting or other floor finish. Figure 6-73 shows a common type of sill.

Place the frame and door in the opening. Adjust it so that there is the same amount of space between the side jambs and rough framing members on both sides. If the sill is not flat on the rough framing members, level it with wedges. The door hanging on its hinges serves to square the frame. Make sure the top of the door and the head jamb are parallel. Drive a nail through the casing into the trimmer studs on each side to hold the frame in position. When

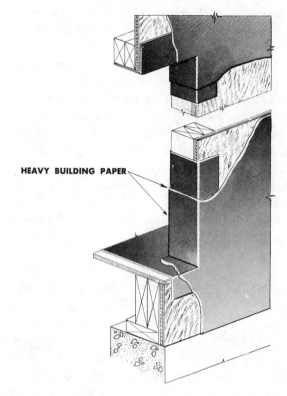

HEAVY BUILDING PAPER

Fig. 6-72. Building paper is placed around the opening before the door frame is installed. A first sheet is placed around the stud and sheathing corner. The outside sheet is turned into the opening.

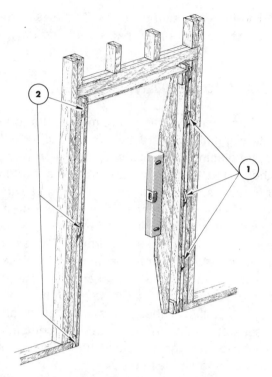

Fig. 6-74. A level and straightedge are used to plumb the door frame. Shingles are used to hold it in place and give solid backing at the butts and the strike plate. On the butt side (1), the shingles are placed on top, bottom, and at butt locations. On the lock side (2), the shingles are placed at the top, bottom, and at the lock height.

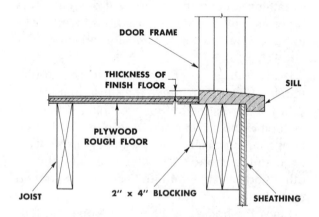

DOOR FRAME

THICKNESS OF FINISH FLOOR

SILL

PLYWOOD ROUGH FLOOR

JOIST

2″ x 4″ BLOCKING

SHEATHING

Fig. 6-73. The sill is supported by the joists and the blocking.

fastening a frame in position, don't drive any of the nails completely into the wood until all the nails have been placed, and a final check has been made to determine if any adjustment is necessary.

The jambs must be plumbed and blocked solid against the rough opening trimmers. Insert wedges or blocks so that they are driven up tight near the top of the frame on one side. Using a level and straightedge, plumb the side jambs. Wedge the top on the opposite side. Place additional wedges at the butt and the strike plate locations and the bottom of each side. Check to see that the jambs remain plumb and square. See Fig. 6-74.

Nail jambs to studs and fasten outside casing to studs with finishing nails spaced about 16 inches apart. If inside trim is included in the package, store it in a safe place until the interior wall finish has been applied.

When nailing trim on the interior or exterior, never drive the nails so far that the hammer hits the surface of the wood. To prevent marring the finished surface of the wood, use a nail set to make the final drive and setting of the nails. Cover the sill by tacking a piece of ¼ inch plywood or other material in place. This protects the sill from excessive wear during the construction of the building.

Setting a door frame in an exterior brick wall. The carpenter is required to set door (Fig. 6-75) and window frames in a brick building. It is the carpenter's responsibility to see that these frames are located according to the working drawing.

Calculate the height of the head jamb above the finished floor and cut off the side jambs to make the door height come out correct in relation to the floor level. Cut a spreader to fit over the threshold to hold the side jambs at the proper dimension and to protect the stone sill. See Fig. 6-75. Check the frame to see that it is square. Renail the corner braces if necessary to make it perfectly square.

Raise the frame up into position. Place it so that it is correct in relation to the masonry wall which is then built up on both sides. See Fig. 6-75. Generally the inside edge of the jamb is to line up with the face of the drywall finish so that

the inside casing lies flat when it is nailed in place. See Fig. 6-71. To determine this location, you must know the thickness of the furring to be placed on the inside face of the wall and the thickness of the gysum board drywall.

Nail short braces to each side of the jamb so that they extend diagonally down to the floor, then nail them to the floor. Nail a brace, which is long enough to reach the floor, on an angle to a top corner of the frame. Plumb the frame parallel to the face of the wall and at right angles to the face of the wall. When the frame is plumb, nail the lower end of the brace to the floor. Fasten a second brace from the other top corner of the frame to the midpoint of the long brace.

The bricklayer builds the wall snug against the brick mold of the door frame and anchors the jamb to the joints in the brick.

The position and openings for doors and windows in concrete walls is determined when the formwork is erected. Rough bucks are placed between the forms. A fuller discussion appears in Chapter 3, *Foundation Formwork*. See pp. 110 to 112.

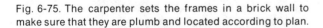

Fig. 6-75. The carpenter sets the frames in a brick wall to make sure that they are plumb and located according to plan.

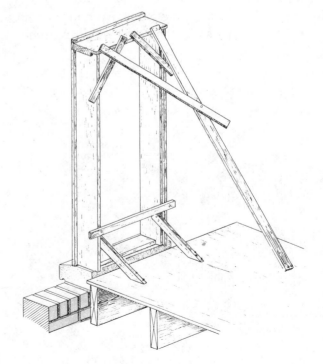

DOORS

Types of Doors

Doors can be categorized in a number of ways. They are either flush or panel doors and either solid or hollow core. They can be finished on the job with locks and butts or packaged, arriving on the jobsite complete with frame and hardware. They can likewise be divided according to their use, such as front entrance doors and patio doors.

Two common classes of doors are the *flush door* and the *panel door*. Flush doors have flat surfaces on both sides. The finish consists of two or three plies of veneer or another type of covering, such as vinyl or hardboard. The construction of a solid core flush door is shown in Fig. 6-76.

Two types of flush doors are made. One called a *solid core door* is made of small blocks glued together to form a core over which surface plies are then placed and glued. See Fig. 6-76. The edges of the door and the places where butts are attached and the location for the lock are all made of long strips or large solid blocks. Solid core doors are generally used for exterior doors. They are heavier and more weather resistant than hollow core doors.

Hollow core doors are made of a honey comb of interlocking strips, confined within a frame-work of members which are equivalent to the side stiles and top and bottom rails of other types of doors. Veneer sheets are glued to the framework to become the face plies. (Note: Hollow core flush doors and the installation of locks and butts are covered in Chapter 7, *Interior Finish*.)

Panel doors are made with vertical members called *stiles* and horizontal members called *rails*. A stile on one side of the door is rabbeted to receive the butts. The stile on the other side is mortised or drilled to receive the lock. The rails at the top, bottom, and at intermediate points serve to provide door strength and to hold it together. Panels, with or without raised surfaces, are held in place by the stiles and rails. Various types of panel doors are shown in Fig. 6-77. Panel doors can be used with many types of architecture but are particularly adapted to traditional styles.

A new approach to making wood panel doors is to press them into the required shape from laminations of wood plies. See Fig. 6-78. Thus though they look like conventional doors made with stiles, rails, and panels, they actually have continuous face plies. This type of construction eliminates some of the old problems with panel doors. In ordinary panel doors, panels tend to pull in and out of the rabbets in stiles and rails as the moisture content of the air changes. Shrinking and swelling of members creates a problem

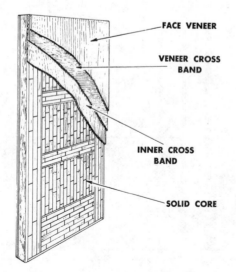

FACE VENEER

VENEER CROSS BAND

INNER CROSS BAND

SOLID CORE

Fig. 6-76. A solid core flush door is made up of many small blocks glued together and covered with layers of veneer.

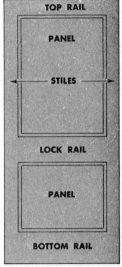

PARTS OF A PANEL DOOR

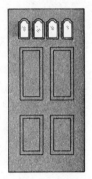   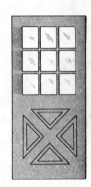 

Fig. 6-77. Exterior wood panel doors are made in many styles.

of keeping the panels painted at the edges. When the air is very dry, the panels shrink and expose a strip of bare wood.

Steel faced doors. Doors made with steel facing have become popular. They are very durable, do not shrink or swell, and have a baked on finish. They are available in a wide variety of styles. See Figs. 6-79 and 6-80. The construction consists of heavy steel sheets formed to make the panels and other door details and placed over wood stiles and rails. The core between the steel sheets (except at the stiles and rails) is made of expanded polystyrene. The front and back facings of the door are formed together to make the edges of the door, then are separated to make a thermal break so that heat or

cold is not transmitted from one side of the door to the other. Metal is a better conductor of heat than other building materials. This can create a problem. Another problem stems from the fact that as the temperature of metal drops, vapor condenses from the air inside the house and forms on the metal surface. However, this problem is eliminated by not having the outside metal face of the door in contact with the inside metal face.

Weather stripping is important at exterior doors. A vinyl or metal strip, which is located at the top and sides and compresses as the door is closed, is used for this purpose. Metal thresholds which include a compressible vinyl strip are generally used. In other cases, the vinyl strip is

Fig. 6-78. Wood panel doors are made from laminations of wood plies with surface plies pressed to resemble panels.

attached to the edge of the door bottom. Some thresholds are adjustable so that the part under the door can be raised or lowered to assure the proper closure and weather seal. Refer to Fig. 6-81.

Special doors and frames. Doors can likewise be classified according to their use. Sliding patio doors provide wide expanses of glass for indoor living areas while at the same time giving access to outdoor living space. Such doors are constructed of wood or metal and are glazed with insulating glass. See Fig. 6-82.

The general procedure for installing patio doors is similar to the procedure for installing other exterior doors. The rough opening is predetermined by the manufacturer. The opening for the same size doors varies from manufacturer to manufacturer. Sills and jambs must be adjusted so that they are level and plumb. Sills require solid blocking, and jambs require wedges to make them tight. The door frame must be adjusted so that the interior face of the jamb is flush with the wall finish and the interior trim lies flat when applied. The wood trim on the outside is part of the frame. Metal frames have a flange used to nail them to rough framing members.

Because of their size and the weight of the glass, patio doors must be adequately supported. Figure 6-83 shows a typical metal gliding door. One pair of doors supported on rollers which ride on a track in the threshold. A typical wood patio door is shown in Fig. 6-84. One of the doors is fixed. The other door rests on a metal glide in the threshold. A metal screen door rides on a track by means of rollers. Some frames are equipped with spring-loaded head jambs. The springs are installed behind the jamb so that

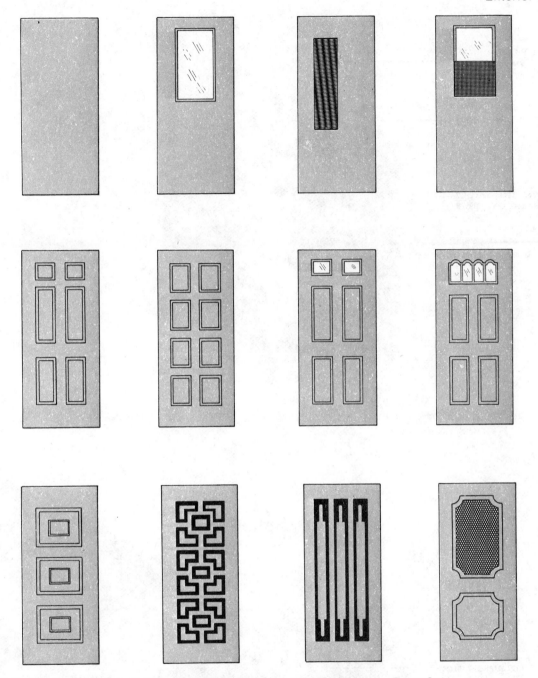

Fig. 6-79. Steel faced doors are available in many styles, with either wood or steel frames. (Pease Co.)

when the door is pressed upward, enough clearance is provided for the door to be removed from the frame.

Front entrance doors and doorways can be very elaborate. They are available with side lights, top lights, and trim members in a variety of architectural styles. See Figs. 6-85 and 6-86. The whole assembly can arrive on the job as a single unit. Some however are broken down into several large parts which are then assembled. The carpenter has the responsibility of providing a rough opening which has a strong header and

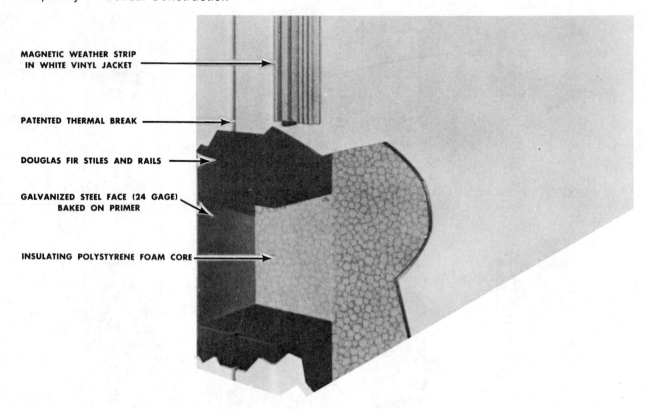

MAGNETIC WEATHER STRIP IN WHITE VINYL JACKET

PATENTED THERMAL BREAK

DOUGLAS FIR STILES AND RAILS

GALVANIZED STEEL FACE (24 GAGE) BAKED ON PRIMER

INSULATING POLYSTYRENE FOAM CORE

Fig. 6-80. Steel faced doors have wood stiles and rails for mounting the butts and the lock. (Pease Co.)

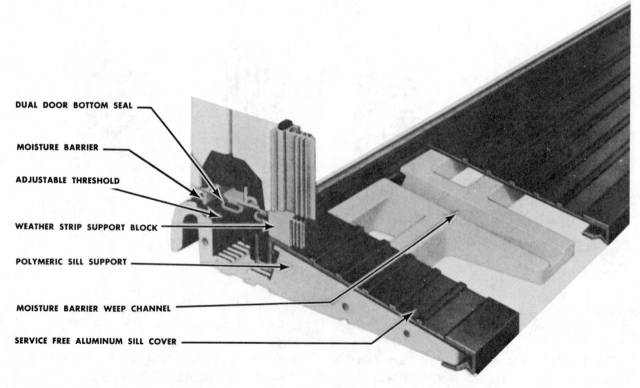

DUAL DOOR BOTTOM SEAL

MOISTURE BARRIER

ADJUSTABLE THRESHOLD

WEATHER STRIP SUPPORT BLOCK

POLYMERIC SILL SUPPORT

MOISTURE BARRIER WEEP CHANNEL

SERVICE FREE ALUMINUM SILL COVER

Fig. 6-81. The threshold at the bottom of a door must provide a weatherproof seal. (Pease Co.)

Fig. 6-82. Wood patio doors provide a wide expanse of glass as well as access to outdoors. (Marvin Windows)

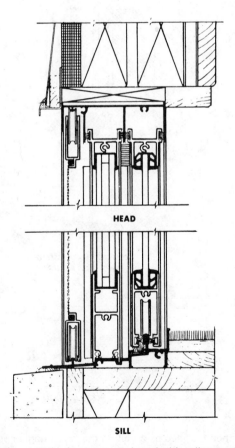

HEAD

SILL

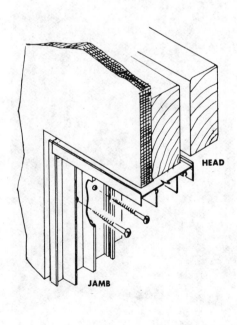

HEAD

JAMB

Fig. 6-83. Aluminum patio doors are equipped with rollers so that they open freely. The doors are fastened in place with screws through the jamb into the side studs. One of the doors is in a fixed position. (Reynolds Metals Co.)

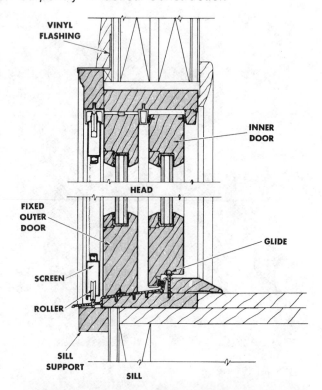

VINYL FLASHING

INNER DOOR

HEAD

FIXED OUTER DOOR

GLIDE

SCREEN

ROLLER

SILL SUPPORT

SILL

Fig. 6-84. The inner door of a pair of wood patio doors moves on glides. The outer door is fixed. (Andersen Corp.)

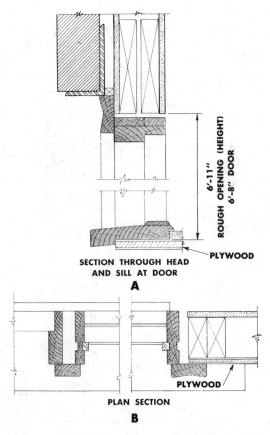

ROUGH OPENING (HEIGHT) 6'-11"
6'-8" DOOR

PLYWOOD

SECTION THROUGH HEAD AND SILL AT DOOR

A

PLYWOOD

PLAN SECTION

B

Fig. 6-85. Front entrance doorways are assembled in a mill and then installed as a unit on the job. A contemporary doorway with sidelights is shown to the left. The view in (A) is a section view through the head and sill at the door. The view in (B) is a plan section (a horizontal slice looking from above) of the door frame. (Morgan Co.)

Fig. 6-86. An elaborate doorway designed for a building with traditional architecture. (Pease Co.)

side jambs. The carpenter may also have to supply blocking or nailing strips to secure the assembly in place. If the building is to be brick, the carpenter must locate and brace the frame for the opening in preparation for the bricklayer.

WINDOWS

The carpenter should know the different types of windows and be familiar with the advantages of each type. He or she must make provisions for installing them as the rough frame of the building is being constructed.

Windows serve several important functions. The purpose or function of a particular window will determine its size, type, and location in the wall. Some of the more common functions of windows include providing light and ventilation and permitting maximum use of room area and wall space. Limits on the amount of light and ventilation (in relation to the floor area) for each room are generally set by the local building codes. In addition to these considerations, the designer must carefully plan the size and placement of the windows to create the best architectural effect.

Types of Windows

The three basic types of windows are categorized according to their method of operation.

The sash for some windows is held in a fixed position. In other types, the sash slides vertically or horizontally. In still other types, the sash is hinged to swing in or out.

Fixed windows. Many contemporary houses are designed so that the living space has large glass areas which provide a view and light. These windows are often called *picture windows*. Insulating glass, made from two panes of glass with a sealed air space in between them, helps compensate for the heat loss normally caused by using such large glass areas. The large fixed sash is usually combined with other types of sash in the same frame. Double hung or casement sash may flank the fixed sash on either side. A hopper or awning sash may be located below the fixed sash. (Note: Insulating glass can be used in all types of windows for all types of climates. This glass compensates for a wide difference between the inside temperature and the outside temperature.)

Sliding windows. The double hung window is considered most practical for cold climates, even though every other type of window is also used. The two sash are arranged in grooves so that they slide past one another in a vertical direction. The horizontal bars where the two sash meet and where the lock is installed are called the *meeting* or *check rails*.

Figure 6-87 is introduced at this point to help you understand how to read section drawings of

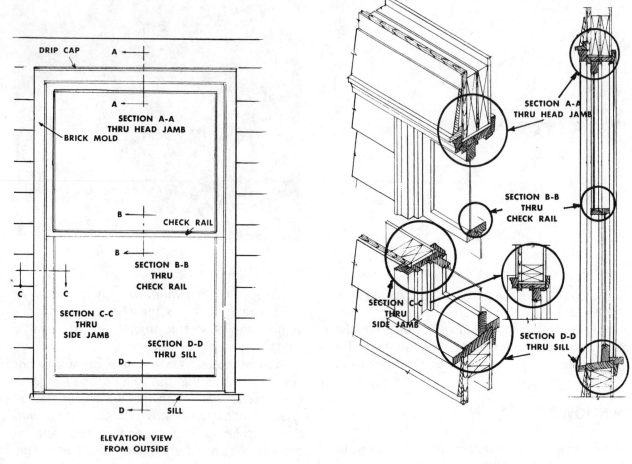

Fig. 6-87. This illustration shows how to read a section drawing of a window. The section lines in the left view are taken through all the parts of a window which require explanations. The isometric drawing, shown in the middle, indicates how the window is assembled. A section view of a window (far right) shows all of the individual section views in relation to each other.

windows, such as are found on shop drawings or architect's working drawings. It is important that you understand the relationship between the various section views, which are generally taken through the head, the sill, and the jamb (in the case of double hung windows, through the meeting rail also). The view at the left shows the window in elevation, viewed from the outside. Notice where the section lines are placed. Also notice the direction of the arrows. The middle view shows the same window with the same cuts in an isometric view. The far right view shows a simplified form of a section view through the window. The two sash are closed to show how the window operates. Figure 6-88 shows the section

view through the window in detail, as it might appear on a working drawing.

If you cut a piece of paper large enough to cover Section C-C, you will notice how Section A-A and Section D-D are related. The lower sash slides up and down in a space between the inside stop and the parting strip. The upper sash slides up and down in a space between the blind stop and the parting strip, extending all the way down to the sill. When you remove the card from Section C-C, you will notice it is a slice taken through the side of the window below the meeting rail. Section C-C is very similar to Section A-A (through the head jamb) except that the rough members framing the opening are the side studs

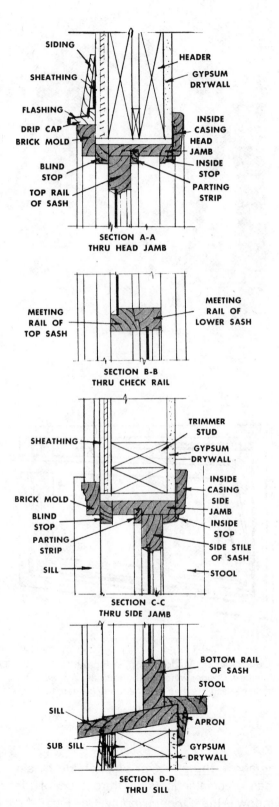

SECTION A-A
THRU HEAD JAMB

SIDING
SHEATHING
FLASHING
DRIP CAP
BRICK MOLD
BLIND STOP
TOP RAIL OF SASH
HEADER
GYPSUM DRYWALL
INSIDE CASING
HEAD JAMB
INSIDE STOP
PARTING STRIP

MEETING RAIL OF TOP SASH
MEETING RAIL OF LOWER SASH

SECTION B-B
THRU CHECK RAIL

SHEATHING
BRICK MOLD
BLIND STOP
PARTING STRIP
SILL
TRIMMER STUD
GYPSUM DRYWALL
INSIDE CASING
SIDE JAMB
INSIDE STOP
SIDE STILE OF SASH
STOOL

SECTION C-C
THRU SIDE JAMB

SILL
SUB SILL
BOTTOM RAIL OF SASH
STOOL
APRON
GYPSUM DRYWALL

SECTION D-D
THRU SILL

Fig. 6-88. A sectional view of a double hung window in a frame wall gives the names of the various parts and shows how they are related.

Fig. 6-89. Metal horizontal sliding windows are popular in warm climates. (Reynolds Aluminum Co.)

instead of the headers. Also no drip cap is shown.

The *horizontal sliding window,* shown in Fig. 6-89, uses the same principle as the double hung window, except that the grooves or tracks are horizontal. It is the most popular window in warm and dry climates. Aluminum horizontal sliding windows are used extensively in some of these areas. They are easy to install and efficient to use. The dry climate and warm temperatures make the condensation on the metal sash and frame an almost negligible problem.

This particular window can be used advantageously in bedrooms. It is located high enough on the wall to insure privacy. This position allows for wall space beneath the window. Both double hung and horizontal sliding types can be provided with spring-loaded jambs which enable the sash to be easily removed for cleaning and painting. By placing pressure on the sash, the jamb compresses enough so that it can be removed from the jamb.

Swinging windows. The most common window that swings on hinges is the *casement window.* See Fig. 6-90. Generally it swings out and is operated by a crank device. Screens and storm sash are located on the room side. One advantage of the casement window is that it provides

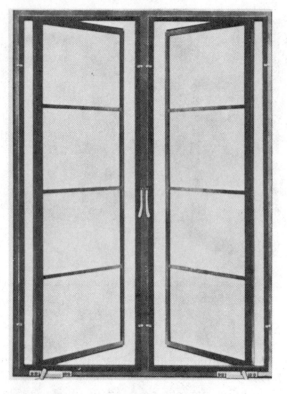

OUTSWINGING AWNING

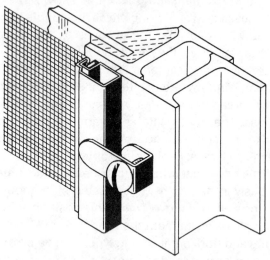

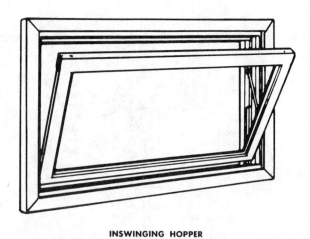

INSWINGING HOPPER

Fig. 6-90. Metal casements, such as shown at the top, prove maximum ventilation and light. The detail at the bottom shows how the sash fit and how the screen is attached. (Republic Steel Corp.)

Fig. 6-91. Awning windows (top) swing out to provide 100% ventilation. Hopper windows (bottom) are hinged at the bottom of the sash to swing into the room. (Andersen Corp.)

100 percent ventilation. Another type of window that is hinged is the *awning window*. See Fig. 6-91. The hinges are on the top and sash swings out. Several awning windows may be stacked one above the other, or placed side by side to create special effects. A third type of hinged window is the *hopper window*. See Fig. 6-91. It is hinged at the bottom and swings into the room. Hopper windows are often used to make up the bottom portion of large combination frames.

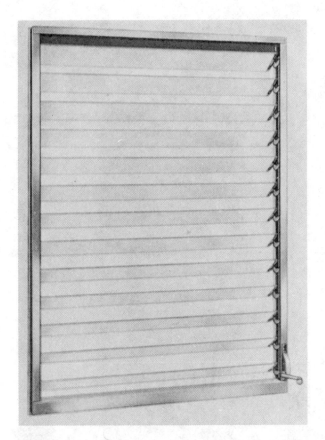

Fig. 6-92. A typical jalousie window. (Andersen Corp.)

Jalousie windows. Jalousie windows are particularly popular in warm climates because they provide 100% ventilation and a view which is unobstructed by the usual wood or metal window parts. They are not used extensively in cold climates because they are not tight enough to prevent air leakage. See Fig. 6-92. Although louvers can be wider, they generally consist of 4 inch wide glass pieces with a ½ inch lap held in place by metal end pieces. A mechanical device operated by a crank opens all of the louvers simultaneously. The screen is fastened on the room side. The frame has a drip cap at the top and flanges on all sides. Screws are used through the flanges to fasten the frame to the sheathing and members of the rough opening. A jalousie is not as airtight as other types.

House Plan A, which appeared at the end of Chapter 1, makes use of casement and awning type windows. House Plan C, shown at the end of Chapter 3-1, makes use of the same types of windows and also uses double hung and fixed sash windows.

Procedure for installing wood windows. Windows are generally delivered to the job with the sash in the frames and the exterior casing nailed on. The procedure for setting window frames in rough openings follows the same routine used for setting door frames.

We begin by assuming that a carpenter has already provided a rough opening for the window which is accurate, with sides plumb and the sill level. The minimum space of ½ inch should be provided on each side of the frame, unless specified otherwise by the manufacturer. This space must be provided in a plumb position. We can also assume that the rough opening is at the correct height above the floor so that the window will appear in the exact position shown on the working drawings. We can now proceed with the steps outlined below.

Drive a nail through the casing to hold the window frame temporarily in place. Place wedge-shaped shims under the sill and adjust the frame to the correct height. At the same time use the shims to level the sill.

The leveling of the sill must be done by placing the shims at points near outside jambs. The sill on narrow windows is usually reliably straight, but the center on multiple windows or very wide windows may sag and require supporting wedges at the center.

After leveling is completed, the frame can be held in position by nails placed near the bottom on each side. Drive the nails through the outside casing and into the sheathing or trimmers.

Plumb both side jambs by testing them with a carpenter's level. Use shims to position the jamb. To hold the frame in position, drive nails through the side casing into the sheathing or framing members near the top of the window.

Check the entire frame with a carpenter's level again to make sure both side jambs are plumb, and the sill is straight and level.

When you are certain the frame is plumb, nail it securely in place against the wall, through the casing into the structural members. Use 16d nails. Space the nails 16 inches apart and ¾ of an inch from the outside edge of the casings. Use a nail set for the final setting of the nails.

Installing metal windows. Certain precautions must be followed in working with metal windows and frames. Openings must be made to the manufacturer's specifications. They must be square, plumb, and level. They must likewise provide sufficient clearance to install the window. Frames must not be distorted, twisted, or crowded in the erection process. The sash should be closed and locked at all times during installation. The window frame members should be caulked where they join siding, masonry, or lintels. Precautions must be taken to keep moving parts, roto-operators, tracks, and runways clear of refuse and trash. Holes intended for drainage must not become plugged. If they become plugged up, they have to be thoroughly cleaned out. Aluminum windows and frame surfaces must be kept clean. Concrete or mortar, if permitted to harden on these surfaces, causes stains. Prefinished vinyl windows and frames should be handled carefully so that the surface is not marred.

A flange which extends around the frame is provided with nail holes to fasten the window to the sheathing and studs. Anchors which become embedded in masonry joints are used for some windows which are to be installed in masonry walls.

Window Construction Details

Packaged windows. Carpenters deal mainly with packaged windows. The term *packaged window* comes from the idea that the window is delivered to the job complete and ready to install. The glass is in the sash, the sash is hung in the frame with all of the operating mechanisms in place, and the exterior trim is securely fastened in place. Interior trim is attached or temporarily screwed in place and finally fastened later. See Fig. 6-93. Some packaged windows come with all the needed parts but in a knocked down manner. These are quickly assembled and installed by the carpenter.

The carpenter should know all of the different types of windows manufactured for use in residential construction and be familiar with procedures used to install them. The manufacturers produce a great variety of windows from the

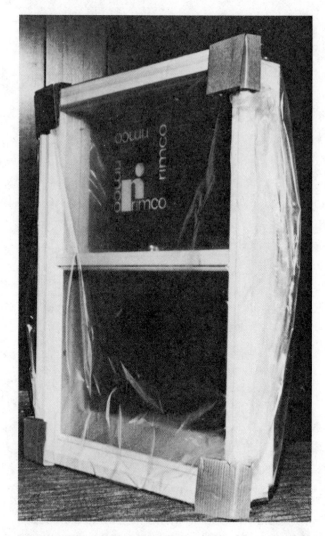

Fig. 6-93. This packaged window is delivered in a polyethylene bag with corner protection. It is ready to be installed in the rough opening. (Rodman Industries, Inc.)

basic operational types, using distinctive and in many cases patented features. The frames for almost every type of window can in most cases be adapted for use in frame, brick, or brick veneer walls. In many instances the same frame can be used for any of a variety of applications.

Wood windows. Wood windows and frames are made of select lumber so that there is a minimum of shrinkage, swelling, or distortion. The more expensive wood windows are clad with acrylic film which protects them from weather and eliminates the need for painting. The film is available in a variety of colors. Wood windows are generally considered superior for use in cold climates. Moisture has less of a tendency to condense on wood sash parts than metal window parts because the wood parts are thicker (usually 1⅜ inches) and because wood compared to metal is a relatively poor conductor of heat.

Metal windows. Aluminum or steel metal windows are available in all of the different window types. They are used in all climates. The condensation problem in areas which have low temperatures is compensated for by various features designed into the windows, careful installation, and the use of ceramic or other non-wood window sills. The windows are prehung and adjusted before they leave the factory. They can be installed rapidly and easily. Because metal windows have narrow frame and sash parts, the glass portion of the window is larger for metal windows than for wood windows.

Aluminum windows are finished in the natural color of the metal or are anodized with black, brown, or gold finish. Steel windows are coated with a baked-on primer. They can be prefinished with vinyl coatings in a variety of colors. Such coatings will last almost indefinitely.

(Note: The following pages are used to show details of the different types of package windows. The carpenter should be introduced to both typical and unusual windows. Within certain broad limits wood and metal units are manufactured in each type and can be used for the same application.)

Double hung windows. Figures 6-94 and 6-95 show how a basic double hung window is adapted for use in either a frame or a brick veneer wall. When used in a brick wall, the brick

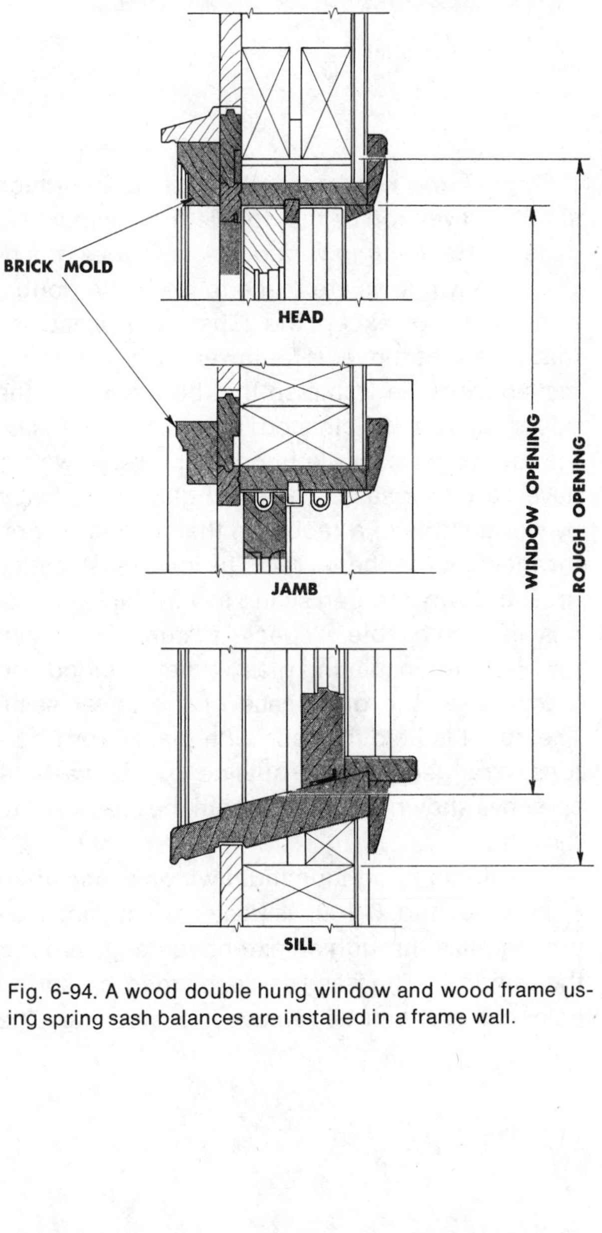

Fig. 6-94. A wood double hung window and wood frame using spring sash balances are installed in a frame wall.

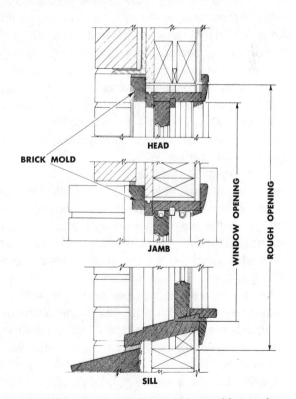

Fig. 6-95. The wood double hung window and frame, shown in Fig. 6-94, are used in a brick veneer wall.

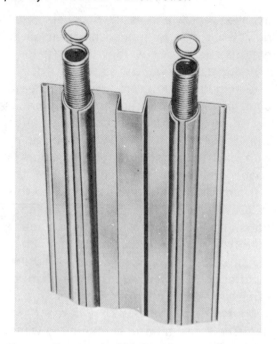 

Fig. 6-96. Sash balances use springs to counterbalance the weight of the sash. (Zegers, Inc.)

mold closes the gap between the wood frame and the steel lintel above and the masonry at the sides. When the frame is used in a frame wall, the brick mold is used as the outside casing. Spring-type sash balances can hold both sash open in any position. Typical sash balances for a double hung window are shown in Fig. 6-96. The springs support the sash when it is opened. The tension on these springs can be adjusted so that the sash operate properly.

A spring-type sash, shown in Fig. 6-97, has two types of sash balances incorporated in one window. The left-hand jamb liner, shown to the left (A) in the elevation view, is made with coil springs behind it. When pressure is placed on the sash toward the left, the sash clears and disengages the jamb liner shown to the right. The sash is removed in this position. The tension spring, shown to the right (B) in the elevation view, adjusts the sash to open at any position. There are several variations of this arrangement. Some adaptations allow the sash to be raised to a certain height, at which point it can be rotated into the room for washing or completely removed from the frame.

Single hung windows. A modified and much simplified version of the double hung window is called a *single hung window.* A few types of single hung windows are made to resemble double hung windows, except the upper sash is made so that it cannot move. (The lower sash can be removed from the frame so that both sides of the upper sash can be cleaned from a position inside the room.) Most single hung windows, however, have no upper sash. The fixed glass is retained by wood stops in a rabbet in the head jamb and the meeting (or check) rail. The lower sash slides up and down and can swing inward for washing. It is also removable. A special feature is shown in Fig. 6-98. An insulating glass panel is added and placed near the outer face of the lower sash. The panel is held in place with plastic turn buttons arranged on the outside of the sash. A screen is shown in position from the check rail to the sill.

The single hung aluminum window, shown in Figs. 6-99 and 6-100, is fastened in place by driving nails through an extended flange around the window. The sash is suspended by spring balances and can be removed for washing. The

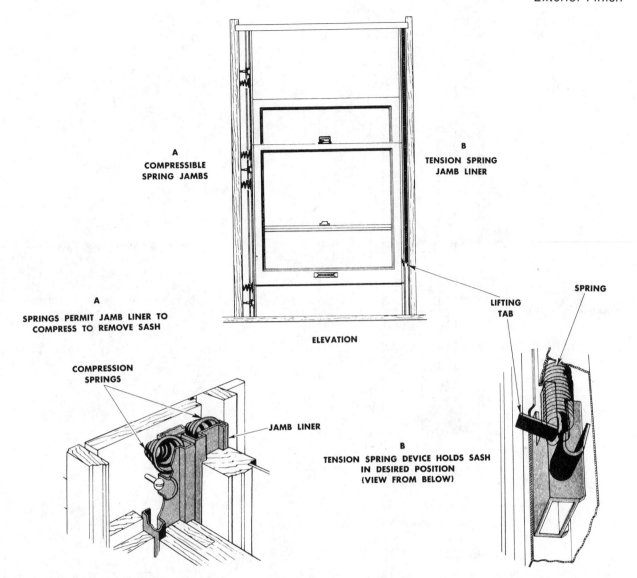

A
COMPRESSIBLE SPRING JAMBS

B
TENSION SPRING JAMB LINER

A
SPRINGS PERMIT JAMB LINER TO COMPRESS TO REMOVE SASH

ELEVATION

LIFTING TAB

SPRING

COMPRESSION SPRINGS

JAMB LINER

B
TENSION SPRING DEVICE HOLDS SASH IN DESIRED POSITION (VIEW FROM BELOW)

Fig. 6-97. This double hung window combines the features of a compression jamb (A) and a spring balance (B). The compression type jamb allows the sash to be removed from the frame. The spring balance is used to hold the window open in any position. (R.O.W. Sales)

usual lifts on the bottom rail are replaced by a horizontal bar. The glass area on single hung windows is slightly larger because there is no top sash.

Horizontal sliding (gliding) windows. Many horizontal sliding windows are popular with builders and homeowners because they are relatively trouble free and give a wide glass area in a horizontal direction. There is little obstruction from window parts. The two sash are held between tracks which allow them to slide. Most types of sash can be taken out of the frames for washing. Some slide on nylon rollers, as shown in Figs. 6-101 and 6-102. Other sash slide on metal or vinyl tracks. Still others have top tracks which are spring loaded to provide some tension and relief for removing the sash. A window designed for warm climates is shown in Fig. 6-101.

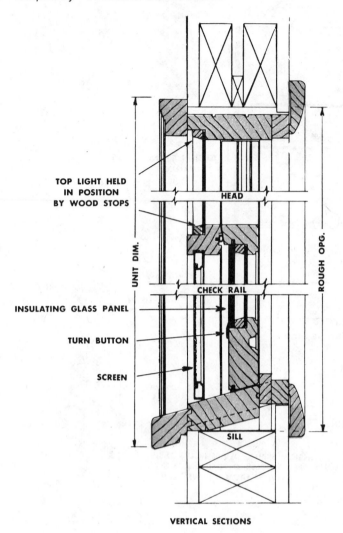

TOP LIGHT HELD
IN POSITION
BY WOOD STOPS

HEAD

UNIT DIM.

ROUGH OPG.

CHECK RAIL

INSULATING GLASS PANEL

TURN BUTTON

SCREEN

SILL

VERTICAL SECTIONS

Fig. 6-98. Single hung windows do not have a top sash. The light is held in place by wood stops. Insulating panels are installed from the outside and held by turn buttons. (Rodman Industries)

The flange on the sash extends over the sheathing to provide nailing space into the studs. Figure 6-102 shows an aluminum window designed for cold climates. It has a removable storm sash which is installed on the inside of the window during cold weather. A vinyl thermal barrier divides the outer part of the aluminum window frame from the inner part so that heat cannot be conducted from one side to the other.

A typical wood horizontal sliding window is shown in Fig. 6-103. The wood parts are covered with a vinyl shield which provides protection from the weather. Steel guides hold the sash in position and permit it to slide easily. Extension

jambs are used to bring the jamb out to the face of the interior wall finish. These extensions can be adjusted or replaced to allow for different wall thicknesses. Insulating glass is used.

An illustration of a single glide window appears in Fig. 6-104. One sash has a vinyl spring-loaded track with adjusting screws to control track pressure. The other sash is fixed. It is cleaned by reaching through the opening after the moveable sash is removed.

Casement windows. Casement windows were among the earliest types of windows in use. See Fig. 6-105. Today they fit into contemporary houses and also houses with a more traditional

Fig. 6-99. An aluminum single hung window. Only the lower sash moves. In place of an upper sash, the lights are held in the window frame itself. The lower sash can be removed for washing. (Keller Industries, Inc.)

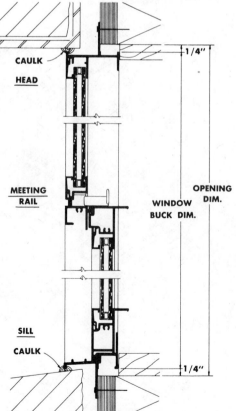

Fig. 6-100. A single hung aluminum window is applied to a brick veneer wall. (Keller Industries, Inc.)

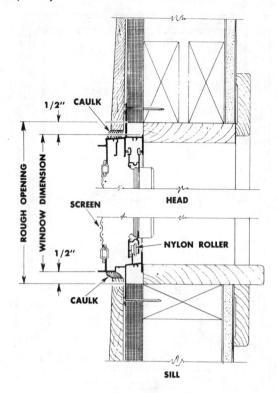

Fig. 6-101. An aluminum sliding window is fastened to a wood wall with nails through the flanges.

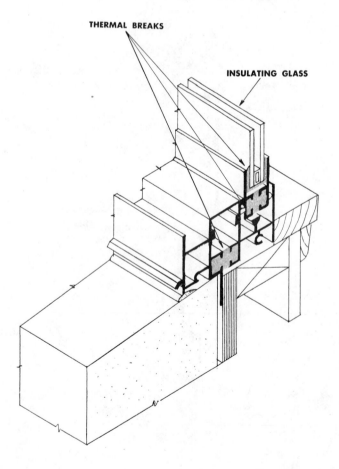

Fig. 6-102. A horizontal aluminum sliding sash window is designed for cold climates. Insulating glass and thermal breaks restrict the conduction of heat. (Reynolds Metals Co.)

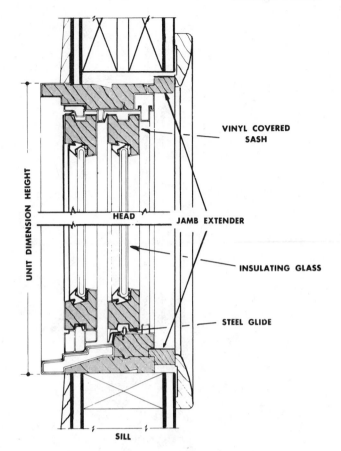

Fig. 6-103. A wood horizontal sliding window rides on steel glides. The exposed parts of the sash and jamb are covered with a vinyl shield. (Andersen Corp.)

VINYL COVERED SASH

HEAD

JAMB EXTENDER

INSULATING GLASS

STEEL GLIDE

UNIT DIMENSION HEIGHT

SILL

Fig. 6-104. Single glide windows have one sash which slides open. The other light (glass) is fixed in the frame. (Marvin Windows)

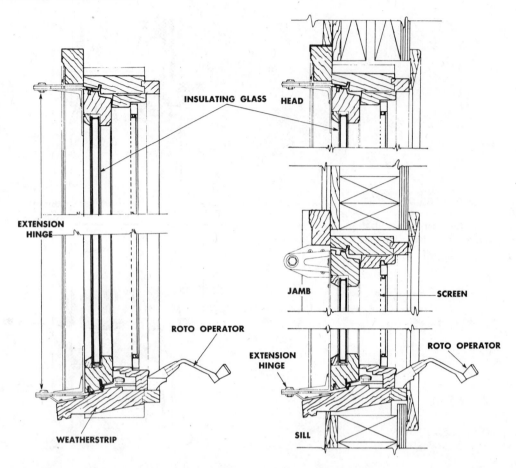

Fig. 6-105. An outswinging wood casement window pivots on extension hinges. A roto-operator keeps the sash in the desired position. The left view shows the window and frame only. The right view shows the window installed in the wall. (Crestline)

design. Except in warm climates, most of the metal windows commonly used in residential construction have casement sash. One advantage casement windows have over certain types is that they permit 100% ventilation. When operated by a roto-operator (a cranking and lever device), they maintain their position regardless of wind. A section through the frame and sash of a typical casement window is shown in Fig. 6-105. (Internal trim and rough framing members are omitted.) Weatherstripping is provided by metal strips which compress against the sash as it closes. They are seen projecting from slots in the head and jamb sections and from the bottom rail in the sill section view. Figure 6-106 shows the position of the roto-operator and the location of the hinges. The hinges are placed so that the sash is extended away from the jamb

and trim. This enables both sides of the window to be washed from the inside. In the view shown in Fig. 6-106, the screen appears on the room side, and a storm panel is installed in the sash.

Awning and hopper windows. Awning windows are windows with roto-operators and hinge arms arranged so that the window projects out from the frames. See Fig. 6-107. Hopper windows are similar to awning windows in that the sash and frames are the same. However hopper windows are hinged on the bottom rail so that they swing into the room. Awning windows are used individually, such as in bedrooms where they can be located above furniture. More often, however, they are used in combination with other types of sash in the same frame. Various sizes and arrangements of awning windows appear in Fig. 6-108. Individual windows are

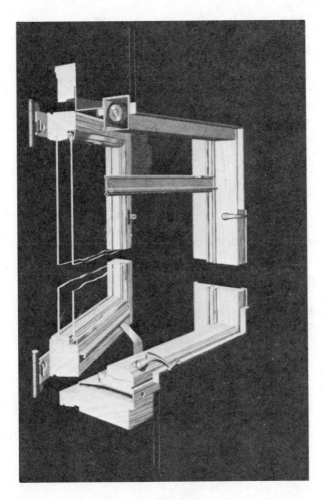

Fig. 6-106. Cutaway view of a casement window. (Rolscreen Co.)

Fig. 6-107. Awning windows are supported on a track and roto-operator device so that they swing out from the bottom. Here the awning windows are used in combination with a fixed sash. (Andersen Co.)

strung out horizontally in an arrangement called *ribbon windows*. Awning windows are frequently used below fixed sash windows to provide ventilation, as shown in Fig. 6-109. Hopper windows are used in a similar fashion.

The details of an awning window, shown in Fig. 6-110, are similar to those of a casement window. A roto-operator projects the window outward. The screen is on the inside, and an insulated panel in this case is fastened with turn buttons to the outside face of the sash. A brick mold is used for an outside casing. Interior casing is the same on all four sides of the window. A lock is placed at the bottom rail for the awning window, and is placed at the top rail for the hopper window.

Bay and bow windows. Bay and bow windows are combinations of standard windows arranged to project from the face of a wall unit. Stock double hung, casement, and fixed sash windows with mullions between each of the units are arranged in a great number of ways. (A mullion is a vertical structural part of the frame holding the sash.) The manufacturer assembles the whole bay or bow window in the shop so that it is delivered to the job in one piece.

Bay windows consist of three windows in one frame arranged so that the whole assembly projects from the wall. Double hung or casement windows are used. A large fixed sash window is

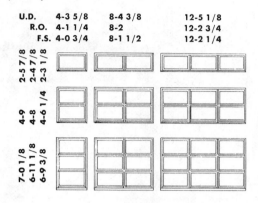

U.D.	4-3 5/8	8-4 3/8	12-5 1/8
R.O.	4-1 1/4	8-2	12-2 3/4
F.S.	4-0 3/4	8-1 1/2	12-2 1/4

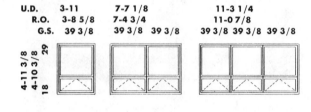

Fig. 6-108. Awning windows are used in combination with each other or below large fixed sash to provide ventilation. The dimensions shown here are samples of stock sizes. U. D. means *unit dimension*. This dimension provides overall dimensions, including casing and sill. R. O. stands for *rough opening*. F. S. stands for *frame size*. (Malta Mfg. Co.)

U.D.	3-11	7-7 1/8		11-3 1/4		
R.O.	3-8 5/8	7-4 3/4		11-0 7/8		
G.S.	39 3/8	39 3/8	39 3/8	39 3/8	39 3/8	39 3/8

Fig. 6-109. Awning type windows provide ventilation. An operating device permits them to swing out. The point of the broken lines indicates the hinge. G. S. stands for *glass size*. (Malta Mfg. Co.)

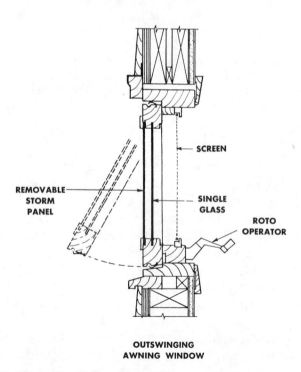

OUTSWINGING AWNING WINDOW

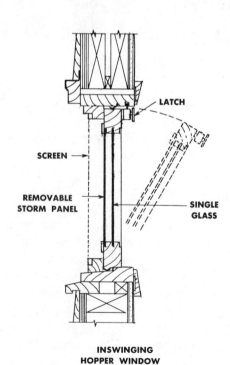

INSWINGING HOPPER WINDOW

Fig. 6-110. Awning (left) and hopper (right) windows have different jambs, sills, and sash. Note the removeable storm panels and the location for the screens.

Fig. 6-111. Bay windows are assembled using double hung windows with 30° or 45° mullions. The side and sash are placed so that they make an angle of 30° or 45° to the plane of the center window. (Anderson Corp.)

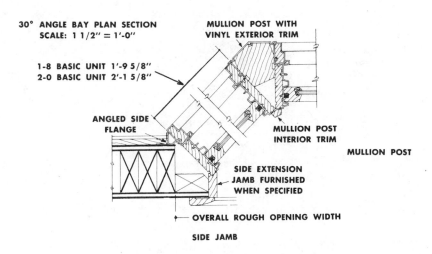

Fig. 6-112. Frame construction details for a bay window show the mullion post, the window jambs, and the interior trim. (Anderson Corp.)

30° ANGLE BAY PLAN SECTION
SCALE: 1 1/2" = 1'-0"

1-8 BASIC UNIT 1'-9 5/8"
2-0 BASIC UNIT 2'-1 5/8"

ANGLED SIDE FLANGE

MULLION POST WITH VINYL EXTERIOR TRIM

MULLION POST INTERIOR TRIM

MULLION POST

SIDE EXTENSION JAMB FURNISHED WHEN SPECIFIED

OVERALL ROUGH OPENING WIDTH

SIDE JAMB

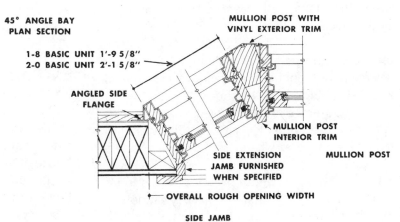

45° ANGLE BAY PLAN SECTION

1-8 BASIC UNIT 1'-9 5/8"
2-0 BASIC UNIT 2'-1 5/8"

ANGLED SIDE FLANGE

MULLION POST WITH VINYL EXTERIOR TRIM

MULLION POST INTERIOR TRIM

MULLION POST

SIDE EXTENSION JAMB FURNISHED WHEN SPECIFIED

OVERALL ROUGH OPENING WIDTH

SIDE JAMB

Fig. 6-113. A bow window is assembled using casement windows with mullions designed to give a curved effect. (Andersen Corp.)

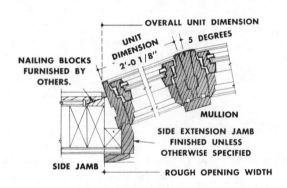

Fig. 6-114. A bow window is an assembly of several casement window frames made into one large frame. The mullion blocks are shaped at a 5° angle on the frame shown to bring the windows into the shape of an arc. (Andersen Corp.)

often used as the center window. The angle between the plane of the center window and the side windows is usually either 30° or 45°. A 45° angle bay is often called an *octagon bay* because the sides form part of an octagon. An octagon bay made with double hung windows is shown in Fig. 6-111. Figure 6-112 shows the manner in which the frames are assembled at the mullion and how the carpenter adjusts the structural members and trim where the bay frame meets the wall.

Figure 6-113 shows a bow window made up of casement sash. It differs from a bay window in that the sash form a part of an arc of a circle. In

this particular window, a 5 degree angle is made at each mullion between one sash and the adjacent one. See Fig. 6-114.

ENTRANCES, PORCHES AND OTHER EXTERIOR FINISH DETAILS

The carpenter is required to build a number of exterior functional and decorative parts of the house other than those that have been described in this chapter. The architect generally covers them with details which appear on the working drawings. They do not fit into any general category because of their specialization. They include decorative treatment at windows, the erection of porches and decks, and hanging shutters. Installing classical and traditional wood work, such as columns, railings, and canopies, has become part of the carpenter's job. Stair construction is discussed in Chapter 7.

SAFETY

The rules for safety apply when you are fastening exterior wall covering, trim, or roofing. The tools involved are mainly of the hammering and cutting variety. Hammers, staplers, the power hand saw, sheet metal tin snips, and cutters for mineral fiber material are some of the more common tools. Each tool has its own potential hazard. Appropriate work clothing should be worn, with special attention given to shoes. Shoes should provide sure and firm footing on ladder rungs, on scaffolds, and on roofs. Some types of roofing are easily marred or broken.

Physical safety includes the correct procedures for lifting. Large sheets of wall covering material must be lifted into place to be fastened. Heavy bundles of shingles must be raised to the roof and placed. Working at heights when applying wall and roof covering material and building cornices requires the safe use of ladders and scaffolds. Ladders must be placed on a substantial base, sloped at a 4–1 pitch, and secured against movement when used. Place ladders so that they make an angle between the ground and wall of 1 foot to 4 feet. (The bottom of a 16 foot ladder, for example, should be placed 4 feet from the wall.)

QUESTIONS FOR STUDY AND DISCUSSION

1. What factors are important when choosing wood for exterior finish? What woods are generally used?
2. What is drop siding and beveled siding?
3. How are corners finished when wood bevel siding is used?
4. What is hardboard? What are its advantages when used for exterior wall covering?
5. How are mineral fiber shingles used for exterior wall covering?
6. Explain how plywood is applied as grooved plywood sheet or plywood sheets with battens.
7. How is metal siding applied?
8. What are the advantages of vinyl siding?
9. How is a story pole used when wood shingles are used as the wall covering?
10. What sizes and shapes are popular for asphalt shingles?
11. Explain *square* and *exposure* as they apply to roofing.
12. How are wood shingles applied to roofs?
13. Describe a Boston ridge.
14. Explain the function of flashing and counterflashing at a chimney.
15. Describe wood shingles and shakes. Material, thickness, and taper.
16. Describe these cornice parts—frieze, fascia, and soffit.
17. What is a closed cornice? An open cornice?
18. How is a metal soffit installed?
19. Describe the purpose of gutters and give the sizes and materials commonly used for them.

20. List the different types of ventilators.
21. How is a door installed in an exterior frame wall?
22. Describe the features of a metal clad entrance door.
23. Describe the operation of a double hung window.
24. What is a single hung window?
25. Describe the operation of a metal horizontal sliding window. How is the sash removed?
26. What are the advantages of horizontal sliding windows?
27. What are the features of a casement window?
28. Explain the different mechanism and operation used in the awning and the hopper windows.
29. Describe a jalousie window.
30. Describe a bay window. A bow window.

CHAPTER 7
INTERIOR FINISH

MATERIALS AND PROCEDURES

The carpenter is responsible for putting many of the finishing touches on the interior of the building prior to the work of the painter and decorator. When this work is done with care, the beauty and value of the house are enhanced greatly. The construction details generally included under *interior finish* are: the wall, ceiling, and floor covering; the installation of windows and doors (also included under *exterior finish*); window and door trim; other trim, such as base and cornice; stairs; and the installation of hardware. In many instances the carpenter also installs the kitchen and built-in cabinets. Special architectural woodwork, such as mantels and room dividers, are also included.

Wood Materials

Some of the wood trim used fifty years ago has been eliminated both because it was costly to install and because styles have changed. Wood however has maintained its popularity as an interior finish material because it is easy to work with and it is attractive. The use of wood for wall covering has increased measurably. The common application is in the form of plywood sheets with a face ply of selected veneer. A variety of beautiful wood, prefinished in several attractive ways, is available. Although less common, board paneling is still used in homes when economy is not the main consideration. The boards usually

selected are without blemish, or at the other extreme, chosen because they have knots and other random imperfections.

Hardboard, which is a reconstituted wood product made by reducing wood to small fibers and then pressing it into sheets, is widely used for wall covering. Hardboard is prefinished in many colors and textures. It is also made to simulate wood paneling, brick, leather, and other material.

Wood floors are specified by some homeowners and architects, especially when they wish to achieve a traditional effect. However, the present trend is toward plywood floors covered over with resilient flooring of carpet, linoleum, or vinyl. Underlayment of hardboard is used over a plywood subfloor to provide a flat smooth surface for the resilient floor finish.

Factory built cabinets, wardrobes, vanities, and bookcases are commonly used almost everywhere. Formerly the carpenter made such wood fixtures on the job. Some carpenters still make special cabinets, but most of this work today is done in mill or cabinet shops. The cabinets are made more efficiently and with greater precision using woodworking machinery. The carpenter has the job of placing and installing the finished pieces.

Non-Wood Materials

Various non-wood materials are today commonly used for interior finish. The adoption of drywall gypsum board for wall and ceiling covering has brought about considerable changes in the carpenter's work. Plaster wall and ceiling finish is rarely specified for residential construction in most areas. (Preparing interior wall surfaces for lath and plaster application has been covered earlier in Chapter 4, *Wall and Floor Framing*.) Nonwood prefinished wall paneling is available in large sheets. It has a gypsum core with paper on each side. One surface is finished with durable vinyl in several colors and textures. Metal moldings in matching colors complete the installation. Moldings for interior trim are available in wood and vinyl, or vinyl-sheathed wood. The vinyl is finished in a color or made to resemble wood. This eliminates the need for additional decorating.

Procedure for Interior Finishing

The work of applying interior finish should not be started until the building is closed in and a satisfactory waterproof roof has been laid. Windows and doors should likewise be installed or the openings at least temporarily closed so that the temperature and humidity can be controlled to a reasonable extent.

Fitting and hanging doors, fitting trim, and installing hardware require both skill and precision. Although many of the doors and windows come to the job prefitted and prehung, carpenters still have the job of installing doors and hardware and trimming the openings for both doors and windows.

Wood members used for trim must not be allowed to absorb excessive moisture. If the conditions at the job are damp, wood members should be delivered shortly before they are used. Some members may be prefinished and require special care in order to protect the finish. Some members have fine moldings which can be nicked or marred.

Operations such as trimming window openings, applying moldings other than the base, fastening stair rails, and installing stairs precede the installation of the finish flooring. The main stairway of a residence is usually built in a shop or factory then installed on the job by the carpenters who specialize in this type of work. Some

stairs are built on the job by the carpenters who finish the house. Carpenters must be familiar with safe stair design and the local code requirements. Other operations such as setting cabinets, installing door trim, fitting doors, and nailing bases usually precede the laying of resilient floor covering. When wood flooring is installed, it is one of the later operations so that it has minimum exposure to damage. It is installed prior to the installation of cabinets and trim members which butt against it.

The last stage of interior finishing is the installation of hardware. Some hardware is fitted but not installed until after the painting is finished. If it is installed beforehand, it is removed for painting then reinstalled.

INTERIOR WALL COVERINGS

New materials providing inexpensive interior wall and ceiling coverings have been introduced and generally accepted into the residential construction field. The most popular material is *drywall*. Drywall can be applied in several ways and provides satisfactory surfaces ready for paint or other types of decoration. Other materials have wood veneer or simulated wood finish. When they are properly fastened and backed, these materials can meet fire rating standards prescribed by national and local building codes and provide walls resistant to damage. The carpenter applies all of these materials. Applying paint or other finish to these walls is the job of another craft.

Drywall

Types of drywall. Several basic features to provide rough framework for drywall have been discussed earlier in Chapter 4. The information covered the structural systems using wood and metal studs for the drywall base and methods for providing fire resistance and reducing sound transmission.

Gypsum wallboard is made with a core of fireproof gypsum molded between two sheets of tough protective paper. The paper is wrapped around and sealed on all edges. The wallboard is made in large sheets, four feet wide, so that

TABLE 7-1. SPECIFICATIONS—GYPSUM PANEL PRODUCTS.

Panel type	Regular		SW and regular		FIRECODE	FIRECODE "C"	
Thickness	1/4"	3/8"	1/2"	5/8"	5/8"	1/2"	5/8"
Lengths	8' & 10'		8' to 14'		8' to 14'	8' to 14'	
Width	48"		48"		48"	48"	
Edges	Tapered		Eased-SW Tapered		Eased-SW Tapered	Eased-SW Tapered	
Face finish	Ivory manila		Ivory manila		Ivory manila	Ivory manila	

Panel type	Foil-back panels			W/R panels		W/R FIRECODE "C"	Exterior ceiling board	
Thickness	3/8"	1/2"	5/8"	1/2"	5/8"	5/8"	1/2"	5/8"
Lengths	8' to 14'			8' & 12'		8' & 12'	8' & 12'	
Width	48"			48"		48"	48"	
Edges	Eased-SW Tapered			Tapered		Tapered	Rounded	
Face finish	Ivory manila			Green manila		Green manila	Gray	

United States Gypsum Co.

it can be quickly installed to cover walls and ceilings. Several types are available for different purposes. There are also various edge treatments. The panels are non-load bearing and require a certain amount of care when handled to avoid cracking or crushing the edges. The installer must use care when he applies it so that the fastening nails and screws do not cut the paper surface. Drywall is not recommended for extremely moist conditions.

Table 7-1 provides data on the different types of gypsum panels available, including the thickness, size, and type of edge treatment. Regular panels are used for most interior walls for the layer which faces the room side. Thin sheets, 1/4 or 3/8 inches thick, are used to cover over old walls which require a new surface. *Firecode* (also called *firestop*) has a higher fire rating than regular gypsum wallboard because it includes expanded vermiculite or other mineral products plus fiberglass in the gypsum core. *Firecode C* is constituted to have even greater fire rating qualities. Other panels called *foil backed panels* are made with a lamination of bright aluminum foil on one side. They are fastened with the foil side toward the studs on outside walls. The foil serves as a vapor barrier to pre-vent moisture inside the building from escaping into the stud space. It also helps prevent outward heat flow in winter and inward heat flow in summer. WR (water resistant) panels, both of the regular and Firecode C types, are used in kitchens and bathrooms as a base for the application of ceramic, metal, or plastic tile. Water resistance is achieved with several layers of chemically treated paper used for the faces, and a gypsum core made water resistant by the addition of an asphalt composition. Exterior ceiling board is weather resistant. It is used for soffits, ceilings in carports, or other places which are covered.

Two types of edges are provided. The long edges of the gypsum board panel are tapered for approximately two inches to permit layers of joint compound and the tape to be embedded so that a smooth wall results. One type *(S.W. Smooth Wall)* has a tapered round edge to reduce ridging in the finished joint. See Fig. 7-1. (When the edge is not rounded, there is a tendency for the joint compound to build up to form a slight ridge at the joint as it dries.)

In addition to the products indicated in Table 7-1, panels are available which are used as coreboard for solid gypsum partitions and for the backing base layer for a two layer application.

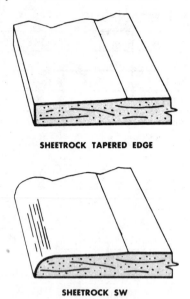

SHEETROCK TAPERED EDGE

SHEETROCK SW

Fig. 7-1. Two types of edges are provided for the application of compound and tape. The SW (smooth wall) type is designed to prevent ridging in the finished joint. (United States Gypsum Co.)

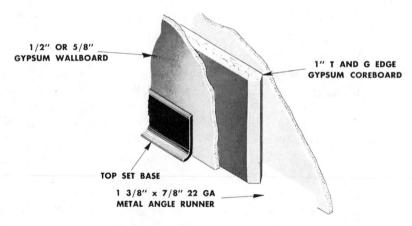

1/2" OR 5/8"
GYPSUM WALLBOARD

1" T AND G EDGE
GYPSUM COREBOARD

TOP SET BASE

1 3/8" x 7/8" 22 GA
METAL ANGLE RUNNER

Fig. 7-2. Solid gypsum partitions are made with one inch thick gypsum coreboard which has gypsum wallboard on each side. (United States Gypsum Co.)

One inch coreboard made with laminations of regular gypsum panels is shown in Fig. 7-2. Backing board is 4 feet wide with thicknesses of 3/8, 1/2, and 5/8 inch. It is less expensive than regular gypsum paneling because the face paper is not intended for exposure to the room side or for decorating. The edges are made square or with tongue and groove. Another product which uses a gypsum core is called *sound deadening board.* It is a low density product (the gypsum is not as compact) used as a base for 2 layer applications.

Measuring and cutting. It is important that the carpenter plan the arrangement of panels so that there is a minimum number of joints. The end joints on panels placed horizontally should meet on wood framing members and be staggered so that the joints are not above one another. The carpenter should check to see that the corners of the room and door frames are plumb and that wall studs are in line. This assures that the gypsum panels will have firm contact and not become bowed. Accurate measurements minimize the need for making corrective cuts and eliminate unnecessary handling of the sheets.

A metal straightedge should be used to mark and cut the panels. A drywall trim knife is best for scoring the face paper. The panel is snapped back to break at the scored line. If the panel does not break easily, the knife is used to cut through the back paper. The rough edge is smoothed out with a piece of coarse sandpaper wrapped around a block. If necessary, the edge can be trimmed with a knife. A rasp made of a piece of metal lath can also be used for this purpose.

Fastening methods. Several types of nails and screws are used to apply single layer gypsum wallboard to wood studs and also to apply the first layer of a two layer application. See Fig. 7-3. Nails are driven using a crowned head or wallboard hammer. The last blow is given so that the head of the nail is driven into the surface to form a slight dimple in the wallboard. Screws are driven in with an electric screw driver, with an

adjustable screw depth control, and a Phillips bit. The depth is adjusted so that the tool stops rotating and the drive head disengages when the screw is deep enough to form a dimple. See Fig. 7-4. Self-tapping screws are used to attach the gypsum wallboard to metal studs. The nails and screws are available in several lengths to accommodate the different thicknesses of wallboard.

A combination adhesive and nailing method reduces the need for nailing by 75%. A 3/8 inch bead of adhesive is run along the face of the studs. The adhesive spreads out to a wide band as the panel is pressed into place. When panels are applied vertically, they require nails spaced 16 inches O.C. at the edges. Temporary nails are placed 24 inches O.C. at the intermediate supports. These fastenings can be removed after 48 hours. When panels are applied to ceilings using the adhesive and nailing method, they require the same nailing used for vertical wall panels. When the panels are applied to walls horizontally, the perimeter nailing requirements are the same but those at intermediate supports can be omitted. A horizontal application of gypsum board is preferred over vertical application because imperfections in the alignment of the studs are less obvious.

Single Layer Method. When the single layer method (shown in Fig. 7-5) is used, the ceilng is applied first with gypsum wallboard sheets placed with the long dimension at right angles to the joists. Sidewall panels are generally applied horizontally so that there is a minimum number of joints to finish and a more uniform wall plane. Nails are spaced at a maximum distance of 8 inches O.C. (screws at 16 inches O.C.) on intermediate joists and studs and on the perimeter of the sheets. The nailing is begun near the center of each sheet. Nails are kept at least 3/8 inch from the edges.

Double Nailing System. A double nailing system requires that the gypsum wallboard be firmly fastened against the wood members with nails or screws spaced 12 inches O.C. See Fig. 7-6. A second nail or screw is driven approximately 2 inches from each of the first fasteners used. This method assures that the wallboard is tight against the wood members. Single nailing is used around the edges, beginning near the cen-

7/8" USG BRAND HI-LO SCREW—TYPE S—BUGLE HEAD

1" USG BRAND HI-LO SCREW—TYPE S—BUGLE HEAD

1 1/4" USG BRAND HI-LO SCREW—TYPE S—BUGLE HEAD

1 1/4" USG BRAND SCREW—TYPE W—BUGLE HEAD

1 7/8" USG BRAND HI-LO SCREW—TYPE S—BUGLE HEAD

1-1/4" USG DRYWALL SCREW—TYPE W—BUGLE HEAD

1 1/2" USG DRYWALL SCREW—TYPE G—BUGLE HEAD

1 1/4" GWB-54 ANNULAR RING NAIL

2 1/2" 7D GYPSUM WALLBOARD NAIL CEMENT COATED

1 7/8" 6D GYPSUM WALLBOARD NAIL CEMENT COATED

1 5/8" 5D GYPSUM WALLBOARD NAIL CEMENT COATED

1 7/8" USG MATCHING COLOR NAIL (STEEL)

Fig. 7-3. Nails and screws are provided to fasten drywall to wood or metal studs. They are available in various lengths to hold different thicknesses of panels. (United States Gypsum Co.)

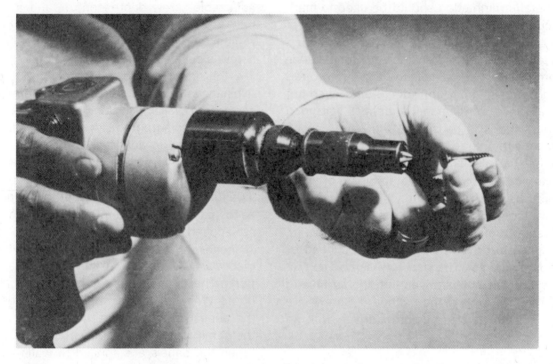

Fig. 7-4. Electric screwdrivers drive drywall screws to a preset depth. A clutch disengages the bit tip at the proper depth.

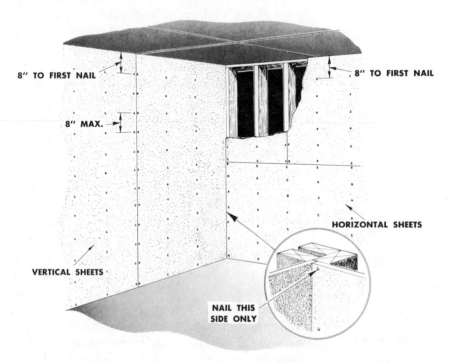

8" TO FIRST NAIL

8" TO FIRST NAIL

8" MAX.

HORIZONTAL SHEETS

VERTICAL SHEETS

NAIL THIS
SIDE ONLY

Fig. 7-5. Single layer application of gypsum drywall shows the nailing pattern for horizontal or vertical sheets. (United States Gypsum Co.)

ter of the panel and proceeding toward the outer ends and edges.

The nailing pattern for a *floating interior angle system* is shown in Fig. 7-5. This simply means that one sheet at each interior angle is not nailed, thus permitting the sheet to move slightly

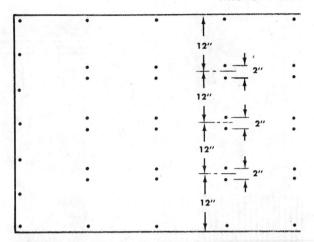

Fig. 7-6. A double nailing procedure insures tight fastening of gypsum wallboard to studs. (United States Gypsum Co.)

when the wood framework of the building dries out.

The double layer method. Two layers of gypsum board provide better fire rating than a single application. The double layer method also helps overcome the nail popping problem and reduces sound transmission.

The first layer may be regular gypsum board, backer board, or sound deadening board nailed (or screwed) to the studs in the conventional manner. The panels are usually placed in a vertical position. See Fig. 7-7A. Special lamination adhesives are spread on the surface of the wallboard with a mechanical spreader. See Fig. 7-7B and C. The finished layer is applied at right angles to the first layer. It is usually applied horizontally to minimize the number of joints. Joints

need not fall on studs but they should be offset at least 10 inches from parallel joints in the base layer. The layer is held in place with duplex nails until the adhesive sets. See Fig. 7-7D. The joints are then spread with compound, and tape is applied in the usual sequence.

Fire restrictive standards. There are a number of gypsum wallboard types listed in Table 7-2. They are used in combination with different partition structures to meet fire code restrictions and sound deadening requirements. Some of the structural members other than conventional wood studs used are metal studs (for several different wall thicknesses), staggered wood studs (such as shown in Fig. 7-8, top), and wood studs with resilient channels. See Fig. 7-8, bottom. Each combination has been tested and rated by

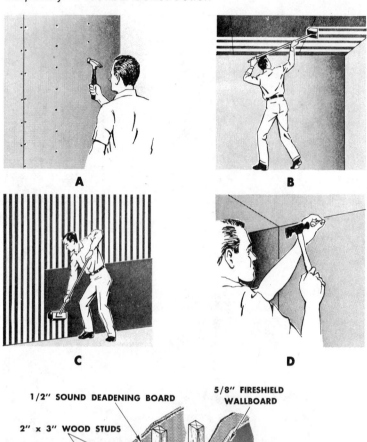

A   B

C   D

Fig. 7-7. Double layer drywall application. In (A), the base sheets are applied vertically in the conventional manner. In (B), the adhesive is applied to the ceiling. The adhesive is applied in bands and then spread over the entire surface. In (C), the adhesive is applied to walls. In (D), the second sheets are temporarily held in place with duplex nails. (National Gypsum Co.)

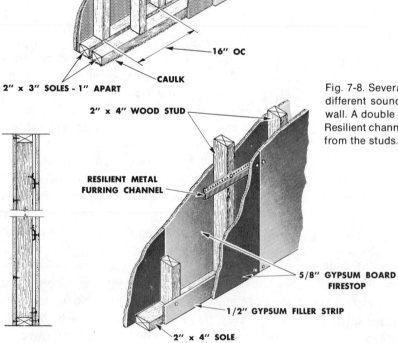

1/2" SOUND DEADENING BOARD   5/8" FIRESHIELD WALLBOARD

2" x 3" WOOD STUDS

16" OC

CAULK

2" x 3" SOLES - 1" APART

2" x 4" WOOD STUD

RESILIENT METAL FURRING CHANNEL

5/8" GYPSUM BOARD FIRESTOP

1/2" GYPSUM FILLER STRIP

2" x 4" SOLE

Fig. 7-8. Several construction methods are used to provide different sound deadening and fire rating qualities for drywall. A double wall (top) separates the two faces of the wall. Resilient channels (bottom) isolate the drywall gypsum board from the studs. (Georgia-Pacific Co.)

TABLE 7-2. TYPICAL FIRE RESISTANCE RATINGS.

1 hr. est	Stag Wd Stud—⅝" SHEETROCK FIRECODE gypsum panels—2x3 16" o.c.—2x3 plates 1" apart—panels att with 1¼" Type W screws 16" o.c.—2" THERMAFIBER ins wool blkts one side—perim caulked wt 8 width 7½"
1 hr.	Wd Stud—⅝" SHEETROCK FIRECODE or W/R FIRECODE "C" gypsum panels—2x4 16" or 24" o.c.—panels nailed 7" o.c.—1⅞" cem ctd nails—joints fin—perim caulked wt 7 width 4⅞"
1 hr.	Wd Stud—2 layers ⅜" SHEETROCK gypsum panels lamin & nailed—2x4 16" o.c.—joints fin wt 7 width 5⅛"
2 hrs.	Met Stud—2 layers ½" SHEETROCK FIRE-CODE "C" gypsum panels ea side—2" USG studs 24" o.c.—panels appl vert & screw att—1½" THERMAFIBER sound attn blkts—joints fin—perim caulked wt 9 width 4"

U.S. Gypsum Co.

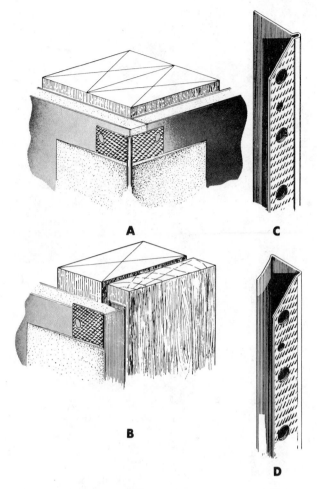

Fig. 7-9. Metal trim helps finish gypsum wallboard corners and edges. A corner bead with wire mesh flanges is shown in (A). An edge protector with mesh flanges is shown in (B). (C) and (D) show metal flanged parts. (United States Gypsum Co.)

qualified private and governmental agencies. The description of a few of the combinations is given in Table 7-2. Builders must check the local building code to determine the fire rating.

Details of assembly. There are several accessories which are useful for completing the job in a professional manner. The most common ones are the external corner bead and the metal trim edge. See Fig. 7-9. A corner bead is applied to the gypsum board corner with drywall nails about 6 inches apart. See Fig. 7-10. The nails are driven opposite each other through the corner bead into the wood member. Several coats of joint compound are applied to finish and protect the corner. The metal trim edge is used to provide a finished end to the panel when it might be exposed, such as at a doorway. See Fig. 7-9.

A window treatment is used to eliminate the usual wood casing. It is done by cutting a piece of the wallboard to return in what is known as a *wrap-around.* See Fig. 7-11. A corner bead can be used to protect the exterior corner, and a trim

edge can be used at the window, although neither of these is commonly used. Insulating tape (or sealant) can be used to close the gap and provide moisture protection for the end of the gypsum panel.

The intersection of partitions is constructed by having the gypsum wallboard panels pass the face layer to engage a steel stud in line with the back of the other face layer. Figure 7-12 shows both a single layer and double layer application. The wall panels are cut short enough to provide space for caulking with sealant before the tape and compound are applied to the internal corner. The intersection of a partition with floor and

Fig. 7-10. A corner bead is applied by driving drywall nails into a structural member. Applications of joint compound finish the corner. (United States Gypsum Co.)

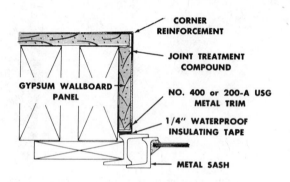

Fig. 7-11. A wrap-around window treatment is used to eliminate wood casings. Metal corner reinforcement and metal end trim protect the gypsum wallboard. (United States Gypsum Co.)

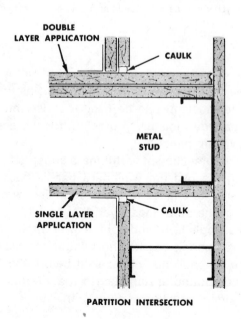

Fig. 7-12. A detail at the intersection of the partitions shows the gypsum wallboard attached to a stud at the opposite side of the main partition. Both double layer (top) and single layer (bottom) applications are shown. (United States Gypsum Co.)

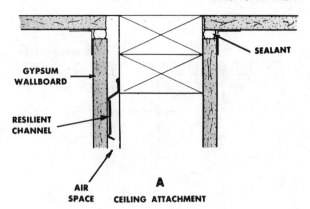

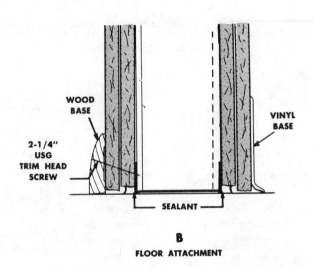

Fig. 7-13. The ceiling attachment detail in (A) shows the installation of the ceiling panel first and a bead of sealant to close the gap between the wall panel and the ceiling panel. A resilient channel is shown on one side at the left. In (B), the floor attachment detail shows a metal stud and runner and a double layer application. Neither panel rests on the floor. One side shows a wood base, and the other shows a vinyl base. (United States Gypsum Co.)

ceiling is shown in Fig. 7-13. Note the use of a resilient channel to provide a space between the gypsum panel and the wood framework. Also note the use of a bead of sealant at the intersection at the ceiling corner. The panels are cut so they do not rest on the floor. Wood trim and vinyl base trim are shown on opposite sides of a wall over two layers of paneling.

Gypsum wallboard paneling can be cut with a keyhole saw or a portable electric jig saw to provide for electrical outlets. Special tools for cutting these openings are also available.

Applying gypsum wallboard to masonry. Metal furring channels are used to attach gypsum wallboard to masonry walls. See Fig. 7-14. The channels are fastened to the masonry with powder actuated fasteners at 24 inch centers. They may have to be shimmed to provide a flat

surface for the wallboard. The wallboard is then fastened with conventional drywall screws on 12 inch centers. Gypsum wallboard can be fastened directly to concrete or brick masonry surfaces above grade with an adhesive. Mortar joints must be cut flush so that the surface is level. Pockets or holes larger than 4 inches in diameter and $\frac{1}{8}$ inch deep should be filled with grout or joint compound. These should be allowed to dry before the wallboard is laminated to the wall.

Water resistant gypsum board panels are made for bathrooms, kitchens, and utility rooms in areas where high humidity could make ordinary gypsum board unsuitable. The paper covering and the gypsum core are both treated so that they are water resistant. These panels are used as backer board wherever ceramic, plastic, or metal tile is applied. Cut edges and nail heads

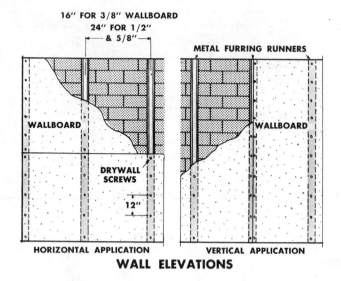

16" FOR 3/8" WALLBOARD
24" FOR 1/2"
& 5/8"

METAL FURRING RUNNERS

WALLBOARD

WALLBOARD

DRYWALL SCREWS

12"

HORIZONTAL APPLICATION · VERTICAL APPLICATION

WALL ELEVATIONS

Fig. 7-14. Metal furring runners are used to fasten gypsum wallboard to masonry walls. (Universal Sections Limited)

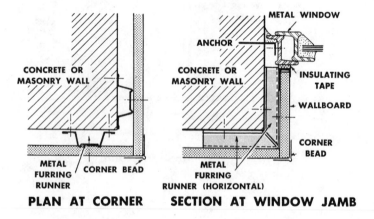

CONCRETE OR MASONRY WALL

METAL FURRING RUNNER / CORNER BEAD

PLAN AT CORNER

METAL WINDOW

ANCHOR

CONCRETE OR MASONRY WALL

INSULATING TAPE

WALLBOARD

CORNER BEAD

METAL FURRING RUNNER (HORIZONTAL)

SECTION AT WINDOW JAMB

are treated with sealant. The tile setter must caulk around tubs and shower stalls carefully.

Taping operation. The taping operation is not usually done by the carpenter. However the carpenter must apply the wallboard in such a fashion that the taper who follows has good joints to work with.

Each joint is buttered with compound. The tape is applied full length. It is pressed into the compound with a broad knife or tape applicator. Internal angles at wall corners and at the ceiling are treated in the same way. Nailheads are coated with compound. After the first coat is dry, a second coat is applied to the joints and nailheads, and the edges are feathered out. When the second coat is dry, it is lightly sanded. A fin-

ish coat is then applied. After 24 hours, surfaces which are not to have a textured finish are sanded smooth.

Vinyl Surface Gypsum Wallboard

Gypsum wallboard is also manufactured with a finished vinyl surface. It is not classed as drywall. Drywall requires that the joints be taped then finished with joint compound. The same gypsum core (or firecode core) with paper faces found in conventional gypsum panels is used. However one surface is covered with a durable vinyl coating. The surface treatment comes in a variety of colors. Several textured finishes, such as stipple, linen, travertine, or wood grain, are likewise available. The wallboard is applied with adhesive

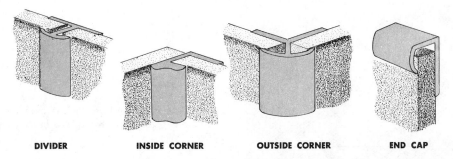

DIVIDER   **INSIDE CORNER**   **OUTSIDE CORNER**   **END CAP**

Fig. 7-15. Moldings for prefinished gypsum panels present a finished appearance. (United States Gypsum Co.)

to a backer layer or studs. The edges are either square or beveled. Some styles leave the edges exposed. Other styles make use of matching color plastic divider moldings. A variety of dividers and moldings is shown in Fig. 7-15. Matching color nails are used in places where the edges of panels cannot be concealed. This type of paneling is usually applied vertically so that it extends from floor to ceiling. It can also be applied horizontally as wainscot. It is not recommended for ceilings because concealing the end joints is too difficult.

Plywood Paneling

The plywood panels used for finished wall covering are generally 3 ply (¼ inch) or 5 ply (⅜ inch). These panels come in several sheet sizes, thus providing greater economy. In general, plywood panels are available in stock sizes of 48 inch widths and in lengths of 7 and 8 feet. Some panels are available in 10 foot lengths. Longer lengths must be ordered. The face plies of panels are chosen so that they are as uniform in grain and color as possible. They are then prefinished to match. Beautiful hardwoods are usually chosen for face plies. Oak, elm, walnut, cherry, and birch are some of the more common native woods used. Exotic and foreign veneers are also available. The carpenter decides on the location for each sheet so that it blends in with the sheets adjacent to it. The carpenter must also see that the best spacing is achieved at corners, doors, and windows. In some cases, the carpenter has several options when placing panels between openings and

Fig. 7-16. Wood paneling is used effectively for wall covering. It has beauty, durability, and requires a minimum of maintenance. (National Forest Products Assoc.)

between openings and corners. For some types, the face ply of the plywood is cut with V grooves or channels in a random plank effect. This feature permits the joints between the 48 inch wide panels to remain unnoticed. See Fig. 7-16.

Hardboard Paneling

Hardboard, shown in Fig. 7-17, is manufactured with a vinyl or plastic coating bonded to the face. This results in a durable surface. It is

Fig. 7-17. Hardboard paneling is made with plastic or vinyl coating to resemble wood and other materials. (Masonite Corp.)

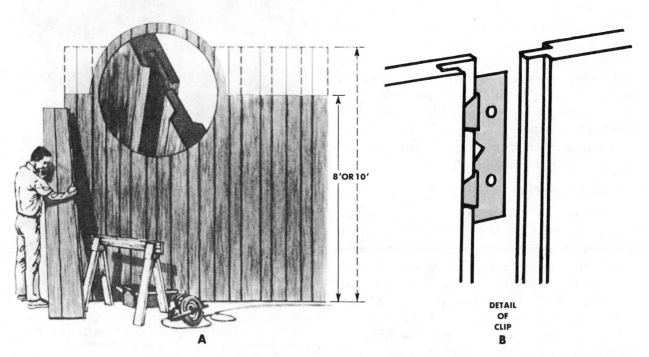

8'OR 10'

A

DETAIL OF CLIP

B

Fig. 7-18. Hardboard planks have tongue and groove edges which slip together to give a V groove effect, as shown in (A). A clip (B) serves to space the planks uniformly. The planks are fastened to the wall with adhesive. (Marlite Div., Masonite Corp.)

available with a variety of effects. Wood patterns used to simulate plywood panels are provided with the same channel effect as planks so that the edges between panels are concealed. High gloss panels used in kitchens and bathrooms are available in a variety of colors and effects. Some panels are made to resemble travertine, or other marble patterns. Other panels are embossed and colored to simulate leather, tile, brick, tapestry, and carved wood. The more common types come in 4 × 8 foot sheets, $\frac{1}{8}$, $\frac{3}{16}$, $\frac{1}{4}$, or $\frac{5}{32}$ inch thick, Vinyl clad wood or vinyl moldings which are the same color as the panels are used to cover joints between the panels. Special nails and sticks of putty can be used to cover nail holes. They are colored to match the panels.

High density tempered hardboard $\frac{1}{8}$ inch thick is used for panels applied in places subject to high humidity, such as around tubs and showers. It is available in 4 × 8, 5 × 5, and 5 × 6 foot sheets. It is applied to backer board with a waterproof adhesive and must be carefully caulked at surrounding and joining moldings to provide a waterseal.

Hardboard is also manufactured in plank form 16 inches wide by 8 feet long and $\frac{1}{4}$ inch thick. The planks have tongue and groove edges. They are applied to existing walls or new walls with a backer board surface, using a combination of adhesive and holding clips. The clips hold the planks in position until the adhesive sets and also serve to space the panel joints correctly. See Fig. 7-18.

Applying prefinished hardboard panels. Some important things to consider are the type of base surface used and its condition, and the trim or other treatment to be used at the floor, the ceiling, and the corners. If the paneling is to be attached to plastered walls or over gypsum backer board, the surface must first be made flat. If the surface is quite irregular, it is made level with grout, drywall compound, or another similar type material. A coat of sealer may be advisable.

When paneling is placed directly over studs, the studs are checked with a straightedge to see if they are in line. Crooked studs should be straightened or faced so that the paneling will have a flat surface. Additional members may be needed to support panel ends at the corners and openings. Adhesive is applied to the face of studs so that the panels will adhere to it firmly. Supplementary nailing is used along the top to hold the panel in position.

Wood furring is commonly fastened to walls when they are badly out of alignment or when panels are placed over masonry or concrete. Wood strips 1 × 2, 1 × 3, or 1 × 4 inches in size are fastened vertically 16 inches O.C. with nails or an adhesive. When the furring is fastened with nails, shims are placed between the furring and the wall to bring the pieces into line. Adhesive used to fasten the furring to the wall is applied with a zig-zag bead. The bead is compressed when the furring strips are positioned and serves the same as shims. Vertical furring strips are placed to provide nailing for panel edges at corners and openings. One method of attaching furring strips, shown in Fig. 7-19, involves cutting horizontal strips to fit between the vertical ones. Thus the horizontal and the vertical furring strips are in the same plane.

Adhesive is spread on all furring strips which come in contact with the perimeter of the panel. It is also spread at intervals along the horizontal furring strips. The panel is raised up from the floor with shims which provide a $\frac{1}{4}''$ space at the bottom. The panel is then plumbed with a level and pressed into place. See Fig. 7-19. A few nails at the top hold the panel in place temporarily. After 15 minutes, the panel is pressed against the wall to make sure that it has adhered at all points. The nails are then removed.

When the panels are fastened directly to walls, a line is snapped at 4 foot intervals, or wherever panels join. Ribbons of adhesive are placed vertically and horizontally at panel edges. They are also placed horizontally at intervals on lines 16 inches O.C. above the floor. The procedure used for furring strips is used for positioning the panel and temporarily fastening it.

Color matching nails are used when the panels are fastened to the furring strips with nails instead of adhesive. The nails are placed approximately 4 inches O.C. on the perimeter of the panel and 8 inches on intermediate furring strips. If the panels are fastened to plastered walls or over gypsum wallboard, the nails are spaced 4 inches O.C. on the perimeter of the panel and 8 inches O.C. at other studs.

Panels faced with plastic laminates. The high durability of plastic laminated surfaces, such as those used for kitchen countertops, is also available for wall paneling. Plastic laminate is a product made up of a sheet, which contains the surface pattern or appearance characteristics, placed over several layers of special kraft paper and then covered with a tough layer of resin coating. It is available in thicknesses of $\frac{1}{32}$, $\frac{1}{20}$ and $\frac{1}{16}$ inch, with several sizes up to 5 feet wide by 12 feet long. One type of this wall covering is manufactured to simulate random width planks. The plastic laminate is bonded to particle board with tongue and groove edges. Fine reproductions of cabinet woods are used

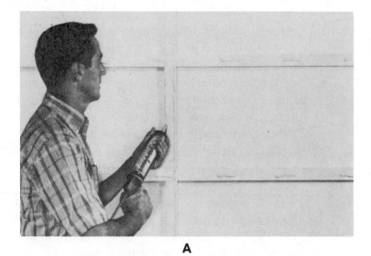

A

B

C

Fig. 7-19. Steps in installing prefinished hardboard panels. (A). Fasten vertical and horizontal furring strips to the wall. Apply a continuous ribbon of adhesive to the vertical furring strips. Apply 3 inch strips (3 inches apart) to the horizontal furring strips. (B). Place the panel in position leaving a ¼ inch space at the floor. Plumb the panel and press it into place. (C). Nail the panel lightly at the top edge. After waiting 15 minutes, press the panel along the points of contact to assure that the panel adheres, then remove the nails. (Masonite Corp.)

for the pattern sheets. This covering is fastened to horizontal furring strips by blind nailing.

Wood Board Paneling

Paneling of pine, fir, cedar and other woods comes in random widths of 4, 6, 8, 10 and 12 inches, with a dressed thickness of approximately ¾ of an inch. The lengths available are from 6 to 16 feet. There are several edge types, but boards with tongue and groove edges provide

the best results for concealing the joints. Some patterns have a V joint or a more elaborate molded edge pattern. See Fig. 7-20.

When the paneling is applied horizontally, no furring strips are required because the paneling is applied directly to the studs. Inside corners are made by trimming the boards to the exact length and butting them flush against the paneling for the other wall. If random width paneling is used, the boards on each wall must have matched

Fig. 7-20. Wood board paneling is available in several patterns with tongue and groove joints.

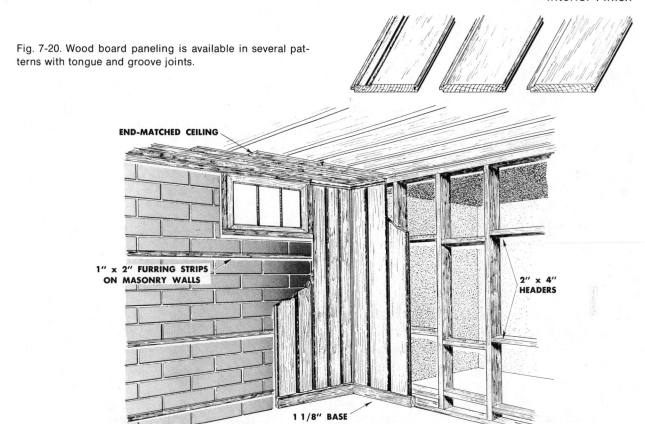

END-MATCHED CEILING

1" x 2" FURRING STRIPS ON MASONRY WALLS

2" x 4" HEADERS

1 1/8" BASE

Fig. 7-21. Board can be nailed to headers placed between the studs. Furring strips are used on masonry walls.

Fig. 7-22. Board paneling is easily installed over furring. Blind nailing is possible through the tongue, as shown at the left. Furring behind the base, shown at the right, provides good nailing.

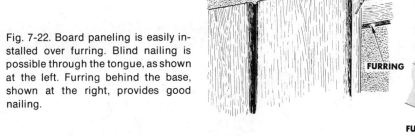

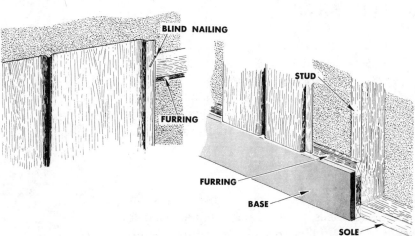

BLIND NAILING

FURRING

STUD

FURRING

BASE

SOLE

width. The tongue and groove allows for blind nailing. When paneling is placed in a vertical position on the wall, furring or solid blocking is required at the top, middle, and bottom of the wall, at the corners, and around doors and windows. Figure 7-21 shows the installation of

Fig. 7-23. Novel effects can be obtained by installing paneling horizontally and vertically.

boards, using solid blocking between studs. The installation of boards with furring placed over the studs is shown in Fig. 7-22. Masonry walls must be furred before applying the boards. A cove or crown molding is used at the ceiling. A number of interesting effects can be produced with board paneling. It can be installed vertically, horizontally, diagonally, or in a combination of these positions. See Fig. 7-23.

CEILING FINISH

Three methods are commonly used to finish ceilings. One method is the conventional drywall system, using gypsum wallboard. The second uses fiberboard (or another composition) tiles which are fastened to furring strips. The third method uses fiberboard, or other panels, suspended in a metal grid.

Drywall

Drywall is applied in one or two layers with screws (or nails) in the same manner used for walls, directly to the joists or the bottom chords of trusses. Resilient channels are also used to provide better sound insulation. The channels are placed on 24 inch centers. The gypsum panels are fastened to the channels with screws. Joints are finished in the same manner as those on side walls.

Fig. 7-24. Acoustical and decorative tile can be fastened by stapling it to furring strips or using runners. (Celotex Corp.)

Tile

Ceilings can be covered with acoustical and decorative tile made from fiberboard or noncombustible mineral fiber. The pieces of tile are 1/2 or 3/4 inch thick and 12 inches square. They have flanges on two of the edges to provide space for stapling. A groove on the other two sides provides a means for interlocking them and concealing the staples. See Fig. 7-24. The furring strips are placed so that the border tiles on opposite sides of the room have equal width. The furring strips should be shimmed so that they all work out to a level plane. Metal channel runners can be used instead of the furring strips. The flanges on the tile engage with the runners of the previously installed tile.

Two methods are used over flat surfaces, such as old drywall or plastered ceilngs. The tiles are stapled to 1 × 3 inch furring strips. The furring strips are nailed to joists 12 inches O.C. The tile can be stapled over a drywall ceiling, or fastened with adhesive to either a drywall or a plastered ceiling. A pat of adhesive is placed at each corner of the tile. Another pat is placed on the ceiling at the center where the title is to be located.

Suspended Ceilings

A suspended ceiling can be used in new homes, if provisions have been made for it. It can also be used for remodeling old homes which have high ceilings. See Fig. 7-25. Wall angles are fastened level around the room at the desired height. A series of T iron members are fastened to the wall angles and hung from the ceiling with wire. The T irons are placed in an inverted position to receive the panels. Cross Tees are inserted between the T irons. These form a grid into which the 2 × 4 foot panels are dropped into place. Figure 7-26 shows an alternative type of grid assembly. Luminous panels made from rigid plastic provide a lighted ceiling. The fixtures and wiring are concealed behind the panel.

Fig. 7-25. T shaped metal members are hung in the form of a grid to support the panels. The steps for installing the suspended ceiling include: (A) leveling the wall angles, (B) placing the grid in place, and (C) placing the panels. (Celotex Corp.)

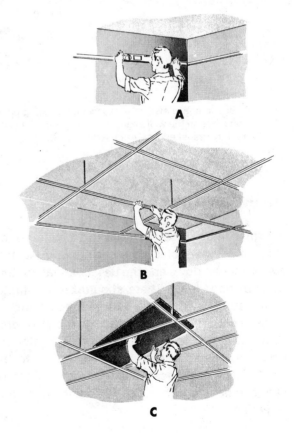

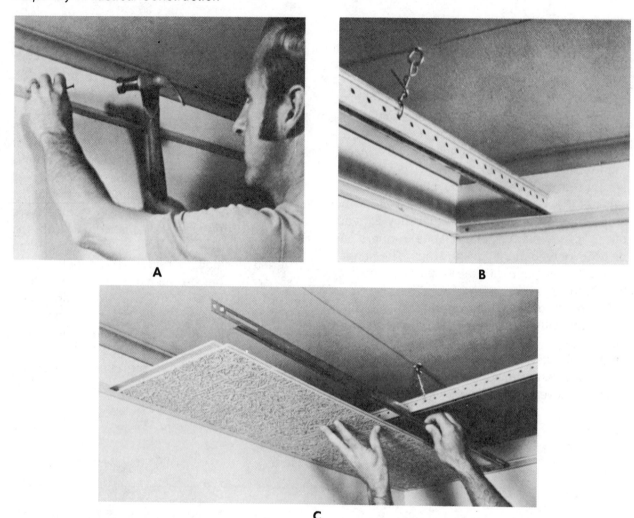

Fig. 7-26. An alternative method for suspending ceiling panels. (A) shows the application of wall angles. In (B), the main runners are placed to rest on the wall angles and hang from the ceiling. In (C), cross tees are snapped on to the main runners and moved over to hold the panel edges. (Armstrong Cork Co.)

INTERIOR TRIM

Trim members used for finishing the interior of most buildings include the casings around the doors and windows and the baseboard along the walls at floor level. There should be a definite architectural relationship between the design of these members and the design for the doors, windows, and the overall decor of the interior. Generally the architect chooses the trim to be used. Stock moldings obtained from a lumberyard can be used when the carpenter makes the selection.

Many varieties of wood are used for interior trim, including birch, oak, mahogany, walnut, white and yellow pine, fir, larch, gum, and tupelo.

The trend today is to make trim members as simple in design as possible. See Fig. 7-27 for details. Trim members are easy to install, stain, or paint, and have a pleasing appearance. The carpenter still must provide the skill to make joints and corners fit exactly.

The carpenter should be familiar with the names, uses, and applications for the many special moldings used in traditional houses or when

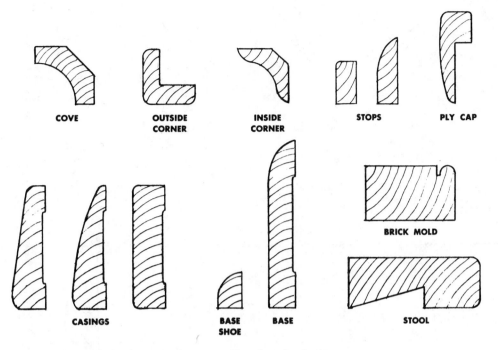

COVE   OUTSIDE CORNER   INSIDE CORNER   STOPS   PLY CAP

CASINGS   BASE SHOE   BASE   BRICK MOLD   STOOL

Fig. 7-27. Moldings and pieces of trim for general use come in a few simple shapes.

special moldings are used to serve a particular purpose. Using the right molding enhances the beauty of the finished home. The highlights and soft shadows which result from the curves and sharp breaks in the smooth surface make them attractive. A great number of moldings are available in different sizes and shapes. Examples of each general classification are provided in Fig. 7-28. Two moldings are often combined to make the final effect more ornate. A recess of about 1/16 inch is cut from the back of some moldings so that they fit tightly against the flat surface of the wall. Note the casing and base moldings shown in Fig. 7-28.

Trim members used with traditional windows include the casing, stool and, when two windows are placed side by side, a mullion casing. See Fig. 7-28 F, G, H, K, and N. A stop is used to hold a sash or a door in place. See Fig. 7-28 L and M. Moldings can be used at the ceiling in houses with traditional architecture, or when wood paneling is used for the wall surface. The ceiling mold can be either a crown mold or a cove mold. See Fig. 7-28 A and C. Those moldings used at the floor are called *base molding*. See Fig. 7-28 I

and J. A base shoe (7-28 E) is often omitted. The quarter round (7-28 D) is used to cover internal angles in special cases.

Some of the moldings shown in Fig. 7-28 are used for exterior trim also. The ceiling molding becomes a crown molding on a cornice, and the bed molding is also used as a cornice trim member. The brick mold (7-28 O) is used with a wood window in a masonry wall. However in many applications it is used as the exterior casing in frame construction. The drip cap (7-28 P) can be used over a window or door or as a member of a water table.

Base Molding

The baseboard is a trim member which extends continuously around a room. The baseboard is the last trim member put in place because it usually has to be fitted against other work, such as door casings and cabinets. The baseboard must be joined in the corners of the room so that the joints are tightly closed and properly aligned. Either the miter or the coped joint is used for joining such trim members.

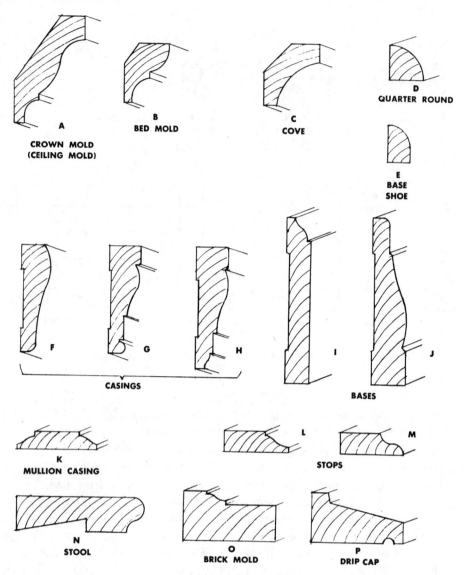

Fig. 7-28. Typical moldings used with traditional architecture are available in many different styles. The crown mold (A) and the bed mold (B) can be used on exterior cornices. The brick mold (O) and drip cap (P) are exterior trim members.

(Very simple types of baseboard can be butted at the corners or terminated on a corner block.)

The internal corners formed by the side walls of a room are either mitered or coped. See Figs. 7-29 A and 7-30. The coped joint is better for interior corners because it does not open while the trim is nailed into place against the wall. Likewise, if the wood on such joints shrinks, the opening is not as noticeable as happens when a mitered joint is used. External corners (B in Fig. 7-29), which are formed at chimneys or any other corners projecting into a room, should have the trim members mitered.

Procedure for coping joints. Coping consists of shaping the ends of a molding or board so that it fits with a butt joint against an adjacent member. See Fig. 7-30. The coped joint is used in fitting moldings on internal or inside corners. To cut a coped joint, follow the procedure given below.

Place the piece of molding in a miter box in the same position that it assumes when in place

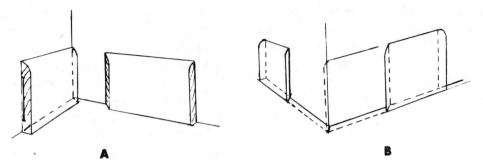

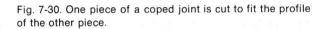

Fig. 7-29. Moldings at internal corners (A) may be mitered. Moldings at external corners (B) are always mitered.

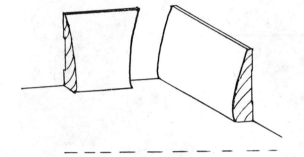

Fig. 7-30. One piece of a coped joint is cut to fit the profile of the other piece.

Fig. 7-31. A miter cut will give the profile or cutting line for a coped joint. (Tony Resendez)

against the wall or other surface. See Fig. 7-31. Place the vertical side of the molding against the back of the miter box. A 45 degree miter

gives the correct outline on the face of the molding for a right angle (90 degree) corner. The outline or profile formed by the miter cut is the cutting line which is followed with the coping saw to obtain the proper cut for making a coped joint. (Note: In case the corner is not a right angle, the angle of the miter must be one half the angle of the corner.)

Place the molding on a sawhorse, with the back in a flat position. See Fig. 7-32. Then using

Fig. 7-32. A coping saw is used to follow the profile of the miter cut to make a coped joint. (Tony Resendez)

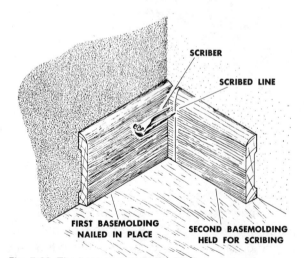

SCRIBER

SCRIBED LINE

FIRST BASEMOLDING NAILED IN PLACE

SECOND BASEMOLDING HELD FOR SCRIBING

Fig. 7-33. The base can be scribed before coping.

a coping saw, closely follow the waste side of the cutting line formed by the miter cut.

Test the joint against the other member which is nailed in place. If necessary, improve the cut by trimming it with a jackknife. Instead of mitering a wide molding to get a profile, scribe it with a pair of scribers or dividers. See Fig. 7-33.

Procedure for mitering joints. The angle of the cut on the molding or board for a miter joint is equal to one half of the angle of the corner around which the molding is fitted. The angle of the miter cut for a square corner is 45 degrees. The two pieces should be carefully cut in a miter box. As they are cut, the pieces must be held in the position they will assume when applied to the wall. When the corner is not 90 degrees, a T bevel should be used and the angle bisected.

Procedure for fitting baseboards. First clean the corners of the room. Also sweep the floor clean along the edges, where the base is to be fitted in place. Locate all wall studs and mark their location on the floor with a light pencil mark.

Square both ends of a piece of base molding to the correct length and fit it into place along the longest wall of the room. A tight fit can be insured by cutting the board 1/16 of an inch longer than the length of the wall, and springing the board into place against the wall. However, be careful not to break the gypsum board (or plaster) in the corners. Nail the board in place with 6d finish nails, holding the board down tightly against the floor. Nail into wall sole unless the base is wide. If the base is wide, nail it into the studs as well. (Note: Sometimes two or more pieces of base molding are needed to complete the length of a wall. If this is the case, the joints should be made on a stud or on solid backing where each of the boards can be securely nailed and the surface block sanded.)

Select a second piece of base molding approximately the same length as the adjacent wall. Scribe the end against the first piece, as shown in Fig. 7-33. Set the scriber for the thickness of the base molding. Cope the end of the board, undercutting it slightly to insure a tight fit on the surface. Cut the second piece to the correct wall length, measuring from the face of the first baseboard to the wall in the other corner.

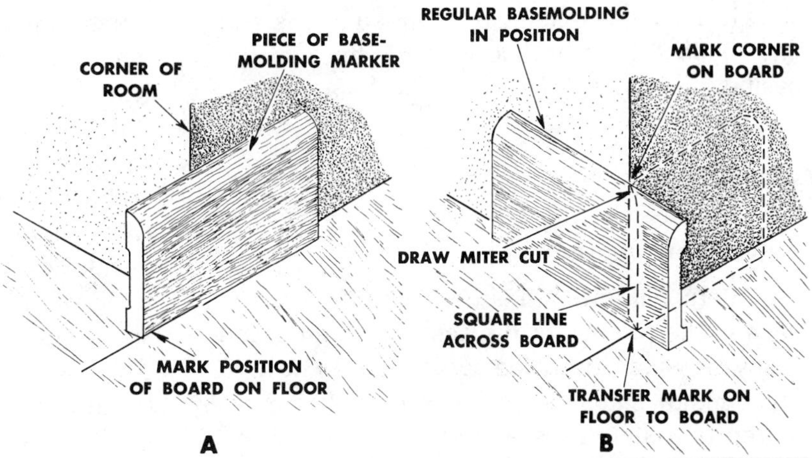

Fig. 7-34. An easy method of laying out a miter cut for the baseboard on the external corner is to mark the face of the board on the floor (or on cardboard placed over the carpeting).

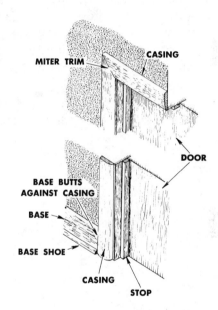

Fig. 7-35. Carpenters can trim an interior door frame with simple moldings. The base shoe is often omitted.

Then nail the second piece in place. Where doorways occur the base molding should be fitted against door casings. Complete fitting base molding pieces around the room, using coped joints. The mitered corners on the base molding for external corners should be laid out on the board when it is in position, as shown in Fig. 7-34. Use the baseboard to draw a mark on the floor, as shown in Fig. 7-34A. Then, with the baseboard fitted in place, draw a mark from the corner to the mark on the floor. See Fig. 7-34B. This procedure takes care of any irregularities in the walls or floor and helps to insure a satisfactory joint.

Wood base trim at doors. Prehung doors and frames are delivered to the job ready to install in the opening. See Fig. 7-35. The trim in many cases is attached on one side and ready to adjust and apply on the other side. When doors are to be hung and the trim applied by the carpenter, the job of trimming involves accurately cutting the casing pieces and nailing them in place. The casings are held back from the face of the jambs 1/4 of an inch. Miters are made at the top two corners, and the casings are made flush with the finished floor. The base is installed after the door casing is in place and is measured so that the base butts against the door casing. If a base shoe is used, it is mitered or rounded on the end and made to stop in line with the back of the casing.

WINDOWS

Adjusting Window Sash.

Modern windows are either delivered with the sash in the frames, or the sash is prefitted and ready for installation. The carpenter may have to make minor adjustments, when he or she is required to install the sash balances, so that the windows operate smoothly. Instructions are supplied by the manufacturer which give the step-by-step procedure. (Note: Patented sash balances and other devices for the efficient operation of windows of all types are discussed in Chapter 6, *Exterior Finish*.)

Trimming a window. Trimming a window can be done in several ways, depending on the type of window and frame and on local practice.

Some windows are finished without a wood casing by returning the drywall into the opening in what is known as a *wrap around* treatment. See P. 417. Other windows have the same casing surrounding the window on all four sides. See Figs. 7-36, 7-37, and 7-38. This method is used for all types of windows, including double hung windows. However, the usual treatment for double hung windows provides casings for the top and sides and a stool and apron below. See Fig. 7-39.

The windows and frames are delivered to the job as a unit and are installed in the wall, as outlined earlier in Chapter 6. Some windows are delivered with jamb extenders and interior trim items included in lengths adequate for the installation. Jamb extenders are pieces which fit into the jambs. They can be cut to provide for different thicknesses of walls. See Fig. 7-36.

Procedure for trimming a double hung window. Set a pair of scribers to the width of the rabbet on the window stool, as shown at A in Fig. 7-40. Place the window stool against the open-

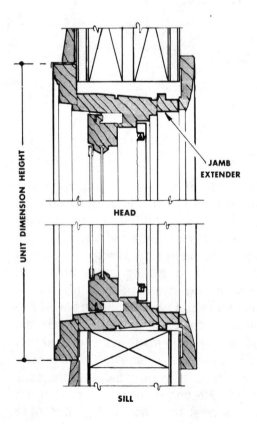

Fig. 7-36. A package casement window sealed in rigid vinyl plastic. Note the absence of exterior casings. The window is operated by a roto-operator through a gear arrangement so that it does not interfere with the inside screen. The jamb extender serves to adjust windows to the various wall thicknesses. (Andersen Corp.)

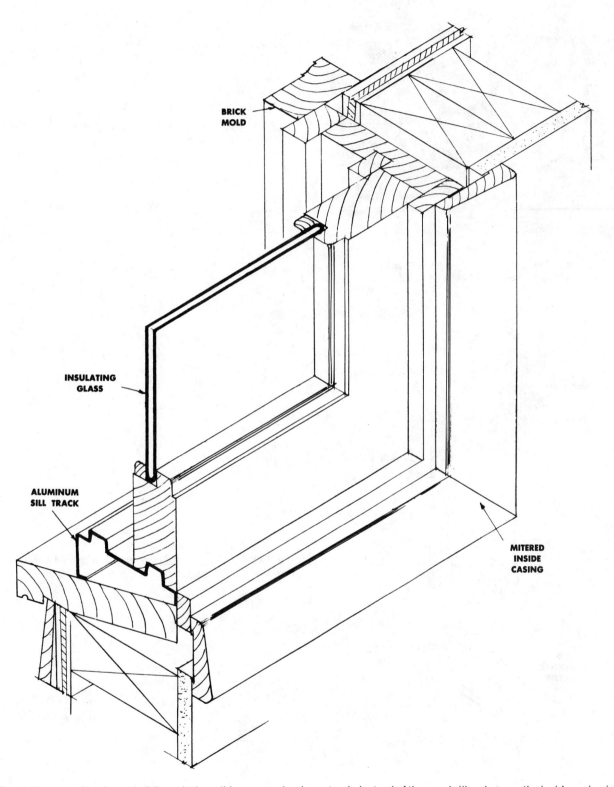

BRICK MOLD

INSULATING GLASS

ALUMINUM SILL TRACK

MITERED INSIDE CASING

Fig. 7-37. A wood horizontal sliding window slides on an aluminum track. Instead of the usual sill and apron, the inside casing is placed on all four sides.

Fig. 7-38. The interior casing is mitered at all four corners. (Georgia-Pacific Corp.)

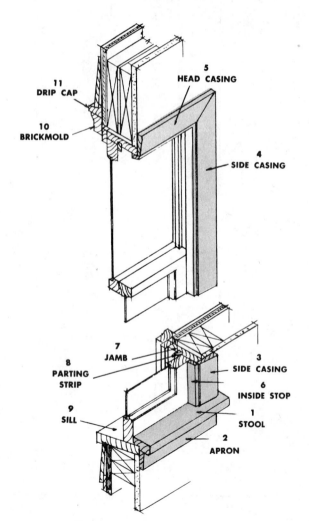

Fig. 7-39. The inside trim members for a double hung window consist of the casings, stops, stool, and apron.

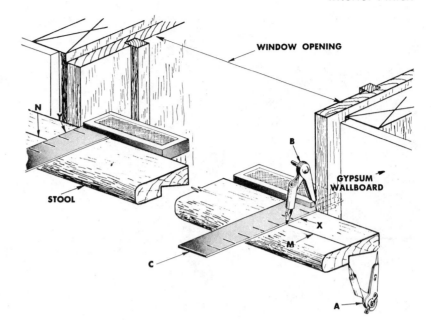

Fig. 7-40. Pictured here are the simple steps suggested for laying out the stool for a double hung window.

ing at the height it will have when fitted. With a square held in the position shown at C in Fig. 7-40, draw lines *X* and *Y*. Scribe the lines *m* and *n* on top of the stool, holding the scribers against the wall, as shown at B. Saw along these lines carefully and remove the wood from the corners of the stool. (Note: When making a finish cut on the trim, cut on the waste side of the cutting line, leaving part of the line. This insures a snug fit against the casing.)

Raise the lower sash and place the stool in its correct position. Bring the sash down and mark a line on the stool along the edge of the sash. Cut and plane to this line and fit the stool with a $1/16$ inch clearance between the stool and the sash.

As shown in Fig. 7-41, the dimension from the casing to the end of the stool *(Y)* is usually made equal to dimension *X*. Sand the stool and nail it in place. The stool when in position is shown at 1 in Fig. 7-39. (Note: When nailing the trim, always drill holes for the nails first, unless the wood is unusually soft. This prevents splitting the wood.)

Fitting the apron under the stool involves several steps. Mark the position of the side casings on the stool. The distance from the outside of one casing to the outside of the other casings gives the length of the apron. In other words, the ends of the apron are in line with the outside edge of the casing. When the apron is a squared piece of stock, cut it to the proper length and finish the ends. When the apron is molded, take a small piece of apron material and draw a pattern for the return of the apron on one end, as shown in Fig. 7-42. Mark both ends of the apron, cut them out with a coping or jig saw, then sandpaper them. Nail the apron in position under the stool with 8d finish nails by nailing into the solid subsill of the rough opening, as shown at 2 in Fig. 7-39.

When starting the casings, leave all pieces about 1 inch longer than required. Put a left and right hand miter on the two side casings and a miter on one end of the head casing. See 3, 4, and 5 in Fig. 7-39. Place the head casing (5 in Fig. 7-39) in position with the end which has the miter on it, $1/4$ inch back of the jamb. Mark the other end, miter it, smooth the cut, put it back in place, and nail it.

The side casings (3 and 4 in Fig. 7-39) have already been mitered. Taking each piece, turn it upside down so that the miter end rests on the stool and mark the other end at the top of the casing. Cut the side casing pieces square on the line and smooth out the cuts. Put the casings in position and nail them, using 4d finish nails to

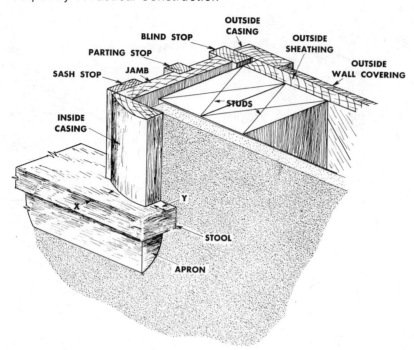

Fig. 7-41. The stool and apron fit snugly. X and Y are made equal.

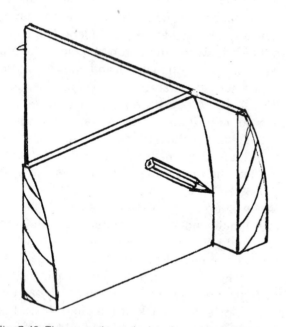

Fig. 7-42. The apron is marked to be cut to return to the wall. The molding is inverted and marked on the back.

fasten the casings to the jamb, and 8d finish nails to fasten the outside of the casings to the wall. Make sure that the 8d nails pass into studs. Drive all the nails home and set them with a nail set. Sandpaper the high point on the joints.

Windows, which are to be trimmed on all four sides with casings, are first finished as described above. The operations are adapted to include a casing across the bottom.

INTERIOR DOORS

Types of Doors

The internal construction and other general information about doors are covered in Chapter 6, *Exterior Finish.* To a certain extent, interior and exterior doors are alike. Exterior doors are usually 1¾ inches thick. Interior doors are either 1¾ or 1⅜ inches. Wood exterior doors require waterproof glue between parts and the veneer plies. The most common interior door is 1⅜ inches thick and has a hollow core construction. Solid core doors are used if fire rating or sound transmission standards require them. Panel doors are used to give the building a more traditional character. Some of the common door styles are shown in Fig. 7-43. Flush doors with louver panels are used to provide ventilation between rooms. See Fig. 7-44.

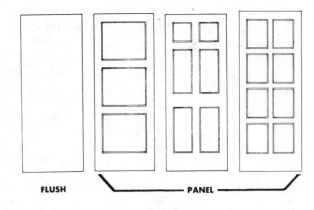

Fig. 7-43. Interior doors include both flush and panel doors.

Fig. 7-44. Interior louvered doors provide ventilation. (Visador Co.)

Setting Inside Door Jambs

The part of a doorway that forms the lining of the door opening is called the *door jamb*. The jambs, together with the casings, provide the finish around the door opening. To close the opening, the door is fitted and hung on these jambs. Most inside door jambs are 7/8 of an inch in thickness and made from the same kind of wood as the interior trim. The inside distance between the side jambs should be equal to the width of the door plus 3/16 inch. The height should be 1/4 inch greater than the height of the door to allow the door to swing freely; Carpet thickness or a threshold will require more space. The head jamb should be housed in the side jambs as shown at A in Fig. 7-45.

Procedure for setting inside door jambs.

Door jambs are usually delivered to the job with all parts assembled and ready for installation. It is advisable to check the width of the jambs with the wall thickness. Likewise check the length of both the head and side jamb with the door and the squareness of the corners. Cut off part of the lugs if necessary. Door jambs are to be assembled with top jamb rabbeted between side jambs. The projections are called *lugs* and may have to be cut so that the door fits in the rough opening.

If the finish floor has been laid, place the jambs in position and verify that the head jamb is perfectly level. If the head jamb is not level, raise the lowest side jamb and cut it off to fit the door. This should bring the head jamb into a level position.

Take a piece of 1 × 6 inch material and cut the ends square so that they fit snugly between the

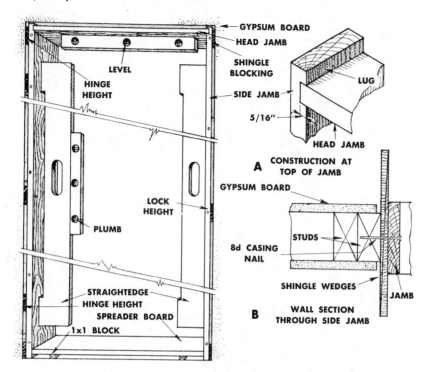

GYPSUM BOARD
HEAD JAMB
SHINGLE BLOCKING
LEVEL
HINGE HEIGHT
SIDE JAMB
LUG
5/16"
HEAD JAMB
A CONSTRUCTION AT TOP OF JAMB
GYPSUM BOARD
LOCK HEIGHT
PLUMB
8d CASING NAIL
STUDS
STRAIGHTEDGE
HINGE HEIGHT
SPREADER BOARD
SHINGLE WEDGES
JAMB
1x1 BLOCK
B WALL SECTION THROUGH SIDE JAMB

Fig. 7-45. The door frame is made plumb and level using shingles driven from both sides.

jambs at the head. This forms a spreader which should be placed on the floor between the jambs at the lower end. See Fig. 7-45.

Draw a faint pencil line on the jamb 7/8 inch beyond the position the door will take. This serves as a center line for nailing through the jamb. The line will later be covered by the stop which hides the heads of the nails.

Set the jambs as near as possible to the center of the opening. Hold the jambs in place with double shims (shingle wedges) on each side at the top so that they are in line with the head jamb. See Fig. 7-45B. Wedge the bottom of each side jamb in this fashion.

Plumb the hinge jamb with a straight edge and level, or place the door itself in the opening. Check the door by placing the level on the door edge while the door it is slightly ajar. Check to see that the edges of the door frame are flat with the wall finish surface. (The casing which is applied later must lie flat.) Fasten the hinge jamb with 8d nails at the top and bottom, placing the nails on the line previously drawn so that they are concealed when the stop is installed. Do not drive the nails home until the lock jamb is adjusted. Check the jamb on the lock side of the door.

When the jamb is plumb, drive the shims tight at top and bottom. Complete the shimming by placing additional pairs of shims in back of each jamb. Place the shims on the hinge jamb, with one pair in the middle and the other two at the hinge location. Prefitted doors are hung so that hinges are a uniform distance from the top and bottom of the door. This is usually 10 inches. However, some carpenters place the top hinge 7 inches from the top and the bottom hinge 10 inches from the bottom of the door. On the lock side, place one pair of shims so that they are centered on the lock position. Place another pair of shims midway between this position and the shims above. Again use a level to check that the jambs are plumb. Slip the door into place to test the straightness of the jambs and the fit of the door. Drive two nails at each shim, staggering them 1/2 inch on each side of the pencil line.

Complete the nailing of the jambs, driving nails into the studs along the pencil line marked on the inside of the jamb.

Trimming a Door Opening

Prehung doors and frames are often delivered with the trim already applied to the jambs on one

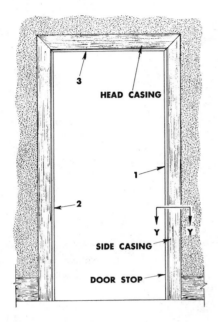

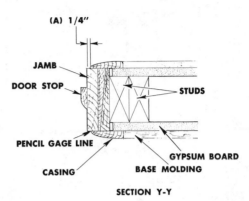

Fig. 7-46. Casings are carefully fitted around the door jambs.

side and loose trim to be applied on the other side. The jambs must be plumbed with the shims, leveled, and fastened securely into studs. These operations are done in a manner similar to those outlined in the preceding section.

A door opening is trimmed by fitting and applying the casings. See Fig. 7-46. This trimming work is done after the jambs have been set and the floor laid, but before the fitting of the baseboards. Many carpenters fit and hang the door before applying the trim so that they can make minor adjustments in the jamb. Such adjustments cannot be done once the casings are in place.

Procedure for trimming door openings. Select the necessary pieces of trim for the door opening and place them at some convenient location near the opening. Gage a light pencil line on the edge of the jamb 1/4 of an inch back from the face, as shown at A in Fig. 7-46. The inside edge of the casings should be set to this line.

Lay out, cut, and fit the top of the side casings and both ends of the top casing. Use the same procedures employed for window trim. Fit the bottom end of the side casings, shown at 1 and 2 in Fig. 7-46, to the finished floor. Nail the casings

in place with 4d nails, nailing into the jambs. Use 6d nails along the outer edges, nailing into the trimmers. Cut, fit, and tack the door stops in place. The stops should not be nailed until after the door has been fitted and hung.

Combined jambs and door trim packages are available. An example of this type of trim package is shown in Fig. 7-47. This type has a split jamb feature which permits quick adjustment to different wall thicknesses. The members are sheathed with vinyl.

Fitting Doors

Doors are delivered to the job in two ways. They may be parts of complete door units. Here the doors are prehung in the frames with some of the casings attached. Other doors are delivered to the job then hung in the frames already in place in the openings.

A door must be handled carefully so that it is not damaged. When delivered to the job, all doors should be carefully stacked on the floor or set against a wall. Doors stored against a wall should be protected against warping. The wall surface should also be protected against marring. The following steps should be observed before the door is fitted or hung.

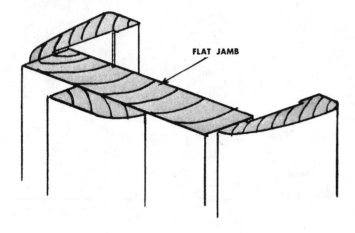

FLAT JAMB

Fig. 7-47. Combined jamb and door trim is provided as a package. The members have a vinyl shield. (Kimberly-Clark Corp.)

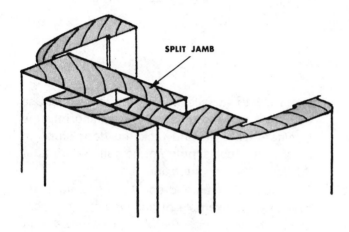

SPLIT JAMB

Consult the working drawings and building specification to determine the type of door required for a particular opening, in addition to any hardware specified and the direction the door is to swing. Mark the door jamb which is to receive the hinges. (Hinges are also called *butts.*) Check the size of the door with the opening size.

Procedure for fitting doors. Mark the hinge stile on the door and fit the door against the hinge jamb, planing the edge if necessary to conform to the shape of the jamb. A power plane is especially useful for this operation. (Note: The hinge edge of the door should usually be square. However a slight bevel at the back provides additional clearance when the door is opened.) Place the door on sawhorses. Measure the width of the door opening (distance between the jambs) at

both the top and bottom, using a rule, then transfer these dimensions directly to the door. Draw a cutting line on the door, making the door $1/8$ of an inch narrower than the width of the opening. (Note: A well-fitted door should fit to the shape of the door opening and have a clearance of $1/16$ of an inch on each side and on the top. This is the thickness of a 4d nail. The bottom clearance should be from $1/8$ to $1/4$ inch. If a threshold is used, a $1/16$ inch clearance should be provided between the door and the threshold. The threshold is fitted into place after the door has been hung. Carpeting also presents a problem. The required door clearance over carpeting is $1/4$ inch.)

Plane the second edge of the door to the cutting line. To insure uniform clearance, check the

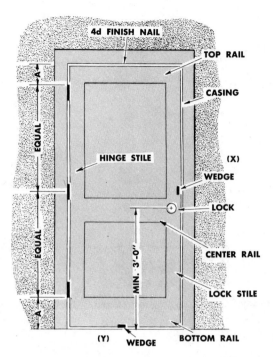

Fig. 7-48. The door must be made to fit the opening with $^1/_{16}$ inch clearance. Dimension A for the hinges is the same at the top and bottom. The lock dimension is a minimum of 36 inches.

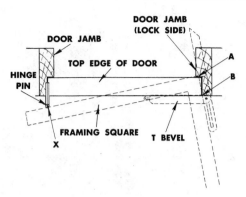

Fig. 7-49. A framing square can be used to determine the amount of bevel needed on the lock stile.

jamb for straightness before planing off the entire amount to be removed.

Place the door in the opening, check the fit, and insert a wedge in place, as shown at X in Fig. 7-48. Scribe the top edge of the door to the top jamb. Scribing can be done with a pair of dividers or a regular scriber. This operation is unnecessary when both the door and jambs are square.

Plane the top edge of the door to fit the top jamb. Bevel the lock edge of the door. The door as it swings forms an arc. The lock side must be beveled so the point on the door (A in Fig. 7-49) clears the edge of the jamb (B in Fig. 7-49). The amount of bevel required depends on the thickness and width of the door and the width of the hinge. The proper bevel can be obtained by drawing a diagonal line on the top edge of the door from one corner to the theoretical pin center of the hinge. In Fig. 7-49, the bevel line, *AB*, is at a right angle to line *AX*. Set the T bevel to line *AB* and to the side of the door to find the correct

angle of the bevel. Plane the lock side edge of the door to this angle. Remove the sharp edges of the door with a plane and sandpaper all edges until they are smooth.

Installing Hinges

The usual types of hinges for door are shown in Fig. 7-50, left and right. They are also called *butt hinges* or *butts.* The round central portion of the butt hinge is known as the *knuckle.* The hinge is ordinarily divided into five sections and called a *five-knuckle butt hinge.* The flat parts of door hinges are known as the *leaves* or *flaps.* The two leaves are held together by a pin running through the knuckle. Door hinges usually are made so that the pin can be removed. This type of door hinge is generally referred to as a *loose-pin hinge* or a *loose-pin butt.* When the pins are not removable, the hinges are called *fast joints.* Hinges with round corners are easy to install if the mortise is cut with a router. See Fig. 7-50, right.

The thickness and weight of the door determine the size of the hinge that should be used. The width of the hinge and the amount of the setback depend on the thickness of the trim. See Fig. 7-51. There must be sufficient clearance so that the door can swing back and be parallel to the wall.

Procedure for installing hinges. Place the door in the door opening, then force the door against the hinge jamb with a wedge, as shown at X in Fig. 7-48. Place a 4d finish nail on top of the door and force the door up against the nail

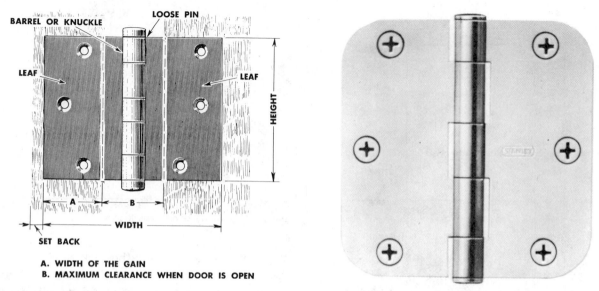

Fig. 7-50. Loose pin mortise hinges (butts) vary in size and shape. (Stanley Power Tools Div., The Stanley Works)

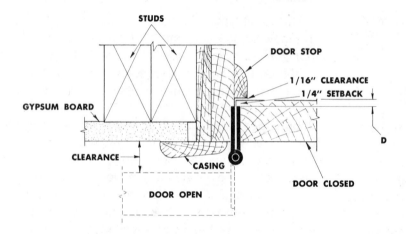

Fig. 7-51. This method of hanging the door provides for clearing the casing when the door is swung wide open.

with a wedge at the bottom. See *Y* in Fig. 7-48. Mark the hinge location on both the jamb and the door stile with a knife or sharp pencil.

The hinge location may vary, depending on the decision of the architect or the contractor. It has become a common practice to locate the top hinge down from the top and the bottom hinge up from the bottom the same distance. This dimension varies between 8 and 10 inches. See Fig. 7-48. However, some carpenters follow an established procedure of placing the top of the upper hinge 7 inches down from the top of the door and the bottom of the lower hinge 10

inches up from the floor. A third hinge should be centered between the top and the bottom hinges, as shown in Fig. 7-48.

Scribe the bottom of the door to the floor, then remove the door. If a threshold is used, measure the thickness of the threshold then add 1/16 of an inch to the amount to allow for clearance. If no threshold is used, then allow a 1/8 inch to 1/4 inch clearance.

Place the door in a door jack. See Fig. 7-52. Use a routing template and router for cutting the gains for hinges. (A *gain* is a cutout or mortise made on one member to receive another mem-

Fig. 7-52. The doors are held in a door jack while the carpenter trims the door edge and installs the lock and hinges. (Stanley Power Tools Div., The Stanley Works)

Fig. 7-53. The template for the leaf of the hinge is positioned in relation to the top or bottom of the door. (Stanley Power Tools Div., The Stanley Works)

ber. In this case, the cutout is made on the door to receive a leaf of the hinge.) See Fig. 7-53. Adjust the template so that the gains are placed at the desired dimensions from both the top and the bottom of the door. Position the template so that a ¼ inch setback is made on the door.

See Fig. 7-54. Choose an appropriate router bit so that you can cut to the same depth as the thickness of the hinge leaf. Cut out the gain with the router. The gains for round corner hinges are quickly made with the router, requiring no additional hand operation. When the gain has been

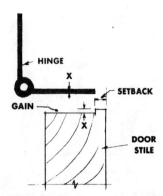

Fig. 7-54. The thickness of the hinge leaf determines the depth of the gain. The setback is made wide enough to prevent the wood from splitting when the gain is cut. The setback is usually ¼".

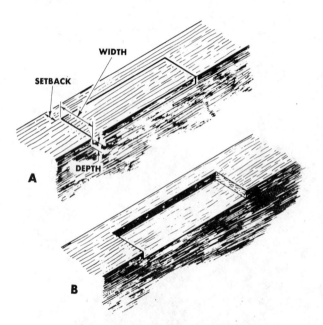

Fig. 7-55. When making a gain for a hinge, the gain is laid out (A) and cleaned out (B) by paring with a chisel.

made on the door, mount the template on the door jamb and repeat the operation. Separate each hinge, then screw one leaf in position on the door and the other leaf on the jamb. Use a drill with a special chuck for driving screws.

If no template is available, mark the gain on the door as shown in Fig. 7-55. Cut the main part with a router and finish the gain with a chisel, or use chisels for the entire operation. (The gain

shown is for a square cornered hinge.) Repeat the operation on the jamb, then mount the hinge leaves.

Hang the door in place on the hinges and check the door for proper clearance. The clearance should be ¹/₁₆ of an inch on each side (Note: If the door has more than ¹/₁₆ of an inch clearance along the hinge jamb, the gain should be deepened slightly. If the door binds against the jamb, place a strip of cardboard behind the butt in the gain. Then drive all screws in place securely.)

Adjust the door stops. There should be ¹/₁₆ of an inch clearance between the stop on the hinge jamb and the door. The stop on the lock side should hold the door flush with the outside edge of the jamb. The top stop should be held in line with the side stops. Nail the door stops securely with 4d finish nails.

Locks

Tubular and cylindrical lock sets. Locks which only require the boring of two holes and the cutting of a small mortise for the face of the lock are used extensively on new jobs because they are easily installed and provide very satisfactory results. They are manufactured for many applications for both exterior and interior doors. The manufacturers provide templates for locating the holes. See Fig. 7-56.

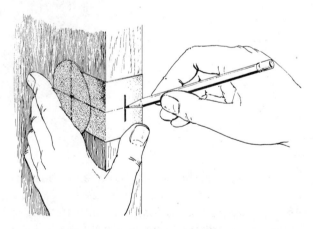

Fig. 7-56. Paper templates are provided by the manufacturers to help the carpenter locate holes accurately. (Yale Lock & Hardware Div., Eaton Corp.)

Fig. 7-57. A jig is fastened to the edge of the door to give accurate location for the holes. (Kwikset Sales and Service Co.)

Fig. 7-58. A router is used to cut the mortise for the lock. The power plane (shown on the floor) is used to trim the door to size. (Stanley Power Tools Div., The Stanley Works)

Jigs are available which clamp on the edge of the door and make the accurate boring of holes a foolproof operation. See Fig. 7-57. A mortising template is also available which cuts the mortise

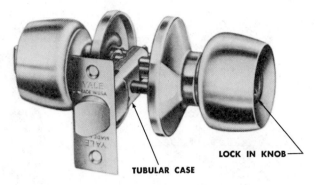

Fig. 7-59. Tubular latch sets are used for interior doors. (Yale Lock & Hardware Div., Eaton Corp.)

Fig. 7-60. Cylindrical lock sets are used for heavy duty on exterior doors. (Yale Lock & Hardware Div., Eaton Corp.)

to the specified depth for the face of the lock. See Fig. 7-58.

Tubular lock sets are used for interior doors on bedrooms, bathrooms, passages, and closets. They are available with outside pin tumbler locks in the knob and turn button or push button locks on the inside. Several variations of this arrangement are manufactured. A typical lock is shown in Fig. 7-59.

Cylindrical lock sets are sturdy heavy duty locks. They are designed for installation in exterior doors and provide high security. See Fig. 7-60 for a typical lock. Manufacturers' instructions supplied with the locks should be followed carefully.

Tubular and cylindrical locks require similar door preparation. They differ in their mechanical construction. Figure 7-61 shows a combination lock with a typical latch bolt in the lower lock

INTERIOR VIEW

EXTERIOR VIEW

**EXTERIOR VIEW WITH
OPTIONAL TRIM**

Fig. 7-61. A combination lock has the features of a latch and a dead bolt. The dead bolt provides high security. The knob on the inside of the dead bolt provides panic safety. (Kwikset Sales and Service Co.)

and a dead bolt in the upper lock. (The bolt of a dead bolt lock has no bevel. Such a lock must be operated by a key from the outside or a knob on the inside.)

Procedure for installing tubular and cylindrical lock sets. Unpack the package containing the lock and examine each part so that you become familiar with the installation requirements. Next open the door to a convenient working location and block it with two wedges under the front edge.

Measure up 36 inches from the floor, which is the usual knob height, and square a line from the edge of the door for the lock spindle. (Note: The knob height can vary from 36 inches up to a height equal to the center of the door. When panel doors are used, the lock is centered on the cross rail.)

Place the template on the edge of the door, as shown in Fig. 7-56, and mark the centers for the two holes. If a jig is available, fasten it to the edge of the door at the proper height.

Drill the two holes to the diameters prescribed in the instructions accompanying the lock. Drill one hole through the door and the other hole to intersect the first hole. See Fig. 7-62. Place the lock in the holes and mark the outline of the mortise for the lock on the edge of the door. Use a chisel to cut the mortise. Use a face plate mortise marker, if one is available. See Fig. 7-63. This device cuts through the wood fibers to make it easy to cut the shallow mortise with a chisel. Install the lock and assemble it according to the manufacturer's instructions.

Close the door and mark the center of the lock on the jamb. Use a strike locator. See Fig. 7-63.

Fig. 7-62. Doors are prepared for locks by drilling two holes and mortising the door edge to receive the lock face. (A) shows the preparation for a tubular lock. The preparation for a cylindrical lock is shown in (B).

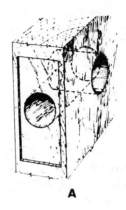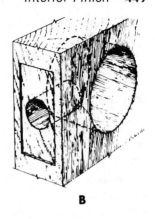

A   B

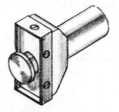

LATCH MARKING CHISEL

Fig. 7-63. Tools help locate and cut the mortises in the door edge and jamb. (Yale Lock & Hardware Div., Eaton Corp.)

STRIKE MARKING CHISEL

STRIKE LOCATOR

Place the strike plate in position and mark the perimeter. The vertical center line must be the same distance from the stop as the center line of the lock case is from the edge of the door. Check the location of the strike plate in relation to the lock. Adjustments cannot be made once the mortise is cut. Use a strike marking chisel. See Fig. 7-63. Carefully cut out the wood for the strike plate in the jamb and cut a recess for the latch bolt itself. Quality locks have strike plates which incorporate a metal box to finish the recess. Fasten strike plate with screws.

Mortise Locks

A large percentage of the locks installed in new buildings are the cylindrical or tubular types. However, mortise locks are still used in sufficient quantity so that carpenters should know how to install them. A modern mortise lock is shown in Fig. 7-64A with a cylinder lock. Mortise locks are generally used in front entrance doors.

Mortise locks are high grade locks and must fit perfectly if they are to operate well. They are usually installed in the door with precision machinery before the door leaves the factory. The door frames are likewise often prefabricated so that the strike is precisely located. Carpenters, however, may still be required to install the locks or the strike plates on the job.

Procedure for installing mortise locks. Unpack the lock and examine each part so that you become familiar with the installation requirements. Open the door on which the lock is to be installed to a convenient working location and block it with two wedges under the front edge. Measure up 36 inches from the floor (or whatever height is specified) and square a line for the lock spindle from the edge of the door. Mark the parts of the door which will be cut away. A template is supplied with the lock. Use it to locate the centers of holes through the door and the mortise for the lock itself on the edge of the door. The location of the holes and the mortise for the body of the lock are shown in Fig. 7-64B.

Using drills of the proper diameter, make the holes through the door. With drill and chisel (or router), cut the deep rectangular hole for the lock. Place the lock in the door and mark the perimeter of the lock face on the edge of the door. Cut the mortise for the lock face with a chisel or a router. Install the lock in the door, following the manufacturer's directions.

Close the door partially and mark the height of the lock bolt and latch bolt on the jamb. Open the door and locate the strike plate position by measuring from the edge of the door to the lock bolt. Duplicate the measurement from the stop in

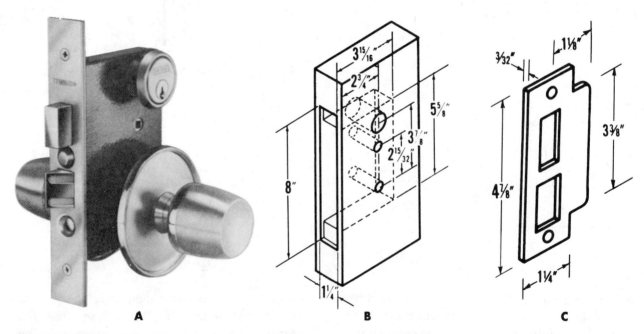

A **B** **C**

Fig. 7-64. A modern mortise lock, shown at the left, provides high security. A strike plate is shown in the middle. The mortise and holes for the spindles are shown at the right. (Sargent and Co.)

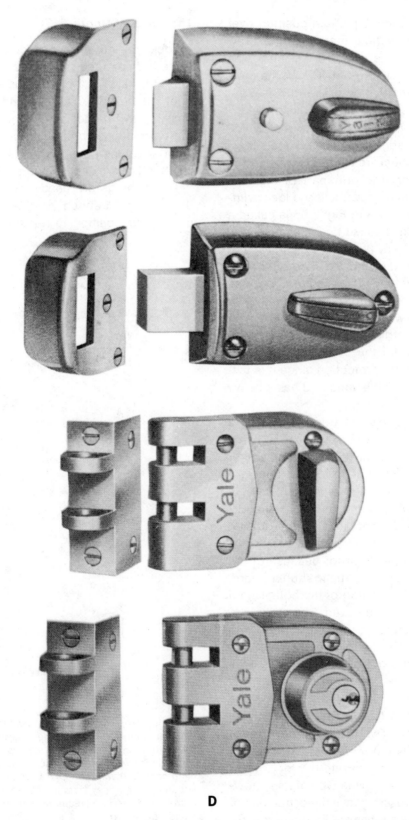

D

Fig. 7-65. Auxiliary latches and locks. (A) spring latch, (B) dead bolt lock, (C) heavy duty jimmy proof dead lock with shutter lock, and (D) heavy duty lock shown in (C) with a lock on both sides. (Yale Lock & Hardware Div., Eaton Corp.)

the jamb. Place the strike plate in position over the horizontal and vertical markings and draw the outline of the strike plate on the jamb. The strike plate is shown in Fig. 7-64C. Cut the mortise for the strike plate, then drill and chisel out the space for the bolts. Install the strike plate.

Several auxiliary latches and locks (also called *rim locks*) used for additional security are shown in Fig. 7-65. All of them are without knobs on the outside of the door. The only part of the lock which is visible from the outside is the lock cylinder. The type shown at A in Fig. 7-65 is called a *spring dead latch.* It is called *dead* because with an extra turn of the key (or the inside knob), the latch bolt is fixed and cannot be pried. The knob on the inside of the door operates the bolt. The knob can be turned so that the bolt is held back within the latch. Two strikes are supplied with the latch. A simple strike plate is mounted in the same manner used for the other locks. The rim strike provides greater protection since it is fastened into the edge of the jamb and has a heavy body to receive the bolt.

The type of lock shown at B in Fig. 7-65 is a lock which operates in a similar fashion to the previous one, except that it has a rectangular bolt (a true dead bolt) and does not act on a spring. It must be operated when the door is closed. The type of lock shown at C, Fig. 7-65, has interlocking bolts operated on the outside with a key and on the inside with a knob. This lock is considered jimmy-proof because of the hardened steel bolt and automatic shutter guard, which prevents the operation of the bolt, even if the outside cylinder is removed. The strike can be either flat (mounted on the face of the jamb) or a rim strike (fastened to the corner of the jamb). The type shown at D (Fig. 7-65) is the same lock as the one shown in C, except it is operated by a key on both sides. This type is called a *double dead bolt lock.* It provides very high security. Even if a would-be intruder breaks a wood panel or the glass in the door, that person cannot reach in to operate the lock, since there is no knob to turn. This type of lock, however, can only be used in a limited number of applications because it cannot be opened without a key. This can cause serious problems in case of an emergency.

The Hand of Doors

Locks are not always made so they can be reversed or changed to suit doors hinged on either the right-hand or the left-hand side, or for doors opening in or out. Therefore, chiefly for the purpose of buying door hardware, it is necessary for the carpenter or builder to be familiar with the standard rules covering locks for right-hand or left-hand doors or casements.

Three things are used to describe whether a door is right-handed or left-handed. These include your position in relation to the door, the

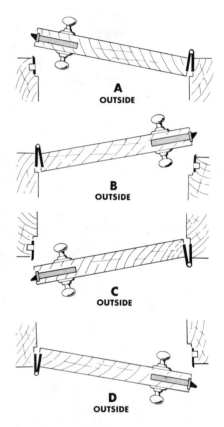

Fig. 7-66. The hand of a door is determined by the position of the observer (standing outside the building or room), the side of the door where hinges are located, and what direction the door swings in relation to the observer. A right-hand door (RH) with a regular bevel bolt lock has hinges on the right and swings away from the observer, as shown in (A). A left-hand door (LH) with a regular bevel bolt has hinges on the left and swings away from the observer, as shown in (B). A right-hand door (RHR) requiring a reverse bevel bolt lock has hinges on the right and swings toward the observer, as shown in (C). A left-hand door (LHR) requiring a reverse bevel bolt lock has hinges on the left and swings toward the observer, as shown at (D).

side of the door on which the hinges are located, and the direction the door swings.

Regarding your position, stand outside the building or room facing the door. Notice the side of the door on which the hinges are located. If the hinges are on the right, the door is a right-handed door. If they are on the left, the door is a left-handed door. If the door swings away from you (the usual situation), it requires a standard bevel bolt lock. If the door swings toward you, it requires a reverse bevel bolt lock. A right-handed door with a regular bevel bolt lock is shown in Fig. 7-66A. To order it you would ask for a RH lock. A left-handed door with a regular bevel bolt lock is shown in Fig. 7-66B. When ordered, this lock is referred to as a LH lock. A right-hand door is shown in Fig. 7-66C. This door requires a reverse bevel bolt lock. This lock is ordered as a RHR lock. Fig. 7-66D shows a left hand door which swings toward you. This requires a reverse bevel bolt lock. It is ordered as LHR lock.

SPECIAL DOORS AND FRAMES

Pocket Doors

A pocket door is hung from a track and operates on rollers. This type of door only requires the regular 3½ inch stud wall thickness. It is a complete sliding door unit with a track and rollers built into one package, as shown in Fig. 7-67. The frame unit is set into the wall opening and covered with the regular interior wall finish. Any door with a thickness of 1⅜ inches can be hung to the rollers when cut to fit correctly. When this door is placed in the wall, any type of interior wall finish can be applied over the horizontal frame members.

By pass doors. By pass doors are hung from a track and slide on glides or rollers. They are arranged in pairs, or combinations of several doors, which slide behind each other on parallel tracks. They are mainly used as closet doors. See Fig. 7-68.

Bifold doors. The most popular door for closets is the bifold door. See Fig. 7-69. The usual arrangement, depending on the width of the closet, is to use two panels as a pair, or four

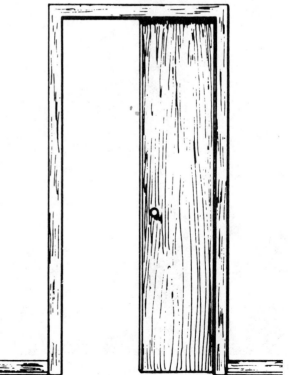

Fig. 7-67. Pocket roll-away door fits into 3½ inch wall space. (Sterling Hardware Mfg. Co.)

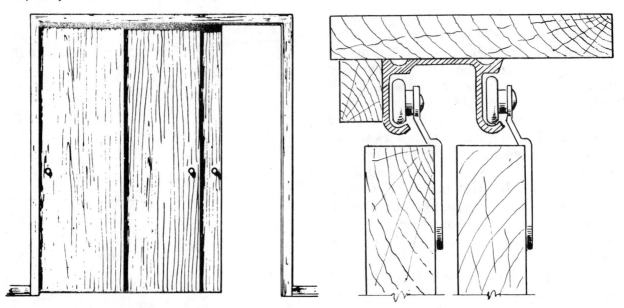

Fig. 7-68. A by pass door (shown at the left) slides in parallel tracks (shown at the right).

Fig. 7-69. Bifold doors are arranged in pairs. (Kinkead Industries)

panels as a double pair. Two panel doors are usually used in jamb openings of 2'-0", 2'-6", 2'-8", or 3'-0". Four panel units are used for 4'-0", 5'-0", or 6'-0" openings. The nominal door height is 6'-8", although other heights are available. The pivot door (door nearest jamb) is suspended from a pivot in the head jamb and has a jamb bracket at the floor. See Fig. 7-70. The guide door is hung from a sliding guide contained in a track across the top of the opening. Hinges permit the doors to fold. Doors are made of plywood, metal, or plastic and are available as prefinished units.

Metal doors and frames. Metal doors used for the interior of a house are generally 1³/₈ inches thick, 6'-8" high, and 1'-6", 2'-0", 2'-4", 2'-6", 2'-8", or 3'-0" wide. The doors are made from steel sheets with stiffeners, reinforcing plates, and a honeycomb core. They are available prefinished, with baked-on enamel, or with a vinyl-laminated surface. The doors have recesses for hinges.

The metal frames which accompany the doors, or which can be used for wood doors, come in different sizes to fit several wall thicknesses. They are shipped knocked-down and can be quickly assembled and fastened to the frame-

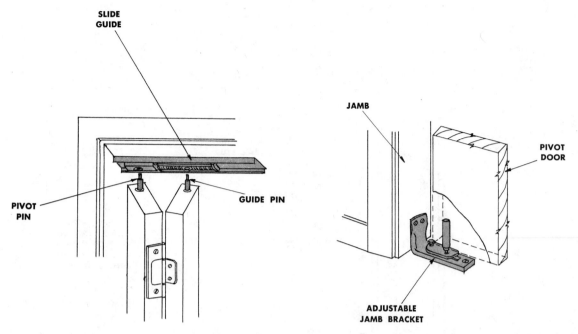

Fig. 7-70. Bifold doors operate on pivots for the door nearest the jamb and on an overhead sliding guide in a track for the inner doors (left). An adjustable jamb bracket holds the bottom edge of the pivot door (right).

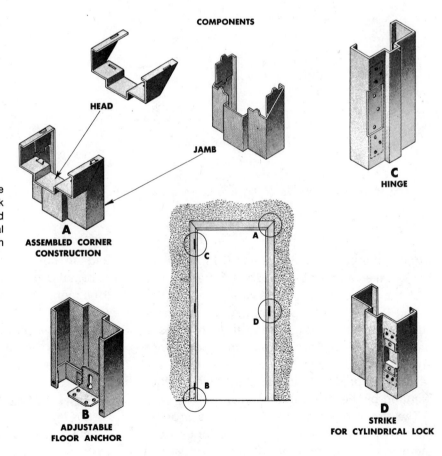

Fig. 7-71. Metal door frames can be assembled on the job. Hinge and lock strikes are positioned accurately and match the hinges and locks on metal doors. Wood doors are hung to match the metal frames. (The Ceco Corp.)

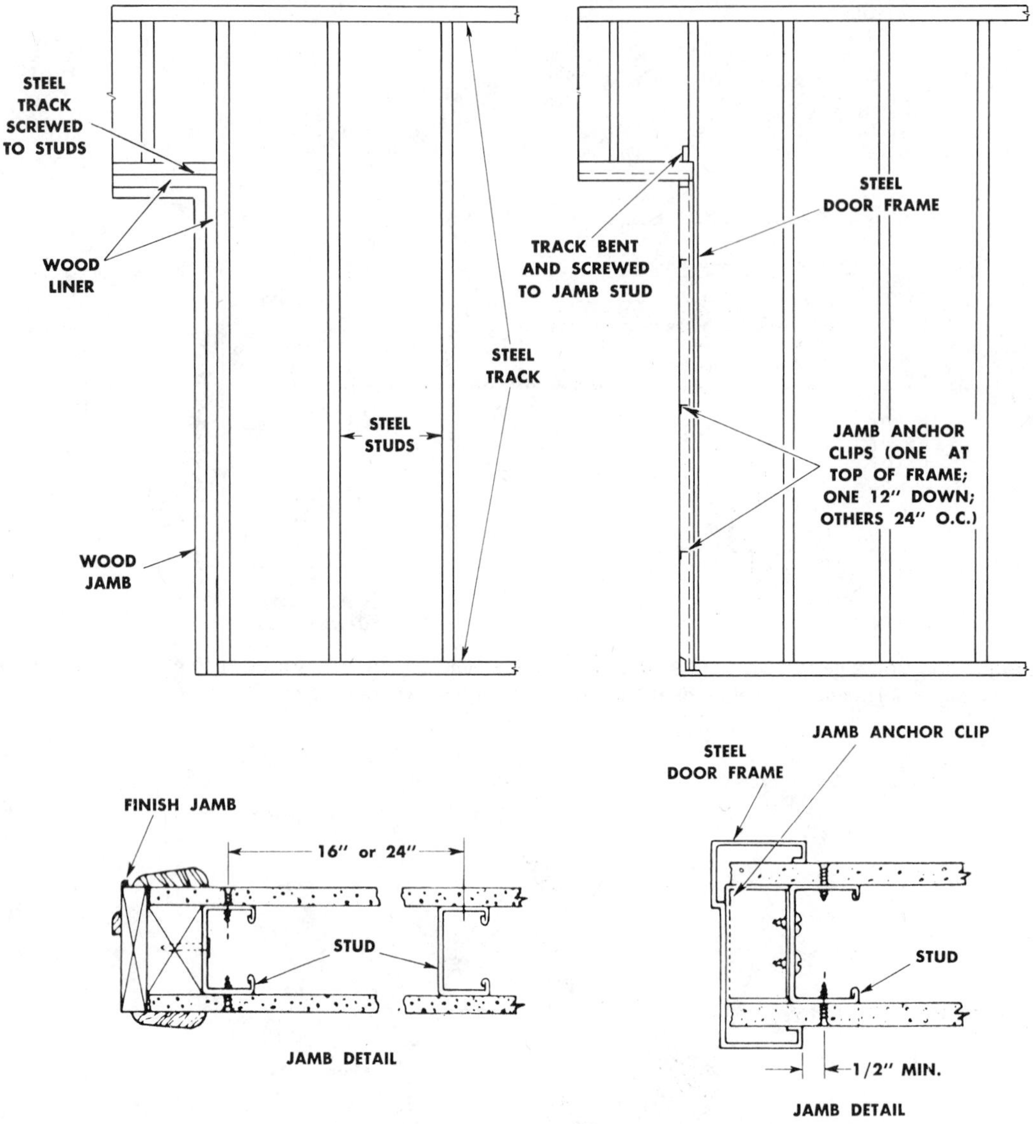

Fig. 7-72. Wood (left) and metal (right) door frames are used with metal studs for drywall application. (National Gypsum Co.)

work of the partition. See A and B in Fig. 7-71. Lock strikes and hinge rabbets are prepared so that the hardware can be easily installed. See C and D in Fig. 7-71. Several devices are available for fastening the frames to different types of partitions. The manner in which the door frames are assembled is shown in Fig. 7-71. There are several minor differences in the frames manufactured by the various companies.

The great emphasis on drywall has brought about the use of metal door frames for use with metal studs. A number of different door frames

and anchoring devices have been developed. Information on wood or metal door frames used for a metal stud partition is provided in Fig. 7-72. A piece of track, the same as that used on the floor for the partition, is fastened over the door to receive the crippled studs. The jamb anchor clips are fastened with screws.

FINISH FLOORING

One of the last jobs the carpenter does before installing the base mold is to install finish flooring or prepare a surface for the installation of resilient flooring by other carpenters who specialize in this field. Resilient flooring is popular in contemporary homes. The variety of resilient flooring includes carpeting, linoleum, vinyl, or vinyl asbestos material. Wood finish flooring is used in more costly residences, or where traditional architecture requires it, and is laid by carpenters who have acquired this skill as part of their general background. Ceramic tile, slate, and terrazzo floors are laid by still other mechanics.

Subfloor and underlayment. The base over which finish flooring is placed is very important. It must be stiff enough to receive the type of finish specified, provide a smooth level surface, and be held fast to the joists so that squeaks, which often develop when the wood dries out and the nails begin to pop, are eliminated. Plywood subfloor is generally used as the base. It is applied as discussed earlier in Chapter 4, *Wall and Floor Framing.*

Structural finish flooring, such as wood strip or plank flooring, can be laid over plywood subflooring without using underlayment (a second layer of material).

When thin vinyl or other types of tile or carpeting are used, two types of subfloor are suggested. One type requires a combination subfloor-underlayment. This material is available in plywood sheets 1/2, 5/8, or 3/4 inch thick, with tongue and groove edges. 6d deformed shank (ring shanked or spiral ring) nails are required to apply it. The nails are spaced 6 inches on panel edges and 10 inches on intermediate joists. This type of plywood is touch sanded to provide an excellent surface for the finish flooring. A type of subflooring with a grade designation of 2-4-1, 1-1/8 inches thick, is also designed for this purpose to span 48" O.C.

Underlayment is often used over the subfloor to provide a smooth base for laying the finish floor. Underlayment provides added protection against nail popping in the subfloor since the nails are covered with the second sheet. It also provides some sound deadening effect. The material used can be plywood made especially for underlayment, particle board, or hardboard. Plywood and hardboard generally have a 1/4 or 3/8

inch thickness, while particle board is 1/2 inch thick. Plywood 3/8 inch thick and hardboard 1/4 inch thick are recommended. Hardboard sheets are supplied in 3 × 4 or 4 × 4 foot sizes. The material is fastened with staples or ring shanked nails, or a combination of nails and adhesive. Fastenings for 3/8 inch plywood are 16 gage staples or 3d ring shanked nails spaced 3 inches O.C. on panel edges and 6 inches O.C. in each direction at intermediate joists. The sheets of underlayment should be spaced 1/16 of an inch apart whenever they meet to avoid buckling. The joints between sheets are filled with a compound made specifically for that purpose. The surface of the sheets at joints and nails is sanded to make a level smooth surface.

Laying resilient flooring. The laying of resilient flooring is the job of specialized mechanics. Such mechanics are part of the carpentry trade in many sections of the country. Space here does not permit an adequate treatment of laying resilient flooring. It is covered elsewhere in printed materials devoted exclusively to it. Cutting carpet to fit in room areas requires skill in measuring accurately, applying the hookless strips which hold the carpeting around the perimeter, and cutting and stretching the carpet. Carpeting adhered to the floor with adhesive requires the same careful measuring, cutting, and seaming. Covering stairs with carpeting requires an especially high degree of skill.

Vinyl, linoleum, and vinyl asbestos tile is applied over a mastic. The tile must be measured carefully so that border tiles at opposite walls are equal in width. The tile is laid beginning with the center so that it is parallel to the walls. Vinyl base is applied to finish the job.

Wide sheets of vinyl and linoleum flooring are used extensively in kitchens and bathrooms. The material comes in 6, 9, or 12 foot widths. When the room dimensions allow it, the whole floor area can be laid with a seamless piece. This type of flooring requires the same skill in measuring and cutting as needed for applying carpet.

Wood Flooring

Materials. Wood for finished floor is selected for its appearance and its ability to withstand wear. The woods most commonly used for finish flooring include oak (both plain and quarter sawn), hard maple, beech, birch, pecan, fir, and yellow pine. Much finish flooring material is edge matched. The boards have a tongue and groove to produce a tight smooth floor. Some are also end matched.

Types of flooring. Three types of wood flooring are used in home building—strip flooring, plank flooring, and block flooring. Parquet flooring is a type of block flooring.

Strip flooring is available in sizes ranging from $1\frac{1}{2}$ to $3\frac{1}{4}$ inches in width and $\frac{5}{16}$, $\frac{3}{8}$, $\frac{1}{2}$ and $\frac{25}{32}$ inch thickness. The thicker material is made with tongues and grooves. The material is edge and end matched and comes in random lengths up to 16 feet. It is blind nailed. See Fig. 7-73. Square edge strip flooring is available $1\frac{1}{2}$ and 2 inches wide and generally $\frac{5}{16}$ inch thick. It is face nailed. Table 7-3, which gives the nailing

Fig. 7-73. Strip flooring is end and edge matched and is blind nailed. (E. L. Bruce Co.)

TABLE 7-3. NAIL SCHEDULE: OAK FLOORING.

Tongue and grooved flooring must always be blind-nailed, square-edge flooring face-nailed.

Size flooring	Type and size of nails	Spacing
(Tongued & Grooved) $\frac{25}{32} \times 3\frac{1}{4}$	7d or 8d screw type or cut steel nail*	10-12 in. apart
(Tongued & Grooved) $\frac{25}{32} \times 2\frac{1}{4}$	Same as above	Same as above
(Tongued & Grooved) $\frac{25}{32} \times 1\frac{1}{2}$	Same as above	Same as above
(Tongued & Grooved) $\frac{1}{2} \times 2$, $\frac{1}{2} \times 1\frac{1}{2}$	5d screw type or cut steel or wire nail	10 in. apart
Following flooring must be laid on wood sub-floor:		
(Tongued & Grooved) $\frac{3}{8} \times 2$, $\frac{3}{8} \times 1\frac{1}{2}$	4d bright casing nail— wire, cut or screw nail	8 in. apart
(Square-Edge) $\frac{5}{16} \times 2$, $\frac{5}{16} \times 1\frac{1}{2}$	1-in. 15 gauge fully barbed flooring brad, preferably cement coated	2 nails every 7 in.

*Machine-driven barbed fasteners of the size recommended by the manufacturer are acceptable.

schedule for oak strip flooring, shows the common sizes.

Plank flooring is made in wide widths up to 9 inches. It is used in homes with traditional architecture. See Fig. 7-74. The third type of wood flooring, *block flooring,* is shown in Fig. 7-75. Square blocks are generally laid in mastic. Some are made of oak strips or veneer with oak face plies. The blocks made of strips resemble parquet

Fig. 7-74. Plank flooring is used in homes with traditional architecture. Simulated wood pegs are spaced along the planks. The planks are blind nailed through a tongue and are fastened with screws at the ends of each plank. (E. L. Bruce Co.)

Fig. 7-76. Examples of parquet flooring. (E.L. Bruce Co.)

Fig. 7-75. Block flooring can be laid in mastic. The blocks have tongue and groove edges. (E. L. Bruce Co.)

floors used in 14th century Europe. See Fig. 7-76. They are laid with the squares in an alternating pattern or to form a herringbone design. Most hardwood flooring is available prefinished.

Installing strip flooring. The flooring should be delivered to the job 24 to 48 hours before it is to be installed. The carpenter should spread it out so that it can adjust to the moisture conditions of the building. Before any finish floor is laid, the rough floor must be thoroughly cleaned, and any protruding nails driven down.

Fifteen pound asphalt felt with a 4 inch lap is placed over the subfloor before the finish floor is laid. Flooring manufacturers generally agree that in ordinary applications, waterproof felt should be used between the rough floor and the finish floor. When the floor is laid above a crawl space, additional moisture protection should be provided by placing polyethylene film over the earth.

Before the actual work of laying the floor is begun, make a careful study of the floor plans to determine the kind of flooring which is to be laid in the different rooms and the relations of the rooms to each other. First, select the key room (usually a living room), which because of its size and importance determines the direction in which the boards of the finish floor should be laid. Next study the relation of the other rooms to the key room to see if the boards of the flooring in the key room are to extend into any other room. See Fig. 7-77. (It is assumed that strip flooring of the same type is used throughout this floor of the building.) Check the walls of the key room to see that the opposite walls are parallel to each other.

Establish a straight line, such as *L* in Fig. 7-77, by snapping a chalk line on the building paper or on the rough floor. This line should be parallel to the longest wall of the key room and should extend into the adjacent room so that the strips of flooring will be continuous.

When laying the first flooring board (shown at *1* in Fig. 7-77.) follow these procedures. Select a long straight piece of flooring, one that is the full length of the room if possible. Place this piece of flooring in position, with the grooved edge toward the wall. A space less than the thickness of the base, usually about 1/4 inch, should be allowed for expansion.

Face nail the board at *A* (Fig. 7-77) with a finish nail, but do not drive the nail home. (Size of nails used depends on the flooring thickness.) Using a tape, measure the distance *X* from the face of the first board to the chalk line *L*. Transfer this distance to *Y* and set a nail at *B*. The board *1* is then parallel to the chalk line *L* and also to the main wall in the key room.

Check the edge of the board *1* for straightness. This can be done with a straightedge, a line, or by sighting. Then face nail the board every 12 inches with finish nails, nailing as near to the wall as possible. Set all nails with a nail set.

Continue to cut, fit, and nail the flooring until the board marked *2* in Fig. 7-77, has been

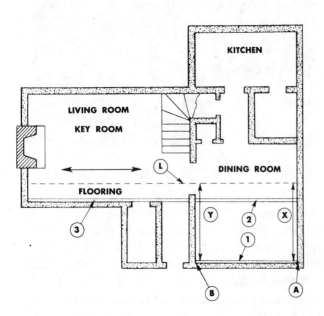

Fig. 7-77. Flooring is laid in relation to the base line (L).

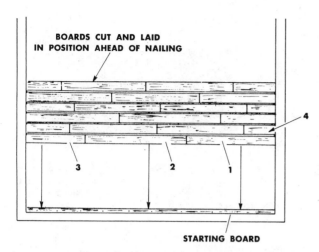

Fig. 7-78. Boards are cut and laid loose, then moved to the starting board.

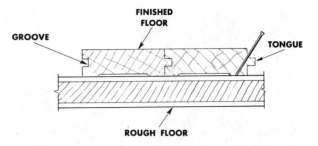

Fig. 7-79. Blind nailing drives the flooring up tight and hides the nails.

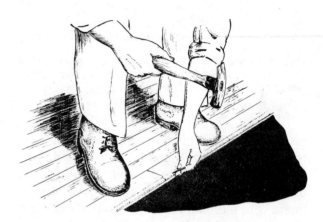

Fig. 7-80. The carpenter must use the correct stance when laying the finish flooring.

reached. At that point proceed as outlined below.

First, cut and fit a number of boards (about 6 or 8) and lay them in sequence on the floor ahead of the nailing, as shown in Fig. 7-78. Use different lengths, matching them so they reach from wall to wall. Never allow the joints in successive courses to come together in line. Begin with piece *1* in Fig. 7-78, then follow with pieces *2* and *3*. The part which is cut off from *3* (unless it is very short) should be used for piece 4 of the next course. In this way, there is little waste of flooring material. Move the loose boards to the starting board for installation.

When tongue and groove flooring is used, blind nail all boards after the first one. Nail three or four pieces, then place a short length of straight edge hard wood against the tongue edge of the outside piece and drive the unit up snugly. This operation is repeated after every three or four pieces are laid. Table 7-3 *(Nailing Schedule)* gives information on nails and spacing for oak flooring. Always drive the nails through the tongue of the board, at an angle of about 50 degrees to the floor. See Fig. 7-79.

When nailing tongue and groove flooring, stand with your feet on the board to be nailed, as shown in Fig. 7-80. The carpenter is standing on the board being nailed. If you follow this pro-

cedure, you can hold the board in position while it is being nailed. By standing you can hammer a straight blow to the nail with an easy motion. When the head of the nail reaches the board, you should raise the handle of the hammer. In this way, the face of the hammer strikes the tongue of the board instead of the edge when the nail is given the last blow. Tongue and groove finish flooring can also be nailed with a manual or pneumatic nailing machine. See Fig. 7-81.

Look back at Fig. 7-77. When the finish floor has been laid up to the line *2*, the starter board *3* in the key room should be laid. The front edge of this board should be the same distance from the chalk line *L* as the front edge of the board *2*. This ensures that the boards will come out correctly at the door opening, where the flooring passes from one room to the next.

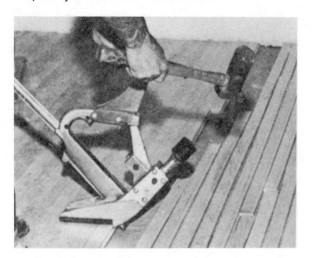

Fig. 7-81. Wood strip flooring is laid using a nailing machine. The nails are blind nailed at the correct angle. (Rockwell International Power Tool Div.)

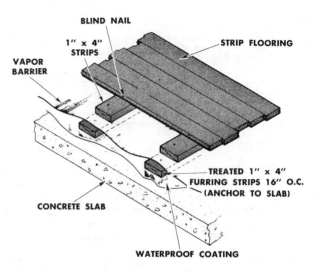

Fig. 7-82. Strip flooring is laid over concrete with furring strips and a vapor barrier. (U.S. Dept. of Agriculture, Forest Service)

Continue laying the finish floor until you have reached within 2 or 3 boards from the opposite wall of the room. Then cut and fit the last few boards to be laid. Open up the groove of each board with a rabbet plane so that the board can be slipped in place. Place the boards in position and draw them tightly together with a pinch bar, then face nail them in place. (Note: In using the pinch bar, be sure to protect the wall finish against marring by placing a strong piece of board between the bar and the wall.)

When there are no projections in the key room, and the finish flooring boards do not extend into any adjacent room, it becomes a simple matter to lay the finish floor. The starting board is nailed as close as possible to one of the two longest walls of the room. Flooring is generally laid parallel to the two longest walls.

Installing strip flooring over concrete. Ordinarily strip flooring is applied over concrete using furring. See Fig. 7-82. Spread waterproof mastic over the floor before fastening the furring in place. Lay pieces of 1 × 4 inch furring strip over the floor on 16 inch centers at right angles to the direction in which the flooring is applied. If the floor is not level, adjust the furring with shims. It is advisable to fasten the furring with powder activated fasteners or masonry nails. Place polyethylene film (4 mil thickness) over the

furring strips and lay a second set of 1 × 4 inch wood strips over the film and furring strips. Nail the two together. Install the strip flooring using the usual procedure.

Installing plank flooring. Modern plank flooring is intended to reproduce the plank flooring used in Colonial times, with wide planks and wood pegs. Originally this type of flooring was fastened with carefully fitted wood pegs instead of nails. Plank flooring today is supplied in widths from 3 to 8 inches, in one inch increments, and $5/16$ or $25/32$ inch thick. The edges and ends of the $25/32$ inch flooring have tongues and grooves. This type is installed with conventional blind nailing, plus screws at the ends of planks. The screws are countersunk $3/16$ inches and covered with a plug of contrasting wood, such as walnut. Other simulated plugs are spaced along the board. The edges of the board are slightly beveled. Allowance must be made for shrinkage because of the width of the boards. See Fig. 7-74 on p. 459.

A thinner material is available. It is $3/8$ inch thick and 3, 5, and 7 inches wide and made of plywood, with a face made to resemble plank flooring. It has tongues and grooves on the sides and ends and is designed to be face nailed into the subfloor. It is supplied in lengths from 12 inches to 5 feet.

Installing block flooring. Two types of block flooring are available. One type is a three ply plywood block, ³⁄₈ inch thick, which has a face veneer of oak resembling one continuous piece of wood. Sizes are 8¹⁄₄ × 8¹⁄₄ inches or 9 × 9 inches. Some types of blocks are made with

Fig. 7-83. Block flooring is provided in large squares held together by face paper or web backing. The squares are laid in mastic, and the paper is peeled off. (Harris Mfg. Co.)

Fig. 7-84. A sanding machine is used to provide a final level surface for block flooring. (Harris Mfg. Co.)

tongue and groove edges. The position of the tongue and grooves automatically positions each square so that the grain is at right angles to those of adjoining blocks. This is shown in Fig. 7-75 on p. 459. Another form of block flooring is made so that it resembles parquet flooring. It is made of several strips of wood glued together, then also held together by two splines which cross the back of the block. Some blocks are held together with a facing paper or web backing. The blocks vary in size, depending on the width of the strips which are used. Typical sizes are 7¹⁄₂, 9, and 11¹⁄₄ inches. The usual thicknesses are ⁵⁄₁₆, ¹⁄₂, and ²⁵⁄₃₂ inch. See Fig. 7-83.

The blocks are laid in mastic spread over underlayment or over a concrete floor. No nails are required. The floor is sanded to make it perfectly smooth and level. See Fig. 7-84.

INSTALLING CABINETS

Cabinets are generally made in a mill or cabinet shop by members of the carpenter trade who specialize in this field. The cabinets are delivered to the job then installed. Kitchen cabinets and vanities are the most common examples of this

Fig. 7-85. Kitchen cabinets are made to fit the architectural style of the house. Counter tops are plastic laminate. (H.J. Scheirich Co.)

Fig. 7-86. The counter tops are solid vinyl. The cabinet faces are vinyl laminated to wood backs. (Du Pont Company)

Fig. 7-87. Vanity cabinets are constructed in a mill and installed on the job. (Brammer Mfg. Co.)

type of work, but a variety of other built-in cabinets are also occasionally used. Such installation work requires skill so that the cabinets are installed plumb and level and fitted tightly against the wall, or other cabinet work. The carpenter must work carefully to avoid damaging these pieces. Many are made of very fine woods. Some are prefinished, others are made with plastic laminate faces. See Figs. 7-85, 7-86, 7-87.

Procedure for Installing Cabinets

(A base kitchen cabinet is used here as an example.) Set the base cabinet in place on the floor in the position where it is to be installed. Level the top in both directions, as shown at 1 and 2 in Fig. 7-88. If the floor is not level, use wedges as shown to bring the cabinet to a level position. After the top has been leveled, plumb the sides as shown at 3 and 4. If the cabinet has been carefully built so that all faces are square, then the sides will be plumb when the top is level.

The variation in level between the floor and the cabinet is usually very small so that trimming the bottom of the cabinet is unnecessary. Vinyl or rubber base or a wood mold when applied will correct and conceal any such irregularities.

Once the cabinet is shoved against the wall as close to its final position as possible and is lev-

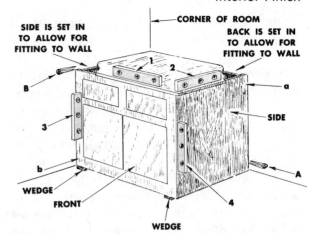

Fig. 7-88. Cabinets are scribed to the wall and trimmed so that they fit perfectly.

eled with wedges, set a pair of scribers (A in Fig. 7-88) to the widest opening between the back of the cabinet and the wall. Scribe a line *(a)* along the end of the cabinet. (Note: Cabinets should be designed with excess material along the edges of ·the outside faces by setting the back and the side of the cabinet from 1/2 to 3/4 inch back from the edges to allow trimming to fit tightly against the wall.)

Set scribers (B in Fig. 7-88) to the widest opening along the front edge and scribe line *b* along the front.

Move the cabinet so that you can use a saw or plane to cut to lines *a* and *b* scribed on the cabinet. When trimming, undercut slightly so that the face surface of the cabinet fits tight against the wall. Place the cabinet back in position, wedge the bottom again, and check all edges for a good fit. Check to see that the top of the base is level.

Fasten the cabinet into place as required by the design. In some cases provisions for fastening the cabinet to the floor and the wall are placed inside the cabinet. Fasten the sink top in place. (Note: In apartment and large housing developments, scribing and fitting cabinets as outlined above is omitted. A simple mold resembling a small base shoe, called a *scribe mold,* is used to cover the intersections of cabinet and wall.)

Other Built-in Units

Contemporary buildings include a number of built-in units other than kitchen cabinets and

Fig. 7-89. A built-in storage unit serves in place of a wall to divide the rooms. (Western Wood Products Assoc.)

bathroom vanities. These units are often made in a cabinet shop and transported to the job completely assembled and installed in the building. Examples of such built-in units include cabinets which serve as room dividers and built-in furniture, such as dressers and wardrobes in bedrooms, china cases in dining areas, and book cases. See Fig. 7-89.

STAIR LAYOUT AND CONSTRUCTION

The staircase, when carefully designed and built, adds dignity and charm to a home. The quality of craftsmanship displayed here affects the character of the entire interior of the building.

In general, stairwork is considered a special field of carpentry. The main stairway and rail may have several parts that are difficult to make on the job. It is usually made in a mill, then assembled in the building under construction. Stairs which are usually built by the carpenter on the job include the porch and other stairs on the outside of buildings and less important stairs within a building. The carpenter must have the necessary information regarding the general principles involved in stair building and knowledge of layout and construction.

Types of stairs. A staircase in a building provides the ordinary means of traveling from one floor to another. The ease with which a stairway can be used depends to a large extend on the proper proportioning of the riser and tread for each step and the number of steps in a single series, or flight. The design of the building and the space allowed for stairs usually limits the type of staircase which can be built.

Straight flight Stairs. A stairway, commonly known as a straight flight stair, is shown in Fig. 7-90A. It is the simplest, and generally the most desirable, stair to build. However, the layout of a building does not always permit the use of a straight flight stairway. A staircase with a long flight (one consisting of more than fifteen steps) is tiring because it doesn't provide a break in the climb (or ascent) for the user. For this reason, a landing should be introduced somewhere in a long flight, usually at the halfway point. See Fig. 7-90 B.

Landings also have the function of changing the direction of the stairs, as shown in Fig. 7-90 C. The staircase returning on itself, as shown in Fig. 7-90D, is economical in space, especially

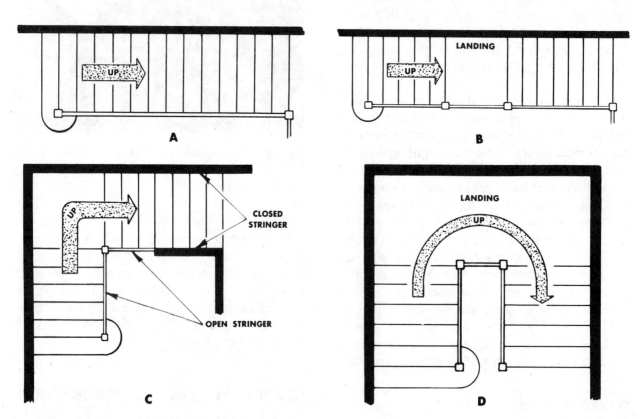

Fig. 7-90. Various types of stairs. Straight-flight stairs, shown in (A), can be made with several arrangements to fit the plan of the house. A straight-flight stair is broken by a landing (B) when the floor to floor height is greater than normal. The landing in (C) serves to change direction in a 90 degree change stair. The open newel stair (D) provides an 180 degree change in direction.

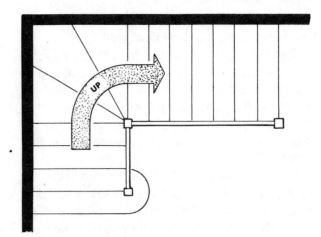

Fig. 7-91. A 90 degree change stair with winders.

Fig. 7-92. Two types of geometrical stairs are shown here. (A) shows the circular stair, and (B) shows the elliptical stair. Geometrical stairs provide an elegant touch to a building.

when there are a number of floor levels to be connected. In this particular stairway, the stairs continue to wind upward.

Winder Stairs. Space limitations frequently demand a staircase which uses winders to change direction. The three-winder stairway illustrated in Fig. 7-91 is frequently used. It is not considered dangerous as long as the treads are approximately the same width on the line of travel, and the stairs meet certain building code limitations. These are discussed later in this chapter under *Winding Stairs.*

Geometrical Stairs. The most complicated and expensive stairways are those that are curved. These stairs are commonly known as a *geometrical stairway.* See Fig. 7-92. The geometrical stairway is a winding stairway. For the sake of

safety, it is designed so that the tread width at the line of travel for all steps is the same. A geometrical staircase can be circular (as shown in Fig. 7-92A) or elliptical (as in Fig. 7-92B). They are often designed with landings so that they can be climbed with ease.

Safety Precautions in Stairway Design

Statistics compiled by the National Safety Council show that stairways are the largest single cause of accidents in the home. These accidents can be attributed to various factors. Some of course are beyond the control of those who build stairs. However, there are many accidents which are directly related to faulty stair design. As a carpenter, you can make a meaningful contribution toward accident prevention by planning and doing your stair work well.

Two factors in stair design are primarily important. All risers should be the same height, and all treads should be the same width in any one flight of stairs. The relationship between the height of the riser and the width of the tread must follow the ratio limitations set up in the local building code.

Two example sets of restrictions on stairs are given in the following pages. They are reproduced to show the basis for the provisions in local codes. Much of the information is self explanatory. Specific items are studied later in this chapter.

The following requirements are taken from the *Minimum Property Standards* of the U.S. Department of Housing and Urban Development (HUD).[1]

402-6   STAIRWAYS

402-6.1 Stairways and landings shall provide for safe ascent and descent under normal and emergency conditions and for the transport of furniture and equipment.

402-6.2 Stairways shall be designed in accordance with the criteria of the table.

Requirements for stairs as found in *Dwelling Construction under the Uniform Building Code*[2] appear below the table under *STAIRS*.

TABLE FOR STAIRWAY DESIGN

	Private Exterior Stairs (attached to dwelling)		Private and Common (1) Interior Stairs	
	Entrance	Basement	Main	Basement
Min. clear headroom	-	-	6'-8"	6'-4"
Minimum width (2)	2'-8"	2'-8"	2'-8"	2'-8"
Minimum run (3)	11"	11"	9"	9"
Minimum nosing	-	-	(4)	(4)
Maximum riser (3) (6)	7½"	7½"	8¼"	8¼"
Winders (5) (6)	Run at point 18" from converging end shall not be less than straight portion			

Notes for Table:

(1) Stairway serving two living units
(2) Clear of handrail
(3) All treads shall be the same width and all risers the same height in a flight of stairs.
(4) Closed riser, 1-1/8" nosing; open riser, 1/2" nosing.
(5) Winders are not permitted in common stairs.
(6) Winders or open risers are not permitted in housing for the elderly.

STAIRS

Rise and Run

The rise of steps in a stairway shall not exceed 7½ inches and the run shall be not less than 10 inches. Variations in the height of risers and the

[1]Publication 4900.1

[2]*Dwelling Construction under the Uniform Code.* International Conference of Building Officials, 5360 South Worman Mill Road, Whittier, Calif. 90601.

width of treads in any one flight shall not exceed ³/₁₆ inch.

EXCEPTION:

In private stairways serving an occupant load of less than 10, the rise may be 8 inches and the run may be 9 inches.

Winders

In residences, winders may be used if the required width of run is provided at a point not more than 12 inches from the side of the stairway where the treads are the narrower, but in no case shall any width of run be less than 6 inches at any point.

Handrails

Stairways shall have at least one handrail. Handrails shall not be required for private stairways having less than four risers.

Handrails shall be placed not less than 30 inches nor more than 34 inches above the nosing of treads. They shall be continuous the full length of the stairs, and except for private stairways at least one handrail shall extend not less than 6 inches beyond the top and bottom risers and ends shall be returned or shall terminate in newel posts or safety terminals.

EXCEPTION:

Stairways 44 inches or less in width and stairways serving one individual dwelling unit in Group 1 Occupanices may have one handrail, except that such stairways open on one or both sides shall have handrails provided on the open side or sides.

Handrails projecting from a wall shall have a space of not less than 1¹/₂ inches between the wall and the handrail.

Headroom

Every required stairway shall have headroom clearance of not less than 6 feet, 6 inches, measured vertically from the tread nosings to the nearest soffit above.

Tread and Riser Relationship

Stairs must be adapted to meet many special requirements to fit into a particular building. Rules have been established to make stairs as comfortable to use as possible. Unfortunately, rules must be occasionally overlooked in order to resolve a problem. This is particularly true in remodeling work. However, it is also true when building a new house that has not been well planned. Carpenters should know how to make choices that provide the best stairs under the circumstances. They should be familiar with the local building code and plan and build stairs accordingly.

Stair ratio is a relationship between the tread run (width of stair) and the riser height so that as one of these increases, the other decreases, and vice versa. A minimum tread run and a maximum riser height keep the stairs from exceeding the critical angle, or slope of the whole stair. See Fig. 7-93. The economical use of materials is also an important consideration. Good design often requires wider boards be used for treads. Such boards are wider than those the carpenter would normally use, if economy were the main consideration.

A formula used by many building authorities for building codes is expressed: the width of a tread plus the height of a riser shall equal not less than 17 inches, nor more than 18 inches. (This calculation does not include the nosing on the tread.) Therefore, if the riser height is 7¹/₂ inches, the tread width would be determined as follows:

$$T + R = 17 \text{ to } 18$$
$$T + 7\tfrac{1}{2} = 17 \text{ to } 18$$
$$T = 9\tfrac{1}{2} \text{ to } 10\tfrac{1}{2}$$

Table 7-4 is based on this formula and gives the minimum and maximum unit run (tread width) for the unit rises of different heights. (Note: These dimensions are the unit tread width and riser height used in laying out the stringer. The nosing is not included in the unit tread width.)

Another common formula found in building codes is stated: the width of a tread and the sum of two risers should not be less than 24 inches, nor more than 27 inches.

Thus if the riser were 7¹/₂ inches, the tread width would be determined as follows:

$$T + R + R = 24 \text{ to } 27$$
$$T + 7\tfrac{1}{2} + 7\tfrac{1}{2} = 24 \text{ to } 27$$
$$T = 9 \text{ to } 12$$

Treads. Material for treads is generally 2 × 10 or 2 × 12 inches (actual size 1¹/₂ × 9¹/₂ or 1¹/₂ by

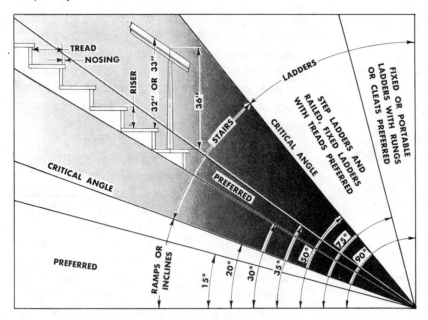

Fig. 7-93. Preferred and critical angles for stairs, ladders, ramps, and inclines. The dimensions for the height of the rail are shown. Ladders should be placed at a 4 to 1 ratio between vertical rise and horizontal run. When interpreting angles into units of rise, 15° is approximately 3 inches of rise per foot, 20° is approximately 4 inches of rise per foot, 30° is approximately 7 inches, 35° is approximately 8 inches, and 50° is approximately 14 inches of rise per foot.

TABLE 7-4. UNIT RISE PLUS UNIT TREAD EQUALS 17 TO 18.

Unit rise	Min. unit tread	Max. unit tread	Unit rise	Min. unit tread	Max. unit tread
6⅝″	10⅜″	11⅜″	7⅜″	9⅝″	10⅝″
6¹¹/₁₆″	10⁵/₁₆″	11⁵/₁₆″	7⁷/₁₆″	9⁹/₁₆″	10⁹/₁₆″
6¾″	10¼″	11¼″	7½″	9½″	10½″
6¹³/₁₆″	10³/₁₆″	11³/₁₆″	7⁹/₁₆″	9⁷/₁₆″	10⁷/₁₆″
6⅞″	10⅛″	11⅛″	7⅝″	9⅜″	10⅜″
6¹⁵/₁₆″	10¹/₁₆″	11¹/₁₆″	7¹¹/₁₆″	9⁵/₁₆″	10⁵/₁₆″
7″	10″	11″	7¾″	9¼″	10¼″
7¹/₁₆″	9¹⁵/₁₆″	10¹⁵/₁₆″	7¹³/₁₆″	9³/₁₆″	10³/₁₆″
7⅛″	9⅞″	10⅞″	7⅞″	9⅛″	10⅛″
7³/₁₆″	9¹³/₁₆″	10¹³/₁₆″	7¹⁵/₁₆″	9¹/₁₆″	10¹/₁₆″
7¼″	9¾″	10¾″	8″	9″	10″
7⁵/₁₆″	9¹¹/₁₆″	10¹¹/₁₆″			

11½ inches). See Fig. 7-94. The run of the unit tread is the distance from the face of one riser to the face of the one which follows it. It is the same dimension as the cut on the stringer. When a 9 inch unit tread width is required, only ½ inch is left for nosing. When a larger unit tread width is required (using the stair ratio formula), a board wider than 2 × 10 inches is used.

The tread is often cut so that it extends the same distance beyond the stringers at each side as the nosing extends in front.

Risers. Risers are usually made of 1 × 8 or 1 × 10 inch boards and ripped to fit. The riser is placed behind the lower tread and snugly fitted against the under side of the upper tread. See Fig. 7-94.

It is very important that all riser heights on a flight of stairs are equal to prevent tripping or misstepping. The riser height should also be limited so that it is reasonably similar to adjacent flights above and below it. The carpenter must adjust the width of the board (not however the

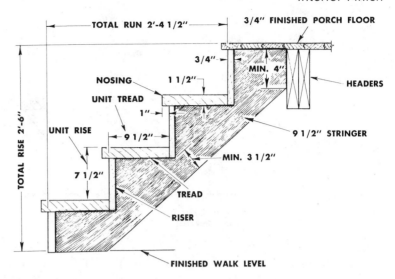

Fig. 7-94. The stringer must be laid out so that the unit rise and the unit tread are the same for each step.

unit rise) for both the top and bottom riser in a flight of stairs to make up for the thickness of the flooring. See Fig. 7-94. This is discussed under *Layout of Stringers* below.

The Stringer. The stringer is the most important stair part. A stringer is the cutout support for treads and risers. If the carpenter has made the layout and the deductions correctly, the stairs will be perfect when installed. The material used for stringers is usually a 2 × 10 or a 2 × 12 inch plank so that a minimum of 3½ inches are left to carry the load after the cuts have been made. See Fig. 7-94. Deductions must be made at the top and bottom of the stringer so that the bottom rise and the top rise of the finished stair are both equal. The thickness of the tread material is deducted at the bottom and is added at the top, unless the flooring and tread thickness are not the same. Further additions and adjustments will be required, depending on the specific application. Adequate bearing (4 inch minimum according to some building codes) against the header must be provided so that the stringer is properly supported.

Laying out a stringer. A simple back porch stair, shown in Fig. 7-94, has a total rise of 2'-6". It is chosen to demonstrate how to lay out a stringer. 2'-6" equals 30 inches and is divided into 4 risers of 7½ inches each. If you use the ratio T + R = 17 to 18, the unit tread width is between 9½ and 10½ inches. Here 9½ inches is

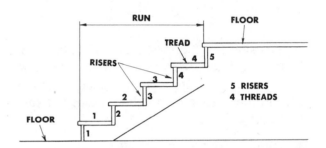

Fig. 7-95. Four treads are needed to go up 5 risers.

used. There are three treads and four risers. This makes the total run equal 28½ inches, or 2'-4½". (Figure 7-95 clarifies why there is one less tread than riser.)

Lay a plank on two saw horses. Lay out three steps, using a framing square with 7½ inches on the tongue and 9½ inches on the blade. See Fig. 7-96. With the same setting on the framing square, draw lines *A* and *B*.

Measure ¾ inch from line *A* and draw a line representing the underside of the porch floor. Measure 7½ inches along line *B* and draw a line which represents the bottom line of the stringer. Measure back 1½ inches from this line, to compensate for the thickness of the treads, and draw the cut line. When the stringer is cut out, the whole stringer drops 1½ inches so that all the riser heights will be equal when the treads are in place.

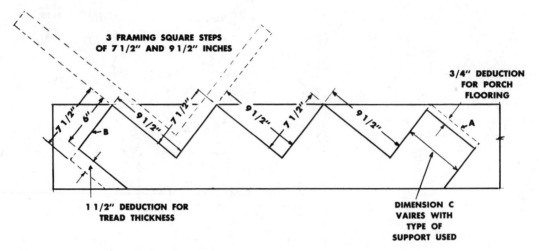

3 FRAMING SQUARE STEPS OF 7 1/2" AND 9 1/2" INCHES

3/4" DEDUCTION FOR PORCH FLOORING

7 1/2" 6" B 9 1/2" 7 1/2" 9 1/2" 7 1/2" 9 1/2" A

1 1/2" DEDUCTION FOR TREAD THICKNESS

DIMENSION C VAIRES WITH TYPE OF SUPPORT USED

Fig. 7-96. The stringer is laid out with a framing square, using the unit riser height and the unit tread width. Deductions are made at the top and bottom to compensate for the thickness of the tread and the flooring.

Dimension *C* in Fig. 7-96 varies with the method used to fasten the stringer at the top. In some cases the headers are brought closer to the risers. In other cases the stringer passes beneath the header.

Designing a Straight Flight Stair

To create a better understanding of how to lay out a stair stringer, we are going to design a stair which must have a total rise of 8 feet, 4 inches. The stairwell already established is 11 feet, 3 inches, as shown in Fig. 7-97.

Before any stringer can be laid out, you must study the plans (or stair location if the building is in progress) to determine the type of stair required. The limitations or restrictions in the area must also be taken into consideration. Typically these might be a beam which cuts down headroom, a door opening at the bottom of the stair, or windows along the stair flight. Frequently, such restrictions determine the place where the stairs start at the bottom and may require that the total run of the stair be shortened, thus changing the standard proportions between the riser and the tread.

When the principles for laying out a simple stair (with no restrictions) are thoroughly understood, then problems, including variations, can be readily solved satisfactorily. The straight flight stair shown in Fig. 7-97 is typical of the type you will run into.

Two methods can be used to find the exact riser height. One method uses a story pole. It is accurate but time consuming. The other method involves some simple mathematics.

Procedure for using the story pole to find the unit rise. Take a story pole (any straight 1 × 2 or 2 × 4 inch piece of material with square ends) and set it on the finished floor. Mark the location of the top of the floor above on the story pole, as shown at *1* in Fig. 7-97. The distance *1 — X* is the total rise of the stair. (In this case the total rise is 8 feet 4 inches.) Then place the story pole on two horses. (Note: If the finish floor has not been laid by the time the measurement is made, put a block of wood equal in thickness to the finish floor in place.)

Set a pair of dividers to 7 inches (an acceptable unit rise per step), and step off the total rise on the story pole, dividing the distance *1 — X* into equal parts. If, after the first trial you find there is a remainder, adjust the dividers and try again. If the remainder is less than 3½ inches, adjust the dividers to a setting larger than 7 inches. If the remainder is more than 3½ inches, adjust the dividers to a setting smaller than 7.

Continue adjusting the dividers and stepping off the distance on the story pole until the last

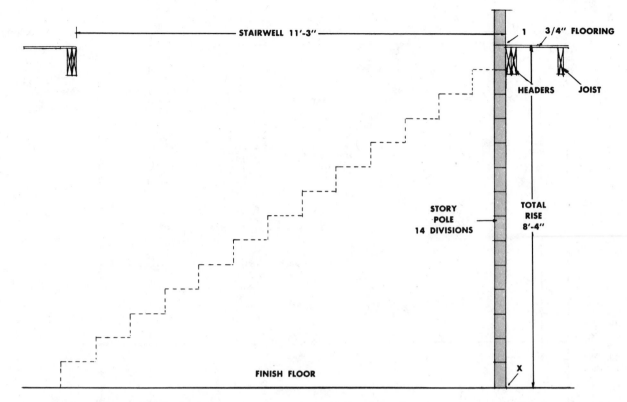

Fig. 7-97. A story pole is used to determine the exact unit rise. The pole is placed in the stairwell in a plumb position, and the two finished floor lines are marked. Then the story pole is removed, and the space is divided into equal parts.

unit is the same as all the others. The dividers are now set to the unit rise. Be careful not to disturb the divider setting.

Using mathematics to find the unit rise. The unit rise per step can also be obtained by dividing the total rise in inches by 7 to find the number of risers. Divide the total rise by the number thusly found to obtain the exact unit rise per step.

Total Rise = 8'-4" or 100 inches

100 divided by 7 = 14²/₇ risers.

(We must choose either 14 or 15 risers.) 100 divided by 14 = 7.143 inches or 7⅛ inches.

Finding the unit tread width. The stair ratio is used to find the unit tread. There is a little leeway permitted the carpenter since the tread width chosen can fall between the limits of the ratio T + R = 17 to 18. The riser height has been determined as 7⅛ inches.

Minimum tread ratio: T + 7⅛ = 17, T = 9⅞ inches.

Maximum tread ratio: T + 7⅛ = 18, T = 10⅞ inches.

A midpoint between the two would be 10⅜ inches.

Referring to the Building Codes, shown on pp. 468 and 469, you will note that the unit rise and unit tread above both fall within the specified limits.

Finding the total run. To find the total run of the stair, multiply the width of the tread by one less than the number of risers. In the stair shown in Fig. 7-98, the width of the tread was determined as 10⅜ inches. The total number of risers was 14. Subtracting 1 from 14 leaves 13, the number of treads. 13 times 10⅜ gives 134⅞ inches, or 11 feet 2⅞ inches, the total run of the stairs. (Approximately 11 feet, 3 inches.)

To find the starting point of the stairs, locate point *X,* shown in Fig. 7-98, on the basement

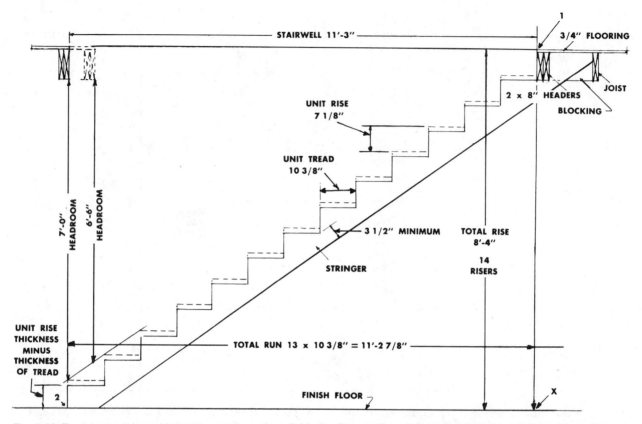

Fig. 7-98. To solve a stair problem using mathematics, divide the floor to floor height (here I-X) to determine the unit tread. Use the ratio (T + R = 17 to 18) to find the unit tread. Then multiply the unit tread by the number of treads to find the total run.

floor by plumbing down from the point *1* in the stairwell. Then lay out the total run of 11 feet, 3 inches ($2^7/_8$ inches) of the stair to locate the starting point. See point *2* in Fig. 7-98.

Finding the headroom. In this particular stair, the length of the stairwell is equal to the total run of the stair. The actual headroom is the finish floor to floor height, minus one unit rise and the thickness of the joist and the thickness of the flooring:

$$
\begin{aligned}
\text{Floor} &= && ^3/_4 \\
\text{Joist} &= && 7^1/_2 \\
\underline{\text{Riser}} &= && \underline{7^1/_8} \\
& && 15^3/_8'' = 1'\text{-}3^3/_8''
\end{aligned}
$$

Floor to Floor Height = 8'-4"

$$8'\text{-}4'' - 1'\text{-}3^3/_8'' = 7'\text{-}0^5/_8''$$

A headroom dimension of 7'-0" is very satisfactory for a residence.

Procedure for laying out a stair stringer. In laying out the stair stringer illustrated in Fig. 7-

99, the method below can be used. (This is the stringer for the stair shown above in Fig. 7-98.)

Select a straight piece of 2 × 12 inch stock of sufficient length and lay it on a pair of sawhorses. The required length is found by taking the unit rise per step ($7^1/_8$ inches) on the tongue of the framing square, and the unit run per step or tread ($10^3/_8$ inches) on the blade. Lay the square at these measurements on the edge of any plank and draw lines *AC* and *CB*, shown in Fig. 7-99. The distance *AB* is the bridge measure per step. Multiply this bridge measure ($12^3/_4$ inches) by the number of risers (14). The result is 14 feet, 10 inches, the approximate length required for this stair stringer. The piece of material should be longer to provide for the layout of the top and the bottom ends.

Begin at the bottom of the stringer. Lay the square in the position shown in Fig. 7-99. (Use framing square clips if available.) Take the unit

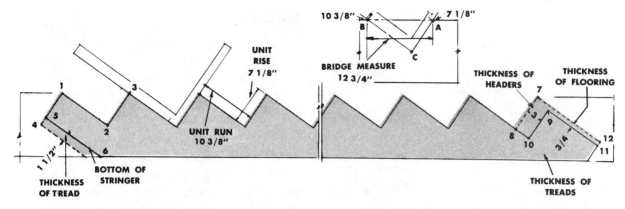

Fig. 7-99. The stringer for the stair shown in Fig. 7-98 is laid out using a framing square set at the unit run and unit rise dimensions. The deductions are marked before the cuts are made so that all the risers have equal height and the stringer itself is well supported at the top end.

rise (7⅛ inches) on the tongue and the unit run or tread (10⅜ inches) on the blade. Draw lines *1-2* and *2-3*. Reverse the square and draw line *1-4* at a right angle to *1-2*. Make the length of line *1-4* equal to the unit rise of the step (7⅛ inches). Shorten the rise of the first step from the bottom an amount equal to the thickness of the tread to be used. In this case it is 1½″. Draw the cutting line *5-6* parallel to line *1-2*.

Continue to lay out the balance of the steps from point *3*, along the edge of the 2 × 12 inch stringer. Great accuracy is needed to lay out the steps. Use a sharpened pencil or knife, and make the lines meet at the edge of the 2 × 12 inch plank.

When point *7* at the top of the stringer has been reached, draw line *9-10* three inches away to allow for the header thickness. Draw line *9-11* three-quarters of an inch from line *7-12* to allow for the floor thickness. Cut the stringer on all the lines marked with solid lines in Fig. 7-99. The stringer is nailed to solid bridging between the joists.

Stairwells and Headroom

The length of the stairwell is usually determined by the architect and shown on the working drawings for the building. If it is not specifically shown in this fashion, the length can be determined from the arrangement of partitions.

The location for the foot of the stairs is also fixed to provide the proper space between the lowest riser and adjacent walls or partitions. The carpenter must work within the limits shown on the working drawings.

In Fig. 7-98, the stairwell and the total run of the stairs works out to the same dimension. It is not always possible to arrange stairs so that the bottom riser is directly below the header. In the case of Fig. 7-98, a 6′-6″ headroom would permit the stairwell to be about 8 inches smaller. Note the hidden lines. Some partition arrangements on the floor above require the use of every bit of space. This results in cutting the stairwell to a minimum.

Headroom should be measured vertically from the front edge of the nosing to a line parallel with the line of nosings. The dimension, according to HUD, should be 6′-8″ minimum for main stairs and 6′-4″ minimum for basement or service stairs. A headroom of 7′-0″ however is preferred. When a soffit develops, as shown on Fig. 7-100, you must provide adequate headroom to dispel the illusion of being boxed in by the ceiling above.

If you run into difficulty making the stairs work out, consult the owner and the architect. If you are on your own doing remodeling work, you might be able to change the header at the end of the stairwell or the unit tread, unit rise, or num-

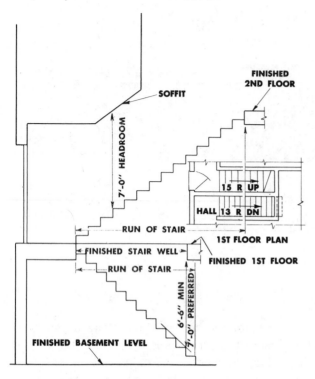

Fig. 7-100. Head room problems can occur in two story dwellings.

ber of risers to achieve the shorter total run. The shorter total run of the stringer increases the headroom between the stair and the soffit.

The width of staircases is determined by the need for two people to be able to comfortably pass on the stairs and the fact that furniture will have to be carried up and down. To allow room for two people to pass, the width of the stair should be 3 or 3½ feet. The minimum set by some codes (such as HUD) for main stairs clear of handrails is 2 feet, 8 inches. The width of stairs needed to allow furniture to be moved up and down depends on the shape of the stairs and the kind of furniture which may have to be moved. The straight flight stairway permits objects to be moved more easily than the winder and platform types do. When winding or platform stairways are open on one side, including open well stairways, they afford a better chance for moving large pieces of furniture. Such objects, unless they are extremely heavy, can usually be lifted over the handrails and the newel posts.

Handrails

Handrails should be provided on one or both sides of a stairs. The height of handrails should be 32 or 33 inches from the edge of the nosing to the top of the rail, or 36 inches from the center of the tread.

Winder Stairs

Winders are not considered as safe as straight stairs which have a platform. Winders are considered less safe because the tread width varies considerably, from almost nothing at the newel to a very wide space at the far end of the tread. To overcome this problem as much as possible, a line of travel is established, with a radius which approximates the line where a person might walk. After the line of travel has been drawn, the risers are placed so that they are spaced equally within the turn and also approximately equal in width to the spacing of the risers in the straight flights above and below. See Fig. 7-101A. (The dimension from *A* to *B* is approximately equal to the line length *B-C*.)

Winding stairs are considered necessary or advantageous in certain situations because they perform various functions. They change the direction of travel of a stair. In some cases they save room because at the same time the direction is changed, a rise is also achieved. Winding stairs are used to provide an interesting architectural effect, particularly when used for finished stairs.

Procedure for building a three winder stair. The carpenter has the problems of cutting out the parts, assembling them, and fastening them to the posts or supports. Stringers can either be nailed to the side of the posts or butt against them. When risers converge at posts, they should be arranged so that good nailing is provided. Blocks can be used to back up or support the risers for nailing on carpenter-made stairs. The newel post (the post on which the narrow end of treads converge) is routed out to receive the risers on mill-made stairs. Outside stringers usually must be made in more than one piece, joining the point where the straight stair ends and the winders begin.

When you build a winder stair, you should first

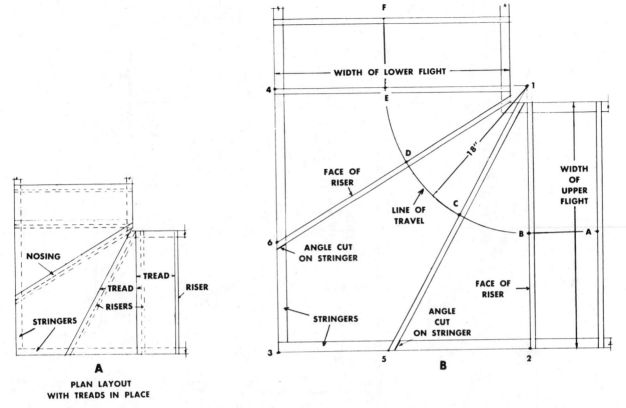

A
PLAN LAYOUT
WITH TREADS IN PLACE

Fig. 7-101. The full size plan layout for winders is used to locate the cuts on stringers. The faces of the risers are equally spaced on the line of travel. Layout with treads in place is shown in (A). The full size plan layout is shown in (B).

draw a full size plan layout on the floor. Show the size and shape of the treads, length of risers, and all angle cuts of both treads and risers. The stringers can also be easily laid out from the plan layout. See Fig. 7-101A. The method below can be used when building winders.

Draw a square (1-2-3-4 in Fig. 7-101B) which is equal in width to that of the stair. Using an 18 inch line of travel, swing an arc with center at *1*. Divide this arc into three equal parts (*BC, CD,* and *DE*) through these points. Draw the riser lines *1-5* and *1-6* through these points. These lines represent the face of the risers. Draw the risers in place.

The width of the treads at their narrowest point is obtained by drawing the full size newel post in position. (The Uniform Building Code requires a minimum tread width at any point of 6 inches. Other codes permit the design shown in Fig. 7-101 A and B.)

Layout of stringer for winders. The winder stringer is in two parts (*7-8* and *9-10* in Fig. 7-102). The layout of each part is different, and both stringers have a different angle of rise from that of the main stair.

The layout of the stringer is made along the edge of the board with the framing square. Information regarding the cuts is obtained from the full size layout shown in Fig. 7-101B. If space permits, the stringer can be laid out, as shown in Fig. 7-102, by projection from the plan layout. The length and end cuts of the risers can also be taken from the plan layout. The rise per step is the same as that of the main stair.

The shape of the treads is obtained by laying out lines representing the nosing on the full size plan layout, 1 to 1½ inches in front of the riser face. The exact size, angles, and cutout at the newel are shown on the plan. Riser lengths and angle cuts are also obtained from it.

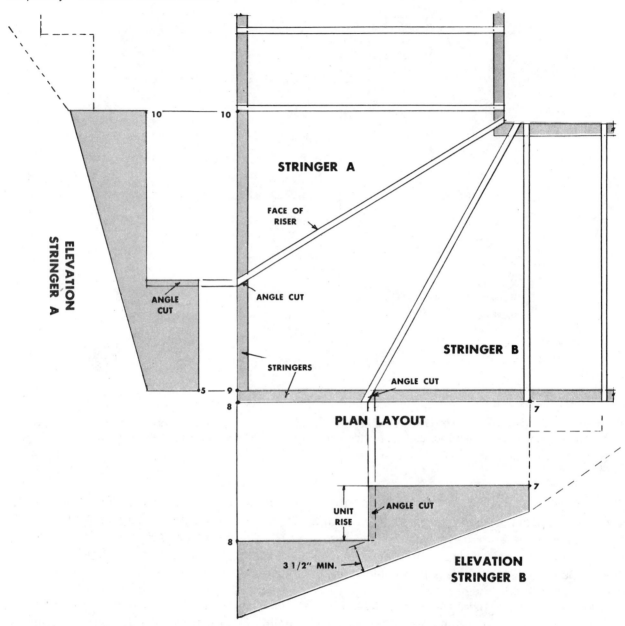

Fig. 7-102. The cuts on the stringers of winding stairs are determined from a full size layout plan. See elevation views. The unit rise for the winders should be the same as the unit rise for the straight flights above and below them.

Sometimes it is possible to rearrange the location of the two straight risers immediately above and below the winders and change the center of the radius to provide a better arrangement of treads. See Fig. 7-103. This particular stair complies with the Uniform Building Code.

Exterior porch and stair construction. Exterior stairs in residential work usually consist of straight flight stairs with a few risers. However, stairs and porches required for two story city dwellings present a challenge for carpenters. They must be constructed to provide maximum safety and stand for many years without deteriorating and rotting out. Figures 7-104 and 7-105 show a stair with many features, including a platform. The first step is to make a plan layout,

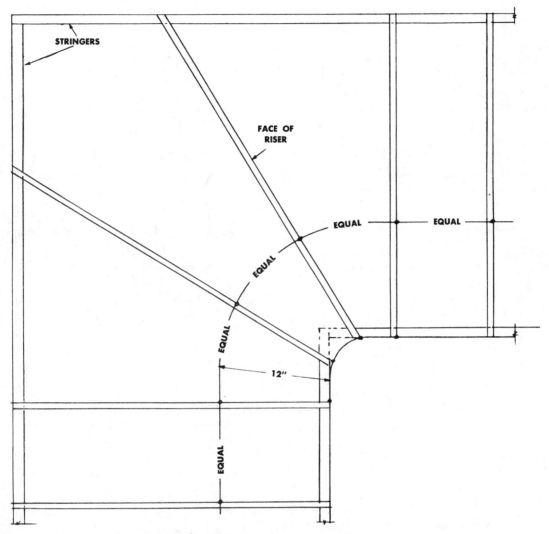

Fig. 7-103. Risers of winder stairs are arranged so that the treads have a safe width at their narrow ends. One building code requires that risers be spaced 6 inches from face to face along the arc at the narrow end of the tread.

either full size (if sufficient space is available) or to scale. The layout locates the posts, headers, and lookouts. The plan layout will show the location of each of the risers and the shape of each of the treads. See Fig. 7-104A. Stringers can be developed as shown in Fig. 7-104B. The carpenter erects the supporting structure for the posts, headers, and lookouts first. The posts rest on a firm bearing, such as a concrete pier, and are notched out where the horizontal members fit in place. These junctions are bedded in white lead or given other preservative treatment. The center newel post is then erected. It extends to a firm

base also on the ground. The porch framing and platform framing is then completed, followed by the installation of the stringers. Fastening the treads and flooring in place and erecting railings complete the job.

Finished Interior Stairs

Making finished interior stairs and rails is generally the job of a specialized carpenter called a *stairbuilder.* However, carpenters should have some knowledge about this phase of building a house since they must prepare the stairwells, supports, and walls for the stairs. Under some

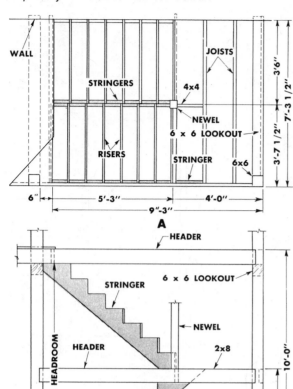

Fig. 7-104. A back porch stair for a two story house, shown in (A), requires a plan layout. Stringers are arranged to be supported by headers and posts, as shown in (B).

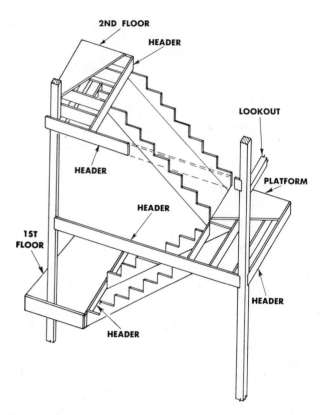

Fig. 7-105. A two story porch. The headers and lookouts are notched into the posts. The joists are hung from the headers.

conditions, they may even be directly involved in assembling a stair. The parts of a stair and their names are provided in Fig. 7-106.

Careful measurements must be made on the job by the stairbuilder. The stair parts are made using accurate woodworking machines and hand tools. When the stair parts have been prepared, the stairbuilder arrives on the job and installs the stairs, making minor corrections along the way, so that the stairs fit perfectly.

Two distinct types of stair are used. One type, the *open* or *mitered stringer stair,* is used when the side of the stair is exposed to view. See Fig.

7-107. The other type, the *closed stringer* or *housed stair* shown in Fig. 7-108, is generally used along a wall.

In first class stairwork, nails are used sparingly. All joints are housed or concealed in some way. Closed stringers are routed out to receive the ends of treads and risers. Treads are rabbeted to receive the top edge of the risers. Risers are rabbeted to receive the end of the treads. Wedges are glued and driven in place to make the stairs solid. Blocks are glued to the underside of the intersection of treads and risers to keep the joint from opening up. See Fig. 7-108.

The plan and elevation views of a stair made up of stock parts which are put together on the job by a carpenter is shown in Fig. 7-109. It is an open stringer stair on the room side. The balustrade is made up of straight and curved parts and turned balusters machined to precision.

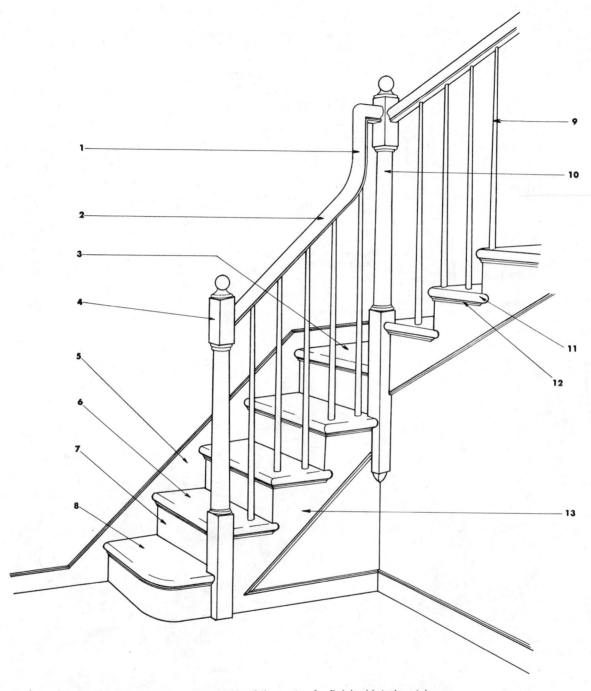

Fig. 7-106. The carpenter should know the names of the parts of a finished interior staircase.

1. Goose neck
2. Handrail
3. Landing
4. Starting newel post
5. Closed stringer
6. Tread
7. Riser

8. Starting step
9. Baluster
10. Landing newel post
11. End nosing
12. Cove molding under nosing
13. Open stringer

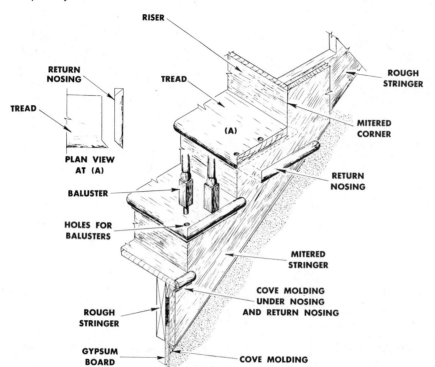

RISER

RETURN NOSING

TREAD

TREAD

PLAN VIEW AT (A)

(A)

ROUGH STRINGER

MITERED CORNER

RETURN NOSING

BALUSTER

HOLES FOR BALUSTERS

MITERED STRINGER

COVE MOLDING UNDER NOSING AND RETURN NOSING

ROUGH STRINGER

GYPSUM BOARD

COVE MOLDING

Fig. 7-107. A finished interior stair is made with a mitered stringer when it is exposed to view.

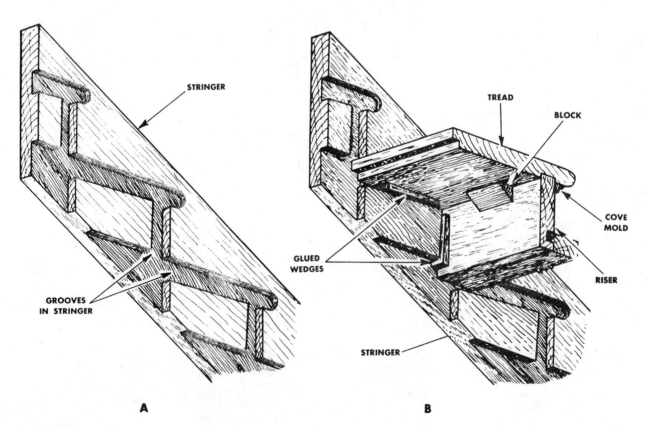

STRINGER

GROOVES IN STRINGER

A

TREAD

BLOCK

COVE MOLD

GLUED WEDGES

RISER

STRINGER

B

Fig. 7-108. The stringer is housed to receive the treads, risers, and wedges. Glued blocks reinforce the center of the treads. The views are from below.

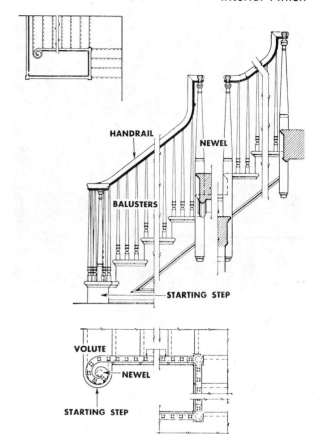

Fig. 7-109. A beautiful stair is made from parts manufactured with precision. It is assembled by the carpenter on the job.

Fig. 7-110. Geometric stairs are made in a stair building shop. The stringers are made of several plies of wood bent into shape and glued together. General carpenters must provide the stairwell and support structure. (Taney Supply and Lumber Co.)

Fig. 7-111. Geometric stairs are made for contemporary buildings. They are built so that they comply with the safety provisions of the various building codes. (Taney Supply and Lumber Co.)

Some of the artistry which goes into designing and building custom circular or elliptical stairs is shown in Fig. 7-110. Stringers are made of single plies of thin wood bent into shape around a huge drum of the proper diameter. Other plies are bent and glued to the first ply. The treads and risers are cut into the stringer as required, either by mitering or housing. See Figs. 7-107 and 7-108. The hand rails require a high degree of skill combining machine and hand work. Figures 7-110 and 7-111 show examples of geometric stairs.

SAFETY PRECAUTIONS

Hazards are present in every area of the construction field, even in the relatively safe environment of the interior of a building nearing completion. The usual potential for physical strains is present wherever workers must handle awkward and heavy objects. Ungainly large panels must be applied to walls and ceilings. Doors must be handled several times while the hinges and locks are installed and they are hung in place. Heavy cabinets and built-in furniture require maneuvering and fitting. The fatigue and unusual position required for laying wood flooring can result in back strain. Makeshift stepladders and platforms should be avoided. Working with adhesives and their solvents often require that the skin be protected.

Perhaps the most important potential for accidents lies with power tools. Power screw drivers, pneumatic staplers, and powder activated drivers are used to fasten material. The sabre saw, portable jig saw, and power miter box are used in trim work. The router and power plane are used in fitting and hanging doors. All of these can cause serious harm or injury if they are not operated properly and safely. Among hand tools, the chisel presents the greatest danger. Specific instructions for the safe handling of all tools are provided in *Fundamentals of Carpentry: Tools, Materials, and Practices.*

QUESTIONS FOR STUDY AND DISCUSSION

1. What is the sequence for installing interior finish in a building?
2. What is drywall finish?
3. In what situations are the following gypsum wallboard products used: backing board, firecode (firestop), foil backed?
4. Explain the procedure for fastening gypsum wallboard to steel studs.
5. Explain the procedure for fastening gypsum wallboard to masonry walls.
6. How are furring strips placed for hardboard application? How is hardboard fastened to furring strips?
7. What is plastic laminate? When and where is it used in interior finish?
8. How are boards used for interior finish over stud walls?
9. How is ceiling tile fastened in place?
10. What are the steps for installing a suspended ceiling?
11. Explain the use for the following moldings: base, casing, cove.
12. What trim members are used at windows?
13. What types of doors are preferred for interior use?
14. Give the steps for hanging a door which is not prehung.
15. What is the difference between tubular, cylindrical, and mortise locks?
16. How is a tubular lock installed?
17. How do locks for right and left hand doors differ?
18. Describe the following doors: pocket, bi-pass, bifold.
19. Give some of the features found on interior metal door frames.
20. What material is used for floor underlayment? Explain how it is applied.
21. Give the basic characterstics of strip, plank, and block flooring.
22. Explain how strip flooring is fastened over a wood subfloor.
23. Explain how block flooring is laid over a concrete floor.
24. What are the steps for installing base cabinets?
25. Explain straight flight stairs, winder stairs, and geometric stairs.
26. What is a stair ratio? What stair ratios are generally suggested?
27. How is the unit riser height determined using a story pole and using simple mathematics?
28. Headroom is measured from what points?
29. What precautions should be followed to make winders more safe?
30. What are closed and open stringer stairs?

CHAPTER 8
INDUSTRIALIZED BUILDING

The term *industrialized building* refers to the construction of homes and other structures in a factory controlled situation. Two methods are used to manufacture industrialized buildings. There are likewise several combinations of these two methods. One method produces panelized (componentized) buildings made from factory-assembled components, such as wall, partition, floor, roof, and truss units. These units are transported to the jobsite and assembled. See Fig. 8-1. The other method produces modular buildings made from three dimensional sections of buildings assembled in a factory. They are transported to the jobsite and joined together to form

Fig. 8-1. Panelized construction consists of two dimensional components which are assembled complete with windows, doors, and interior and exterior wall finish. (Windsor Homes, Inc.)

Fig. 8-2. Modular buildings are made from three dimensional building units which are positioned on the foundation. (National Homes Corp.)

the building. See Fig. 8-2. Both panelized and modular houses are designed to meet the same construction requirements as conventional on-site buildings.

In this chapter we are not concerned with components in the usual sense, that is as individual building parts. Components are factory-built parts of houses which are brought to the job and incorporated into a conventional building. They are used as a part of conventional building practice, as well as industrialized building. These components have already been discussed earlier in the text. A quick listing of this coverage appears below.

Chapter 4 discussed wall and floor framing components, including floor truss systems. Plywood components, such as built up headers and stressed skin panels, were also discussed. Chapter 5 discussed roof trusses and trussed rafters. Chapter 6 included package windows, pre-hung exterior doors, and exterior soffit systems. Chap-

ter 7 covered pre-hung interior doors, kitchen cabinets, bookcases, vanities, and other built-in cabinet items. In industrialized building, the idea of using components is extended to whole wall, floor, and roof sections.

The expansion in industrialized housing over the last ten years has been so extensive that it now accounts for about one third of the housing units built each year. There are several reasons for the success of this type of construction. The buildings have generally proved to be as satisfactory in construction as conventionally built structures. People living in these units have found them an attractive solution to their housing needs. The manufacturing process and overall careful scheduling both make it possible to build housing with a minimum delay. The cost of the buildings is at least competitive with houses built on site and often lower. Builders provide a number of options on exterior finish, window and door styles, and interior finish so that each building has its own individual distinct flavor. However, the variety of house plans offered by any one manufacturer is generally limited. To use production line methods and have more than a few basic plans with a number of minor variations would defeat the purpose of the system. Some few industrialized builders, using computer techniques, are equipped to build very distinctive homes in almost any style.

Several basic problems confront the home manufacturer. One is the high cost of equipment and the plant building itself. A large factory type building is carefully planned for material storage, material flow, and assembly line operation. A high level of production on a continuing basis is necessary to return a profit to the business. This demands a sales organization and a setup of distributorships which is highly efficient. Another problem is transporting the units to the jobsite by truck. Distance adds enormously to this expense. Most states have restrictions covering the width and length of a truck load and a daily time period when such wide load trucks may travel on highways. The height of units is restricted by overpass clearances. Manufacturers must protect the units against the wracking effect of rough roads by strengthening and stiffening the building parts. Some building sites are inaccessible to trucks. Once the units arrive at

the site, they must be lifted into place with cranes. Room must be available for the crane to lift the large building assemblies.

The savings involved in using industrialized buildings units over conventional methods is largely in the area of labor cost. When considering the overall cost of a home (including land, materials, and furnishings), the average labor cost is nearly one third of the total. The ratio of skilled workers, such as carpenters, electricians, and plumbers, to semiskilled or unskilled workers varies throughout the industrialized building industry. Some production plants use skilled labor exclusively. On the other extreme, certain plants, largely because of automated machinery, use only unskilled or semiskilled help. Once the building units arrive at the jobsite, the work calling for skilled mechanics sometimes requires 30 days and sometimes only a few days, depending on the type of job and the amount of work already done in the factory. Preparing the jobsite and building the foundation are done in the conventional manner by qualified local mechanics.

PANELIZED CONSTRUCTION

With panelized construction, the house is divided into parts for manufacturing purposes, such as wall and partition units. These units are

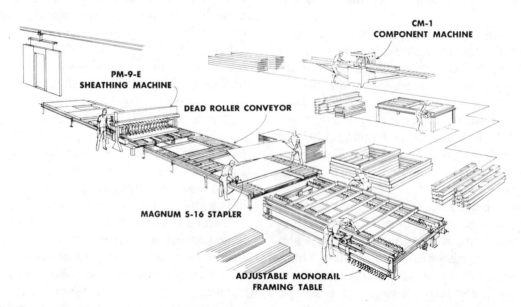

Fig. 8-3. Panels are assembled on mechanized tables which have equipment for automatic positioning and nailing. (Paslode Co.)

Fig. 8-4. Closed panel construction. At left, the electrician finishes the wiring, while another worker installs the baseboard heating unit. At right, the finished wall panel is loaded on a truck for delivery to the jobsite. (Wausau Homes Inc.)

constructed on long mechanized tables which enable the unit to move from station to station. See Fig. 8-3. At one point during the assembly, pneumatic nailing machines fasten the plates to studs. Openings for windows and doors are provided where required. At another station along the table, sheathing is stapled in place.

Open Panel Construction

Some plants are equipped to fasten windows and doors in place and apply sheathing to the outside of the unit. Such units are called *open panels* because the back is open so that the electrical, plumbing, and interior finishing can be done later on the job. The wall panel units, along with floor and roof units, are then loaded on a truck in sequence so that they are in proper order for removal and assembly. Bulk items, such as roof shingles, flooring, interior finish, and miscellaneous items, such as hardware, are included. Everything needed to assemble the house is delivered at once except the electrical, plumbing, and heating equipment. These items are delivered separately and are installed by local contractors.

Closed Panel Construction

Closed panel construction is an extension of open panel construction. In closed panel construction, the walls and roof panels are finished on both sides, with insulation in place. See Fig. 8-4. Windows are installed, doors are prehung and in place, including the hardware. Wiring and plumbing are done at the factory before the panels are closed. The work required on the job includes placing the panels, completing some of the interior finish (particularly where panels join), hooking up plumbing fixtures, finishing wiring for light and power, and installing the heating plant.

The computer plays an important part in the manufacture of panelized buildings. Some manufacturers have computers which print out all of the information needed for cutting stock and assembling wall panels. See Fig. 8-5. A draftsman or other technician, following the architectural floor plans, punches the proper keys on the computer which prints a diagram of the panel stud arrangement for each section. The computer also prints a diagram to tell the installers the exact position for each panel in the construction of the building. The printout of the panels is then fed to the assembly line. Material is cut to the correct lengths and is assembled automatically, including bracing, with provisions for door and window openings.

Fig. 8-5A. By operating a computer keyboard, accurate shop drawings are printed on a drum plotter. The plotter subsequently produces detailed drawings of panels and components. The computer makes a tape which is used to automatically arrange the framing members on the assembly table. (PanelTech Corp.)

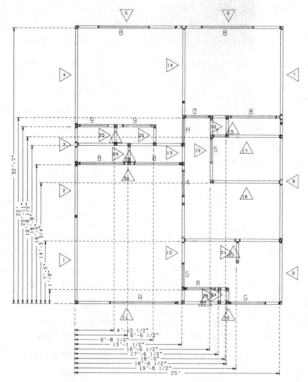

Fig. 8-5B. The shop drawing of the building shows how panels are divided and gives dimensions for their size and location. PanelTech Corp.)

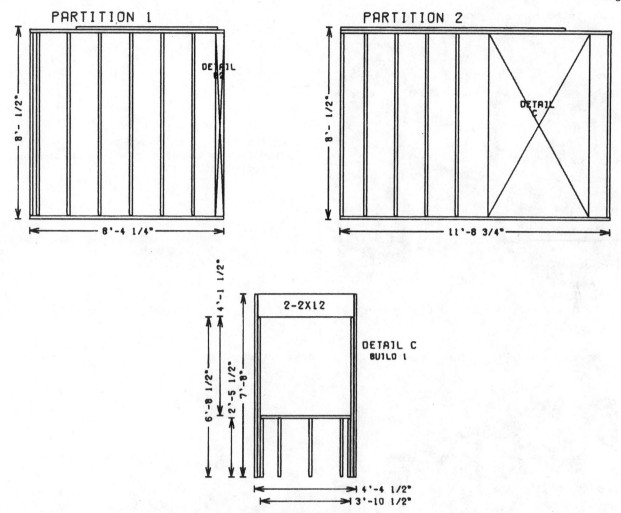

Fig. 8-5C. The computer provides printouts showing the stud arrangement for each panel. Components, such as Detail C, are dimensioned to drop into place in the panel. (Panel Tech Corp.)

MODULAR CONSTRUCTION

Modular construction in housing applies to the manufacture of three dimensional cubes or box-like structures which are fabricated and assembled in a factory. They are then transported to the job for assembly to form a complete building. See Fig. 8-6. They are completed to different degrees by various manufacturers. Some modules have siding and roofing installed. Others have wood wall paneling or interior finish, such as gypsum wallboard and even carpeting. The particular arrangement can be a two piece single story home. Three or more piece

homes and complete apartments designed to stack two high are other arrangements. See Fig. 8-7. The modular building method is therefore not limited to individual home building. It has been successfully used to provide schools and apartments.

Mechanical Core

One type of module has received a great deal of attention because it reduces the on-site labor for plumbing, electricity, and even heating. This is the mechanical core, often called a *wet core* or *utility core*. See Fig. 8-8. The mechanical core usually consists of a bathroom complete with

Fig. 8-6. Modules are complete with exterior finish and roofing. They are stacked to make two story structures. (American Plywood Assoc.)

Fig. 8-7. The completed modular building is a multi family dwelling. (American Plywood Assoc.)

fixtures and finished walls. All of the plumbing pipes are in the wall and floor and only need to be connected to supply and waste lines when in place on the job. Some mechanical cores consist of one and one half baths. Others have the "wet" part of the kitchen and kitchen cabinets installed on an outside wall of the bathroom core and are arranged to become a wall of the kitchen when in place. Still other mechanical cores have parallel walls back of the plumbing wall arranged to provide a space for water heater, the heating plant, and ductwork. This is a true mechanical core because it includes all of the piping for the plumbing and heating. Mechanical cores are available for use in residential construction.

The mechanical core has been successfully

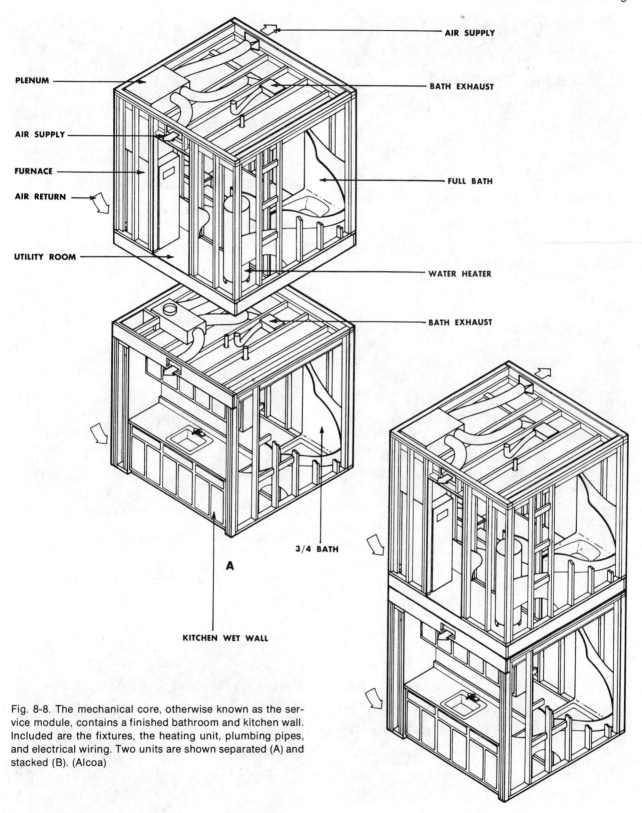

AIR SUPPLY

PLENUM

BATH EXHAUST

AIR SUPPLY

FURNACE

FULL BATH

AIR RETURN

UTILITY ROOM

WATER HEATER

BATH EXHAUST

3/4 BATH

A

KITCHEN WET WALL

B

Fig. 8-8. The mechanical core, otherwise known as the service module, contains a finished bathroom and kitchen wall. Included are the fixtures, the heating unit, plumbing pipes, and electrical wiring. Two units are shown separated (A) and stacked (B). (Alcoa)

Fig. 8-9A. This photo and the following sequence of photos show the steps in the erection of a four section modular home. This view shows that an accurate foundation is built and braced well in advance of the delivery of the units. The tracks are supported on posts and the wall. (Dickinson Homes)

Fig. 8-9D. The first unit is raised using jacks to the level of the track. The open end is protected with polyethylene film.

Fig. 8-9B. The unit is transported to the jobsite on a low bed trailer.

Fig. 8-9E. Carpenters stationed at the far wall use a winch to pull the unit across on the track.

Fig. 8-9C. The unit arrives at the jobsite. The overhang and the cornice have been left off because of restrictions on the width of loads.

Fig. 8-9F. The unit is in place and is carefully lowered to rest on the foundation wall and steel posts.

Fig. 8-9G. The second unit is placed in the same manner as the first. In this view, it has not yet been dropped to be level with the first unit.

Fig. 8-9H. The third unit is in place. Temporary diagonals stiffen the inside partition. Units one and two are bolted together where they join.

Fig. 8-9I. Carpenters will finish the siding, overhang, and other odd jobs. The electrical and plumbing work is hooked up at this stage.

Fig. 8-9J. A garage is added. It is built using conventional framing.

used in a number of hotel and apartment applications. The cores are built away from the jobsite, delivered, lifted by crane, and slid into place at the proper floor level.

Figure 8-9 shows a series of photographs of a typical modular home in the process of construction. It is built in four modules because of load and other restrictions. The foundation is built at the jobsite to exact dimensions, and the ground is prepared so that the heavy units can be brought into position along side of the foundation wall.

The construction of such a modular home begins in the plant, with the building of a sturdy girder, joist, and floor assembly. Such assembly is built for each module. It is designed to resist the stress of travel on the road and handling during erection on the job. The pre-built panels are added as each module moves from station to station in the assembly line operation. These include the wall, partition ceiling, and truss parts. Windows and doors are then fitted and fastened into place. Plumbing and electrical work are added. The plumbing stubs extend below the floor. Sheathing is next applied to walls and roof. Insulation is put in place. Exterior and interior wall finish are installed. Interior doors are hung, cabinets are placed, and interior trim is fitted and nailed. Each module is braced where large openings occur. The open sides are protected by sheets of polyethylene film. The trip to the job site is made on a large flat bed trailer, as shown in B and D of Fig. 8-9.

The module is brought alongside the foundation wall and lifted by four jacks on to a track which is placed across the foundation and resting on posts and the foundation wall. A winch and chain device is then used to draw the module across to its position over the opposite wall (E). Jacks are again used to lift the module until the tracks are removed. The module is then lowered into place on the foundation wall and posts (F). The second module is delivered in the same manner and slid into position on tracks. The two modules are then bolted together (G). The two remaining modules are delivered and installed in the same fashion (H). In the course of a day, the entire building is delivered, placed, and fastened together. The plumbing and electrical work can be completed in a few additional days. A carpenter finishes (I) the siding and roofing where the modules join and completes minor trim work inside. The completed building is then ready for the owners to move in. A garage (J) is sometimes added on. It can be a modular unit, built in the same manner as the other modules, or built using conventional framing methods.

QUESTIONS FOR STUDY AND DISCUSSION

1. What are some typical components used in building homes, using conventional construction methods?
2. List some of the advantages of using industrialized methods for building.
3. What are some of the problems involved in manufacturing industrialized homes?
4. What is panelized construction?
5. What work is done at the jobsite before a panelized home is delivered?
6. Explain the difference between an open panel and a closed panel.
7. How is a computer used in panelized construction?
8. What is a module?
9. What kind of buildings are adaptable to be constructed with modules?
10. Describe the function of a mechanical core.

CHAPTER 9

METRIC MEASUREMENT

The monetary system in the United States is essentially based on multiples of ten. We are all familiar with the ease with which transactions can be carried out using this system. Several European countries (notably France) have used a system of measure, which is likewise based on multiples of ten, for over one hundred years. They have found this system very convenient and efficient. This system of measure is called the *metric system.* The metric system of measurement has since been adopted by all industrial nations of the world with the exception of the United States. Great Britain spent the decade from 1965 to 1975 bringing about a complete conversion to metrics.

The economic pressures of world trade and the need to remain competitive in foreign markets has created renewed interest in the United States in metrics and stimulated progress toward the eventual and inevitable conversion of the U.S. to this system. Several industries, such as the pharmaceutical, photographic, and to some extent the machine and electrical industries, have already converted to metric measure. Other major industries and labor groups are seriously considering the mechanics of changeover and what impact it will have on their particular field.

The United States has had a long standing interest in the metric system of measure. As early as 1866, Congress passed legislation which made the metric system legal in this country. In 1875, representatives of our government signed the Metric Convention, an agreement between nations, which defined metric standards for length and mass and established the International Bureau of Weights and Measures. The United States is currently a member of this group. An intensive study of the metric system was contained in a report, in 1971, conducted jointly by the Bureau of Standards and the Department of Commerce. The Metric Conversion Act of 1975 has added further stimulus toward the adoption of a conversion program.

WHAT IS THE METRIC SYSTEM?

The metric system is a standardized world wide system of measurement known as SI (Systéme International) for (1) length, (2) mass, (3) time, (4) electrical current, (5) thermodynamic temperature, (6) amount of substance, and (7) luminous intensity. The basic units chosen by this international body are multiplied or divided on the base of ten, with a prefix used to identify the new unit. See Fig. 9-1. The meter (abbreviated m) is equivalent to 39.37 inches. This is the base for length. The meter is equal to 1000 millimeters, 100 centimeters, and 10 decimeters. (A millimeter is abbreviated mm, a centimeter is abbreviated cm, and a decimeter is abbreviated dm.) Terms for lengths larger than a meter include decameter (10 meters), hectometer (100 meters), and kilometer (1000 meters), which is the standard measure for long distances in met-

PREFIX	VALUE		MEANING	SYMBOL
MILLI	THOUSANDTHS	= ÷	1000	m
CENTI	HUNDREDTHS	= ÷	100	c
DECI	TENTHS	= ÷	10	d
DECA	TENS	= X	10	da
HECTO	HUNDREDS	= X	100	h
KILO	THOUSANDS	= X	1000	k

Fig. 9-1. In the metric system, the meter is the standard of linear measure. Prefixes, such as those appearing above, denote the divisions or multiples of the meter. The most common unit used in the construction industry is the millimeter.

ric countries. Thus distances between cities are measured in *kilometers* (abbreviated km) instead of *miles*.

The construction industry simplifies metric measure by using only two of the units. The millimeter (0.03937 inch) is used for material sizes and dimensions on working drawings. Large dimensions, such as dimensions for lots and other tracts of land, are given in meters. Carpenters are mainly concerned with linear (length) measure. However, they may also find themselves required to work with areas and volumes. The concepts of area and volume are actually based on meters (and millimeters). Area is measured in terms of a square meter (or millimeter) and volume in terms of cubic meters (or millimeters). Carpenters occasionally work with weights (mass) and liquid measure. The kilogram is the metric measure for weight. It is equivalent to 2.205 pounds. The liter, the measure for liquid, is equivalent to 1.057 quarts. Tables 9-1 and 9-2 show equivalent factors used for converting from our present foot-pound-quart system (generally called the *English System*) to the metric system.

USING THE METRIC SYSTEM

If the metric system were suddenly adopted and all materials were converted to metric sizes and working drawings were dimensioned in metric measurements, the main problem for the mechanics would be to adapt to a new measuring tool. This measuring tool is a rule divided into millimeters. Each ten millimeters is numbered to make a centimeter, and 100 centimeters are marked as a meter. Finding units on the rule

for laying out members before cutting or assembling them is relatively easy. Adding and subtracting quantities in the metric system is very simple, especially when compared to our present method of adding fractions of inches, and inches and feet.

Even with careful planning, a changeover to metrics (called *metrication*) requires a span of 10 years or more before architects, engineers, manufacturers, etc., could be expected to completely coordinate their efforts. A period of transition, during which the carpenter must be able to work with either system of measure or both, is inevitable.

The carpenter can get a firm grasp on how the metric system of measurement works by acquiring a rule divided into metric units. The carpenter should then use the rule to practice on common objects and see how easy it is to read and interpret metric units. The carpenter can prepare for the inevitable changeover to metrics by learning how to convert from the present English system to the metric system. (There may be some need, at a later date, to convert from the metric system back to English measure.)

Various aids which resemble slide calculators are available. These provide rough approximations between the English system and the metric system. Tables, such as Table 9-1 provided here, give direct metric equivalents for fractions of an inch, inches, and feet. Converting from the English system to metric measure often requires nothing more than a simple multiplication, using the conversion factors provided in Table 9-2.

The following examples show how conversions are made.

Using Table 9-1. When only one type of unit, such as inches, is involved in the conversion the equivalent is read directly. When feet, inches, and fractions of inches are involved, it requires taking the equivalents from the applicable sections of the table and adding them together.

Problem: To find the equivalent metric size of a 2 × 4 (1½ × 3½ inches) inch piece.

$1\frac{1}{2}$ inches = 25.4 + 12.7 = 38.1 mm
$3\frac{1}{2}$ inches = 76.2 + 12.7 = 88.9 mm

A 2 × 4 inch piece = 38.1 × 88.9 millimeters.

TABLE 9-1. FRACTIONAL INCH—MILLIMETER AND FOOT—MILLIMETER CONVERSION TABLES.
(Based on 1 inch = 25.4 millimeters, exactly)

Fractional inch to millimeters

In.	Mm.	In.	Mm.	In.	Mm.	In.	Mm.
1/64	0.397	17/64	6.747	33/64	13.097	49/64	19.447
1/32	0.794	9/32	7.144	17/32	13.494	25/32	19.844
3/64	1.191	19/64	7.541	35/64	13.891	51/84	20.241
1/16	1.588	5/16	7.938	9/16	14.288	13/16	20.638
5/64	1.984	21/64	8.334	37/64	14.684	53/64	21.034
3/32	2.381	11/32	8.731	19/32	15.081	27/32	21.431
7/64	2.778	23/64	9.128	39/64	15.478	55/64	21.828
1/8	3.175	3/8	9.525	5/8	15.875	7/8	22.225
9/64	3.572	25/64	9.922	41/64	16.272	57/64	22.622
5/32	3.969	13/32	10.319	21/32	16.669	29/32	23.019
11/64	4.366	27/64	10.716	43/64	17.066	59/64	23.416
3/16	4.762	7/16	11.112	11/16	17.462	15/16	23.812
13/64	5.159	29/64	11.509	45/64	17.859	61/64	24.209
7/32	5.556	15/32	11.906	23/32	18.256	31/32	24.606
15/64	5.953	31/64	12.303	47/64	18.653	63/64	25.003
1/4	6.350	1/2	12.700	3/4	19.050	1	25.400

Inches to millimeters

In.	Mm.	In.	Mm.	In.	Mm.	In.	Mm.	In.	Mm.	In.	Mm.
1	25.4	3	76.2	5	127.0	7	177.8	9	228.6	11	279.4
2	50.8	4	101.6	6	152.4	8	203.2	10	254.0	12	304.8

Feet to millimeters

Ft.	Mm.	Ft.	Mm.	Ft.	Mm.	Ft.	Mm.	Ft.	Mm.
100	30,480	10	3,048	1	304.8	0.1	30.48	0.01	3.048
200	60,960	20	6,096	2	609.6	0.2	60.96	0.02	6.096
300	91,440	30	9,144	3	914.4	0.3	91.44	0.03	9.144
400	121,920	40	12,192	4	1,219.2	0.4	121.92	0.04	12.192
500	152,400	50	15,240	5	1,524.0	0.5	152.40	0.05	15.240
600	182,880	60	18,288	6	1,828.8	0.6	182.88	0.06	18.288
700	213,360	70	21,336	7	2,133.6	0.7	213.36	0.07	21.336
800	243,840	80	24,384	8	2,438.4	0.8	243.84	0.08	24.384
900	274,320	90	27,432	9	2,743.2	0.9	274.32	0.09	27.432
1,000	304,800	100	30,480	10	3,048.0	1.0	304.80	0.10	30.480

Problem: To convert a dimension of 14'-9½" to millimeters.

½ inch =	12.7 mm
9 inches =	228.6 mm
10 feet =	3048.0 mm
4 feet =	1219.2 mm
Total	4508.5 mm

Problem: To find the size of a 4 × 8 foot sheet of plywood, refer to the table.

A 4 × 8 foot sheet = 1219.2 × 2438.4 millimeters.

Using Table 9-2. Multiply the English unit by the constant given to derive the metric quantity with a single step. However, the English units must be converted to their decimal equivalent first. The following rules involve the use of constants shown in Table 9-2.

TABLE 9-2.
CONVERSION OF ENGLISH TO METRIC UNITS.

Lengths

1 inch (in)	= 2.540 cm or = 25.40 mm
1 foot (ft)	= 30.48 cm or = 304.8 mm
1 yard (yd)	= 91.44 cm or = 0.9144 m
1 mile	= 1.609 km

Areas

1 square inch (in^2)	= 6.452 cm^2 or = 645.2 mm^2
1 square foot (ft^2)	= 929.0 cm^2 or = 0.0929 m^2
1 square yard (yd^2)	= 0.8361 m^2

Volumes

1 cubic inch (in^3)	= 16.39 cm^3
1 cubic foot (ft^3)	= 0.02832 m^3
1 cubic yard (yd^3)	= 0.7646 m^3

Weights

1 ounce (oz) (AVDP)	= 28.35 grams (g)
1 pound (lb)	= 453.6 g or = 0.4536 kilogram (kg)
1 (short) ton	= 907.2 kilograms (kg)

Liquid measurements

1 (fluid) ounce	= 0.02957 liter or = 28.35 grams
1 pint (pt)	= 473.2 cm^3
1 quart (qt)	= 0.9463 liter
1 (U.S.) gallon (gal)	= 3785 cm^3

Power measurements

1 horsepower (hp)	= 0.7457 kilowatt

Temperature measurements

To convert degrees Fahrenheit to degrees Celsius (Centigrade), use the following formula:
$$°C = 5/9 \ (°F - 32)$$

TABLE 9-3.
CONVERSION OF METRIC TO ENGLISH UNITS.

Lengths

1 millimeter (mm)	= 0.03937 in or = 0.003281 ft
1 centimeter (cm)	= 0.3937 in
1 meter (m)	= 3.281 ft or = 1.0937 yd
1 kilometer (km)	= 0.6214 miles

Areas

1 square millimeter (mm^2)	= 0.00155 in^2
1 square centimeter (cm^2)	= 0.155 in^2
1 square meter (m^2)	= 10.76 ft^2 or = 1.196 yd^2

Volumes

1 cubic centimeter (cm^3)	= 0.06102 in^3
1 cubic meter (m^3)	= 35.31 ft^3 or = 1.308 yd^3

Weights

1 gram (g)	= 0.03527 oz (AVDP)
1 kilogram (kg)	= 2.205 lb
1 metric ton (t)	= 2205 lb

Liquid measurements

1 cubic centimeter (cm^3)	= 0.06102 in^3
1 liter (= 1000 cm^3)	= 1.057 quarts or = 2.113 pints or = 61.02 in^3

Power measurements

1 kilowatt (kw)	= 1.341 horsepower (hp)

Temperature measurements

To convert degrees Celsius (Centigrade) to degrees Fahrenheit use the following formula:
$$°F = (9/5 \ °C) + 32$$

Some important features of the SI are:

1 cubic centimeter of water = 1 gram. Pure water freezes at 0 degrees Celsius and boils at 100 degrees Celsius.

TABLE 9-4. DECIMAL EQUIVALENTS OF AN INCH.

4ths	8ths	16ths	32nds	64ths	to 2 places	to 3 places	4ths	8ths	16ths	32nds	64ths	to 2 places	to 3 places
				1/64	0.02	0.016					33/64	0.52	0.516
			1/32		0.03	0.031				17/32		0.53	0.531
				3/64	0.05	0.047					35/64	0.55	0.547
		1/16			0.06	0.062			9/16			0.56	0.562
				5/64	0.08	0.078					37/64	0.58	0.578
			3/32		0.09	0.094				19/32		0.59	0.594
				7/64	0.11	0.109					39/64	0.61	0.609
	1/8				0.12	0.125		5/8				0.62	0.625
				9/64	0.14	0.141					41/64	0.64	0.641
			5/32		0.16	0.156				21/32		0.66	0.656
				11/64	0.17	0.172					43/64	0.67	0.672
		3/16			0.19	0.188			11/16			0.69	0.688
				13/64	0.20	0.203					45/64	0.70	0.703
			7/32		0.22	0.219				23/32		0.72	0.719
				15/64	0.23	0.234					47/64	0.73	0.734
1/4					0.25	0.250	3/4					0.75	0.750
				17/64	0.27	0.266					49/64	0.77	0.766
			9/32		0.28	0.281				25/32		0.78	0.781
				19/64	0.30	0.297					51/64	0.80	0.797
		5/16			0.31	0.312			13/16			0.81	0.812
				21/64	0.33	0.328					53/64	0.83	0.828
			11/32		0.34	0.344				27/32		0.84	0.844
				23/64	0.36	0.359					55/64	0.86	0.859
	3/8				0.38	0.375		7/8				0.88	0.875
				25/64	0.39	0.391					57/64	0.89	0.891
			13/32		0.41	0.406				29/32		0.91	0.906
				27/64	0.42	0.422					59/64	0.92	0.922
		7/16			0.44	0.438			15/16			0.94	0.938
				29/64	0.45	0.453					61/64	0.95	0.953
			15/32		0.47	0.469				31/32		0.97	0.969
				31/64	0.48	0.484					63/64	0.98	0.984
1/2					0.50	0.500	1					1.00	1.000

Rules for converting from English to metric linear units:

1. If the English unit is a fraction of an inch, use Table 9-4 to find its decimal equivalent and then multiply by the constant 25.40.

Example: To find the equivalent of 3/4 inch.

$$.75 \times 25.40 = 19.05 \text{ mm}$$

2. If the English unit is in feet, inches, and fractions of an inch, use Table 9-5 to find the decimal

TABLE 9-5. DECIMAL EQUIVALENTS OF A FOOT.

0″	.0000	1″	.0833	2″	.166667	3″	.2500
$1/16$	.0052	$1\,1/16$	.0885	$2\,1/16$	.171875	$3\,1/16$	.2552
$1/8$	.0104	$1\,1/8$	.09375	$2\,1/8$	.1771	$3\,1/8$	.2604
$3/16$	.015625	$1\,3/16$	.0990	$2\,3/16$	.1823	$3\,3/16$	.265625
$1/4$	.0208	$1\,1/4$	.1042	$2\,1/4$	.1875	$3\,1/4$	.2708
$5/16$	.0260	$1\,5/16$	.109375	$2\,5/16$	.1927	$3\,5/16$	.2760
$3/8$	.03125	$1\,3/8$	.1146	$2\,3/8$	.1979	$3\,3/8$	.28125
$7/16$	.0365	$1\,7/16$	.1198	$2\,7/16$	.203125	$3\,7/16$	.2865
$1/2$	.0417	$1\,1/2$	.1250	$2\,1/2$	.2083	$3\,1/2$	.2917
$9/16$	.046875	$1\,9/16$	.1302	$2\,9/16$	.2135	$3\,9/16$	.296875
$5/8$	.0521	$1\,5/8$	.1354	$2\,5/8$	.21875	$3\,5/8$	.3021
$11/16$	.0573	$1\,11/16$	.140625	$2\,11/16$	.2240	$3\,11/16$	.3073
$3/4$	.0625	$1\,3/4$	.1458	$2\,3/4$	.2292	$3\,3/4$	.3125
$13/16$	.0677	$1\,13/16$	.1510	$2\,13/16$	.234375	$3\,13/16$	.3177
$7/8$	.0729	$1\,7/8$	.15625	$2\,7/8$	.2396	$3\,7/8$	.3229
$15/16$	.078125	$1\,15/16$	.1615	$2\,15/16$	.2448	$3\,15/16$	.328125
4″	.3333	5″	.416667	6″	.5000	7″	.5833
$4\,1/16$	.3385	$5\,1/16$	.421875	$6\,1/16$	.5052	$7\,1/16$	.5885
$4\,1/8$	.34375	$5\,1/8$	.4271	$6\,1/8$	.5104	$7\,1/8$	.59375
$4\,3/16$	.3490	$5\,3/16$	.4323	$6\,3/16$	.515625	$7\,3/16$	.5990
$4\,1/4$	.3542	$5\,1/4$	.4375	$6\,1/4$	.5208	$7\,1/4$	.6042
$4\,5/16$	.359375	$5\,5/16$	.4427	$6\,5/16$	.5260	$7\,5/16$	.6093
$4\,3/8$	.3646	$5\,3/8$	.4479	$6\,3/8$	.53125	$7\,3/8$	.6146
$4\,7/16$	.3698	$5\,7/16$	.453125	$6\,7/16$	.5365	$7\,7/16$	.6198
$4\,1/2$	.3750	$5\,1/2$	.4583	$6\,1/2$	.5417	$7\,1/2$	.6250
$4\,9/16$	.3802	$5\,9/16$	.4635	$6\,9/16$	.546875	$7\,9/16$	.6302
$4\,5/8$	.3854	$5\,5/8$	.46875	$6\,5/8$	.5521	$7\,5/8$	.6354
$4\,11/16$	.390625	$5\,11/16$	.4740	$6\,11/16$	.5573	$7\,11/16$	.640625
$4\,3/4$	.3958	$5\,3/4$	.4792	$6\,3/4$	.5625	$7\,3/4$	.6458
$4\,13/16$	.4010	$5\,13/16$	.484375	$6\,13/16$	.5677	$7\,13/16$	.6510
$4\,7/8$	.40625	$5\,7/8$	.4896	$6\,7/8$	.5729	$7\,7/8$	.65625
$4\,15/16$	.4115	$5\,15/16$	.4948	$6\,15/16$	.578125	$7\,15/16$	.6615
8″	.666667	9″	.7500	10″	.8333	11″	.916667
$8\,1/16$	.671875	$9\,1/16$	.7552	$10\,1/16$	.8385	$11\,1/16$	.921875
$8\,1/8$	.6771	$9\,1/8$	.7604	$10\,1/8$	.84375	$11\,1/8$	.9271
$8\,3/16$	.6823	$9\,3/16$	.765625	$10\,3/16$	.8490	$11\,3/16$	.9323
$8\,1/4$	.6875	$9\,1/4$	.7708	$10\,1/4$	.8542	$11\,1/4$	.9375
$8\,5/16$	.6927	$9\,5/16$	.7760	$10\,5/16$	.859375	$11\,5/16$	.9427
$8\,3/8$	.6979	$9\,3/8$	.78125	$10\,3/8$	.8646	$11\,3/8$	.9479
$8\,7/16$	.703125	$9\,7/16$	.7865	$10\,7/16$	.8698	$11\,7/16$	.953125
$8\,1/2$	.7083	$9\,1/2$	.7917	$10\,1/2$	.8750	$11\,1/2$	.9583
$8\,9/16$	.7135	$9\,9/16$	.796875	$10\,9/16$	.8802	$11\,9/16$	.9635
$8\,5/8$	.71875	$9\,5/8$	.8021	$10\,5/8$	.8854	$11\,5/8$	.96875
$8\,11/16$	.7240	$9\,11/16$	.8073	$10\,11/16$	.890625	$11\,11/16$	.9740
$8\,3/4$	.7292	$9\,3/4$	.8125	$10\,3/4$	.8958	$11\,3/4$	.9792
$8\,13/16$	.734375	$9\,13/16$	.8177	$10\,13/16$	.9010	$11\,13/16$	.984375
$8\,7/8$	.7396	$9\,7/8$	.8229	$10\,7/8$	.90625	$11\,7/8$	.9896
$8\,15/16$	.7448	$9\,15/16$	.828125	$10\,15/16$	.9115	$11\,15/16$	.9948

equivalents of a foot and multiply by the constant 304.8.

Example: 14'-9½" equals 14.79 feet
14.79 × 304.8 = 4508.50 mm

(Two decimal places are adequate for carpentry calculations.)

Other quantities, such as those involving areas, volumes, weights, and liquid measure, are converted to metric units by using the conversion factors shown in Table 9-2.

Using Table 9-3. When quantities are given in metric units, the English units are obtained by multiplying the quantity by the proper constant shown in the table.

Example: To convert 4000 mm to feet and inches. 4000 × .003281 = 13.124 feet.
By referring to Table 9-5, you find that
.124 feet = 1½ inches
4000 mm = 13'-1½".

WHAT THE METRIC SYSTEM MEANS FOR CARPENTERS.

Carpenters find the period of transition a time for adjusting to new concepts. These include using new measuring tools and materials supplied in new sizes. Carpenters first have to become acquainted with new measuring tools. They eventually will need new cutting tools also. Bit and drill sizes and saw blade diameters used will be available exclusively in metric units. Power hand tools and machines will be manufactured with metric measure, thus making the previous tools largely obsolete when parts must be replaced.

When the size of materials changes, the spacing of studs and other building parts also changes accordingly. A 2 × 4 inch piece (38.1 × 88.9 mm) is changed to a size which is more convenient in terms of the metric system. A 4 × 8 foot sheet of plywood, for example, now

TABLE 9-6. METRIC SIZES OF SAWN SOFTWOOD.

Thickness in mm	75	100	125	150	175	200	225	250	300
16	X	X	X	X					
19	X	X	X	X					
22	X	X	X	X					
25	X	X	X	X	X	X	X	X	X
32	X	X	X	X	X	X	X	X	X
38	X	X	X	X	X	X	X		
44	X	X	X	X	X	X	X	X	X
50	X	X	X	X	X	X	X	X	X
63		X	X	X	X	X	X		
75		X	X	X	X	X	X	X	X
100		X		X		X		X	X
150				X		X			X
200						X			
250								X	
300									X

Timber Research and Development Association, Huhenden Valley, High Wycome, Buckinghamshire, England

has the equivalent of 2.97 square meters and dimensions of 1219 × 2438 millimeters. Under the metric system, plywood is sold by the square meter and has dimensions which conveniently fit the new stud spacing. The size of materials is yet to be determined.

Perhaps the most difficult adjustment for carpenters is acquiring a new concept of space and learning to think metric, and using the new linear units. A carpenter knows how large a 12 × 15 foot room is but must have a whole new space concept to understand the equivalent dimensions expressed as 3700 × 4600 millimeters. (These numbers are roughly equivalent.)

TABLE 9-7. STANDARD METRIC LENGTHS OF SOFTWOOD ARE AVAILABLE FROM 1.8 M, INCREASING BY INCREMENTS OF 300 MM.

Metric lengths	Equivalent in feet and inches
1.8 m	5'-10$\frac{7}{8}$"
2.1 m	6'-10$\frac{5}{8}$"
2.4 m	7'-10$\frac{1}{2}$"
2.7 m	8'-10$\frac{1}{4}$"
3.0 m	9'-10$\frac{1}{8}$"
3.3 m	10'- 9$\frac{7}{8}$"
3.6 m	11'- 9$\frac{3}{4}$"
3.9 m	12'- 9$\frac{1}{2}$"
4.2 m	13'- 9$\frac{3}{8}$"
4.5 m	14'- 9$\frac{1}{8}$"
4.8 m	15'- 9"
5.1 m	16'- 8$\frac{3}{4}$"
5.4 m	17'- 8$\frac{5}{8}$"
5.7 m	18'- 8$\frac{3}{8}$"
6.0 m	19'- 8$\frac{1}{8}$"
6.3 m	20'- 8"

Timber Research and Development Association, Huhendon Valley, High Wycome, Buckinghamshire, England

WHAT THE METRIC SYSTEM WILL MEAN TO THE CONSTRUCTION INDUSTRY

A straight conversion of the conventional sizes of material used today to metric units would result in very awkward descriptive units. On the other hand, converting machinery to provide new sizes of lumber, plywood, and all other material is extremely costly. A ripple effect is created as various components, such as windows, doors, and trusses, are changed to meet new space requirements.

Tables 9-6 and 9-7 show how Great Britain solved the problem of establishing new size standards for softwood lumber. Table 9-6 provides various sizes, beginning with a thickness of 16 mm (approximately $\frac{5}{8}$ inch) and proceeding to timbers 300 mm (approximately 12 inches) thick. Widths begin with 75 mm (approximately 3 inches). Table 9-7 shows the lumber lengths available, beginning with 1.8 m (equal to 5'-10$\frac{7}{8}$") and increasing in 300 mm (approximately 12 inch) increments.

Architects and engineers have special problems when they begin planning structures using a new concept of space measurements. In addition to thinking in metrics, they are forced to use material in both the old and new sizes, especially during the early stages of the transition period. Calculating loads is very simple from a mathematical standpoint because of the base of ten which is used. However, all the basic engineering books or tables have to be revised.

READING METRIC WORKING DRAWINGS

Figure 9-2 shows a floor plan drawn to metric measure. A carpenter should find this plan easy to follow and have little difficulty using the metric rule. The same floor plan converted to English units is shown in Fig. 9-3. Hopefully, a clear distinction between the two systems can be made possible. Most buildings would then be

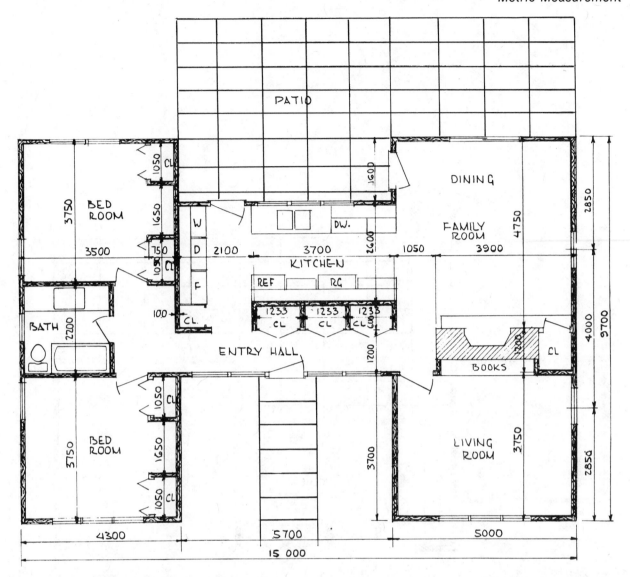

Fig. 9-2. This floor plan is designed to use metric measure. The plan shows the ease of working with metric dimensions.

planned using the conventional English dimensions and present building material sizes until a certain time when the metric dimensions would be used with metric sized material. Realistically, however, the transition period usually requires the use of dual dimensioning. With dual dimensioning, all dimensions are provided in both English and metric units, as shown in Figs. 9-4 and 9-5.

All metric dimensions on working drawings are given in millimeters. A dimension of 10'-5" is shown as 3175 mm. No comma is used to set off the thousand place. A space may be used instead as shown in 3 175 mm.

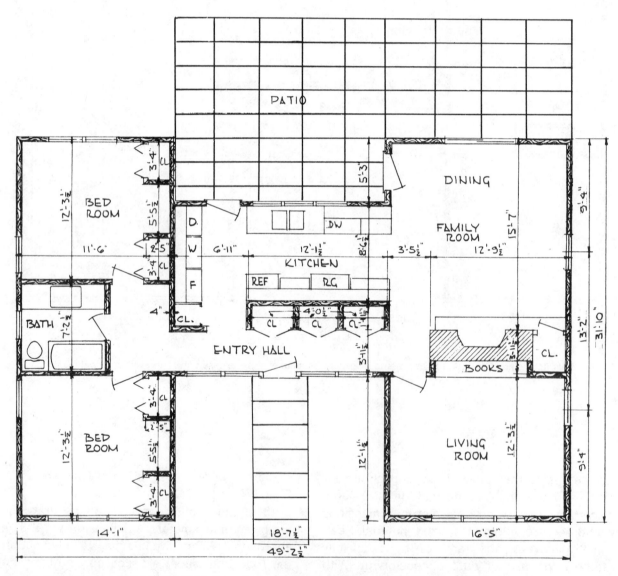

Fig. 9-3. The floor plan shown in Fig. 9-2 is here dimensioned with English units (to the nearest ½ inch).

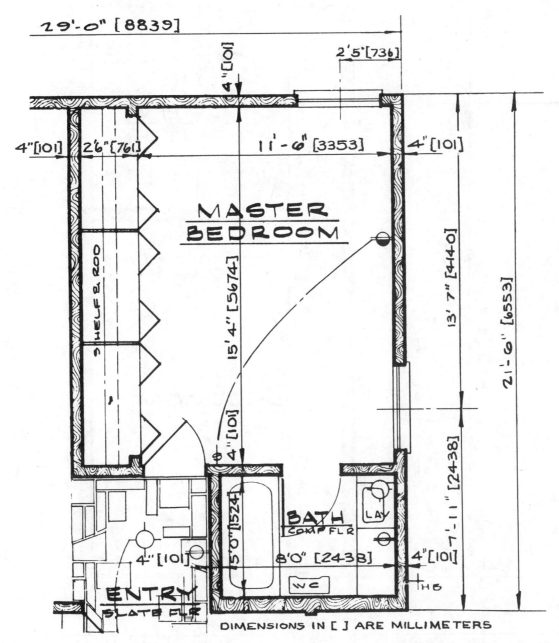

29'-0" [8839]

2'5' [736]

4" [101]

4" [101]   2'6" [761]   11'-6" [3353]   4" [101]

MASTER
BEDROOM

SHELF & ROD

15'4" [5674]

13'7" [4140]

21'-6" [6553]

4" [101]

φ 4" [101]

5'0" [1524]

BATH
COMP FLR

LAV

7'-11" [2438]

ENTRY
SLATE FLR

4" [101]

8'0" [2438]

WC

4" [101]

HB

DIMENSIONS IN [ ] ARE MILLIMETERS

Fig. 9-4. An example of dual dimensioning. This portion of a floor plan shows the conventional English measure with metric measurements in brackets.

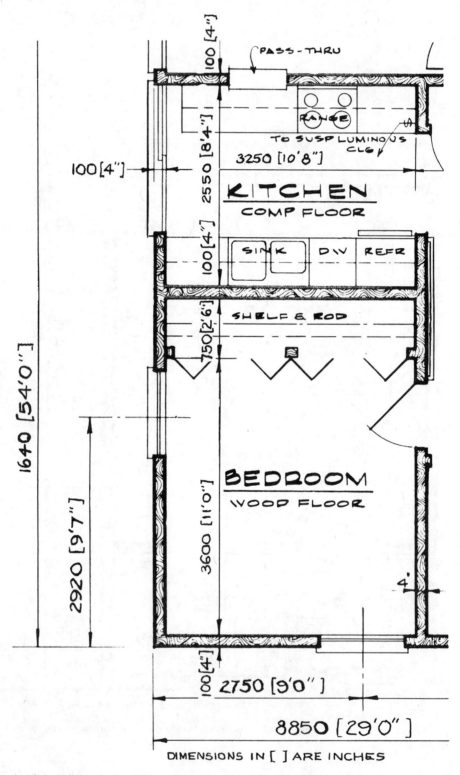

Fig. 9-5. An example of dual dimensioning. This portion of a floor plan shows metric measure with the conventional English measurements in brackets.

QUESTIONS FOR STUDY AND DISCUSSION

1. What are the advantages of converting to the metric system of measurement?
2. Which types of measurement directly apply to the construction industry?
3. What are the metric sizes for lumber of the following sizes: 1 × 10, 4 × 6, 2 × 8?
4. What direct effect on the carpenter does a changeover to the metric system have?
5. How does a changeover affect the manufacturer of building material?
6. How does a changeover affect the architect and engineer?
7. How are metric measurements indicated on working drawings?

GLOSSARY

A

admixture: An ingredient added to a concrete batch to give the concrete special properties.

aggregate: The gravel, crushed stone, or other hard inert material used in concrete.

anchor bolt: A metal bolt used to tie down a wood sill to a masonry or concrete foundation wall.

annular ring nail: Nails with fine rings on the shanks which give them strong holding power.

apron: A piece of inside window trim placed under the stool, flat against the wall.

areaway: An opening adjacent to a basement window or door to allow air and light to enter.

asbestos cement: A compound of Portland cement and asbestos. Used as the basis for shingles and siding.

awning window: A type of window in which the sash opens outward with hinges placed at the top of the sash.

B

backfill: Coarse dirt or other material, used to build up the ground level around the foundation wall. Provides a slope for drainage away from the foundation.

backing board: Gypsum wallboard used as a first layer in a two layer drywall application.

baffle: A construction of boards used to restrain the flow of concrete.

baluster: Small vertical members of a railing used between a top rail and the stair treads or a bottom rail.

balustrade: A railing made up of balusters, top rail, and bottom rail, or with balusters extending to treads.

band iron: Strips of metal which have perforations for nailing. Used to hold the foundation forms and structural building parts together.

base: A finish board covering the wall where it meets the floor.

batt insulation: Blanket insulation cut into short lengths so that it is easy to handle and apply.

batten: A narrow strip of wood used to cover the joint between two vertical pieces of siding.

batterboard: A construction of stakes and horizontal boards from which lines are strung to define the building lines.

beam: A steel structural member or heavy wood member used to support floor joists.

bearing partition: A partition which supports a vertical load in addition to its own weight.

bed molding: A molding applied where two surfaces come together at an angle. Commonly used in cornice trim.

bench mark: In surveying, a mark on some object firmly fixed in the ground from which distances and elevations are measured. It is usually a mark established by the city as a local point of reference.

bifold door: Doors arranged in double pairs so that a pair of doors are hung on each side of an opening. The inner doors hang from a track.

bird's mouth: A cutout near the bottom of a rafter which fits over the rafter plate.

blade: The longer of the two extending sides of a framing square.

blind valley: A valley of a roof in which the main roof is sheathed and the jack rafters are framed on top of the sheathing.

bow window: Several casement sash in a combination frame in the form of an arc of a circle.

brick molding: A molding for window and exterior door frames. Serves as the boundary molding for brick or other siding material.

bridging: Bracing between joists or studs to add stiffness to the floor or walls.

buck: Rough framing around a door or window opening in a concrete or masonry wall.

builders' level: An instrument with an optical telescope used to lay out building lines and establish elevations.

butt: A hinge for a door.

by pass door: Doors hung on tracks and arranged to slide past each other.

C

camber: A slight arch in a beam or other horizontal member which prevents it from bending into a downward or concave shape due to its own weight or load it must carry.

cant strip: An angular board installed at the intersection of a roof deck and a wall to avoid a sharp right angle where the roofing is to be installed.

cantilever: A projecting beam or construction supported only at one end and projecting from a wall.

cap molding: A molding used to finish the top of a wainscot.

casement window: A window in which the sash opens outward on hinges located at the side of the sash.

casing: A wood trim member covering the space between the plaster or drywall and the jamb at windows and doors.

casing nail: A wire nail with a flared head used for outside finish and also for nailing flooring.

cement: The binding part of a concrete mix.

chamfer: A beveled edge.

channel: An assembly of studs where partitions and walls intersect.

chord: A horizontal member of a truss.

collar: A horizontal tie beam connecting two opposite rafters.

column: A vertical free standing support member.

common rafter: A simple rafter extending from the ridge to the plate.

component: A part of a house assembled before delivery to the building site.

concrete: A mixture of sand, cement, and gravel with water in varying amounts according to the use of the finished product.

contour lines: Lines on a plot plan drawn to pass through points which have the same elevation.

control joint: A formed, cut, or tooled groove in concrete structures to regulate the location of cracks caused by shrinkage.

coped joint: A juncture between pieces of molding in which one piece is cut away to receive the molded part of the other piece.

corner bead: A metal protective strip placed at the outside corners of drywall finish.

cornice: A horizontal molded projection which crowns or finishes the eaves of a building.

course: A level layer of brick, stone, or other material.

cove molding: A concave molding used on inside corners.

crawl space: The space between the ground and the bottom of the joists in a house without a basement.

crimp: To fold the edges in a pinching action.

Used to fasten certain metal studs to the floor and ceiling channels.

crown molding: A molding which is part of a cornice. A molding to fit an internal angle, such as occurs at the intersection of a wall and ceiling.

cut: The relation between the rise and the run of a roof expressed in inches of rise per foot of run.

D

datum point: A point of reference established by the city from which levels and distances are measured.

dead bolt: A rectangular lock bolt controlled by a key.

dead load: The weight of the building itself, including the permanent equipment.

detail: A drawing, usually at a large scale, used to show some feature of construction.

dormer: A projection, usually with a window, built out from a sloping roof.

double hung window: A window with upper and lower sash which slide up and down in the grooves of the window frame.

dowel: A vertical reinforcing and positioning steel bar which is embedded in a footing and extends into a column.

downspout: A vertical pipe to carry rainwater from the gutter to the ground or sewer. (Same as a *leader*.)

drain tile: Rough unglazed ceramic tile laid with open joints around the footing. Used to conduct water to a point from which it is diverted away from the building.

drip cap: A molding placed above the top of a window or door casing to provide a means for water to run off.

drip edge: A protective edge for the roof which is installed below the shingles.

drywall: A system of interior wall finish which uses sheets of gypsum board. Finished with tape and compound.

duplex nail: Double headed nail used in temporary constructions, such as formwork, so that it can be easily removed.

E

eaves: The portion of the roof which overhangs the side wall.

elevation: (A) Drawings of buildings or parts of buildings made as though the observer were looking directly at the building.
(B) The height above a horizontal measuring point.

exposure: Amount of the shingle exposed to the weather.

F

fascia: A flat vertical board located at the outer face of a cornice.

fiberboard: A broad term used to describe sheet material of widely varying densities. Manufactured from wood, cane, and other vegetable fibers.

fiberglass: Material made from spun glass.

fire cut: A beveled cut made on the end of a joist which is anchored in a masonry wall.

firestop: Any blocking of air passages to prevent the spread of fire in a building. A block of wood closing off a space between studs.

flashing: Sheet metal used in roof and wall construction to keep out water.

footing: The spread portion at the base of a foundation wall or column which distributes the weight over a larger area.

frieze: The part of a cornice which is the lower vertical board at the wall.

frost line: The depth to which frost penetrates the earth. The particular depth is established for each area.

furring strip: Narrow strips of wood fastened to a wall or ceiling. Serves as a leveling device and provides a means for fastening the finishing materials.

G

gable: The triangular end of a house with a gable roof.

gable roof: A roof that slopes up from two sides.

gain: Notch or mortise cut to receive the end of another structural member, or a hinge and other hardware.

gambrel roof: A roof with two different slopes on each of its sides.

girder: A large supporting horizontal member used to support joists or beams.

girder pocket: A notch in a concrete or masonry wall made to receive the end of a girder.

grade: The level of the ground around a building.

grade beam: A concrete wall foundation in the form of a strong reinforced beam which rests on footings or concrete columns spaced at intervals.

ground: A piece of wood fastened to the rough framework which serves as a stop and thickness gage for plaster.

gusset: A flat surface of plywood or metal used to reinforce a joint of a truss.

gutter: A horizontal trough at the edge of a roof for carrying away rain water or melting snow.

gypsum: A calcium product used in plaster and as a core in sheets of drywall.

gypsum board: An air-entrained core of gypsum between two layers of fibrous absorbent paper.

H

hardboard: A reconstituted board made by exploding wood chips into a fibrous state and then pressing them in heated hydraulic presses to form dense rigid board.

header: A joist or joists placed at the ends of an opening in the floor used to support side members. The top rough framing members over a window or door opening.

headroom: The clear, vertical space between the sloped line passing over the nosing of stair treads and the ceiling or stairs above.

heel: The corner of a framing square.

hip jack rafter: A rafter which fits against a hip rafter.

hip rafter: A rafter at the junction of two slopes of a roof where an exterior angle is formed.

hip roof: A roof sloping up from all sides of a building.

hopper window: A type of window in which the sash opens inward with hinges at the bottom of the sash.

horizontal sliding window: Windows in which the sashes slide horizontally in grooves in the jambs

I

insulating glass: Two sheets of glass separated by an air space and sealed around the edges with either a glass or a metal closure.

insulation board: Structural building board made of wood, cane, or other cellulose fiber.

isolation joint: Asphalt-impregnated material used to isolate one part of a concrete slab from another, or a slab from a wall.

J

jack rafter: A rafter similar to a common rafter with one or both ends cut on a diagonal to fit against hip or valley rafters.

jalousie window: A window consisting of narrow pieces of glass arranged horizontally which open outward under the control of a mechanical operator.

jamb: The main members of a window or door frame which form the sides and top.

joist: The framing members which directly support the floor.

K

kerf: A cut made with a saw.

keyway: A groove formed in one placement of concrete to provide an interlock with a subsequently placed part.

L

lally column: A metal pipe filled with concrete used to support beams or girders.

lap siding: Siding in which one piece is applied to lap over the one below it.

laser: A device in which atoms stimulated by focused light waves are amplified and concentrated and then emitted in a narrow very intense beam. Surveying instruments which use the laser beam.

latch: A spring lock for a door.

lath: Metal mesh or wood strips which are fastened to structural members to provide a base for plaster.

ledger: (A) A horizontal part of a batterboard from which building lines are strung. (B) A member fastened to the side of a wood beam to help support joists.

level: (A) Horizontal. (B) A builder's instrument used to transfer points in laying out foundations. Revolves only in a horizontal plane. (C) A carpenter's or mason's tool used to level building parts in the course of building. (D) To adjust into a horizontal position.

light: A pane of glass.

liner: A vertical member used to keep walers in line. Also called a *stiffback*.

lintel: A horizontal member supporting the wall over an opening.

live load: The total of all moving and variable loads that may be placed on a building, including people, furniture, etc.

louver: A slatted opening used for ventilating attics and other roof spaces.

M

mansard roof: A roof with two slopes on all four sides.

mil: A unit of thickness of one thousandth of an inch (.001 inch equals 1 mil).

mineral fiber: A compound of asbestos and Portland cement formed into sheets and used in roof and exterior wall coverings.

miter joint: The joining of two pieces at an angle requiring 45° cuts on each piece (for a 90° corner).

module: (A) A unit of measurement established at 4 inches. (B) A complete part of a building assembled in a shop, such as a bathroom or kitchen.

monolithic concrete: Term used for concrete construction placed and cast in one unit without joints.

mortise: A slot cut in a plank or board (usually on the edge) to receive a tenon of another piece of wood, or (in the case of a door) to receive the lock.

mudsill: The lowest horizontal member of a frame wall anchored to the foundation.

mullion: The structural member between windows which come in pairs or in series.

muntin: The small members dividing the glass lights in a window sash.

N

nosing: The extension on a stair tread beyond the face of the riser.

P

package door: A factory assembled door complete with frame, lock, and hinges.

package window: A factory assembled window complete with frame, hardware, and trim.

parquet: Narrow strips of wood flooring arranged to form a geometric pattern. Strips may be glued or fastened together to form squares.

particleboard: A processed wood panel consisting of wood fiber, flakes, or shavings bonded together by synthetic resin, or some other binder.

partition: An interior wall which separates a space into rooms.

pennyweight: Term used to indicate nail length for common, box, casing and finishing nails. The symbol for pennyweight is the letter *d*.

perspective: A pictorial representation drawn so that it has the same appearance as when viewed from a particular location.

phillips head screw: A type of screw having a cross slot cut in the head for driving or removing the screw.

pier: A masonry or concrete column.

pilaster: A rectangular masonry column attached to a wall to provide stiffening.

pitch: (A) The slope of a floor toward a drain expressed in inches per foot. (B) A ratio between the rise of a roof and the span.

plancier: A horizontal board which is the underside of an eave or a cornice.

plaster: A pasty composition of lime, sand, and water which hardens on drying. Used for coating wall and ceiling surfaces.

plastic laminate: Sheet material made of several layers of strong kraft paper impregnated with synthetic resins. An upper cellulose sheet is used to give it color and pattern. The top covering consists of a protective resin sheet.

plate: The top horizontal structural member of a frame wall. Also known as a *top plate.*

plot plan: A part of the working drawings which shows information about the lot and the location of the building on the lot.

plumb: (A) Vertical. (B) To adjust into a vertical position.

plumb bob: A weight attached to a line used to locate a point.

polyethylene membrane: A sheet of plastic used for waterproofing.

point of beginning: A point on or near the lot from which horizontal and vertical (elevations) measurements are made.

prints: Reproductions of the architect's final working drawings. Commonly called *prints* or *working drawings.*

projections: When referring to roofs, the horizontal distance from the face of a wall to the end of the rafters.

R

rabbet: A groove cut in the surface or on the edge of a board to receive another member similarly cut.

rafter: A sloping roof member which supports the roof covering.

rail: A horizontal member of a panel door or a window sash.

random: A manner of laying stones or other units so that they do not follow regular patterns or courses.

reinforcing bar: Steel bars manufactured for the purpose of reinforcing concrete structural members.

ribbon: A narrow strip of board cut to fit into the edge of studding to help support joists.

ridge: The top horizontal member of a roof framed with rafters.

rise: On a roof built using simple rafters, the rise is the vertical (straight up and down and plumb) distance measured from the highest point (the ridge) of the rafter to the lowest point at the other end of the rafter.

riser: A vertical board at the edge of a stairway step.

roto-operator: Mechanical device used to operate casement and awning windows.

run: A horizontal measurement. In relation to roofs, the horizontal measurement from the center line of the ridge to the outside of the plate.

S

saddle: A small gable roof placed behind a chimney on a sloped roof to shed water.

sash: The frame in which the window lights (glass) are set.

sash balance: A device designed to counter-balance the window sash in a double hung window.

scab: A short piece of lumber fastened to overlap a butt joint to add strength.

screed: Grade level forms set at the desired elevation so that the concrete can be leveled by drawing a straightedge over their surface. Also the straightedge.

scribe: Fitting wood work to an irregular surface.

scriber: A tool resembling a compass with a pencil on one leg and a metal point on the other. Used to mark the irregularities in fitting cabinets or trim members to the walls or floor.

sealant: A substance used between members to provide a waterproof seal.

seat cut: A horizontal cut on a rafter where it rests on the plate.

setback: A specified minimum distance that a structure must be placed from a lot line.

shake: Handsplit wood shingles.

sheathing: Fiberboard, gypsum board, plywood, or rough boards that cover up the rough framing and rafters.

shed roof: A roof with only one slope.

shim: A thin piece of wood used to fill a gap between furring strips and an irregular masonry surface.

shiplap: Lumber which has been rabbetted along each edge to provide a close lapped joint by fitting two pieces together.

shoe mold: The small molding covering the joint between flooring and baseboard.

shortening: The amount of the rafter end that is cut off to allow for the thickness of the intersecting members.

sidelap: The amount of material overlapping at sides in the laying of vapor barrier film, roll roofing, building paper, or similar material.

siding: Outside wall finish made in long narrow units.

sill: (A) The bottom rough structural member which rests on the foundation. (B) The bottom exterior member of a window or door.

single hung window: A window in which the lower sash slides up and down in a groove in the jambs. The upper sash is fixed.

site: The piece of ground on which the building is located.

slope: Incline of a roof, expressed as rise in inches per foot of run.

snap tie: Concrete wall form tie. The end can be twisted or snapped off after the forms have been removed.

soffit: A lower horizontal surface, such as the underface of eaves, cornice, or beam.

sole: The horizontal member of a frame wall or partition which rests on the floor.

span: The distance between the outside of wall supports.

specifications: A typed set of instructions prepared by the architect, covering materials, procedures, quality of workmanship, and guarantees.

spreader: A piece of wood cut the exact thickness of the foundation wall. Used to keep the forms apart.

square: The amount of roofing which can cover 100 square feet when laid.

stiffback: A vertical member used to keep walers in line. Also called a *liner*.

stile: Side vertical parts of a panel door or window sash.

stop: The inside molding or piece of trim fastened to the jamb which holds the sash in place on a window.

story pole: A strip of wood used to lay out and transfer measurements for door and window openings, siding and shingle courses, and stairway risers.

stress: Force exerted on a body that tends to strain or deform its shaped.

strike plate: A metal piece mortised into, or fastened to, the face of a door frame side jamb to receive the latch or bolt when the door is closed.

stringer: The member on each side of a stair which supports the treads and risers.

stucco: A cement exterior coating.

stud: Vertical structural member which makes up the walls and partitions in a frame building.

subfloor: Boards or panels laid directly on floor joists over which a finished floor is laid.

survey: A plan or map of a lot prepared by a licensed surveyor, showing dimensions, elevations, etc.

T

tail cut: Cut at the lower end of a rafter.

tail piece: The part of the rafter which extends beyond the plate. Also called the *overhang*.

termite: Wood devouring white ants.

terrazzo: A floor material consisting of broken marble chips bedded in concrete which is polished to a smooth surface after it is laid.

threshold: A piece of material over which a door swings.

tie: In masonry veneer, a metal strip used to tie the masonry wall to the wood sheathing. In concrete formwork, devices used to tie the two sides of a form together.

toenailing: To drive a nail at a slant with the initial surface in order to permit it to penetrate into a second member.

tongue: The shorter leg of a framing square.

tongue and groove: Boards or planks machined in such a manner that there is a groove on one edge and a corresponding tongue on the other.

toplap: The amount of material overlapping at the ends in the laying of vapor barrier film, roll roofing, building paper, or similar material.

transit: A surveyor's instrument used by builders to establish points and elevations. The transit operates in both the horizontal and vertical planes.

travertine: A type of soft limestone.

tread: The horizontal board in a stair.

truss: A structural member with primary and secondary members arranged to form a triangular assembly.

trussed rafter: A roof truss which serves to support the roof and ceiling construction.

U

underlayment: Floor covering of plywood or fiberboard used to provide a level surface for carpet or other resilient flooring.

V

valley: The internal angle formed by two inclined sides of a roof. (A valley rafter is the rafter supporting the valley.)

valley jack rafter: A rafter which fits against a valley rafter.

valley rafter: A rafter which forms the intersection of an internal roof angle.

vanity: A cabinet lavatory.

vapor barrier: A material used to prevent the

passage of moisture or water vapor into or through structural elements (floors, walls, ceilings, etc.).

veneer: In masonry, a facing of stone, brick, or other masonry material placed over a frame superstructure. In carpentry, a thin layer of wood.

vermiculite: A mineral closely related to mica which expands when heated to form a lightweight material with insulating qualities. Used as bulk insulation, also as aggregate in insulating and acoustical plaster, and in concrete.

vernier: A graduated scale used to read fractions of a degree on a leveling instrument.

vibrator: A mechanical device used to compact freshly placed concrete in the forms.

vinyl: Noncrystalline thermoplastic material with tough and strong properties.

volute: A spiral portion at the end of certain types of handrail.

W

waler: Horizontal member which holds concrete forms in line and provides a stiffening effect. Also called a *wale*.

water table: The members of wood trim at the bottom of exterior siding. Designed both for finish and a means to keep water from running down the foundation wall.

web member: Secondary member of a truss placed between chords.

welded wire fabric: Heavy steel wire welded together in a grid pattern. Used for reinforcing concrete slabs.

wet core: A module which contains the plumbing for a building. Can be a bathroom or a bathroom/kitchen combination.

working drawing: The architect's final drawings used to build the structure. Reproduced in the form of prints for use on the job.

INDEX

Numbers appearing in **bold type** refer to illustrations.